Proceedings in Life Sciences

Hearing and Sound Communication in Fishes

Edited by
William N. Tavolga
Arthur N. Popper
Richard R. Fay

With 196 Figures

Springer-Verlag
New York   Heidelberg   Berlin

William N. Tavolga
Mote Marine Laboratory
1600 City Island Park
Sarasota, Florida 33577, U.S.A.

Richard R. Fay
Parmly Hearing Institute
 and Department of Psychology
Loyola University of Chicago
6525 N. Sheridan Road
Chicago, Illinois 60626, U.S.A.

Arthur N. Popper
Department of Anatomy
Georgetown University
 Schools of Medicine and Dentistry
3900 Reservoir Road N.W.
Washington, D.C. 20007, U.S.A.

Production: Kate Ormston

The figure on the front cover shows the brain and auditory apparatus of a shark (*Squalus*), redrawn and modified, after E. H. Weber, De aure et auditu hominis et animalium. Pars I. De aure animalium aquatilium. Lipsiae, 1820, plate 10, figure 88.

Library of Congress Cataloging in Publication Data
Main entry under title:
Hearing and sound communication in fishes.
 (Proceedings in life sciences)
 Papers originally presented at a meeting in Apr. 1980 at Mote Marine Laboratory, Sarasota, Fla.
 Includes bibliographies and index.
 1. Fishes–Physiology–Congresses. 2. Hearing–Congresses. 3. Sound production in animals–Congresses. 4. Lateral line organs–Congresses. I. Tavolga, William N., 1922-   . II. Popper, Arthur, N. III. Fay, Richard R. IV. Series.
QL639.1.H4   597'.01825   81-5653
             AACR2

ISBN 0-387-90590-1 Springer-Verlag New York Heidelberg Berlin
ISBN 3-540-90590-1 Springer-Verlag Berlin Heidelberg New York

This volume is dedicated to
Dr. Sven Dijkgraaf
and to the memories of
Dr. Willem A. van Bergeijk and *Dr. Gerard G. Harris,*
three investigators whose contributions to fish acoustics pervades
this volume, and all work leading to it.

Preface

This volume is a compilation of the papers presented at a meeting that took place in April 1980 at the Mote Marine Laboratory, Sarasota, Florida. The meeting and this volume are outgrowths of two earlier international meetings on marine bio-acoustics that occurred in 1963 and 1966 (Tavolga 1964, 1967). The first meeting took place at the Lerner Marine Laboratory of the American Museum of Natural History, while the second meeting was at the American Museum itself, and was under the sponsorship of the Department of Animal Behavior. It is apparent that these two volumes have had immense impact on the current study of marine bio-acoustics, and particularly on fish audition. In a preliminary conference in Sarasota in 1979 we decided that it was time for another such meeting, to bring together as many as possible of the investigators interested in fish acoustics in order to assess the current state of our knowledge and predict directions for research for the next several years. Such a meeting appeared particularly timely, since over the past four or five years there have been many new studies that have provided new empirical and theoretical work on basic mechanisms of fish audition. Furthermore, it became evident, as we made up preliminary lists of possible participants, that few of the currently active workers were in the field back in 1966. In fact, of the current participants, only Drs. Enger, Myrberg, Popper and Tavolga were at the 1966 meeting, and only Tavolga is on the list of attendees of the 1963 meeting.

One of the great pleasures in planning this meeting was the extraordinary cooperation of virtually the entire staff at the Mote Marine Laboratory, the site of the 1980 meeting. We are particularly indebted to the president and director of the laboratory, Dr. William H. Taft, who did everything possible to make the meeting the success that it was. We were particularly impressed with the way that Dr. Taft saw to it that the Marine Sciences building at the laboratory was completed in time (electricity was turned on for the first time the day before the meeting) for our use during the meeting. We cannot name all of the people at the Mote Marine Laboratory who gave assistance at the meeting, but we do wish to express special thanks to Mr. William R. Mote, Chairman of the Board of the Mote Marine Laboratory, for his generous aid and interest in the meeting. We are also most appreciative of the time and effort spent by Ms. Patricia Morrissey for deciphering the contents of the tape recordings of the discussions and of the extraordinary cooperation given us by all of the participants in the meeting in getting the completed manuscripts to us as rapidly as they did.

We are grateful for the financial support for the meeting given through a grant (BNS-79-17024) from the National Science Foundation. We thank Dr. Terrence Dolan, director of the program in Sensory Physiology and Perception, for his advice and guidance in the preparation of our grant application. Finally, we cannot begin to express our appreciation to our wives, Margaret C. Tavolga, Helen A. Popper and Catherine H. Fay, for their unfailing support and help throughout the planning and execution of this meeting.

William N. Tavolga
Arthur N. Popper
Richard R. Fay

Contents

List of Participants

H. DAVID BALDRIDGE   Mote Marine Laboratory, 1600 City Island Park, Sarasota, Florida 33577, U.S.A.

CURTIS C. BELL*   Neurological Sciences Institute, Good Samaritan Hospital, Portland, Oregon 97209, U.S.A.

JOHN H. S. BLAXTER*   Scottish Marine Biological Association, Dunstaffnage Marine Research Laboratory, P. O. Box 3, Oban, Argyll PA3 4AD, Scotland.

ROBERT BOORD   School of Life and Health Sciences, University of Delaware, Newark, Delaware 19711, U.S.A.

J. ROBERT BOSTON*   Department of Anesthesia/CCM, 1060E Scaife Hall, University of Pittsburgh School of Medicine, Pittsburgh, Pennsylvania 15261, U.S.A.

MARK BRAFORD   Department of Anatomy, Georgetown University Schools of Medicine and Dentistry, 3900 Reservoir Rd. N.W., Washington, D.C. 20007, U.S.A.

THEODORE H. BULLOCK*   Department of Neurosciences A-001, School of Medicine, University of California, San Diego, La Jolla, California 92093, U.S.A.

ROBBERT J. A. BUWALDA*   Laboratory of Comparative Physiology, Jan Van Galenstraat 40, Utrecht, The Netherlands.

PHYLLIS CAHN   National Oceanographic and Atmospheric Administration, Office of Marine Pollution Assessment, RD/MP1, 11400 Rockville Pike, Rockville, Maryland 20852, U.S.A.

*Presenting papers.

SHERYL COOMBS* Department of Zoology, University of Hawaii, Honolulu, Hawaii 96822 and Department of Anatomy, Georgetown University Schools of Medicine and Dentistry, 3900 Reservoir Rd. N.W., Washington, D.C. 20007, U.S.A.

JEFFREY T. CORWIN* Department of Neurosciences A-001, School of Medicine, University of California, San Diego, La Jolla, California 92093, U.S.A.

LEO S. DEMSKI* School of Biological Sciences, University of Kentucky, Lexington, Kentucky 40506, U.S.A.

ERIC J. DENTON* Marine Biological Association of the United Kingdom, The Laboratory, Citadel Hill, Plymouth PL1 2PB, England.

PER S. ENGER* Institute for Zoophysiology, University of Oslo, P. O. Box 1051, Blindern, Oslo 3, Norway.

RICHARD R. FAY* Parmly Hearing Institute and Department of Psychology, Loyola University of Chicago, 6525 N. Sheridan Road, Chicago, Illinois 60626, U.S.A.

MICHAEL L. FINE* Department of Biology, Virginia Commonwealth University, 901 West Franklin St., Richmond, Virginia 23284, U.S.A.

P. J. FRASER* Zoology Department, University of Aberdeen, Aberdeen, Scotland.

PERRY W. GILBERT Mote Marine Laboratory, 1600 City Island Park, Sarasota, Florida 33577, U.S.A.

GRANT GILMORE Harbor Branch Foundation, Inc., Link Port, RR 1, Box 196, Fort Pierce, Florida 33450, U.S.A.

CHARLES R. GORDON 525 South River Rd., New Port Richey, Florida 33552, U.S.A.

JOHN A. B. GRAY* Marine Biological Association of the United Kingdom, The Laboratory, Citadel Hill, Plymouth PL1 2PB, England.

SAMUEL GRUBER School of Marine and Atmospheric Science, University of Miami, 4600 Rickenbacker Causeway, Miami, Florida 33149, U.S.A.

LORRAINE HALL* Department of Psychology, Loyola University of Chicago, 6525 N. Sheridan Road, Chicago, Illinois 60626, U.S.A.

A. D. HAWKINS* Marine Laboratory, P. O. Box 101, Victoria Road, Torry, Aberdeen AB9 8DB, Scotland.

KATHLEEN HORNER* Marine Laboratory, P. O. Box 101, Victoria Road, Torry, Aberdeen AB9 8DB, Scotland.

DAVID B. JENKINS*   Department of Anatomy, School of Medicine, University of North Carolina, Chapel Hill, 102 Medical Research Building D, Chapel Hill, North Carolina 27514, U.S.A.

SHEN JUN   Department of Biology, Shandong College of Oceanology, Qingdao, Shandong, China and (1979-1980) Department of Anatomy, Georgetown University Schools of Medicine and Dentistry, 3900 Reservoir Rd. N.W., Washington, D.C. 20007, U.S.A.

JAMES I. KENDALL   1543 Babcock Road #1001, San Antonio, Texas 78229, U.S.A.

DAVID KOESTER   School of Life and Health Sciences, University of Delaware, Newark, Delaware 19711, U.S.A.

CATHERINE A. McCORMICK*   Department of Anatomy, Georgetown University Schools of Medicine and Dentistry, 3900 Reservoir Rd. N.W., Washington, D.C. 20007, U.S.A.

GLORIA MEREDITH   Department of Anatomy, Georgetown University Schools of Medicine and Dentistry, 3900 Reservoir Rd. N.W., Washington, D.C. 20007, U.S.A.

HIN KIU MOK*   Harbor Branch Foundation, Inc., Link Port, RR 1, Box 196, Fort Pierce, Florida 33450, U.S.A.

WILLIAM R. MOTE   Mote Marine Laboratory, 1600 City Island Park, Sarasota, Florida 33577, U.S.A.

T. J. MUELLER   Department of Biological Sciences, University of Southern California, University Park, Los Angeles, California 90007, U.S.A.

ARTHUR A. MYRBERG, JR.*   School of Marine and Atmospheric Science, University of Miami, 4600 Rickenbacker Causeway, Miami, Florida 33149, U.S.A.

R. GLENN NORTHCUTT*   Division of Biological Sciences, University of Michigan, Ann Arbor, Michigan 48109, U.S.A.

JANET OTT   Department of Biological Sciences, University of Southern California, University Park, Los Angeles, California 90007, U.S.A.

BRIAN L. PARTRIDGE*   Department of Biology, University of Miami, P. O. Box 249118, Coral Gables, Florida 33124, U.S.A.

MARIAN PATRICOSKI*   Department of Psychology, Loyola University of Chicago, 6525 N. Sheridan Road, Chicago, Illinois 60626, U.S.A.

CHRISTOPHER PLATT*   Department of Biological Sciences, University of Southern California, University Park, Los Angeles, California 90007, U.S.A.

ARTHUR N. POPPER*   Department of Anatomy, Georgetown University Schools of Medicine and Dentistry, 3900 Reservoir Rd. N.W., Washington, D.C. 20007, U.S.A.

ROY RICHARDS   Mote Marine Laboratory, 1600 City Island Park, Sarasota, Florida 33577, U.S.A.

ROBERT RIGGIO   School of Marine and Atmospheric Science, University of Miami, 4600 Rickenbacker Causeway, Miami, Florida 33149, U.S.A.

BARRY L. ROBERTS*   Marine Biological Association of the United Kingdom, The Laboratory, Citadel Hill, Plymouth PL1 2PB, England.

OLAV SAND*   Department of Physiology, Veterinary College, P. O. Box 8146 DEP., Oslo 1, Norway.

ARIE SCHUIJF*   Laboratory of Comparative Physiology, Jan Van Galenstraat 40, Utrecht, The Netherlands.

W. G. SOKOLICH*   Division of Otolaryngology, School of Medicine, University of California at Los Angeles, Los Angeles, California 90024, U.S.A.

DAVID STRELIOFF*   Division of Otolaryngology, School of Medicine, University of California at Los Angeles, Los Angeles, California 90024, U.S.A.

WILLIAM H. TAFT   Mote Marine Laboratory, 1600 City Island Park, Sarasota, Florida 33577, U.S.A.

WILLIAM N. TAVOLGA*   Mote Marine Laboratory, 1600 City Island Park, Sarasota, Florida 33577, U.S.A.

A. O. D. WILLOWS   Division of Neurobiology, National Science Foundation, Washington, D.C. 20051, U.S.A.

PART ONE

Form and Function

Over the past several years there has been considerable new interest in the structure of the auditory system in fishes and the implications of such structure for the processing of acoustic information. This interest has grown with the advent of scanning electron microscopy, enabling investigators to study inner ear ultrastructure that was heretofore beyond the scope of light microscopy. In particular, these studies have provided a far better understanding of the ear's sensory epithelia and their ultrastructural features which are likely to be intimately involved in sound localization and other aspects of acoustic processing. More broadly, structural analyses of the auditory system of fishes are not only of intrinsic interest, but provide the basis for understanding and interpreting data obtained from physiological and behavioral experiments, such as those described in Parts Two and Three of this volume.

In Chapter 1, Platt and Popper provide a survey of the structure of the teleost ear. They demonstrate the extensive variation seen in the ear among teleost fishes and suggest the presence of interspecific variation in inner ear function. In Chapter 2, Blaxter, Denton, and Gray provide an extension of the functional principles discussed in the first chapter by describing the organization of the highly specialized auditory system in the clupeoid fishes. Blaxter (Chapter 3) then discusses the function of the swimbladder, while Jenkins (Chapter 4) describes the morphology of the utricle in a catfish. Finally, Corwin (Chapter 5) gives a detailed analysis of the auditory system in elasmobranchs and provides evidence for the involvement of the macula neglecta in hearing.

Chapter 1

Fine Structure and Function of the Ear

CHRISTOPHER PLATT* and ARTHUR N. POPPER**

1 Introduction

One of the most striking features in the auditory system of fishes is the extensive structural diversity in the inner ear and its peripheral accessories. In this chapter we will summarize this diversity in the ear, from the gross structure to the ultrastructure of the sensory epithelia, and suggest some of the possible functional meanings for these structural differences. We hope that this discussion will stimulate interest in pursuing direct experimentation on the function of the fish ear in order to fill in the gaps in our understanding of peripheral auditory mechanisms. Two major points will be stressed throughout this chapter. First, we feel that dividing up of auditory and vestibular functions between the different otolithic organs of the ear may not be as absolute as has been often implied, so it may be necessary to reconsider some of the basic "classical" assumptions of auditory organ functions, at least with regard to the teleost ear. Second, we suggest that the notion of a functionally or structurally "typical teleost ear" is no longer tenable, since the breadth of interspecific structural variation in teleost ears may imply significant functional variation.

1.1 The Inner Ear

Both vestibular and auditory functions in fishes are mediated by the inner ear, which consists of several mechanosensory end-organs that are located in interconnected fluid-filled chambers (Fig. 1-1). The two major types of end-organs are the semicircular canals and the otolith organs. The ears in many species may also have a nonotolithic accessory sensory area, the macula neglecta (see Retzius 1881).

*Department of Biological Sciences, University of Southern California, Los Angeles, California 90007
**Department of Anatomy, Schools of Medicine and Dentistry, Georgetown University, Washington, D.C. 20007

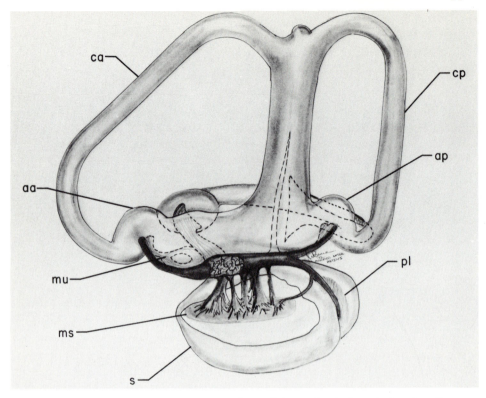

Figure 1-1. Medial view of the right ear from *Salmo salar* (salmon) showing the sensory maculae and innervation by portions of the eighth nerve. aa, anterior ampulla; ap, posterior ampulla; ca, anterior semicircular canal; cp, posterior semicircular canal; ms, saccular macula; mu, utricular macula; pl, lagena; s, sacculus. (Redrawn from Retzius 1881.)

Functionally, the inner ear is responsive to a variety of mechanosensory stimuli. In addition to oscillations at auditory frequencies, physiological responses have been obtained from gravistatic, acceleratory, and vibrational stimuli. Classically, the ear has been divided into the pars superior, consisting of the semicircular canals and one otolith organ, the utricle, and the pars inferior, consisting of two other otolith organs, the saccule and the lagena. Behavioral experiments have suggested that the pars superior mediates postural responses, while the pars inferior mediates acoustic behavior (von Frisch 1938, von Holst 1950). However, while such evidence may suggest major roles for such organs, they do not necessarily demonstrate the full capabilities of each individual otolithic end-organ in mechanoreception (see Lowenstein 1971).

1.2 Structure

The inner ear (Fig. 1-1) is a membranous system of contiguous ducts and pouches, containing a liquid endolymph with special viscous and ionic properties (Fänge, Larsson, and Lidman 1972). Each semicircular canal is a duct extending from the rest of the inner ear, forming a ring of endolymph, with a single dome-shaped swelling, the ampulla, at the origin of the canal. The sensory crista of the ampulla is a saddle-shaped ridge lying across the base of the ampulla. Above the crista is a gelatinous cupula, which is deformed by fluid movements within the canal (Steinhausen 1933, Oman, Frishkopf, and Goldstein 1979). In contrast, the three otolith organs are pouches containing dense calcifications (otoliths or statoconia) of crystalline calcium carbonate (Carlström 1963) embedded in a gelatinous otolithic membrane. This otolithic membrane contacts the sensory macula which lies on a wall or floor of the pouch. Some areas of otolith organ maculae as well as the whole macula neglecta are covered only by gelatinous membrane without calcification.

The sensory epithelia each contain mechanoreceptive hair cells which are innervated by fibers of the eighth cranial nerve. The sensory hair cells in fishes are all of the cylindrical form known as Type II (Wersäll 1961), based upon their internal ultrastructure and synaptic contacts. Each hair cell has an apical bundle of hairlike sensory processes (Figs. 1-2 and 1-3) protruding into the lumen of the end-organ. These ciliary bundles are made up of a single true cilium, the kinocilium, arising at one end of an array of microvilli-like stereocilia. The stereocilia are anchored in a cuticular plate, while the kinocilium sends microtubular rootlets down to the basal body with the cell (Fig. 1-2). The stereocilia usually decrease in height with increasing distance from the kinocilium (Fig. 1-3). Each hair cell in the epithelium is usually surrounded by thin projections from several supporting cells. The apical surfaces of the supporting cells are covered by short microvilli, while the base of these cells lie beneath the sensory cells on a thin basement membrane (Wersäll, Gleisner, and Lundquist 1967, Hama 1969, Dale 1976, Popper 1979a). Tight junctions between the sensory and supporting cells presumably electrically isolate the high-potassium endolymph from the intracellular fluid and cranial fluid outside of the ear (Hama and Saito 1977, Popper 1979a).

1.3 Auditory and Vestibular Input

Fibers from the eighth nerve penetrate the basement membrane and their terminals make synaptic contacts with the bases of the hair cells (Fig. 1-2). Chemical synaptic transmission is suggested by the response latencies, fatigue, and ultrastructural features of the synapses (Furukawa and Ishii 1967a, Hama 1969, Nakajima and Wang 1974, Popper 1979a). Both afferent and efferent synapses are distinguishable not only by the position of synaptic vesicles, but also because the afferent synapse from hair cell to fiber characteristically contains a large presynaptic body associated with synaptic vesicles, while efferent endings have typical synaptic bars (Fig. 1-2) (Hama 1969, Nakajima and Wang 1974, Popper 1979a). Though in the mammalian organ of Corti there are more afferent nerve fibers than sensory cells (Spoendlin 1973), some fish maculae may have up to ten times as many hair cells as innervating fibers to it (Flock 1964, Popper and Northcutt in prep.), and in the goldfish, at least, a single fiber may innervate widely separated hair cells (Furukawa 1978).

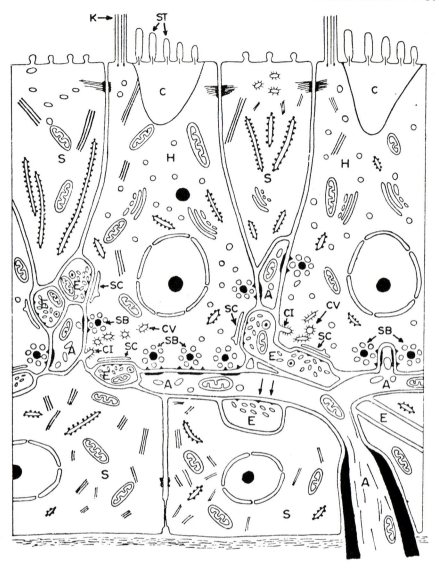

Figure 1-2. Schematic drawing showing a cross-sectional view of sensory and supporting cells from the goldfish saccular macula. The sensory cells sit on a layer of supporting cells which also send projections to the apical surface of the sensory epithelium
and surround each of the sensory cells. The supporting cells lie on a basement membrane. A, afferent nerve endings; C, cuticle; CI, coated invaginations; CV, coated vesicles; E, efferent nerve endings; H, sensory hair cells; K, kinocilia; S, supporting cells;
SB, synaptic bodies; SC, subsynaptic cistern; ST, stereocilia. (From Nakajima and
Wang 1974.)

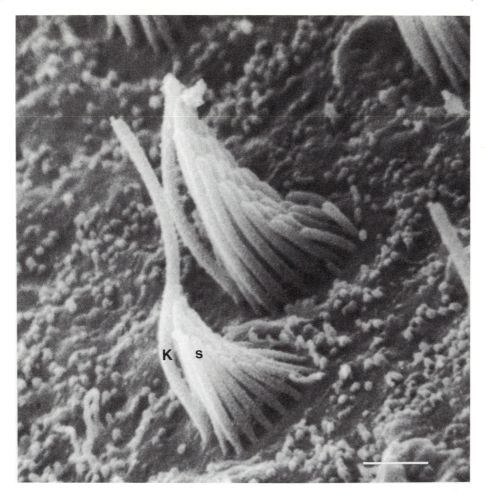

Figure 1-3. Scanning electron micrograph of apical ciliary bundles on the epthelial sensory surface of an otolith organ in the goldfish, *Carassius auratus*. Two types of ciliary bundles are illustrated, showing different relative lengths of the kinocilia (K) and stereocilia (S). The kinocilium arises eccentrically at one end of the ciliary array, and the tallest stereocilia are closest to it.

1.4 Cellular Mechanisms

Transduction from mechanical to electrochemical energy by the hair cells in fish ears, as in fish lateral line organs and the ears of other vertebrates, results from the bending of the apical bundle of stereocilia by lateral shearing forces (Flock 1965, Wersäll, Flock, and Lundquist 1965, Hudspeth and Jacobs 1979, Strelioff and Sokolich, Chapter 24). When the bundle is bent toward the kinocilium, the inside of the cell becomes depolarized relative to its resting potential, and the deflection in the opposite direction produces hyperpolarization (Hudspeth and Corey 1977). These transmembrane potential changes depend on the opening of membrane channels for cations and increase flow of Na^+ inward and K^+ outward across the cell membrane (Corey and Hudspeth 1979). The presence of the eccentrically placed kinocilium means that each cell can be assigned a morphological orientation that is a correlate of this directional sensitivity (Wersäll et al. 1965). Orientation patterns of cell groups thus have considerable functional significance (see Section 4.3).

Many afferent eighth nerve fibers are continuously active, with discharge rates in fishes ranging from less than 1 to over 100 spikes per second (Enger 1963, Lowenstein 1971, Platt 1973, Fay 1978a, 1978c, Highstein and Politoff 1978, Horner, Hawkins, and Fraser, Chapter 11). Depolarization is believed to be coupled to increased release of neurotransmitter from the hair cell to the afferent fiber terminals, causing a change in the firing rate of the fiber (Furukawa and Ishii 1967a, Highstein and Politoff 1978). The continuous "spontaneous" firing allows either an increase or decrease in resting rate to occur, providing another mechanism for directional sensitivity.

While comparable data are not available for any fish ear, the hair cell response in the bullfrog saccule is a sigmoid function of the extent of the displacement of the ciliary bundle and is nearly linear over part of its displacement range. Although the response curve saturates when the bundle is deflected more than $2°$, this curve can be shifted by an adaptive process that restores sensitivity during maintained displacements of up to $30°$ (Eatock, Corey, and Hudspeth 1979). The frequency response of these hair cells in amphibians appears to depend on displacement, not velocity, for stimuli of up to 150 Hz (Hudspeth and Corey 1977, Strelioff and Honrubia 1978). Given the wide band-pass and broadly linear response dynamics of the receptor cells, it is important to consider how accessory structures associated with fluid displacement or interactions between components of the maculae might act as peripheral filters to "tune" particular cell populations. For example, displacement stimuli can be considered to produce a wide spectrum of frequencies for relative motions between otolith and macula. Oscillating stimuli from acoustic signals produce high frequency (to over 1,000 Hz) relative motions, and gravistatic stimuli from postural changes produce low frequency (to less than 0.01 Hz) relative motions. As will be discussed (Section 4.1), this filtering problem is particularly relevant to studies of teleost fishes, where the gross structure of the presumed gravistatic and auditory otolith organs are very similar to one another.

2 Auditory Mechanisms and Behavior

The evolutionary radiation of fishes into a wide range of habitats has been accompanied by a diversity in auditory anatomy that may indicate some phylogenetic relationships, as has been shown in reptiles (Wever 1978, Miller 1980). Although data on acoustic behavior of fishes are limited, there also are likely to be some correlations between structural and functional aspects of the auditory system, as shown in both amphibians and reptiles (Weiss, Mulroy, Turner, and Pike 1976, Lewis and Levernez 1979).

2.1 Phylogenetic Diversity

The structural variability found in the teleost auditory periphery rivals that found in terrestrial vertebrates. Strikingly, while such diversity is great, some taxonomically distinct groups of teleosts show considerable similarity in the structure of the auditory system, including the ear (particularly the pars inferior), swimbladder, and other accessory structures. It is not yet feasible to make broad generalizations about the taxonomic significance of any but a very few aspects of the auditory system. In fact, it seems more likely that there are cases of apparent convergence, as a variety of different teleosts have been able to exploit only a few possible different mechanisms for detection and processing of similar signals.

The variety of teleosts in which the auditory system has been studied by electron microscopy and/or modern behavioral techniques is shown in Table 1-1, with evolutionary relationships based upon a major proposed phylogenetic scheme (Greenwood, Rosen, Weitzman, and Myers 1966).

2.2 Acoustic Behavior

Interspecific variation in the detection and production of sounds is likely to have evolved under selective pressures involving the presence of both biologically and non-biologically produced sounds in the environment (Marler 1955, Morton 1975, Fine, Winn, and Olla 1977, Michelsen 1978, Popper and Coombs 1980, 1981, Myrberg, Chapter 20). Such selection would have affected how fishes use different parameters of sound, such as the frequency spectrum or the pattern of the sound, and would have affected the nature of the detection and processing systems. Data relating the tuning of the hearing system to different acoustic parameters is not as widely available for fishes as it is for birds and other terrestrial vertebrates (Marler 1977, Dooling 1980), but it is clear that most of the acoustic signals produced by fishes in communication involve lower frequencies (below 1,000 Hz), and that these are the frequencies most readily detected by the species that have been studied (see Tavolga 1971, 1976, 1977a, Demski, Gerald, and Popper 1973, Popper and Fay 1973, Fine et al. 1977, Fay and Popper 1980, Hawkins, Chapter 6; Myrberg, Chapter 20).

Acoustic sensitivity has been experimentally tested in only a few fish species (see Tavolga and Wodinsky 1963, Jacobs and Tavolga 1967, Chapman and Hawkins 1973, Popper and Fay 1973, Chapman and Sand 1974, Fish and Offut 1972, Tavolga 1976, Fay 1978b, Fay and Popper 1980, Popper 1981), and shows considerable interspecific variation. Figure 1-4 demonstrates this auditory diversity for eight of these species. It becomes imperative to point out that earlier generalizations relating hearing capabilities to the presence or absence of certain structures may not necessarily be completely valid. For example, because the Ostariophysi have a chain of bones, the Weberian ossicles, coupling the swimbladder and inner ear, many investigators have proposed that they have the best hearing sensitivity among fishes (von Frisch 1936, 1938, Jacobs and Tavolga 1967, Lowenstein 1971). But it is now known that at least one nonostariophysine, the squirrelfish (Beryciformes) *Myripristis kuntee,* has hearing capabilities exceeding that of an ostariophysine, the goldfish (*Carassius auratus*) (Coombs and Popper

Table 1-1. List of fish taxonomic groups (generally at familial level) for which ultrastructural data are available regarding inner ear structure

Taxonomic group	Reference	Hearing data	Other references	Comments
Holostei	Popper and Northcutt in prep.		McCormick, Chapter 18, CNS	*Amia* (bowfin)
Acipenseriformes	Popper 1978a			Sturgeon
Polypteriformes	Popper 1978a			Reedfish
Anguilliformes	Popper 1979a			Moray eel
Clupeiformes	Popper and Platt 1979	Blaxter et al., Chapter 2		Herrings, etc.
Osteoglossomorpha			Greenwood 1973, taxonomy	
Mormyridae	Popper 1979b	Stipetić 1939	Bell, Chapter 19, CNS	Elephant nose
Notopteridae	Coombs and Popper in prep.	Coombs and Popper in prep.		Knifefish
Osteoglossidae	Popper 1979b	Coombs in prep.		
Pantodontidae	Popper 1979b			Butterfly fish
Salmoniformes				
Salmonoidei	Popper 1976 Popper 1977	Hawkins and Johnstone 1978		Salmon and relatives
Myctophoidei	Popper 1977		Marshall 1967, on acoustic behavior and sound production	Lanternfish
Stomiatoidei	Popper 1980			
Argentoidei	Popper 1980			
Ostariophysi				
Cyprinidae	Hama 1969 Platt 1977	Jacobs and Tavolga 1967	Fay, Chapter 10	Goldfish
Siluriformes	Jenkins 1979a, 1979b Popper and Tavolga in prep.	Poggendorf 1952 Popper and Tavolga in prep.	Tavolga 1964, 1977b, on sound production	Catfishes

Taxon				
Bericiformes				
Holocentridae	Popper 1977	Coombs and Popper 1979	Salmon 1967, on sound production	Squirrelfish
Perciformes				
Cichlidae	Popper 1977	Tavolga 1974	Myrberg, Kramer, and Heinecke 1965, on sounds	Cichlids
Chaetodontidae	Popper 1977			Butterfly fish
Pomacentridae	Popper 1977	Tavolga and Wodinsky 1963	Myrberg and Spires 1972	
Sparidae	Popper 1977	Tavolga 1974		
Centrarchidae	Popper 1977	Sand 1974	Gerald 1971, on sound production	Sunfishes
Percidae	Enger 1976			Perch
	Popper 1977			
Goodeidae	Popper 1977	Iversen 1967		
Scombridae	Popper 1978b			Tunas
Acanthuridae	Popper 1977	Schneider 1941		Soldierfish
Anabantidae	Wegner 1979			Bubblenest builders
Scorpaenidae	Popper 1979b			
Scorpaeniformes				
Scorpaenoidei	Popper 1980	Chapman and Sand 1974	Platt 1973, on vestibular responses	Flatfish
Pleuronectiformes	Jørgensen 1976			
Gadiformes				
Bregmaceroidei	Popper 1980	Chapman and Hawkins 1973	Brawn 1961, on sound production	Cod
Gadoidei	Dale 1976			

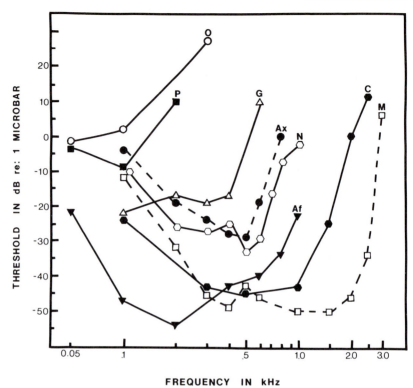

Figure 1-4. Behavioral audiograms for a several teleost species showing hearing sensitivity at different frequencies. Af, *Arius felis* (marine catfish) (Popper and Tavolga in prep.); Ax, *Adioryx xantherythrus* (squirrelfish) (Coombs and Popper 1979); C, *Carassius auratus* (goldfish) (Popper 1972); G, *Gadus morhua* (codfish) (Chapman and Hawkins 1973); M, *Myripristis kuntee* (squirrelfish) (Coombs and Popper 1979); N, *Notopterus chitala* (clown knife fish) (Coombs and Popper in prep.); O, *Opsanus tau* (toadfish) (Fish and Offutt, 1972); P, far-field pressure sensitivity *Limanda limanda* (flatfish) (Chapman and Sand 1974).

1979). Still, generalizations about the Beryciformes cannot yet be made since another species in this group, *Adioryx xantherythrus,* has poor hearing sensitivity (Coombs and Popper 1979, Coombs, Chapter 8). Hearing sensitivity in *Adioryx* in fact resembles that for several other species, including an osteoglossid, *Notopterus chitala* (Coombs and Popper in prep.) which has a connection between swimbladder and the ear that is very like that in *Myripristis*; a cod (*Gadus morhua*) which appears to use the swimbladder in sound detection (Chapman and Hawkins 1973); a toadfish (*Opsanus tau*) (Fish and Offutt 1972) which has, but might no use, the swimbladder; and *Limanda limanda,* a pleuronectid flatfish (Chapman and Sand 1974), which does not have a swimbladder.

Substantially less is known about phylogenetic differences in the behavioral uses of sound, particularly with regard to different types of signals being used by different species. In general, teleost sounds are low frequency (below 1,000 Hz), broad-band pulses, and evidence is accumulating to support the hypothesis that the pulse characteristics rather than the frequency spectrum of the sounds carried information for communication (Winn 1972, Fine et al. 1977, Myrberg, Spanier, and Ha 1978, Myrberg, Chapter 20). The data, however, are limited to species living in shallow inshore waters where the acoustic environment of a soft bottom, rocks, coral heads, and a close air-water interface, may strongly determine the most useful types of signals (Banner 1971, Tavolga 1971, Schuijf, Chapter 14). In fact, literature on bird acoustics (see Marler 1955, Morton 1975) suggests that in areas of high acoustic absorption and scattering, acoustically similar to these shallow water environments, sounds used by birds have some common characteristics to sounds used by fishes (Popper and Coombs 1980a). Thus, acoustic selective pressures may overwhelm interspecific differences related to phylogeny of fishes.

2.3 Morphological Diversity of Acoustic Accessories

Morphological phylogenetic diversity in the peripheral auditory system involves the structure of each portion of the system, as well as the interconnections between structures of the whole system (see Blaxter, Chapter 3). This variation may markedly affect diversity in acoustic sensitivity and processing, though data are still quite limited. Structural diversity includes the gross morphology of the swimbladder which may have single or multiple chambers, and the presence or absence of anterior diverticulae toward the auditory bullae. Furthermore, a compressible gas chamber may be acoustically coupled to the ear, either by attachment of the swimbladder to the ear as in notopterids and some beryciforms; by a bony chain between the bladder and ear as in the Ostariophysi; or by small gas-filled bullae or bubbles that may or may not be swimbladder extensions, as in clupeids, mormyrids, and anabantids. Such connections to a compressible mechanism may significantly affect auditory sensitivity. However, it should be noted that sometimes the presence of the swimbladder (as in cichlids), a direct connection of the swimbladder to the ear (as in notopterids), or coupling of the ear to the swimbladder by Weberian ossicles (as in the ostariophysine *Arius felis*) does not necessarily mean an extended hearing range for a particular species (see Fay and Popper 1975, Coombs and Popper in prep., Popper and Tavolga in press).

3 Structural Organization

Anatomical studies reveal that structural variability of the ear among different fish species exists from the gross to the ultrastructural level. A few organizational principles have appeared that are common to all species, but the functional significance of many structural features remains unclear.

3.1 Gross Features

The position of the parts of the inner ear within the head is one notable variable. The canals and otolith organs lie next to the brain in many teleosts (Furukawa and Ishii 1967a, Dale 1976, Popper 1976). In contrast, the inner ear may be located more laterally in some deep-sea fishes (Popper 1980) and in some chondrosteans and holosteans (Popper 1978a, Popper and Northcutt in prep.). There is also substantial variability in the adherence of the inner ear membranes to the cranial bones. In some cichlids, the whole membranous labyrinth is only loosely adherent to the bones (Popper 1977), while in *Myripristis* and *Notopterus* (Popper 1977, Coombs and Popper in prep.) the adhesion is extremely close, and in holosteans the ears are almost totally embedded in the lateral wall of the cranium (Popper and Northcutt in prep.). Tight adhesion should cause a far better coupling of the maculae to head movements, in comparison to a "floating" labyrinth within the cranial cavity.

The shape of the organs, and the relative sizes of the different organs, also vary considerably (see Retzius 1881). The canal system is far larger than the otolithic system in the flying fish (Exocetidae) and angler fish (Lophiidae), the saccule is particularly large in the gobies (Gobiidae), the lagena is particularly large in the holostean *Amia calva*, while the utricle is particularly large in the marine catfish (Ariidae) (Chardon 1968, Popper and Tavolga in prep.) and lungfish (*Protopterus*) (Retzius 1881).

Perhaps the most consistent gross structural feature of the ear is seen when comparing the Ostariophysi to other teleosts. In Ostariophysi the lagenar otolith is substantially more massive than the saccular otolith, although the total area of the actual sensory epithelium in each macula may be relatively similar in both organs. In contrast, in the nonostariophysine teleosts, the saccular otolith and macula generally are far larger than the lagenar otolith and maculae. The relationship in holosteans and chondrosteans resembles that in Ostariophysi, although the saccular and lagenar otoliths in these groups are less different in relative size than in the Ostariophysi (Popper 1978a, Popper and Northcutt in prep.).

3.2 Otolith Sculpture

The shapes of the otoliths found in teleost ears are species-specific. There is wide interspecific variation in these shapes, although there appears to be less variation in the utricular otolith than in the lagenar otolith. The most extensive variation is found in the saccular otolith (see Figs. 1-5 and 1-6).

The utricular otolith in most species that have been examined tends to be an oblate sphere which sits on top of the utricular macula. The lagenar otolith in nonostariophysine teleosts is generally small and round or triangular in shape, while that in the Ostariophysi is roughly circular and laterally compressed, with scalloping at the edges. The medial side of the lagenar otolith in the Ostariophysi has a sulcus or groove which is the shape of the lagenar sensory epithelium and which contains the gelatinous membrane that covers the epithelium. There is no clear sulcus in the lagenar otolith in most nonostariophysines, although there may be a shallow indentation in the general region of the sensory epithelium.

The saccular otolith in the Ostariophysi is long and delicate with thin fluted longitudinal vanes (see von Frisch 1936a, Adams 1940, Furukawa and Ishii 1967a, Jenkins 1977, 1979a, 1979b) (Fig. 1-5). The shape and size of the saccular otolith is substantially more variable in the nonostariophysines (Fig. 1-6) where it may partially or totally

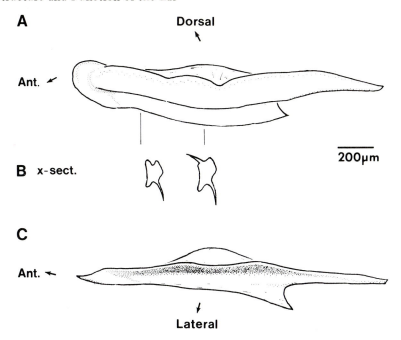

Figure 1-5. Sketches of the saccular otolith from an ostariophysine, the goldfish *Carassius auratus* (5 cm). (A) Medial view, of surface facing the macula. (B) Cross sections at levels indicated by lines, showing thinness and orientation of vanes. (C) Ventral view, showing medial-lateral flattening, fluting, and medial extension of dorsal vane almost orthogonal to ventral vane. Note that in situ the long axis of this otolith is tilted roughly 30° to the anterior-posterior axis of the fish (see Fig. 1-9A).

fill the otolithic sac. Most commonly, the saccular otolith is a laterally flattened elipsoid, with a deep medial sulcus into which the sensory epithelium closely fits (Fig. 1-6A,B). Variants on this shape are seen among most groups of teleosts; extremes range from a condition where one end of the otolith is massive and the other very thin and delicate in the deep-sea salmoniform *Opisthoproctus soleatus* (Fig. 1-6C) (Popper 1980) to the presence of gentle sculpting on the ventral side of the otolith in the gonorynchoform *Chanos chanos* (Fig. 1-6D) (Popper 1981).

There is also substantial interspecific variation in the relationship between the otoliths and the sensory epithelia. Again, the variation seems to be least extensive in the utricle and most extensive in the saccule. The utricular otolith often does not cover the anterior striolar, or transition, region of the macula, but a thin gelatinous extension of the otolith membrane may extend from under the otolith to cover this area (Werner 1929). Similarly, the lagenar otolith often does not cover the anterior or dorsal tip of its macula, while the otolith membrane does cover this region (Platt 1977, Popper 1977).

The relationship between saccular otolith and macula is most variable in the non-ostariophysines. The otolith completely covers the macula and fills, or almost fills,

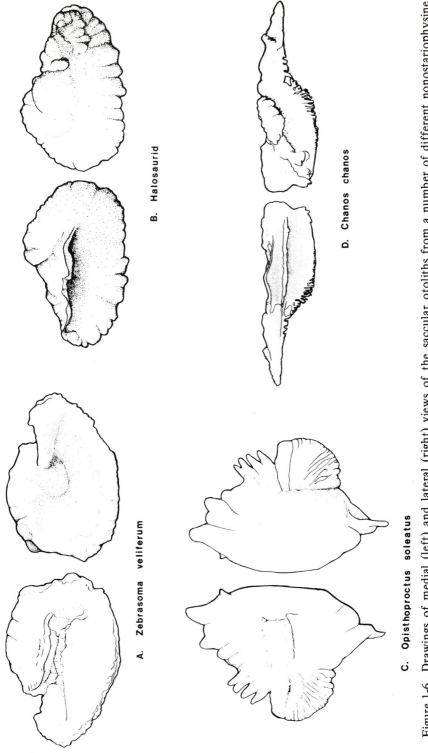

B. Halosaurid

D. Chanos chanos

A. Zebrasoma veliferum

C. Opisthoproctus soleatus

Figure 1-6. Drawings of medial (left) and lateral (right) views of the saccular otoliths from a number of different nonostariophysine fishes. *Zebrasoma* (A) and the Halosaurid (B) are fairly typical of many nonostariophysines in being generally ellipsoid in shape and having a deep groove (sulcus) on the medial side into which the sensory epithelium fits. *Opisthoproctus* (C) (from Popper 1980) and *Chanos chanos* (D) (from Popper 1981) are divergent forms showing some of the possible extremes in sculpturing found among fishes.

the saccular chamber in some species such as in gobiids, holocentrids, and many perciforms (Retzius 1881, Popper 1977). In other species, such as the salmonids (Popper
1976, 1977), the rostral end of the macula is covered only by extensions of the otolith
membrane. In these species the otoliths either are, in some dimensions, smaller than
the maculae or they curve away from the maculae. There are also species where the
saccular otolith may contact as little as half of the macula, as in the myctophids and
the tetraodontids (Popper 1977 unpublished).

3.3 Macular Shape

The overall shape of the sensory epithelia of each of the otolithic organs, as well as
the structure of the otoliths, shows the least interspecific variation in the utricle,
more in the lagena, and the most in the saccule. With a few notable exceptions [the
chondrostean *Scaphirhynchus,* clupeids, the ariid catfish *Arius* (Popper 1978a, Popper
and Platt 1979, Popper and Tavolga in press)] the utricular macula is dish-shaped and
it often has a lateral extension, the lacinia (Fig. 1-7B,C,E,F) (Werner 1929, Flock
1964, Dale 1976, Platt 1977, Popper 1978a, Best and Gray 1980, Popper and Northcutt in prep.). The ostariophysine lagenar macula is broadly curved while in nonostariophysines it is shaped approximately like a fat crescent (Fig. 1-8). The saccular macula
in Ostariophysi is quite narrow across the dorsal-ventral axis and quite elongate (Fig.
1-9A) (Platt 1977, Jenkins 1977, 1979a, 1979b), while in nonostariophysines the
macula is also elongate, but generally wider at the rostral than caudal end (Fig. 1-9B,E).
In many nonostariophysines the macula somewhat resembles the shape of a tadpole,
although exceptions are seen in the complex patterns of the notopterids, anabantids,
gobiids, chondrosteans, and other diverse species (Fig. 1-9F) (Popper 1977, 1978a,
1979a, 1979b, 1980, Wegner 1979).

It is particularly significant, especially when considering the patterns of sensory
cells on the otolithic maculae (Section 3.5) that the actual shapes of each of the otolithic maculae are far more complex than can be shown in a two-dimensional illustration. In fact, the maculae often have curves and bends that cause different macular
regions to be oriented in different planes somewhat off of the main "plane" of each
macula. Nonplanar epithelia are evident in both ostariophysine and nonostariophysine
saccular maculae. It is also significant that in most species the two ears are not aligned
parallel to each other in parasagittal planes; instead, the bilateral mirror symmetry of
each macular pair often involves divergence of the anterior ends away from the midline (Sand 1974, Popper 1976, Wegner 1979).

3.4 Ciliary Bundle Diversity

Use of the scanning electron microscope (SEM) has tremendously increased the
amount of data available regarding the nature of the ciliary bundles in teleost sensory
epithelia. Variability in ciliary bundle length occurs among agnathans (Lowenstein,
Osborne, and Thornhill 1968, Lowenstein 1970, Hoshino 1975) and elasmobranchs
(Lowenstein, Osborne, and Wersäll 1964, Barber and Emerson 1980, Corwin, Chapter
5), as well as in teleosts (e.g., Dale 1976, Platt 1977, Popper 1977, 1978a, 1978b).

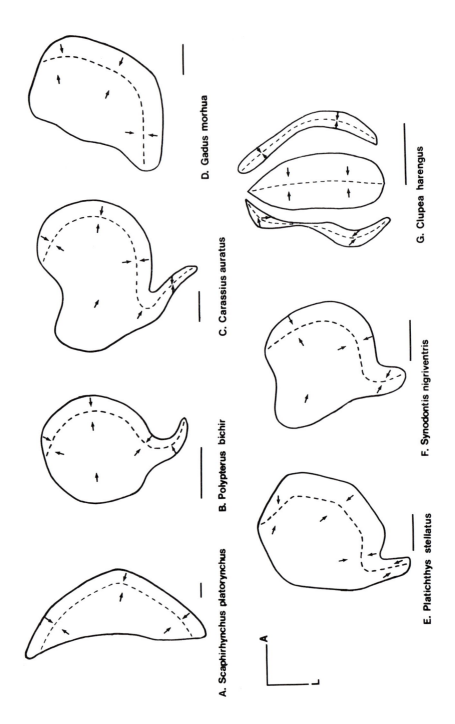

A. Scaphirhynchus platorynchus

B. Polypterus bichir

C. Carassius auratus

D. Gadus morhua

E. Platichthys stellatus

F. Synodontis nigriventris

G. Clupea harengus

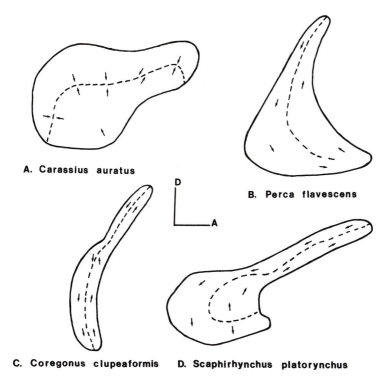

A. Carassius auratus

B. Perca flavescens

C. Coregonus clupeaformis   D. Scaphirhynchus platorynchus

Figure 1-8. Lagenar maculae from four different fish species. (A) *Carassius auratus* (Platt 1977). (B) *Perca flavescens* (Perch) (Popper 1977). (C) *Coregonus clupeaformis* (lake whitefish) (Popper 1976). (D) *Scaphirhynchus platorynchus* (sturgeon) (Popper 1978a).

Several studies of bony fishes show the presence of different forms of ciliary bundles. Summarized data suggest these bundles can be grouped into up to seven distinguishable types based upon the lengths and relative sizes of the stereocilia and kinocilia (Fig. 1-10). These ciliary bundle types have regionally local macular distributions, and most bundle types are found in each species studied (Dale 1976, Enger 1976, Popper 1976, 1977, 1978b, 1979a, Platt 1977, Jenkins 1979a, Popper and Northcutt in prep.). Bundles with very short cilia (K2s1, Fig. 1-10A) consistently occur at the margins of almost all maculae. At least in elasmobranchs these small bundles are on

Figure 1-7. Schematic drawings of the hair cell orientation patterns for the utricular maculae from different fish species. In this and succeeding figures the tips of the arrows indicate the side of the sensory cells in each region at which the eccentrically placed kinocilium is located. The dashed lines indicate the approximate dividing lines between groups of differently oriented hair cells. (A) *Scaphirhynchus platorynchus* (sturgeon) (from Popper 1978a). (B) *Polypterus bichir* (reedfish) (Popper 1978a). (C) *Carassius auratus* (goldfish) (Platt 1977). (D) *Gadus morhua* (codfish) (Dale 1976). (E) *Platichthys stellatus* (starry flounder) (Platt 1976). (F) *Syndontis nigriventris* (upside down catfish) (Platt unpublished). (G) *Clupea harengus* (herring) (Popper and Platt 1979). Scale bar equals 250 μm.

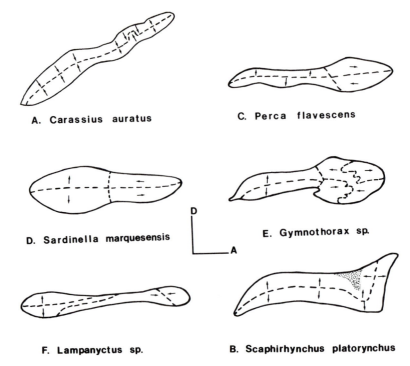

A. Carassius auratus

C. Perca flavescens

D. Sardinella marquesensis

E. Gymnothorax sp.

F. Lampanyctus sp.

B. Scaphirhynchus platorynchus

Figure 1-9. Saccular maculae from different fish species. (A) *Carassius auratus* (Platt 1977). (B) *Perca flavescens* (Popper 1977). (C) *Sardinella marquesensis* (Platt and Popper 1981). (D) *Gymnothorax* sp. (moray eel) (Popper 1979). (E) *Lampanyctus* sp. (lantern fish) (Popper 1977). (F) *Scaphirhynchus platorynchus* (Popper 1978a).

the newest cells in the growing macula (Corwin 1977a), and the very similar marginal ciliary bundles in the bullfrog (*Rana catesbeiana*) are physiologically responsive (E. R. Lewis, pers. comm.). A form of bundle with a very long kinocilium and short stereocilia (K10s3, or type F2 of Popper 1977, Fig. 1-10B) occurs in many fish maculae in a rather restricted band located between the small marginal bundles and much larger central bundles. These bundles also occur as the sole central form in the lacinia of the utricle and in the macula neglecta of goldfish, two areas covered only by nonotholithic membrane (Platt 1977).

While four forms of large ciliary bundles often are encountered (Fig. 1-10C-F), it is not clear how distinct they may be. All four bundle types frequently have overlapping distributions, but some clearly are the preponderant form in certain macular regions. There are a short and a tall version of a large bundle in which the tallest stereocilia are roughly half as long as the kinocilium (K6s3, K8s4, or type F1 of Popper 1977, 1978b, Fig. 1-10C,D). Similarly, there are a short and a tall version of a large bundle in which the tallest stereocilia nearly reaching the tip of the kinocilium (K6s5, K8s7, or type F3 of Popper 1978b, Fig. 1-10E,F). At present, the only clearly defined distribution appears to be the consistent occurrence of large numbers of the tall bundles with tall stereocilia in some macular regions believed to have an auditory function (Platt 1977, Popper 1977, Popper and Platt 1979, Platt and Popper 1981).

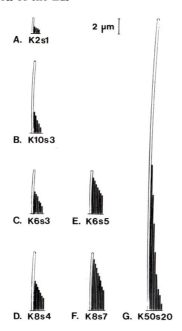

Figure 1-10. Schematic drawings of different types of ciliary bundles found on various sensory maculae in fishes. Ciliary bundles A-F are generally found on the otolith maculae while G is only found on the cristae of the semicircular canals. (After Platt 1981).

One additional type of ciliary bundle is found only in the semicircular canals and is extremely tall, with a kinocilium that is more than twice as long as the longest stereocilia (K50s20; Fig. 1-10G). These bundles, in fishes as in other vertebrates, are the only type found on the cristae (Wersäll and Bagger-Sjöbäck 1974, Dale 1976). It is important to note that cilia with bulbed tips have not been observed in any fish macula, although they are common in amphibian and reptilian auditory maculae (Lewis and Li 1975, Weiss et al. 1976, Miller 1980).

3.5 Orientation Patterns

Each otolithic macula in all vertebrates can be divided into a number of regions within which all the ciliary bundles are morphologically polarized in the same general direction. A dividing boundary line can be drawn between the regions having directly opposing hair cell orientations, while there is a more gradual transition between regional cell groups that are oriented less than 90° from one another. Although all of the ciliary bundles in each region are oriented in the same general direction, individual cells may lie 5° to 10° from the mean orientation of their group, and there may be a gradual change of orientation across the entire population, particularly when the macula is curved. The utricular macula in most species lies roughly horizontally and extends out laterally, but the saccular and lagenar maculae lie approximately vertically, with the long axis in a generally rostrocaudal direction. It must be kept in mind that the macular surfaces are rarely planar, that there is a three-dimensional pattern of orientations for each macula, which is not readily shown by planar projection maps (Figs. 1-7-1-9).

The hair cell orientation patterns in the end-organs of the pars superior are fairly stable throughout the fishes, while there is more interspecific variation in the end-organs of the pars inferior. In the cristae of each semicircular canal all of the sensory cells are oriented in the same direction. In the horizontal canal the kinocilium is always on the side of the bundle facing the utricle, while in the two vertical canals the kinocilium is on the side of the bundle facing away from the utricle (Lowenstein and Wersäll 1959, Lowenstein 1971).

There is little interspecific variation in the orientation patterns in the utricular macula. Usually an anteriolateral band of cells faces posteriomedially, and a more central population faces anterolaterally (Fig. 1-7 and Jenkins, Chapter 4). The line dividing these populations curves to lie along the midline of the region of large hair cell bundles called the striola (Werner 1929, 1960), and continues along the midline of the lacinia when it is present. Most bony fishes studied including nonteleosts (Fig. 1-7A,B), and nonostariphysan teleosts (Fig. 1-7D,E) and Ostariophysi (Fig. 1-7C,F), share this pattern with other vertebrates, but a few exceptions have now been found among teleosts. The most striking variation on the "standard" utricular pattern is found in the herrings and related clupeids (Clupeiformes) which have a tripartite utricular macula with a dividing line along each part (Fig. 1-7G) (Popper and Platt 1979, Best and Gray 1980, Platt and Popper 1981). A second atypical pattern occurs in the marine catfish *Arius felis,* which has a slim equatorial macula running around about three-fifths of a large otolith, with the dividing line horizontally bisecting the two opposing hair cell populations (Popper and Tavolga in press). Even in both of these unusual maculae though, the opposition is between cell groups with the kinocilia facing each other.

Data on the macula neglecta are very limited because it is extremely small in most species, but in several species it is in two distinct patches (Retzius 1881). In goldfish, at least, one patch has its cells oriented rostrally, while the other patch has cells oriented caudally (Platt 1977) and a two-patch neglecta with opposing hair cells has also been found in the one nonostariophysine that has been studied, the anabantid *Colisa labiosa* (Wegner 1979). These are the only reports of orientation pattern in the macula neglecta for bony fishes, but in elasmobranchs comparative data on several species shows a gradation from a single patch with oppositely oriented cells to the two cell groups of opposite polarization making up two separate patches; this gradation may be related to increasing auditory sensitivity (Corwin 1977b, 1978, 1979, Chapter 5).

In comparison to the utricular maculae, there is more interspecific variaton in the hair cell orientation patterns of teleost lagenar maculae and extensive variation in the patterns of the saccular maculae. The hair cells of the lagenar maculae in almost all bony fishes are divided into two major populations, opposing each other across a curving dividing line (Fig. 1-8) (Popper 1977, 1979a, 1980). In some species (Fig. 1-8A) the angle of the opposition may change along the macula, so that some macular regions contain opposing hair cell groups oriented parallel to the opposition boundary line, while in other regions the opposing orientations face each other across the boundary, as in the utricular macula (see Fig. 1-7). In other species only one of the two opposition patterns is found (Fig. 1-8C,E). In each case, however, the opposition boundary divides the macula into approximately equal areas. Furthermore, the primary interspecific differences seen in the lagenar macula are related to differences in macula shape and curvature, rather than to changes in the relative positions of the oriented groups as seen in the saccular macula (see below).

The saccular macula in fishes shows far more variability in hair cell orientation pattern than any other macula. Two primary patterns are found, one having two hair cell orientation groups and the other having four groups. In addition, there are a substantial number of major variants on the basic four-group pattern. The pattern in the Ostariophysi and in nonteleost bony fishes that have been studied is similar to that found in tetrapods (see Spoendlin 1964, Lindeman 1969) which have two oppositely oriented populations; but unlike the utricular and lagenar maps, the cells in these saccular populations are oriented away from a single longitudinal dividing line (Hama 1969, Saito 1973, Lowenstein 1974, Platt 1977, Popper 1978a, 1978b, Jenkins 1979a, 1979b, Popper and Northcutt in prep., Popper and Tavolga in prep.). The pattern in Ostariophysi (Fig. 1-9A) is quite consistent, the variation being limited to slight differences in the relationship of the two orientation groups. For example, the goldfish has a tongue of ventrally oriented cells extending into the dorsal anterior macular region (Platt 1977) and similar patterns have been seen in other members of the same family (Cyprinidae) but not in members of other osatriophysine groups including characids (Popper unpublished) or the Siluriformes (catfishes) (Jenkins 1979a, 1979b, Popper and Tavolga in press). Several, but not all, catfishes have alternating rows of oppositely oriented hair cells at the anterior tip of the macula (Jenkins 1979a, 1979b). The holostean and chondrostean saccular macula shows bidirectionality as in the Ostariophysi, but because the macular shape is more bent there is a curving gradation of orientations relative to the vertical plane (Fig. 1-9F) (Popper 1978a, Popper and Northcutt in prep.).

A saccular pattern that is unusual among vertebrates has been found in almost all other teleosts. In all major taxonomic divisions of fishes other than the Ostariophysi (see Section 2.1), with one known exception, the rostral region of the macula has groups of cells oriented horizontally in at least two discrete opposing groups, while the caudal region of the maculae is similar to the general vertebrate pattern with two groups showing opposite dorsal-ventral orientations (Fig. 1-9B,E) (Dale 1976, Enger 1976, Jørgensen 1976, Platt 1976, Popper 1976, 1977, 1979a, 1979b, 1980, Wegner 1979, Platt and Popper 1981). When variation does occur in the caudal region of the macula, it generally involves the presence of small groups of horizontally oriented cells on the most caudal part of the macula, isolated from the horizontal cells at the rostral end of the macula. This pattern has been seen in a number of species including two gadiforms, the cod *Gadus morhua* (Dale 1976) and a deep-sea form, *Bregmacerous* sp. (Popper 1980), as well as in a scarid (Popper unpublished). Whereas there is little interspecific variation in the relationships between the dorsally and ventrally oriented caudal groups of hair cells, there is substantial variation among the horizontally oriented rostral groups (Popper 1977). This variability is in the group boundaries, while in each case the pattern retains the horizontal orientations. The most common pattern found in many, though not all, perciforms (Popper 1977) and some other species has the posteriorly oriented cell group located above the anteriorly oriented group (Fig. 1-9B, C). Variants on this pattern include an alternation of position between the two cell groups, so that both polarizations occur on both the dorsal and ventral side of the macula (Fig. 1-9D), as has been seen in the Elopormopha (moray eel and some relatives) (Popper 1979a) and in an osteoglossid, *Pantodon buchholzi* (Popper 1979b). Another pattern of the horizontally oriented cells shows that one of the two hair cell groups may be anterior to the other. For example, in the Myctophidae the posteriorly oriented cells are found rostral to the anteriorly oriented cells (Fig. 1-9E) (Popper 1977), while the opposite pattern occurs in one species of squirrelfish (Holocentridae), *Myripristis kuntee,* but not in another member of the same family, *Adioryx xanthyrethrus* (Popper 1977). The one known exception to the presence of horizontally oriented cells in the nonostariophysine teleosts is found in a mormyrid (*Gnathonemus* sp.), where only vertically oriented cells have been observed (Popper 1979b). This pattern is particularly striking since no other member of the same taxonomic group (Osteoglossomorpha) that has been examined lacks horizontally oriented cells, although members of this group show some extraordinarily diverse saccular hair cell orientation patterns (Popper 1979, Coombs and Popper in prep.). While earlier reports, based on work with the transmission electron microscope, indicated a similar bidirectional pattern in a gadid, *Lota lota* (Wersäll et al. 1965), recent examination of the saccular macula in *Lota* with the scanning electron microscope has shown the presence of horizontally oriented cells on the rostral region of the macula (Popper unpublished).

4 Form and Function?

The extensive data now available on the structure of the inner ear in fishes have resulted from the cascading increase in the use of the scanning electron microscope to study the sensory surfaces. Consequently, much of the following discussion on the function of fish ears is speculative, and based largely on ultrastructure of the teleost ear, along with results from physiological studies in other vertebrates in which there are similar sensory structures. There are three major questions to be considered. First, what effect might the variety of coupling mechanisms between the acoustic medium and receptors have on the processing of acoustic input, both between receptor organs and at different regions within a receptor? Second, what is the functional role of the form and orientation of the ciliary bundles? Finally, what, in fact, are the roles of each of the end-organs found in the teleost ear?

4.1 Coupling and Intramacular Functional Variation

In considering coupling between the various portions of the auditory system in fishes, it is necessary to include the mechanism for getting sound from the water to the ear, both directly or via the swimbladder, the coupling of the swimbladder to the ear, and from the otolith to the sensory cells. Also worth considering are the coupling of the ear itself to the rest of the fish's body and the different types of coupling mechanisms that might occur between the otoliths and the sensory maculae.

   Recent studies of fish auditory mechanisms have demonstrated that sound gets to the ear by at least two possible pathways (Chapman and Sand 1974, Fay and Popper 1974, 1975, 1980, Fay and Olsho 1979, Hawkins, Chapter 6, Buwalda, Chapter 7).

Furthermore, it is now clear that no matter what the pathway, the ultimate mechanism for stimulation of the sensory hair cell is a shearing action (de Vries 1950, von Békésy 1960, Hudspeth and Corey 1977) which, in the teleost otolith organs, results from relative motion between the otoliths (or otolithic membrane) and the sensory epithelium (see Dijkgraaf 1960). One of the two pathways involves direct stimulation of the inner ear by the particle displacement component of the sound and, as a result, there is no intervening function of the swimbladder. The second pathway involves the swimbladder, which converts impinging pressure changes to particle displacement, and this output is sufficient for stimulation of the ear (see Fay and Popper 1974, 1975, 1980). The physical coupling of the swimbladder to the ear should then have an effect on the sound transmission from the swimbladder to the ear. Until recently, it had been suggested that the proximity of the swimbladder to the ear, whether through the presence of Weberian ossicles or of anterior diverticula from the swimbladder to the bulla, would affect auditory sensitivity. However, as discussed in Section 2.2, recent data from *Arius felis* (Popper and Tavolga in press) and *Notopterus chitala* (Coombs and Popper in prep.) shows that this hypothesis must be reexamined. In fact, these data lead to questioning the whole nature of coupling between the outside and the inner ear. It is necessary to ask questions regarding the functional significance of the coupling and the specific role this coupling plays in auditory detection.

Within the ear it is necessary to consider the coupling between displacement signals and the otoliths, between displacement signals and the sensory epithelium, and the coupling between the otoliths and the sensory epithelium. During acoustic stimulation the otolith and the fish's body, as a result of their markedly different densities, move at different amplitudes and phase (see Dijkgraaf 1960, Wever 1969, Fay 1978b, Corwin 1979, Fay and Popper 1980, Popper 1981), resulting in a shearing action on the sensory epithelia. However, movement of the otoliths also would be affected by the frictional forces set up when the otolith moves within the endolymphatic fluid of the ear. Since the saccular otolith is highly sculpted it is likely that the drag set up by the various otolith surfaces moving against the fluid of the ear would affect the patterns through which the otoliths move (Popper 1978b, 1981). Thus, the movement of the otoliths relative to the sensory epithelium could involve more than a single direction and, in fact, might be a complex rocking motion (Sand and Michelsen 1978).

As a result of this potentially complex motion, it is likely that different macular regions would be stimulated under different, as yet undefined, acoustic conditions. Perhaps such different conditions might include different frequencies or sounds from different directions or at different distances. In addition to differences imposed by the otolith shape, it seems likely that regional differentiation might occur, depending on whether the different sensory cell regions are closely coupled to the otoliths or whether the ciliary bundles are in a region covered only by a gelatinous membrane. In the latter case, such as the macula neglecta, it might be expected that the motion imposed upon the ciliary bundles would be far different, possibly in phase relations as well as magnitudes, from those cases where the otolith lies in close proximity to the sensory cells (see Platt 1977, Popper 1978b, 1981, Corwin 1979). Although it may be a substantially different system, the basilar papilla of the alligator lizard has regions with differences in coupling between the sensory cells and the tectorial mass, and these differences are associated with different responses to acoustic signals (Weiss et al.

1976). Observations in fishes lead to the suggestion of a possible "placelike" mechanism in teleosts for signal processing in the ear. While data are still very limited, recordings from single fibers from the goldfish sacculus show that neurons presumably innervating different regions of the macula respond best to different frequencies (Furukawa and Ishii 1967a, Fay 1978a), indicating that different frequencies may be providing maximum stimulation to different regions (see Fay, Chapter 10). Similarly, microphonic recordings from the ear in the perch (*Perca*) show that different regions along the macula are most responsive to different frequencies (Sand 1974, Sand and Enger 1974).

Whereas the mechanism of stimulation of the sensory cells has been "modeled" as a relatively simple system, with the otolith and macula moving relative to one another as a result of amplitude and phase differences in their responses to the impinging signals, it is possible that the differences seen in the coupling of the membranous labyrinth to the bones of the skull also would affect motions of the sensory epithelia. While it would be difficult to indicate how this will affect motion of the systems, it is possible to envision four, rather than three sets of moving objects: the sensory cells, the gelatinous membranes, the otoliths, and the cranium. Each of these would have their own motional characteristics, and since they are interconnected, motions of one would affect motions of the other.

4.2 Ciliary Bundle Form

There has been considerable speculation that there might be a clear correlation between ciliary bundle length and frequency sensitivity, but only a few direct tests have been done in tetrapods. In the alligator lizard the basal portion of the basilar papilla is innervated by fibers with high "characteristic frequencies" and contain hair cells with tall bundles, while fibers with low-frequency sensitivity come from the apical portion where hair cells have short bundles (Weiss et al. 1976).

In agnathans, tall ciliary bundles appear to be localized to regions of known gravistatic sensitivity, with short bundles in regions of vibrational sensitivity (Lowenstein 1970, Hoshino 1975). In the goldfish, on the other hand, tall ciliary bundles occur most densely in the posterior part of the saccular macula (Platt 1977), where the best low-frequency auditory responses are found (Fay 1978a). In goldfish, the utricular lacinia and the macula neglecta both have bundles with very long kinocilia, but it is unknown whether these regions are sensitive to sound or vibration in teleosts, as they are in elasmobranchs (Lowenstein and Roberts 1951, Corwin 1979).

Very short small bundles that are widespread on the utricular floor in goldfish (Platt 1977) are very similar to those found in the bullfrog utricle, a region that is innervated by purely gravistatic afferent fibers (Baird 1979). If short cilia are associated with low-frequency gravistatic function, however, it becomes remarkable that the very tallest bundles, those of the semicircular canal cristae, also are in low-frequency detectors. Further complicating the situation is the presence of tall "saccular-like" bundles in the utricular macula in herring (Popper and Platt 1979, Platt and Popper 1981) where the utriculus is believed to have an auditory function (Blaxter and Denton 1976, Denton, Gray, and Blaxter 1979, Blaxter, Denton, and Gray, Chapter 2, Best and Gray 1980).

Our understanding of the functional significance of the different bundle types is still confusing and contradictory, particularly when comparing results from fishes and the alligator lizard. More direct data are needed involving measurement of the frequency response characteristics from specific macular regions with known ciliary bundle types, or directly from sensory cells with different bundles. However, even in the latter case, the natural response characteristics of the cells may not be determinable from direct cellular recordings unless the ear is intact since it seems likely that the response of the different cells depends on coupling characteristics to the other components of the ear.

4.3 Hair Cell Orientation Patterns

Orientation patterns of the sensory hair cells presumably give clues to the direction of maximum sensitivity for a given population in each macula (Dale 1976, Enger 1976, Platt 1977, 1981, Popper 1977, 1978b, 1981). The orientation and curvature of the maculae within the head determines the specific orientation of the sensory cell populations relative to the world outside of the fish. In the utricle, for example, the fact that the dished macula curves up the lateral walls as well as covering the floor (in most species examined) means that there is no position that is really one of zero gravistatic shear for all the cells at once (Flock 1964).

There has been considerable speculation regarding the mechanisms for sound localization by fishes (van Bergeijk 1964, Sand 1974, Tavolga 1977b, Fay 1978b, Fay and Popper 1980, Popper 1981, Popper and Coombs 1980a, 1980b, Schuijf and Buwalda 1980, Schuijf, Chapter 14). It is becoming apparent that not only are mechanisms significantly different from those encountered in any terrestrial vertebrate, but that localization is likely to involve the individual maculae, as well as binaural effects. While these mechanisms will not be discussed here (but see Schuijf, Chapter 14), it is worth asking questions regarding the functional significance of the different hair cell orientation patterns of the auditory maculae encountered in different species. Bidirectional populations allow discrimination of both compression and decompression (rarefaction) phases of sound, so they can presumably give phase and timing cues (Furukawa and Ishii 1967a, 1967b, Denton and Gray 1979, Fay, Chapter 10, Hawkins and Horner, Chapter 15). The presence of sensory cells oriented in widely different directions could enhance central sensitivity to particle displacement signals from different directions. However, it is not clear why in nonstariophysines there is such great interspecific variation in the patterns of the horizontally oriented patterns of cells on the rostral end of the saccular macula, but not in the vertically oriented patterns on the caudal end of the macula.

4.4 End-Organ Function

Comparative study of the shape and orientation patterns in all three otolithic maculae in fishes suggests that the pars superior contains patterns that are more evolutionarily stable, and that the pars inferior contains patterns that are more variable. Experiments suggesting separate major roles for these different maculae provide some explanation for this variability among the different sensory epithelia. In the pars superior the canal cristae pattern is constant throughout the cartilagenous fishes, the bony fishes, and all other vertebrates. The utricular hair cell orientation pattern shows some changes, from the patchy pattern in elasmobranchs (Lowenstein et al. 1964, Barber and Emerson 1980) to the two opposed facing populations in a broad or ovoid macula in most bony fishes (see Flock 1964, Dale 1976, Platt 1977, Popper 1978a) and other vertebrates (Lindeman 1969, Spoendlin 1964). The circular equatorial macula of the marine catfish, *Aris felis* (Popper and Tavolga in press), is probably a derivation from this form. The tripartite macula of the clupeid utricle, while unique, still retains the facing opposition of hair cell populations (Denton and Gray, 1979, Popper and Platt 1979, Best and Gray 1980, Platt and Popper 1981), presumably derived from a more generalized utricular macula. We have no comparative data on teleost macula neglecta patterns, although such data would be of interest and value, particularly in light of findings with elasmobranchs that this organ is an auditory receptor (Corwin 1979, Chapter 5) whose size and shape changes in correlation to the ecology of the animals (Corwin 1978).

The canals mediate responses to head rotation in all three planes. Such stimuli should be essentially identical, except for the magnitude of rotational acceleration, in all vertebrates. The lack of variability in orientation of the cristae sensory cells might well reflect this lack of variety in the stimulus. The utricles largely, though probably not exclusively, mediate postural responses in teleosts. Here, the habits of some species normally swimming nose-down, nose-up, upside-down, or rolled to one side lead to an expectation of variability. However, the utricular maculae of nearly all teleosts examined, including such species (Fig. 1-7E,F) have a remarkable similarity in pattern, except for the marine catfish and herring mentioned above; the patterns appear derivable from a simple form like that in the sturgeon (Popper 1978a), a chondrostean (Fig. 1-7). It should be noted that vibration sensitivity is found in the utricle of elasmobranchs (Lowenstein and Roberts 1951, Budelli and Macadar 1979), and at least in the sculpins (Cottidae) (Enger 1963) and turbots (Pleuronectidae) (Platt 1973) among teleosts. In the case of the herring, it is possible that the unique pattern is related to this group of fishes using the utricle for sound detection (Denton, Gray, and Blaxter 1979).

In the pars inferior, the lagenar cell orientation pattern shows some variation, with facing opposed populations as in the utricle, in some cases, and in others the opposition is better described as "antiparallel." The saccular macula has by far the widest variability in all aspects of its structure and ultrastructure. The hair cell orientation pattern is particularly variable, and ranges from the rather simple bidirectional divergence of two orientations in the Ostariophysi and tetrapods, through the "quadrant" orientation patterns of many teleosts, to the elaborate pattern of several patches found in several taxonomically diverse forms.

There is behavioral and physiological evidence that the lagena participates in both gravistatic (von Holst 1950) and acoustic (Furukawa and Ishii 1967a, Fay 1978a, 1978b, 1978c, Chapter 10, Fay and Olsho 1979) function. Its macula shows interspecific variability over a wider range of species than found for the utricular macula. Finally, the major function of the saccule in most teleosts is believed to be acoustic reception, although some very low frequency responses may occur (Furukawa and Ishii 1967b).

Neuroanatomical data show that there are differences between anterior (pars superior) and posterior (pars inferior) eighth nerve branches in their distribution to brainstem nuclei in all fish classes (Bell, Chapter 19, McCormick, Chapter 18, Northcutt, Chapter 16, Roberts, Chapter 17). It is not yet clear whether this general distinction holds at the finer level of individual fibers, or alternatively whether, for example, acoustic and gravistatic fibers from a single end-organ may segregate into different brainstem target nuclei.

From these initial correlations, it seems reasonable to suggest that acoustic input can be more complex than gravistatic input, so that portions of the sensory macula with complex orientation patterns may reflect complexity in that portion for peripheral auditory processing. A "phylogenetic experiment" testing this hypothesis and lending support is the implied acoustic role of the utricle in the herrings (Denton and Gray 1979, Denton et al. 1979, Blaxter et al., Chapter 2). In this case, where the utricle rather than the saccule is coupled to a gaseous compressible system, the utricular macula has become far more complex than usual. A converse case occurs in the flatfishes, which lie on one side as an adult, and utilize the saccule rather than the utricle as the major gravistatic organ (Schöne 1964, Platt 1973). However, in these fishes the saccular macula has a quadrant pattern which is similar to that found in most other non-ostariophysines and not unique as originally supposed (Jørgensen 1976, Platt 1976).

5 Meanings of Diversity?

We have described the ultrastructural diversity of each macula in the fish ear. Structural variability in coupling systems and in ciliary bundles may reflect functional differences in the auditory system or alternatively, the variability may reflect, to a greater or lesser degree, phylogentic relationships among fishes. The particular significance of the variability is not known, although two preliminary suggestions can be made. One suggestion is that the variability reflects real functional differences in the ears of different species as adaptations to handle different types of acoustic detection and processing chores. A second possibility is that different species have evolved a variety of different mechanisms to analyze similar acoustic signals to derive similar data. In either case, it can be suggested that each macula should be considered as a receptor mosaic (Platt 1981). In structural and ultrastructural features, and in functional innervation, it is clear that there are distinguishable differences in different regions of the sensory epithelia and, as a consequence, it would not be overly surprising to find multiple roles, particularly in the saccule and lagena, within a single macula in a single species.

Further, and more specific, suggestions as to the functional and evolutionary significance of the diversity seen in the teleost auditory system seem premature at this time. In spite of our now considerable knowledge of the structure of the auditory system, there is still a real paucity of data regarding function (see Fay, Chapter 10, for a further discussion of these data). In doing the appropriate physiological experiments, it is imperative that experimenters keep in mind the two points developed since the beginning of this chapter: (1) It is not feasible to talk just of "the fish ear" since the structural data now suggests that there may be striking variability in the way that the ears of different species analyze sounds; and (2) There should not be fixation upon certain types of functions for any particular sensory epithelial region in fish ears, since even the limited evidence now indicates that various organs may have different functions in different species, and there is the potential for multiple functions even within a single organ.

Acknowledgments. Portions of the work reported here was supported by U.S. Public Health Service Grant NS-13946 from the National Institute of Neurological and Communicative Diseases and Stroke to C. P. and by NINCDS Grant NS-15090, National Science Foundation Grant BNS 78-22441 and NINCDS Research Career Development Award Number NS-00312 to A.N.P.

References

Adams, L. A.: Some characteristics of otoliths of American Ostariophysi. J. Morphol. 66, 497-523 (1940).

Baird, R.: Correspondences between structure and function in the bullfrog utricle and lagena. Soc. Neuroci. Abstr. 5, 15 (1979).

Banner, A.: Propagation of sound in a shallow bay. J. Acoust. Soc. Am. 49, 373-376 (1971).

Barber, V. C., Emerson, C. J.: Scanning electron microscopic observations on the inner ear of the skate, *Raja ocellata*. Cell Tissue Res. 205, 199-215 (1980).

van Bergeijk, W. A.: Directional and nondirectional hearing in fish. In: Marine Bio-Acoustics. Tavolga, W. N. (ed.). Oxford: Pergamon Press, 1964, pp. 281-299.

Best, A. C. G., Gray, J. A. B.: Morphology of the utricular recess in the sprat. J. Mar. Biol. Assoc. U.K. 60, 703-715 (1980).

Blaxter, J. H. S., Denton, E. J.: Function of the swimbladder-inner ear-lateral line system of herring in the young stages. J. Mar. Biol. Assoc. U.K. 56, 487-502 (1976).

Brawn, V. M.: Sound production by the cod (*Gadus callarias* L.). Behaviour 18, 239-255 (1961).

Budelli, R., Macadar, O.: Statoacoustic properties of utricular afferents. J. Neurophysiol. 42, 1479-1494 (1979).

Carlström, D.: A crystallographic study of vertebrate otoliths. Biol. Bull. 125, 441-463 (1963).

Chapman, C. J., Hawkins, A. D.: A field study of hearing in the cod, *Gadus morhua* L. J. Comp. Physiol. 85, 147-167 (1973).

Chapman, C. J., Sand, O.: Field studies of hearing in two species of flatfish, *Pleuronectes platessa* (L.) and *Limanda limanda* (L.) (Family Pleuronectidae). Comp. Biochem. Physiol. 47A, 371-385 (1974).

Chardon, M.: Anatomie Comparée de l'Appareil de Weber et des Structures Connexes chez les Siluriformes. Musée Royal de l'Afrique Centrale, Annales Ser. 8 Sciences Zoologisches N°· 169 (1968).

Coombs, S., Popper, A. N.: Hearing differences among Hawaiian squirrelfish (family Holocentridae) related to differences in the peripheral auditory system. J. Comp. Physiol. 132A, 203-207 (1979).

Corey, D. P., Hudspeth, J.: Ionic basis of the receptor current in a vertebrate hair cell. Soc. Neurosci. Abstr. 5, 18 (1979).

Corwin, J. T.: Ongoing hair cell production, maturation and degeneration in the shark ear. Soc. Neurosci. Abstr. 3, 4 (1977a).

Corwin, J. T.: Morphology of the macula neglecta in sharks of the genus *Carcharhinus*. J. Morph. 152, 341-362 (1977b).

Corwin, J. T.: The relation of inner ear structure to the feeding behavior in sharks and rays. Scanning Electron Microscopy/1978, 1105-1112 (1978).

Corwin, J. T.: Parallel channels for sound detection in the fish ear. Soc. Neurosci. Abstr. 5, 18 (1979).

Dale, T.: The labyrinthine mechanoreceptor organs of the cod *Gadus morhua* L. (Teleostei: Gadidae). Norw. J. Zool. 24, 85-128 (1976).

Demski, L. S., Gerald, J. W., Popper, A. N.: Central and peripheral mechanisms of teleost sound production. Am. Zool. 13, 1141-1167 (1973).

Denton, E. J., Gray, J. A. B.: The analysis of sound by the sprat ear. Nature 282, 406-407 (1979).

Denton, E. J., Gray, J. A. B., Blaxter, J. H. S.: The mechanics of the clupeid acoustico-lateralis system: Frequency responses. J. Mar. Biol. Assoc. U.K. 59, 27-47 (1979).

de Vries, H.: The mechanics of the labyrinth otoliths. Acta Otolaryngol. 38, 262-273 (1950).

Dijkgraaf, S.: Hearing in bony fishes. Proc. R. Soc. London Ser. B 152, 51-54 (1960).

Dooling, R.: Psychophysical determinations of hearing capabilities in birds. In: Comparative Studies of Hearing in Vertebrates. Popper, A. N., Fay, R. R. (eds.). New York: Springer-Verlag 1980, pp. 261-288.

Eatock, R. A., Corey, D. P., Hudspeth, A. J.: Adaptations in a vertebrate hair cell: Stimulus-induced shift of the operating range. Soc. Neurosci. Abstr. 5, 19 (1979).

Enger, P. S.: Single unit activity in the peripheral auditory system of a teleost fish. Acta Physiol. Scand. 59, Suppl. 210, 1-48 (1963).

Enger, P. S.: On the orientation of haircells in the labyrinth of perch (*Perca fluviatilis*). In: Sound Reception in Fish. Schuijf, A. and Hawkins, A. D. (eds.). Amsterdam: Elsevier, 1976, pp. 49-62.

Fänge, R., Larsson, Å., Lidman, V.: Fluids and jellies of the acousticolateralis system in relation to body fluids in *Coryphaenoides rupestris* and other fishes. Marine Biol. 17, 180-185 (1972).

Fay, R. R.: Coding of information in single auditory nerve fibers of the goldfish. J. Acoust. Soc. Am. 63, 136-146 (1978a).

Fay, R.: Sound detection and sensory coding by the auditory systems of fishes. In: The Behavior of Fish and Other Aquatic Animals. Mostofsky, D. (ed.). New York: Academic Press, 1978b, pp. 197-236.

Fay, R. R.: Phase-locking in goldfish saccular nerve fibers accounts for frequency discrimination capacities. Nature 275, 320-322 (1978c).

Fay, R. R., Olsho, L.: Discharge patterns of lagenar and saccular neurons of the goldfish eighth nerve: Displacement sensitivity and directional characteristics. Comp. Biochem. Physiol. 62A, 377-386 (1979).

Fay, R. R., Popper, A. N.: Acoustic stimulation of the ear of the goldfish (*Carassius auratus*). J. Exp. Biol. 61, 243-260 (1974).

Fay, R. R., Popper, A. N.: Modes of stimulation of the teleost ear. J. Exp. Biol. 62, 379-387 (1975).

Fay, R. R., Popper, A. N.: Structure and function in teleost auditory systems: In Comparative Studies of Hearing in Vertebrates. Popper, A. N., Fay, R. R. (eds.). New York: Springer-Verlag, 1980, pp. 1-42.

Fine, M., Winn, H., Olla, B.: Communication in fishes. In: How Animals Communicate. Sebeok, T. (ed.). Bloomington: Indiana University Press, 1977, pp. 472-518.

Fish, J. F., Offutt, G. C.: Hearing thresholds from toadfish, *Opsanus tau*, measured in the laboratory and field. J. Acoust. Soc. Am. 51, 1318-1321 (1972).

Flock, Å.: Structure of the macula utriculi with special reference to directional inter-play of sensory responses as revealed by morphological polarization. J. Cell Biol. 22, 413-431 (1964).

Flock, Å.: Electron microscopic and electrophysiological studies on the lateral line canal organ. Acta Otolargyngol. Suppl. 199, 1-90 (1965).

Furukawa, T.: Sites of termination on the saccular macula of auditory nerve fibers in the goldfish as determined by intracellular injections of procion yellow. J. Comp. Neurol. 180, 807-814 (1978).

Furukawa, T., Ishii, Y.: Neurophysiological studies on hearing in goldfish. J. Neuro-physiol. 30, 1377-1403 (1967a).

Furukawa, T., Ishii, Y.: Effects of static bending of sensory hairs on sound reception in the goldfish. Jpn. J. Physiol. 17, 572-588 (1967b).

Gerald, J. W.: Sound production during courtship in six species of sunfish (Centrar-chidae). Evolution 25, 75-87 (1971).

Greenwood, P. H.: Interrelationships of osteoglossomorphs. In: Interrelationships of Fishes. Greenwood, P. H., Miles, R. S., Patterson, C. (eds.). London: Academic Press, 1973, pp. 307-332.

Greenwood, P. H., Rosen, D. E., Weitzman, S. T., Myers, G. S.: Phyletic studies of teleost fishes, with a provisional classification of living forms. Bull. Am. Mus. Nat. Hist. 131, 339-456 (1966).

Hama, K.: A study on the fine structure of the saccular macula of the gold fish. Z. Zellforsch. 94, 155-171 (1969).

Hama, K., Saito, K.: Gap junctions between the supporting cells in acoustico-vestibular receptors. J. Neurocytol. 6, 1-12 (1977).

Hawkins, A. D., Johnstone, A. D. F.: The hearing of the Atlantic salmon, Salmo salar. J. Fish Biol. 13, 655-673 (1978).

Highstein, S. M., Politoff, A. L.: Relation of interspike baseline activity to the spon-taneous discharges of primary afferent fibers from the labyrinth of the toadfish, Opsanus tau. Brain Res. 150, 182-187 (1978).

Hoshino, T.: An electron microscopic study of the otolithic maculae of the lamprey (Entosphenus japonicus). Acta Otolaryngol. 80, 43-53 (1975).

Hudspeth, A. J., Corey, D. P.: Sensitivity, polarity, and conductance change in the re-sponse of vertebrate hair cells to controlled mechanical stimuli. Proc. Nat. Acad. Sci. 74, 2407-2411 (1977).

Hudspeth, A. J., Jacobs, R.: Stereocilia mediate transduction in vertebrate hair cells. Proc. Nat. Acad. Sci. 76, 1506-1509 (1979).

Iversen, R. T. B.: Response of the yellowfin tuna (Thunnus albacares) to underwater sound. In: Marine Bio-Acoustics II. Tavolga, W. N. (ed.). Oxford: Pergamon Press, 1967, pp. 105-121.

Jacobs, D. W., Tavolga, W. N.: Acoustic intensity limens in the goldfish. Anim. Behav. 15, 324-335 (1967).

Jenkins, D. B.: A light microscopic study of the saccule and lagena in certain catfishes. Am. J. Anat. 150, 605-629 (1977).

Jenkins, D. B.: A transmission and scanning electron microscopic study of the saccule in five species of catfishes. Am. J. Anat. 154, 81-101 (1979a).

Jenkins, D. B.: Anatomical investigation of the saccule in Clarius batrachus. Scanning Electron Microscopy/1979, 949-954 (1979b).

Jørgensen, J. M.: Hair cell polarization in the flatfish inner ear. Acta Zool. 57, 37-39 (1976).

Lewis, E. R., Li, C. W.: Hair cell types and distributions in the otolithic and auditory organs of the bullfrog. Brain Res. 83, 35-50 (1975).

Lewis, E. R., Leverenz, E. L.: Direct evidence for an auditory place mechanism in the frog amphibian papilla. Soc. Neurosci. Abstr. 5, 25 (1979).

Lindeman, H. H.: Regional differences in structure of the vestibular sensory regions. J. Laryngol. Otol. 83, 1-17 (1969).

Lowenstein, O.: The electrophysiological study of the responses of the isolated labyrinth of the lamprey (*Lampetra fluviatilis*) to angular acceleration, tilting and mechanical vibration. Proc. R. Soc. London Ser. B 174, 419-434 (1970).

Lowenstein, O.: The labyrinth. In: Fish Physiology, Vol. 5. Hoar, W. S., Randall, D. J. (eds.). New York: Academic Press, 1971, pp. 207-240.

Lowenstein, O.: Comparative morphology and physiology. In: Handbook of Sensory Physiology, Vol. VI/1. Kornhumber, H. H. (ed.). New York: Springer, 1974, pp. 75-120.

Lowenstein, O., Osborne, M. P., Thornhill, R. A.: The anatomy and ultrastructure of the labyrinth of the lamprey (*Lampetra fluviatilis* L.). Proc. R. Soc. London Ser. B 170, 113-134 (1968).

Lowenstein, O., Osborne, M. P., Wersäll, J.: Structure and innervation of the sensory epithelia of the labyrinth in the thornback ray (*Raja clavata*). Proc. R. Soc. London Ser. B 160, 1-12 (1964).

Lowenstein, O., Roberts, T. D. M.: The localization and analysis of the responses to vibration from the isolated elasmobranch labyrinth. A contribution to the problem of the evolution of hearing in vertebrates. J. Physiol. (Lond.) 114, 471-489 (1951).

Lowenstein, O., Wersäll, J.: A functional interpretation of the electron-microscopic structure of the sensory hairs in the cristae of the elasmobranch *Raja clavata* in terms of directional sensitivity. Nature 184, 1807-1808 (1959).

Marler, P.: Some characteristics of some animal cells. Nature 176, 6-7 (1955).

Marler, P.: Structure of animal communication sounds. In: Recognition of Complex Acoustic Signals. Bullock, T. (ed.). West Germany: Dahlem Konferenzen, 1977, pp. 17-35.

Marshall, N. B.: Sound-producing mechanisms and the biology of deep-sea fishes. In: Marine Bio-Acoustics II. Tavolga, W. N. (ed.). Oxford: Pergamon Press, 1967, pp. 123-133.

Michelsen, A.: Sound reception in different environments. In: Sensory Ecology. Ali, M. A. (ed.). New York: Plenum Press, 1978, pp. 345-373.

Miller, M. R.: The reptilian cochlear duct. In: Comparative Studies of Hearing in Vertebrates. Popper, A. N., Fay, R. R. (eds.). New York: Springer-Verlag, 1980, pp. 169-204.

Morton, E.: Ecological sources of selection on avian sound. Am. Nat. 109, 17-34 (1975).

Myrberg, A. A., Jr., Kramer, E., Heinecke, P.: Sound production by cichlid fishes. Science 149, 555-558 (1965).

Myrberg, A. A., Jr., Spanier, E., Ha, S. J.: Temporal patterning in acoustical communication. In: Contrasts in Behavior. Reese, E. S., Lighter, F. (eds.). New York: Wiley, 1978, pp. 137-180.

Myrberg, A. A., Jr., Spires, J. Y.: Sound discrimination by the bicolor damselfish, *Eupomacentrus partitus*. J. Exp. Biol. 57, 727-735 (1972).

Nakajima, Y., Wang, D. W.: Morphology of afferent and efferent synapses in the hearing organ of the goldfish. J. Comp. Neurol. 156, 403-416 (1974).

Oman, C. M., Frishkopf, L. S., Goldstein, M. H., Jr.: Cupula motion in the semicircular canal of the skate, *Raja erinacea*. Acta Otolaryngol. 87, 528-538 (1979).

Platt, C.: Central control of postural orientation in flatfish. I. Postural change dependence on central neural changes. J. Exp. Biol. 59, 491-521 (1973).

Platt, C.: Asymmetry of semicircular canal-extraocular muscle function in flatfish. Soc. Neurosci. Abstr. 2, 1060 (1976).

Platt, C.: Hair cell distribution and orientation in goldfish otolith organs. J. Comp. Neurol. 172, 283-298 (1977).

Platt, C.: The peripheral vestibular system in fishes. In: Fish Neurobiology and Behavior. Northcutt, R. G., Davis, R. E. (eds.). Ann Arbor: Univ. of Michigan Press, 1981 (in press).

Platt, C., Popper, A. N.: Otolith organ receptor morphology in herring-like fishes. In: Vestibular Function and Morphology. Gualtierotti, T. (ed.). New York: Springer-Verlag, 1981 (in press).

Poggendorf, D.: Die absoluten Hörschwellen des Zwergwelses (*Ameiurus nebulosus*) und Beiträge zur Physik des Weberschen Apparate der Ostariophysen. Z. Vergl. Physiol. 34, 222-257 (1952).

Popper, A. N.: Auditory threshold in the goldfish (*Carassius auratus*) as a function of signal duration. J. Acoust. Soc. Am. 52, 596-602 (1972).

Popper, A. N.: Ultrastructure of the auditory regions in the inner ear of the lake whitefish. Science 192, 1020-1023 (1976).

Popper, A. N.: A scanning electron microscopic study of the sacculus and lagena in the ears of fifteen species of teleost fishes. J. Morphol. 153, 397-417 (1977).

Popper, A. N.: Scanning electron microscopic study of the otolithic organs in the bichir (*Polypterus bichir*) and shovel-nose sturgeon (*Scaphirhynchus platorynchus*). J. Comp. Neurol. 181, 117-128 (1978a).

Popper, A. N.: A comparative study of the otolithic organs in fishes. Scanning Electron Microscopy, II, 405-416 (1978b).

Popper, A. N.: The ultrastructure of the sacculus and lagena in a moray eel (*Gymnothorax* sp.). J. Morphol. 161, 241-256 (1979a).

Popper, A. N.: Inner ear auditory receptors in Osteoglossomorph fishes. Soc. Neurosci. Abstr. 5, 29 (1979b).

Popper, A. N.: Scanning electron microscopic study of the sacculus and lagena in several deep sea fishes. Am. J. Anat. 157, 115-136 (1980).

Popper, A. N.: Organization of the inner ear and auditory processing. In: Fish Neurobiology and Behavior. Northcutt, R. G., Davis, R. E. (eds.). Ann Arbor: Univ. of Michigan Press, 1981, (in press).

Popper, A. N., Coombs, S.: Auditory mechanisms in teleost fishes. Am. Sci. 68, 429-440 (1980a).

Popper, A. N., Coombs, S.: Acoustic detection by fish. In: Environmental Physiology of Fish. Ali, M. A. (ed.). New York: Plenum Press, 1980b, pp. 403-430.

Popper, A. N., Fay, R. R.: Sound detection and processing by teleost fishes: A critical review. J. Acoust. Soc. Am. 53, 1515-1529 (1973).

Popper, A. N., Platt, C.: The herring ear has a unique receptor pattern. Nature 280, 832-833 (1979).

Popper, A. N., Tavolga, W. N.: Structure and function of the ear in the marine catfish, *Arius felis*. J. Comp. Physiol. (in press).

Retzius, G.: Das Gehörorgan der Wirbelthiere: morphologisch-histologische Studien, Vol. 1. Das Gehörorgan der Fische und Amphibien. Stockholm: Samson and Wallin, 1881, 221 pp.

Saito, K.: Fine structure of macula of lagena in the teleost inner ear. Kaibogaku Zasshi. Acta Anatomica Nipponica 48, 1-18 (1973).

Salmon, M.: Acoustical behavior of the menpachi, *Myripristis berndti,* in Hawaii. Pacific Sci. 21, 364-381 (1967).

Sand, O.: Directional sensitivity of microphonic potentials from the perch ear. J. Exp. Biol. 60, 881-899 (1974).

Sand, O., Enger, P. S.: Possible mechanisms for directional hearing and pitch discrimination in fish. Rheinisch-Westfal. Akad. Wiss. 53, 223-242 (1974).

Sand, O., Michelsen, A.: Vibration measurements of the perch saccular otolith, J. Comp. Physiol. 123A, 85-89 (1978).

Schneider, H.: Die Bedeutung der Atemhöhle der Labyrinthfische für ihr Hörvermögen. Z. Vergl. Physiol. 29, 172-194 (1941).

Schöne, H.: Über die Arbeitsweise der Statolithenapparate bei Plattfischen. Biol. Jahresh. 4, 135-156 (1964).

Schuijf, A., Buwalda, R. J. A.: Underwater localization—A major problem in fish acoustics. In: Comparative Studies of Hearing in Vertebrates. Popper, A. N., Fay, R. R. (eds.). New York: Springer-Verlag, 1980, pp. 43-78.

Spoendlin, H.: Organization of the sensory hairs in the gravity receptors in the utricle and saccule of the squirrel monkey. Z. Zellforsch. 62, 701-716 (1964).

Spoendlin, H.: The innervation of the cochlear receptor. In: Basic Mechanisms in Hearing. Moller, A. M. (ed.). New York: Academic Press, 1973, pp. 185-230.

Steinhausen, W.: Über die Beobachtung der Cupula in den Bogengangsampullen des Labyrinths des lebenden Hechts. Pfluegers Arch. Ges. Physiol. 232, 500-512 (1933).

Stipetić, E.: Über das Gehörorgan der Mormyriden. Z. Vergl. Physiol. 26, 740-752 (1939).

Strelioff, D., Honrubia, V.: Neural transduction in Xenopus laevis lateral line system. J. Neurophysiol. 41, 432-444 (1978).

Tavolga, W. N.: Sonic characteristics and mechanisms in marine fishes. In: Marine Bio-Acoustics. Tavolga, W. N. (ed.). Oxford: Pergamon Press, 1964, pp. 195-211.

Tavolga, W. N.: Sound production and detection. In: Fish Physiology, Vol. 5. Hoar, W. S., Randall, D. J. (eds.). New York: Academic Press, 1971, pp. 135-205.

Tavolga, W. N.: Signal/noise ratio and the critical band in fishes. J. Acoust. Soc. Am. 55(6), 1323-1333 (1974).

Tavolga, W. N.: Recent advances in the study of fish audition. In: Sound Reception in Fishes. Benchmark Papers in Animal Behavior, Vol. 7. Tavolga, W. N. (ed.). Stroudsburg, Pa.: Dowden, Hutchinson and Ross, 1976, pp. 37-49.

Tavolga, W. N.: Recent advances in the study of sound production in fishes. In: Sound Production in Fishes. Benchmark Papers in Animal Behavior, Vol. 9. Tavolga, W. N. (ed.). Stroudsburg, Pa.: Dowden, Hutchinson and Ross, 1977a, pp. 45-53.

Tavolga, W. N.: Mechanisms for directional hearing in the sea catfish (Arius felis). J. Exp. Biol. 67, 97-115 (1977b).

Tavolga, W. N., Wodinsky, J.: Auditory capacities in fish. Pure tone thresholds in nine species of marine teleosts. Bull. Am. Mus. Nat. Hist. 126, 177-239 (1963).

von Békésy, G.: Experiments in Hearing. New York: McGraw-Hill, 1960.

von Frisch, K.: Über den Gehörsinn der Fische. Biol. Rev. 11, 210-246 (1936).

von Frisch, K.: Über die Bedeutung des Sacculus und der Lagena für den Gehörsinn der Fische. Z. Vergl. Physiol. 25, 703-747 (1938).

von Holst, E.: Die Arbeitsweise des Statolithenapparates bei Fischen. Z. Vergl. Physiol. 32, 60-120 (1950).

Wegner, N. T.: The orientation of hair cells in the otolithic organs and papilla neglecta in the inner ear of the Anabantid fish Colisa labiosa (Day). Acta Zool. (Stockholm) 60, 205-216 (1979).

Weiss, T. F., Mulroy, M. J., Turner, R. G., Pike, C. L.: Tuning of single fibers in the cochlear nerve of the alligator lizard: Relation to receptor morphology. Brain Res. 115, 71-90 (1976).

Werner, C. F.: Experimente über die Funktion der Otolithen bei Knochenfischen. Z. Vergl. Physiol. 10, 26-35 (1929).

Werner, C. F.: Das Gehörorgan der Wirbeltiere und des Menschen. Leipzig: G. Thieme, 1960.

Wersäll, J.: Vestibular receptor cells in fish and mammals. Acta Otolaryngol. Suppl. 163, 25-29 (1961).

Wersäll, J., Bagger-Sjöbäck, D.: Morphology of the vestibular apparatus. In: Handbook of Sensory Physiology, Vol. VI/1. Kornhuber, H. H. (ed.). New York: Springer-Verlag, 1974, pp. 123-170.

Wersäll, J., Flock, Å., Lundquist, P.-G.: Structural basis for directional sensitivity in cochlear and vestibular sensory receptors. Cold Spring Harbor Symp. Quant. Biol. 30, 115-132 (1965).

Wersäll, J., Gleisner, L., Lundquist, P.-G.: Ultrastructure of the vestibular sense organs. In: Myotatic, Kinesthetic and Vestibular Mechanisms. de Reuck, A. V. S., Knight, J. (eds.). Boston: Little, Brown, 1967, pp. 105-120.

Wever, E. G.: Cochlear stimulation and Lempert's mobilization theory. Principles and methods. Arch. Otolaryngol. 90, 68-73 (1969).

Wever, E. G.: The Reptile Ear. Princeton: Princeton Univ. Press, 1978, 1024 pp.

Winn, H. E.: Acoustic discrimination by the toadfish with comments on signal systems. In: Behavior of Marine Animals, Vol. 2. Winn, H. E., Olla, B. L. (eds.). New York: Plenum Press, 1972, pp. 361-385.

Discussion

WILLOWS: Back in 1958, Morris and Kittleman [Science, 158, 368-370 (1958)] reported that macular otoliths were strongly piezoelectric, but I have not heard a word about this since. Could there be a piezoelectric mechanisms for response to pressure changes in fish otoliths?

POPPER: Offutt [J. Aud. Res. 10, 226-228 (1970)] theorized about the possible piezoelectric function of otoliths, but nothing has ever come of it.

BULLOCK: I don't know of any work on this but it does not seem plausible since the piezoelectric effect depends on distortion. The otoliths are, as it were, held at one end and pushed at the other. Under such conditions I doubt that any piezoelectric energy could appear.

WILLOWS: Some sort of anchoring of the otoliths may allow for distortion or, at least, the possibility of differential movements. One of the interesting properties of piezoelectric materials is that it takes very little distortion to elicit an effect. I think it might be worth while testing.

PLATT: If one considers the cellular ultrastructure of the end-organs together with the evidence that we now have on the physiology of the primary afferent, we have a situation that I think can be described as a mosaic of end organs. There is evidence that an individual end organ may have discrete populations of hair cell bundles, and also may have individual afferents that are either auditory or gravistatic. The behavioral evidence from the work of the Germans in the thirties has suggested that there is a pars superior of the utricle and semicircular canals that is gravistatic, and a pars inferior that is auditory. With the current technical developments, I think it should be interesting to look at the central connections of the primary afferents. The afferents from the auditory organs converge on auditory nuclei as distinct from vestibular nuclei. It seems that one group is suggesting that the ear is divisible into auditory and gravistatic portions. Yet there is physiological and ultrastructural evidence that each of the end-organs may be doing the same things. We should be aware of our own tunnel vision. I see the ear primarily as a vestibular sense. Popper sees it primarily as an auditory sense.

I notice that both of us have been up here for forty minutes or so, and we have not once mentioned the macula neglecta that Corwin will be talking about. Here is another patch of end-organs about which we know very little. It is very difficult to record from in teleosts because it is so small.

SCHUIJF: Concerning the hypertrophy of different parts of the inner ear that you describe, would it be useful to make some grouping of fishes as to acoustical structures?

POPPER: It would be useful, but it just has not worked out. An example is the comparison of the two squirrelfishes. *Myripristis* and *Adioryx* have very different ear structures. Deep sea fishes seem to have some aspects in common, and so do many catfish.

NORTHCUTT: The same problem exists in the comparative study of photoreceptors. We examine species after species, and try to correlate a particular photopigment with a particular habitat, and the conclusion is that there is no simple relationship. In the case of audition, the problem is even more complex. We have to know what the fish are listening to or if they listen in the same way.

FAY: In the face of all this variation, the characteristics of the ear organization in Ostariophysi seems most interesting. How much variation is there among the Ostario-physi? We really don't know, but here is an auditory system that is adapted for responding to input from the swimbladder in a very special way.

Chapter 2

Acousticolateralis System in Clupeid Fishes

J. H. S. Blaxter*, E. J. Denton** and J. A. B. Gray**,†

Clupeoid fish such as *Clupea harengus* (herring) and *Sprattus sprattus* (sprat) live in schools. Such schools change direction repeatedly and suddenly. Although the distance between the fish in the schools is small the fish never seem to collide in the course of fast maneuvers in tight formation, and it seems likely that the acousticolateralis system plays a role (see Partridge, Chapter 26). Another feature of the behavior of many clupeid species is that they migrate between the surface and quite considerable depths; herring may migrate as much as 200 m, which is equivalent to a change of 20 atm in pressure. Such a change means that any gas used as a pressure-displacement converter in the auditory system could be reduced to 1/20 of its surface volume during each migration unless there are special adaptations. We have used a number of techniques to study acousticolateralis systems in larval and juvenile herring and sprat. This account considers: first, behavioral experiments because these can indicate some of the requirements for sensory information which the acousticolateralis must supply; second, the mechanics of the liquid flow in the system and of the adaptation to change in depth; third, the detailed structure of the utriculus and the coupling of its sense organs to the auditory bulla; and finally, the electrical mass responses of the utricular receptors to a variety of pressure stimuli.

1 Experiments on Behavior

These species cannot be restrained successfully and some of the conditioning techniques for plotting audiograms are not possible. However, fish in a small school respond to taps on the walls of their tank; usually the fish give a sharp turn and then accelerate. Some of these responses show characteristics of the "startle" response associated in other fish with the Mauthner cells. We have not attempted to categorize these different

*Dunstaffnage Marine Laboratory, Oban, Argyll, Scotland.
**Marine Biological Association of the United Kingdom, The Laboratory, Citadel Hill, Plymouth, England.
†Member of the external scientific staff of the Medical Research Council.

responses, but have used them to obtain some information about the sensory mechanisms of the fish.

A school of about 400 young herring 12 cm long was allowed to swim round a circular tank containing a circular island of netting so that at one point in the cycle they had to pass through a gap 23 cm wide × 30 cm high (Blaxter, Gray, and Denton in prep.). In the side of the tank on the edge of this gap a vibrator was attached to a water-filled metal bellows. Stimuli were applied through this vibrator. After each set of behavioral observations an array of three transducers was lowered into the gap to measure the particle velocity as well as one for pressure (Fig. 2-1A). The responses were graded and were scored on a scale relating to the nature of the response and the number of fish involved. Consistent scores were obtained on a steep response/stimulus relation (Fig. 2-1B). Furthermore, scores were consistent from day to day as well as within one experimental run. The stimuli were sine waves which always started at the same point in the cycle and continued for the number of cycles desired. Varying the number of cycles showed that a stimulus lasting only one cycle was as effective as a long stimulus. Similar results were obtained with sine waves of frequencies between 26 and 400 Hz (with a single cycle, change of frequency implies only a change in the rate of rise to the peak). Stimulus against frequency curves for constant responses were prepared taking either pressure or resultant velocity as the stimulus. The curve obtained using pressure fitted very closely with the pressure vs. frequency curve for a constant electrical response from the utricular receptors and, over the range tested, sensitivity was nearly constant (see Fig. 2-6B). In other experiments the burst of sinusoidal waves was made to grow over a few cycles rather than starting abruptly. The dashed line in

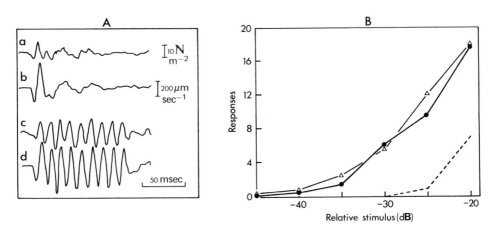

Figure 2-1. The "startle" response. Clupeids subjected to a sufficient noise make an "escape" reaction. Herring swimming round a circular tank were subjected to stimuli as shown in (A). (A) a, one cycle, pressure recording; b, one cycle, recording of velocity in the direction of the tank radius—the major component; c, several cycles, pressure; d, several cycles, velocity. Responses were measured on an arbitrary scale and (B) shows such measurements (means of runs at various frequencies) plotted against size of stimulus in arbitrary units. ●, one cycle only; △, several cycles; ---, shows the fall in sensitivity found in experiments in which the full amplitude of stimulus was only reached after four cycles.

Fig. 2-1B shows the effect of taking four cycles to reach full amplitude. Even when a stimulus took as little as two cycles to reach its full amplitude the threshold was raised significantly (by a factor of about two). A striking example of this kind of adaptation is that a school of herring will swim quietly past a very intense source of steady sinusoidal stimulation. Further experiments using a high speed video camera have shown that the overwhelming majority of responses near the sound source were away from the stimulus and that there was no difference in this respect if the polarity of the stimulus was reversed. The latency was 20-25 msec between the beginning of the stimulus and the bending of the tail. These results showed that, in this simple response, a single cycle is sufficient to enable the fish to determine the amplitude with considerable accuracy and possibly also to decide the direction, though in the case of the latter there could have been other clues. Other behavioral studies suggest that the auditory bullae play an important role in these responses. Thus, they only appear in the larval fish when the bullae first fill with gas and bursting the bulla membrane in the juvenile or adult raises the threshold of the response.

Larval herring become much more sensitive to hydrostatic pressure changes once the bullae fill with gas. They show this by moving upward in response to small pressure increases and downward in response to small pressure decreases (Blaxter and Denton 1976). This experiment suggests that sensitivity to larger and slower changes than occur in acoustic stimuli need to be looked for in the system.

2 The Gas-Filled Structures and the Liquid Flows in the Inner Ear

Figure 2-2 shows after a number of earlier authors and our own observations some of the main features of the acousticolateralis system of an adult sprat (see Allen, Blaxter, and Denton 1976). The swimbladder, inner ear, and lateral line system are all linked. The central feature is a pair of prootic auditory bullae. Each is divided into gas-filled and liquid-filled parts by a membrane mainly composed of elastin under tension (8 N m^{-1} in the sprat). The upper part contains perilymph connected with that of the labyrinth via a fenestra in the upper wall of the bulla (Fig. 2-2C). Lateral to the fenestra is a very compliant membrane which presents little reactance to flow in the working range. This is the lateral recess membrane and is positioned in the wall of the skull at the back of the lateral recess from which all the lateral line canals radiate (Figs. 2-2D and 2-3A). The auditory bullae, because of the compliance of the gas, allow pressure changes to generate flows of liquid which reach the sense organs of the inner ear and lateral line (Fig. 2-3A).

The gas-filled part of each bulla is connected to the swimbladder by a precoelomic duct, a strong cartilaginous tube containing living tissue in the center of which there is a gas-filled duct whose internal diameter is only about 7 μm (Fig. 2-2B). In an adult herring the precoelomic duct is almost a centimeter long. When a fish moves downward in the sea the main part of the swimbladder wall, which is compliant, quickly collapses under the increase of pressure and the gas in the swimbladder follows the changes in ambient pressure. The gas in the bulla is compressed less since the bulla membrane is much less compliant and a pressure difference develops between the gas in the bulla and that in the swimbladder so that gas passes along the precoelomic duct

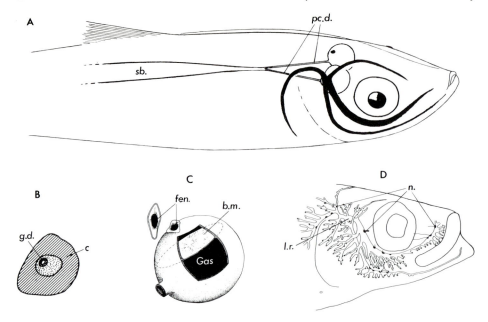

Figure 2-2. The main feature of the clupeid acousticolateralis system. The special features of this system are the existence of the bullae, as pressure-displacement converters, the hydrodynamical connections between the ear and lateral line, and the gas connections between the bullae and swimbladder which allow adaptation to depth. (A) The position of the two bullae, the main lateral line canals, and the connections between bullae and swimbladder; sb., swimbladder; pc.d., precoelomic ducts. (B) A cross section of the precoelomic duct between a bulla and the swimbladder; the small gas-filled central part (g.d.) has a diameter of only about 7 μm; c, cartilage. (C) A bulla and (in plan) its fenestra; b.m., bulla membrane; fen., fenestra; the elastic thread is not shown. (D) The lateral line system of the sprat; l.r., lateral recess; n, neuromasts.

Figure 2-3. Liquid flow in the inner ear and lateral line. Volume changes in the gas in ▶ the bulla accompanying pressure changes cause liquid to flow in the acousticolateralis system. (A) Diagram of the pathways in which this flow occurs: a, swimbladder; b, precoelomic duct; c, bulla; d, bulla membrane; e, fenestra; f, utricular macula; g, utriculus; h, lateral recess membrane; i, lateral line canal; j, sacculus; k, saccular macula; l, auditory foramen. The dynamics of the flow indicated by the arrows in (A) were often investigated by recording light reflected from one of the membranes d or h (from Gray and Denton 1979). (B) A record from h with the corresponding pressure. Records such as (B) allowed the linearity of the system to be plotted as in (C) and the frequency response as in (D) (B, C, and D from Denton, Gray, and Blaxter 1979).

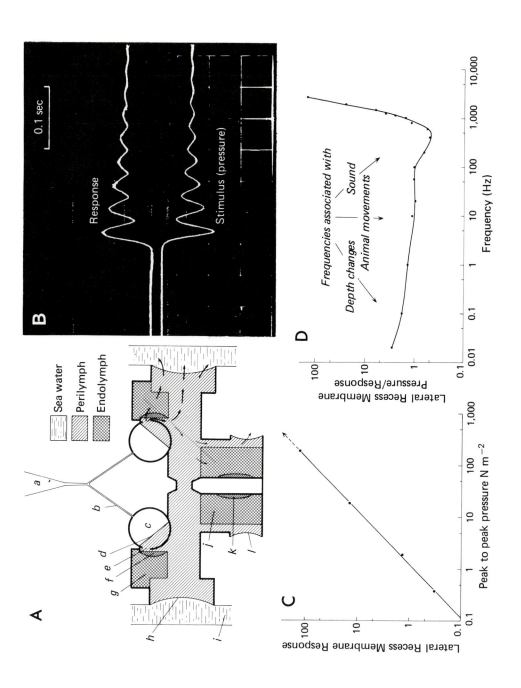

restoring the bulla membrane to the flat resting state. After a discrete change in depth it takes about one minute for the membrane to become completely flat again (time constant ca. 15 sec). When the fish moves upward the reverse processes occur and gas passes from the bulla to the swimbladder. We have concluded that the swimbladder cannot be an effective buoyancy device and mainly acts as a reservoir of gas for the auditory bulla system allowing the *volume* of the gas-filled parts of the bulla to remain constant as a fish changes depth (Denton and Blaxter 1976, Blaxter, Denton, and Gray 1979).

For all but very low frequencies, such as are given by changes in depth, the bulla can be considered independent of the swimbladder. Pressure changes acting on the fish cause displacements of the elastic bulla membrane which lead to liquid displacements of the perilymph. These movements are possible because the bulla contains gas, but the *extent* of the movements is limited by the restoring force exerted by the elastic bulla membrane as well as that of the gas which the bulla contains. Several sets of sense organs, i.e., those of the utriculi, the sacculi and possibly the lagenae, and the neuromasts of the lateral line systems are linked mechanically by flows of liquid to the auditory bullae. These mechanical links are shown diagrammatically on Fig. 2-3A. By direct observation and by measurement it has been shown that the utricular macu-lae are mechanically very responsive to displacements of perilymph produced by the auditory bullae (Table 2-1), whereas the sense organs of the sacculus are very little affected. This is not surprising for the utricular maculae lie closely above the fenestra through which all of the liquid displaced by movements of the bulla membrane must flow while only a small fraction of the flow through the fenestra passes by the saccu-lus. The major fraction of this flow is dissipated along the lateral line canals which radiate from the lateral recess (Fig. 2-2D). The lateral line system itself is a very open one and communicates through many pores directly to the sea. As shown on Fig. 2-4A there is also an elastic thread between the bulla membrane and utriculus.

Some of the main mechanical properties of the sprat and herring bulla membrane, utricular maculae, lateral recess membrane, and lateral line have been elucidated. The application of large pressures at low frequencies enabled absolute measurements to be made, while an optical method enabled the relative responses to small pressure changes over a wide range of frequencies to be found (Gray and Denton 1979, Denton, Gray, and Blaxter 1979). The main features of the displacements of liquids in the system are:

1. The amplitude of response is almost perfectly linearly related to pressure over a very wide range up to about 0.5 atm (peak to peak) (5×10^4 N m^{-2}) (Fig. 2-3C).
2. The frequency response changes little from about 0.01-1,000 Hz, but above 1,000 Hz the mechanical sensitivity falls rapidly (Fig. 2-3D).
3. The phase lag behind pressure increases with increasing frequency (Fig. 2-6A) but the *absolute* time delay is small (ca. 0.5 msec) and changes little.

A consequence of these simple relationships is that complicated waveforms of pres-sure containing a mixture of frequencies are reproduced, with relatively little change, in the displacements of liquid produced by the bulla. The mechanical responses of the bulla can be predicted reasonably well by treating it as a forced oscillator for

Table 2-1. Quantitative results for sprats standard length ca. 8 cm.

	Mean	Range
Bulla		
Volume	1.67 mm^3	0.86-1.91
Displacement of membrane ÷		
membrane radius	1.37×10^{-5}/N m^{-2}	1.26-1.54
Volume displacement	6.03×10^3 μm^3/N m^{-2}	3.16-6.95
Volume displacement ÷		
bulla volume	3.61×10^{-6}/N m^{-2}	3.48-3.67
Liquid displacements		
Linear displacement		
at fenestra (Fig. 2-3A.e)	300 nm/N m^{-2}	
in large lateral-line canals		
(Fig. 2-3A.i)	30 nm/N m^{-2}	
at ends of canals	3 nm/N m^{-2}	
Utriculus		
Displacement of floor	5 nm/N m^{-2}	1.2-8.7
Flow through utriculus	ca. 20% of total at 50 Hz	
Force on macula through		
elastic thread	2×10^{-10} N/N m^{-2}	
Endolymph potential	+ 10 mV	4-14
Thresholds		
Herring medulla	5.6×10^{-3} N m^{-2}	Enger (1967)
Herring school	3×10^{-2} N m^{-2}	Olsen (1976)
Human threshold intensity		
(10^{-12} W m^{-2}) in water	3.5×10^{-3} N m^{-2} [a]	
"Startle" response (1st		
pressure peak)	17 N m^{-2}	12-23

[a] At 10^{-12} W m^{-2} human cochlea partition displaces 0.4×10^{-2} nm (Johnstone 1968) and the sprat utricular floor 1.7×10^{-2} nm.

which the restoring force is given by the elastic bulla membrane and the gas which the bulla contains while the relevant mass is that of the perilymph within the bulla. The natural resonant frequency of the bullae of herring and sprat is around 500-600 Hz and the damping is usually but not always less than critical. The mechanical responses of the utricular maculae follow those of the bulla membrane closely except that below about 30 Hz their mechanical responses fall. The responses of the lateral recess membrane (Fig. 2-3D), and hence of the liquid within the lateral line canals, follow those of the bulla membrane well although the phase delays between pressure and response tend to be greater at the higher frequencies. We may note that displacements in the medium outside the fish have very little effect on the bulla and utriculus but they are, as we shall see later, accompanied by displacements in the lateral line canals. Table 2-1 gives, for an 8 cm sprat, some numerical values relating to the mechanical properties described above.

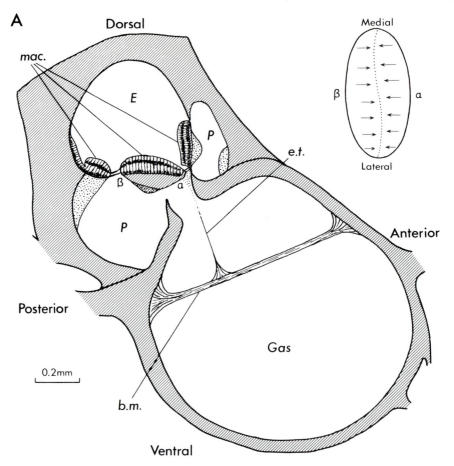

Figure 2-4. The coupling between the sprat bulla and the utricular receptors. Movements of the bulla membrane are relayed to the floor of the utriculus mainly by the flow of perilymph in and out of the fenestra; there is also a coupling through an elastic thread. (A) A sagittal section, showing relation of the maculae to the fenestra: the hatched areas are tough collagenous or calcified tissue; b.m., bulla membrane; e.t., elastic thread; E., endolymph; P., perilymph; mac., maculae (αβ is the minor axis of the middle macula); inset, view of the middle macula from above to show orientation of hair cells (from Denton and Gray 1979). (B) A perspective diagram to show the suspension of the maculae in the alternating liquid flow from the bulla: m.mac., middle macula; a.mac., anterior macula; p.mac., posterior macula; sus., suspension of anterior and middle maculae; fen., fenestra; ot, otolith (from Best and Gray 1980).

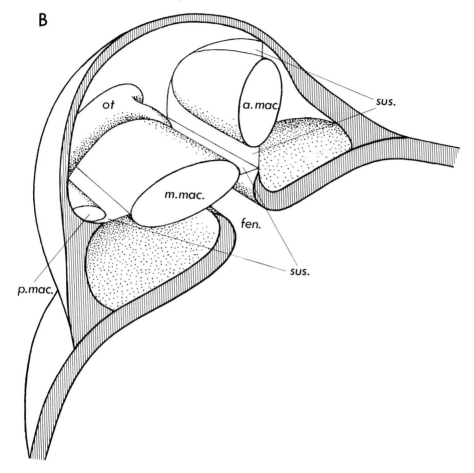

3 The Structural Relationships between the
Bullae and the Utricular Sense Organs

In Fig. 2-4 we give (after Best and Gray 1980, also cited by Denton and Gray 1979)
diagrams showing how the sprat utricular maculae are related to the auditory bullae.
The lateral parts of the middle maculae are backed by strong connective tissue and
suspended by thin membranes containing elastic fibers; the anterior maculae are also
suspended and backed by connective tissue. On decompression the middle maculae
move dorsally as a single unit and the anterior maculae move rostrally; on compression the movements are reversed. The medial parts of the middle maculae have less
connective tissue, have shorter suspensions, and are associated with a small otolith,
the lapillus. The cupulae of the maculae, which are not shown in the figure, lie in the
endolymph of the utriculus and those of the middle and anterior maculae are near to
each other and probably interact mechanically. A very thin elastic thread connects
the center of the bulla membrane to the floor of the utriculus through the medial

corner of the fenestra of the bulla under the otolith. It seems likely that this thread is associated with the detection of slow pressure changes. In the sprat the hair cells of the middle and anterior maculae are arranged so that all cells have their kinocilia on the sides which are nearest to lines down the middle of the long axes of the maculae (Fig. 2-4A insert). In this they resemble those of the herring described by Popper and Platt (1979). In the sprat the hair cells of the posterior macula are orientated along rather than across its length and in this they differ from the herring where all three maculae have a closely similar disposition of hair cells. Each sprat utriculus has about 13,000 hair cells, about 60% on the middle macula and slightly more than 20% on the anterior macula.

4 Electrophysiological Responses

Electrical responses (microphonics) were obtained from the utriculus of the sprat during stimulation by bursts of oscillatory pressure ranging in frequency from 2.5 to 880 Hz and in amplitude from 0.12 to 220 N m^{-2} (30-90 dB above threshold) (Denton and Gray 1979, 1980). We give examples of such responses in Fig. 2-5. Responses to simple sinusoidal pressures characteristically consisted of two negative waves rising from a "base line" for each cycle of stimulus. For frequencies between about 40-200 Hz one of the waves was associated with the compression, this we have called the c-response, and the other, which accompanied decompression, the d-response. At any given frequency of stimulation the timing of the peaks of the c- and d-responses relative to pressure was independent of the intensity of stimulation (Fig. 2-5B). With change of frequency the *waveform* of response did not change but the phase relationship between the stimulating pressure and the response did change (Fig. 2-5C). In the 80-160 Hz range the peaks of compression and the c-responses almost coincided, as did the peaks of decompression at the d-responses. At lower frequencies the peaks of response were generally phase-advanced on the peak of pressure; at higher frequencies they were phase-delayed until, at about 800 Hz, the c-response accompanied the decompression and the d-response the compression (Fig. 2-6A). The phase differences between stimulus and response become, however, very small if we compare the electrical responses not directly with pressures but with the displacements of the bulla membrane (and hence the utricular maculae) produced by pressures. On Fig. 2-6A we plot for a range of frequencies the angular phase differences between the stimulating pressure and the electrical responses of the utriculus and between the pressure and the mechanical responses of the bulla membrane. This figure shows that the two kinds of response are alike in phase both have, e.g., an angular phase lag of about 180° at 880 Hz. It is thus clear that the electrical responses are, as we might have expected, much more closely related to *displacements* of the utriculae than to pressure changes. In Fig. 2-6B we show the frequency characteristic of the electrical responses described here. If we take account of the differences between the pressure and the liquid displacements and relate the utricular macular responses to the latter it becomes evident that (at least between 30-880 Hz), in response to sinusoidal stimulation, the amplitudes of the c-responses depend almost entirely on the *amplitudes* of the displacements of the utricu-

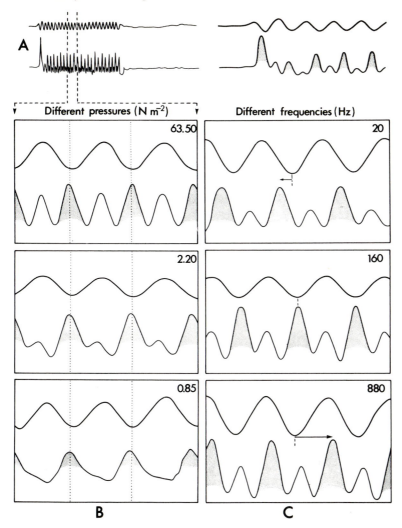

Figure 2-5. Responses to pressure from sprat utricular receptors. An electrode in the endolymph above the maculae records a potential change which is the resultant of the currents set up by the activity of the receptors: in these pairs of records the upper traces are stimulus pressure (compression up), and the lower traces are electrical responses (negative upward, d-responses stippled). (A) Response to a stimulus burst at 80 Hz and 7.8 N m^{-2} peak to peak pressure; left traces show the whole burst and the right traces the beginning of the burst on a faster sweep. The records in (B) and (C) come from the middle of bursts such as (A). (B) 80 Hz with different pressures. The peak to peak pressures and responses were: for 63.5 N m^{-2}, 480 μV; for 2.2 N m^{-2}, 52 μV; for 0.85 N m^{-2}, 19 μV. (C) Different frequencies with stimuli giving about the same response size (range 100 to 139 μV). This shows constancy of shape but phase differences with respect to pressure.

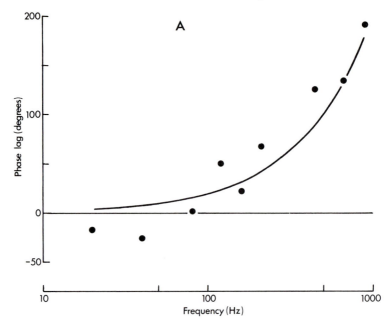

Figure 2-6. Characteristics of the responses of sprat utricular receptors. This figure shows that above about 30 Hz d-responses are closely related to upward movements of the utricular floor and c-responses to downward movements. (A) The phase delays of the receptor responses relative to pressure (●) are closely similar to the average delays of the perilymph displacement from the bulla behind pressure (line) (from Denton and Gray 1980). (B) Full line shows the pressure required to give constant receptor activity at different frequencies (dashed line joins means of experimental points and stippled area indicates range); above 30 Hz this is like the corresponding relation for a constant liquid displacement (Fig. 2-3D).

The filled circles in (B) link these observations on the mass responses of utricular receptors with behavior. They are for the thresholds of the "startle" responses such as Fig. 2-1, plotted with the result at 80 Hz (ringed) made equal to one.

lar floor away from its rest position toward the fenestra of the bulla while the amplitudes of the d-responses depend almost entirely on the *amplitudes* of the displacements of the utricular floor away from the fenestra. Except possibly at frequencies below 30 Hz the *rate* of displacement becomes unimportant in determining the amplitude of response.

In some further experiments using the same method of recording two frequencies of pressure stimulation were combined as for example in Fig. 2-7 where 200 and 30 Hz waves were mixed. In this and similar experiments we have always found that the timing and relative magnitudes of the response peaks could be predicted very well on the simple hypothesis that displacements of the utricular floor toward or away from the mean utricular position produce responses in one or other of two major groups of receptors giving, respectively, and c- and d-responses. We may note that while the mechanical responses of the utriculus to combined frequency stimulation are simply

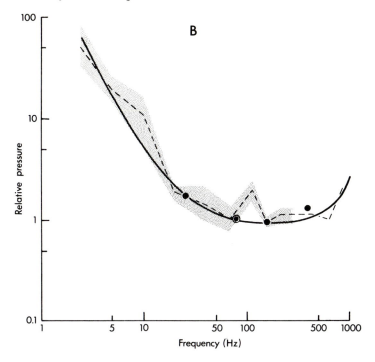

the added responses to the separate frequencies this is certainly not true for the electrical responses (Fig. 2-7B). Furthermore, the electrical responses were such that the peaks were exaggerated and information about polarity preserved. The responses of these receptors to the first cycle of a burst of sinusoidal pressure waves was usually larger than that to subsequent cycles (Fig. 2-5A).

5 Discussion

To have an understanding of an animal's hearing it is evidently very desirable to link closely results on a range of studies: on sounds in the sea; on behavior; on electrophyiology; on gross and fine structures; and on mechanical responses of tissues. We have tackled only a limited number of problems in herring and sprat, and have concentrated our efforts on studies of the effects of pressure changes on the bulla system and the structures associated with this system.

A striking feature of the mechanical response of the bulla is the relatively flat frequency response between 0.01 and 1,000 Hz (about 20 octaves) (see Fig. 2-3D). At the lower end of this frequency range it seems probable that these responses are, in life, concerned with pressure changes with depth. This is made more likely by the finding that larval herring became sensitive to small pressure changes (centimeters of water) at the time when their bullae are first filled with gas and also by the discovery of the fine elastic thread shown on Fig. 2-4A which links the bulla membrane with the utriculus;

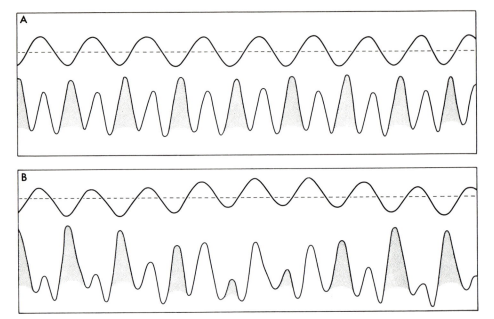

Figure 2-7. Mixed sinusoidal waveforms. If a low frequency (30 Hz) pressure wave is mixed with a 200 Hz wave, the resultant electrical activity of the utriculus is considerably changed and is not the sum of the responses to the two frequencies taken alone. (A) 200 Hz, 12.5 N m^{-2} (peak to peak) alone; top trace pressure, compression is up and dotted line is mean pressure; bottom trace response, negative is up and d-responses are stippled. (B) Mixed waveform stimulus (200 and 30 Hz) with the 200 Hz wave the same size as (A); note that the envelopes of both the d-responses and the c-responses follow the lower frequency. (From Denton and Gray 1979).

the thread is also found in the larval stage of herring. With respect to higher frequencies we know far too little about the sounds to which clupeoids are exposed in life (see Denton, Gray, and Blaxter 1979) but we can be certain that in the frequency band between 3-30 Hz there will be large particle displacements given by the swimming activity of a fish itself and of its neighbors.

Watching a closely knit school of herring, especially if under intense stimulation in a tank, one is impressed by the ability of the fish to avoid the walls and each other at such high speeds and when so confined. Although we know that schools break up at night and that vision is important for reforming the schools and keeping them together, it does seem likely that "acoustic" signaling is playing a part during rapid maneuvers like escape reactions. We know that herring can respond to high frequencies (Enger 1967, Olsen 1976). It may be that the perception of the rapid transients given out by neighbors in the school during fast maneuvers is the secret to the coordinating ability by clupeoids. The only experimental evidence which relates to this question comes from the "startle" experiments. These show that thresholds are independent (over a wide range) of rate of rise within single peaks, a finding that accords well with the flat frequency response of the system. The startle response is obviously important, but this limited type of response uses only a small part of the information that can be transmitted by the clupeid acousticolateralis. We have little idea how the full range of acoustic information is used in relation to the other senses and more complex behavior patterns.

Let us now consider our results on the utriculus for these are the ones which are most complete. Mechanically the movements of the floor of the utriculus follow the waveforms of stimulating pressures over a range and variety of simple and mixed frequencies between 30 and at least 800 Hz. Below 30 Hz the electrical responses (see Fig. 2-6B) fall off sharply with decreasing frequency and it is clear that this decline is at least in part due to the mechanics of the utricular floor. In one experiment on a sprat, electrical responses (like those of Fig. 2-5) were obtained during slow sinusoidal (0.25 Hz) changes of pressure of up to 6500 N m^{-2} peak to peak. These slow pressure waves mimicked pressure changes due to changes in a fish's depth. The results led to the conclusion that (as far as the utriculus was concerned) normal rates of fish movements (up and down) would have little effect on the reception of vibratory signals emitted by other organisms.

There can be little doubt that the electrical responses recorded inside the utriculus (Fig. 2-5) are derived from the hair cells of the utricular maculae. Following Lowenstein and Wersäll (1959) and Flock (1965) it has commonly been found that hair cells are preferentially excited when their ciliary processes are displaced toward their kinocilia and it seems almost certain that the two peaks (c and d), given in response to each cycle of a sinusoidal stimulus, correspond to the excitation of the two major populations of oppositely orientated hair cells in the two larger maculae. Because of the strong connective tissue backing to the main parts of these maculae (Fig. 2-4A) each must behave *mechanically* as a single unit. It does not follow automatically, however, that the electrical responses of, for example, the d-receptors lying on the middle maculae will all have the same flat frequency response. From our knowledge of the mammalian cochlea (see e.g., Russell and Sellick 1978) and the turtle cochlea (Fettiplace and Crawford 1978) it would be quite reasonable to expect that there might be a number of groups of d-receptors which would have different frequency responses although they were all lying on a platform which was displaced as one unit. Results like those of Fig. 2-7 give no support to this idea for if it were true we should expect that the electrical response to a mixture of 30- and 200-Hz pressure waves would be in large part the addition of the two separate responses and this is never the case. Much the simplest explanation of our results is that all the receptors respond in the same way over a wide band of frequencies.

Although most work on hearing has been concerned with tones it is generally recognized that, even in humans, noises play an important role in auditory discriminations. Helmholtz (1877) emphasized that in music we depend very much on irregular noises such as the scratching and rubbing of the violin bow to distinguish between instruments and as Pippard (1978) writes "the ear (human) in fact, occupies an intermediate position between the tape recording or disc which responds to an immediate sound pressure and a sound spectrograph designed to sort out sinusoidal components." Fish can discriminate between different frequencies of stimulation but, as Fay (1978) has argued, they do have auditory systems which are to a greater degree than our own, organized to give rapid information about the amplitude and temporal characteristics of brief noises. If a brief noise is to be recognized sufficient information must be transmitted in the time available to make all necessary discriminations. Our results on the sprat utriculus lead us to conclude that, in the anterior macula and in the lateral half to two-thirds of the middle macula, the amplitude of each pressure peak, appropriately modified, will with a short delay be represented by the total activity of the receptor population, the timing by the time of the mean, or other measure of location, and the polarity by the half of the population involved; since the receptor population is large the precision could be high. This kind of representation is virtually instantaneous in contrast to one dependent on resonances since, if a resonant system is highly tuned and sensitive, the stimulating vibration needs to act for several cycles to reach its full amplitude. It remains of course possible that other parts of the clupeid auditory system are more capable of Fourier analysis and we may note that, even in the utriculus, a great deal of information about frequencies is retained and our results on the utriculus are not inconsistent with those of Enger (1967) on the medulla of the herring.

The utriculus could act as a reference based on acoustic pressure (Schuijf 1976) whose activity might be compared with that of organs responding to directional stimuli. The utriculus is an organ which is likely to be involved in the "startle" response but, since this response seems to have directional properties, it seems almost certain that some other auditory structure will also be involved (it is possible that some directional information can be provided by the action of the otolith on the middle macula of the utriculus). We have done practically no work on the sacculus but we have worked on the lateral lines of herring and sprat and two of us (E. J. D. and J. A. B. G.) have made some recent studies on the sprat. These showed: that (1) when a fish and the surrounding seawater were made to vibrate together the liquid throughout the lateral line canals moved, as might be expected, in exactly the same way as the fish; and (2) when a fish was held still and the seawater outside the fish made to vibrate, the liquid in the center of the canals could move relative to the fish and so presumably stimulate its neuromasts in the canals. These movements were maximal in the parts of the main canals which were parallel to the axis of motion and very small in the parts of the canals which were perpendicular to the motion. In the canals for which the movements were greatest the amplitude of liquid movements inside the canals, relative to that outside the fish, depended (for sinusoidal vibrations) on frequency. As the frequence was raised from 10 to 80 Hz the relative movement rose from about 0.1 to about 1 and was approximately proportional to frequency. Above 80 Hz (at least as far as 160 Hz) the ratio was about 1.

There is the possibility that the clupeid may be able to estimate the distance as well as the direction of a vibrating source by determining the ratio of pressure to particle displacement. We have no direct experiments on this but, since we know a good deal about the mechanics of the clupeid ear, we can calculate the relative stimuli which are likely to be given by simple external sources of sound to the sense organs of the utriculus and lateral line by pressure and particle displacement respectively. We have given the results of such calculations in an earlier paper (Denton, Gray, and Blaxter 1979) and these gave support to the idea. Our recent observations on lateral lines, described briefly above, seem to us to make the hypothesis a more plausible one. If, as our experiments indicated, the particle displacement inside the lateral line system is (up to

80 Hz) approximately proportional to frequency, then the ratios of the liquid movements in the canals caused on the one hand by the pressure component and on the other by the particle displacements of an acoustic wave will be less dependent on frequency than was previously supposed.

References

Allen, J. M., Blaxter, J. H. S., Denton, E. J.: The functional anatomy and development of the swimbladder-inner ear-lateral line system in herring and sprat. J. Mar. Biol. Assoc. U.K. 56, 471-486 (1976).

Best, A. C. G., Gray, J. A. B.: Morphology of the utricular recess in the sprat. J. Mar. Biol. Assoc. U.K. 60, 703-715 (1980).

Blaxter, J. H. S., Denton, E. J.: Function of the swimbladder-inner ear-lateral line system of herring in the young stages. J. Mar. Biol. Assoc. U.K. 56, 487-502 (1976).

Blaxter, J. H. S., Denton, E. J., Gray, J. A. B.: The herring swimbladder as a gas reservoir for the acoustico-lateralis system. J. Mar. Biol. Assoc. U.K. 59, 1-10 (1979).

Denton, E. J., Blaxter, J. H. S.: The mechanical relationships between the clupeid swimbladder, inner ear and lateral line. J. Mar. Biol. Assoc. U.K. 56, 787-807 (1976).

Denton, E. J., Gray, J. A. B.: The analysis of sound by the sprat ear. Nature 282, 406-407 (1979).

Denton, E. J., Gray, J. A. B.: Receptor activity in the utriculus of the sprat. J. Mar. Biol. Assoc. U.K. 60, 717-740 (1980).

Denton, E. J., Gray, J. A. B., Blaxter, J. H. S.: The mechanics of the clupeid acoustico-lateralis system: Frequency responses. J. Mar. Biol. Assoc. U.K. 59, 27-47 (1979).

Enger, P. S.: Hearing in herring. Comp. Biochem. Physiol. 22, 527-538 (1967).

Fay, R. R.: Sound detection and sensory coding by the auditory systems of fishes. In: The Behavior of Fish and Other Aquatic Animals. Mostofsky, D. I. (ed.). New York: Academic Press, 1978, pp. 197-231.

Fettiplace, R., Crawford, A. C.: The coding of sound pressure and frequency in cochlear hair cells of the terrapin. Proc. R. Soc. London Ser. B 203, 209-218 (1978).

Flock, Å.: Electron microscopic and electrophysiological studies on the lateral line canal organ. Acta Otolaryngol. Suppl. 199, 1-90 (1965).

Gray, J. A. B., Denton, E. J.: The mechanics of the clupeid acoustico-lateralis system: Low frequency measurements. J. Mar. Biol. Assoc. U.K. 59, 11-26 (1979).

Helmholtz, H. L. F.: On the Sensations of Tone as a Physiological Basis for the Theory of Music, 2nd English ed. New York: Dover (1877).

Johnstone, B. M.: General discussion. In: Hearing Mechanisms in Vertebrates. de Reuck, A. V. S., Knight, J. (eds.). London: Churchill, 1968.

Lowenstein, O., Wersäll, J.: A functional interpretation of the electron-microscopic structure of the sensory hairs of the cristae of the elasmobranch *Raja clavata* in terms of directional sensitivity. Nature 184, 1807-1808 (1959).

Olsen, K.: Evidence for localization of sound by fish in schools. In: Sound Reception in Fish. Schuijf, A., Hawkins, A. D. (eds.). Amsterdam: Elsevier, 1976, pp. 257-270.

Pippard, A. B.: The Physics of Vibration, Vol. 1. Cambridge: Cambridge University Press, 1978.

Popper, A. N., Platt, C.: The herring ear has a unique receptor pattern. Nature 280, 832-833 (1979).

Russell, I. J., Sellick, P. M.: Intracellular studies of hair-cells in the mammalian cochlea. J. Physiol. 284, 261-290 (1978).

Schuijf, A.: The phase model of directional hearing in fish. In: Sound Reception in Fish. Schuijf, A., Hawkins, A. D. (eds.). Amsterdam: Elsevier, 1976, pp. 63-86.

Discussion

BULLOCK: For the record, it should be emphasized here that the auditory system in fishes is highly suitable for recording the polarity of the stimulus. Their system is much better than ours at discriminating the initial compression versus the initial rarefaction.

BELL: I note that the frequency response of the microphonics to displacements is much flatter at the low end of the spectrum. Presumably the animal will not be sensing hydrostatic changes. Is that a reason for the difference in frequency response?

DENTON: That is certainly one reason. In fact, the main parts of the utricular maculae do not follow low frequencies very well.

There must be reasons for such a fast sensitivity drop at low frequencies. I think it would be undesirable for the system to remain sensitive in the low frequency region, otherwise it would be thrown out of gear by small depth changes. Even if you apply a large pressure change at a very low frequency, say the equivalent of 6,000 Pascals over one cycle, it hardly affects the responses to high frequencies at all. The system works quite well independently of the fish moving up and down. Responses to depth change are more the concern of the medial parts of the utriculus, and not of the main system we have described here.

BULLOCK: To follow up on my earlier account, I should like to point out that the ability to discriminate between rarefaction and compression, and, therefore, phase, may go with another ability of these animals that we may not possess. The fish auditory system may look at complex waveforms not in terms of Fourier or frequency analysis, but in the time domain. Piddington [J. Exp. Bio. 56, 403-410 (1972)] was the first to present evidence for this. I think it is very likely that fish are good at discriminating mixtures of frequencies, even with the power spectra being the same. The relative phase of the component harmonics may be critical. We find that this is true in electroreception, and here is where the importance of the time domain as distinct from the frequency domain has been suggested. If you clip a sine wave in one polarity, it may be quite different from clipping it in the other polarity. The frequency content is the same, the power spectrum is the same, but the phase relations of the harmonics have been changed. The fish system should be very good at such a task. Even if it may not be a great frequency analyzer, in the sense of discriminating Beethoven from Mozart, it may be very good at analyzing the phase structure.

Assuming such abilities, the fish may not be able to tell very much about the quality of the sound source. However, it should be able to tell if the source is approaching or receding, by determining the proportion of the rarefaction cycle to the compression cycle. An analysis of the fine structure of harmonics seems like a great deal more than we ever thought the fish ear could do, but, in fact, it may be very good at this kind of analysis.

PARTRIDGE: I think that the reason fish have chosen the time domain over frequency domain is in order to do the tasks that Bullock has suggested when they school. It is much more important to know whether your neighbor is coming closer or moving away than to recognize the individual, as by the tail flips, for example.

DENTON: Tone analysis is very important to us because we are interested in speech, while the sound quality may not be so important to fish. They may be more concerned with transients generated by other organisms. They also do not have a great deal of time to make up their minds as to what frequencies are coming in and what to do about them. For a frequency of, say, 20 Hz, if you have to wait three or four cycles to recognize it, it is too late—you're caught. The fish must get its information within the first cycle or half cycle, and the kind of a system which fish possess is very good at getting a lot of information as quickly as possible. For a startle response, the fish gets all the needed information in the first half cycle, makes its response, and it may not care what happens later.

WILLOWS: That is a beautiful description of the transformation that takes place from an energy source through the mechanics of the sensory system. What evidence is there on the transformation at the next step, i.e., from the sensory-mechanical to the neural level? There may be still another story at the central level.

DENTON: We are fairly well convinced of what happens. The fish makes use of information obtained simultaneously from the lateral line and utriculus. This is organized centrally. The sacculus is involved to a smaller degree in the response to pressure. Most of the energy flow goes from the bulla, through the utriculus, then out to the lateralis system. The proportion that goes through the sacculus is small. However, I am oversimplifying.

GRAY: We have not recorded at all from the primary nerve fibers, and some interesting properties should be found through such techniques. The utriculus alone has some thirteen thousand hair cells. We are talking about the same order of magnitude as in the human cochlea. We have also counted afferent nerve fibers in one specimen, but I think the numbers may be too low. The middle macula has about a thousand nerve fibers, and about seven thousand hair cells. However, the middle macula has about forty really big fibers, a couple of hundred middling ones, and about eight hundred little ones. The anterior macula has almost entirely small fibers, and the ratio of hair cells to fibers is less. The big fibers in the middle macula may be concerned with rapid connections, while the smaller fibers should provide a more detailed analysis of the stimulus.

BLAXTER: It seems from the literature that the startle responses are randomized. They are escape reactions, and to best avoid a predator, it is desirable that they should be random, so that the predator cannot predict how the prey will move. We may or may not be looking at Mauthner's cell responses. From the viewpoint of the school, however, it is desirable that responses, especially startle responses, should be carefully coordinated among the members of a school to avoid collisions. There seems to be a paradox here.

PARTRIDGE: If you cut the lateral line and then startle them, the fish will indeed collide. In fact, they bump so violently that they settle and float to the surface.

DENTON: Were these herring?

PARTRIDGE: No, these were saithe.

TAVOLGA: I am familiar with the startle response that you get from herring and other clupeids, but in my experience they habituated quite rapidly. How did you manage to keep them responding without getting tired of all this?

DENTON: We used a tank about two meters across, with a central island. Three or four hundred herring swim round the island. We provided stimuli at one point in the tank about every five seconds. The chances of getting the same herring with a stimulus twice were very small. Herring are like Ford cars, in that their responses are all alike.

GRAY: When we used only a few fish, we did observe some habituation, but with a large population we were able to get consistent results quickly.

BULLOCK: I should like to discuss the question of what the brain may be getting out of all this. Recorded microphonics can give information about local effects and local differences, but if you are interested in what the fish can hear and what its brain does with the information, there is another way of investigating this besides the behavioral approach. There is a technique of doing this electrophysiologically that may not be familiar to everybody here. You don't have to hunt for single fibers, cast magic spells to enhance your luck, and occasionally get the right fibers. There are electrophysiological techniques that are now being used in mammals. They are used extensively, they are reliable and consistent. This is the method of averaging action potentials. You can look at responses low in the brain stem and see just how much analysis may be taking place there, or you can look at later stages. It is astonishing how simple this method is and it can tell you whether the brain is making a discrimination between this or that stimulus. I recommend it as a method. We are just beginning to get some experience with it in fish.

BOSTON: I should like to add a caution. I have worked both with microphonics and action potentials. Some of the problems in interpreting the average response of potentials is just as difficult as trying to interpret microphonics.

FAY: I should like to propose the question of this relation between the ear and lateral line. Is this unique to this species, or is it found in other groups?

GRAY: It is found in all the clupeoids, i.e., herrings and anchovies.

FAY: I once injected India ink into the perilymphatic system of the goldfish, and the ink just popped right out into the head lateral line canals. I wonder if there are other groups we should be looking at for relations between the lateral line and the ear.

GRAY: I just want to point out that there is no trunk lateral line in the clupeids, just the head canal system.

FAY: Your analysis of this fish ear is remarkably complete, especially as compared to analyses of other fish ears. Your analysis includes the notion of a "release window." Weaver called this a fluid mobilizing system, and he identified the mechanism in many animal groups. From your description, it seems clear how such a fluid circuit allows movement to take place. In most other species, it is not known where this mechanism is and what its characteristics are.

Chapter 3

The Swimbladder and Hearing

J. H. S. BLAXTER*

1 Introduction

There is adequate experimental evidence to show that the hearing ability of fish is enhanced by the presence of a swimbladder. Sand and Enger (1973) found that removing gas from the swimbladder of *Gadus morhua* (cod) reduced the sensitivity at 300 Hz by about 20 dB, while Fay and Popper (1975) found a decline in the sound pressure sensitivity of *Ictalurus punctatus* (catfish) above 100 Hz after swimbladder deflation. In a neat experiment on the dab *Limanda limanda,* a species without a swimbladder, Chapman and Sand (1974) found that the presence of a small air-filled balloon below the head increased the sensitivity by about 3 dB at 30 Hz and by nearly 20 dB at 200 Hz; the upper limit of the frequency response also increased from about 200 Hz to 350 Hz. A gas-filled structure very close to the ear, such as the otic bullae of the clupeoids which lies near the utriculus, also improves hearing (Blaxter, Denton, and Gray, Chapter 2). A coupling mechanism between the swimbladder and the ear enhances hearing still further as shown by comparing the audiograms of the Ostariophysi, which possess Weberian ossicles, with nonostariophysine fish (Popper and Fay 1973). The auditory sensitivity of different holocentrid (squirrelfish) species can also be related to the degree of connection between the swimbladder and labyrinth (Coombs and Popper 1979).

2 Resonance Frequency and Hearing Ability

A number of workers have measured the audiograms of fish and compared the result with the resonance frequency of the swimbladder since it might seem desirable for the swimbladder to resonate within the frequency range of hearing. In the cod, Chapman and Hawkins (1973) found a frequency range of 30-470 Hz but a peak of swimbladder resonance around 2 kHz (Sand and Hawkins 1973). Depending on the conditions,

*Dunstaffnage Marine Research Laboratory, Oban, Argyll, Scotland.

this peak can be as much as 20 dB higher outside the frequency range of hearing than within it. Resonance frequencies are inversely related to the volume of the swimbladder; hence larger fish have lower resonance frequencies (Fig. 3-1). The resonance frequency is only matched to the hearing range if the fish are quite large or the frequency range is wide (as in the Ostariophysi). There may, in a sense, be a conflict between the need to keep the swimbladder about 5% of the body volume in a marine fish for neutral buoyancy and the possible desirability of having a larger swimbladder with a lower resonance frequency.

3 The Effect of Depth Change

The volume of the swimbladder changes if fish move vertically. Substantial vertical migrations, toward the surface at dusk and toward the seabed at dawn, are a regular behavioral feature of many species. *Pollachius virens* (coalfish), for example, is known to migrate over at least 100 m (Schmidt 1955). The gas in the swimbladder expands or contracts during a pressure change according to Boyle's law, although the swimbladder, in most species being attached at various points over its surface, does not necessarily behave like a spherical bubble.

The swimbladder is the main structure returning the echo from sonar devices. The strength of the returning echoes from fish is important in fisheries for echo-sounding estimates of fish biomass. If fish make substantial vertical migrations the change in volume of the swimbladder, and indeed the change in resonance frequency, causes a problem in calibrating fish counters. Surprisingly, few acousticians have considered this problem. Dunn (1978) reported that the target strength of various gadoid fish [cod, *Melanogrammus aeglefinus* (haddock) and *Merlangius merlangus* (whiting)] was reduced initially by about 6 dB after subjecting the fish to enforced pressure increases in the range 1.75-5.0 atm. The attenuation appeared to depend on the relative change

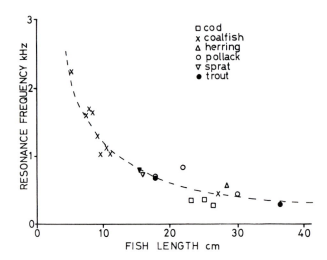

Figure 3-1. Resonance frequency of swimbladders of different species related to fish length (redrawn from Løvik and Hovem 1979).

of pressure rather than the absolute pressure which is consistent with the swimbladder volume changing in accordance with Boyle's law. Løvik and Hovem (1979) measured the immediate change in resonance frequency of *Clupea harengus* (herring), *Sprattus sprattus* (sprat), and coalfish (Fig. 3-2) after enforced depth increases from 4 to 50 m (1.4-6.0 atm). The increase in resonance frequency was consistent with theory, the resonance frequency being proportional to $p^{5/6}$ where p was the ambient pressure.

4 Acclimation to Depth Change

Dunn (1978) found that if cod, haddock and whiting were allowed to adapt to the new depth, equivalent to pressures of 1.75 to 5 atm, the target strength was restored after 20-100 hr (Fig. 3-3), suggesting that gas secretion had returned the swimbladder to its previous volume. Løvik and Hovem (1979) plotted the change in resonance frequency of a coalfish during acclimation after a sequence of enforced depth changes from 4 to 32 to 44 to 12 m (Fig. 3-4). The fish was not able to restore its resonance frequency in the 12- to 24-hr period allowed. If the fish had been given time to regulate its swimbladder volume, the resonance frequency should have been proportional to $p^{1/2}$ where p is the ambient pressure.

It should be possible to relate acclimation time to the rate of gas secretion or resorption in the swimbladder. The secretion rate for coalfish is about 2 ml STP/kg/hr and the resorption rate about 8 ml STP/kg/hr (Tytler and Blaxter 1973). The time required to adapt fully after an ascent or descent is shown in Fig. 3-5). In most daily vertical migrations of tens to a hundred meters the swimbladder will not have adapted fully before the arrival time for the next phase of the vertical migration. The fact that rates of gas secretion "lag" after the dawn ascent means that the fish has a greater flexibility to move up again on the dusk ascent without bursting the swimbladder (Tytler and Blaxter 1973).

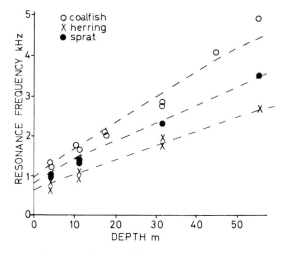

Figure 3-2. Resonance frequencies at different depths for nonadapted fish. Dashed line is theoretical result where resonance frequency follows $p^{5/6}$ power law (redrawn from Løvik and Hovem 1979).

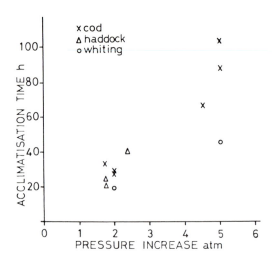

Figure 3-3. Time for target strength to be restored after enforced pressure increases between 1.75 and 5 atm (redrawn from Dunn 1978).

5 Swimbladder "Behavior"

The latency of swimbladder adaptation after a change of pressure should affect hearing and other possible swimbladder functions. Yet some unpredicted changes take place after an enforced descent. In the cod, simple changes of resonance frequency (Fig. 3-6A) do not occur after a change of depth (Sand and Hawkins 1973, Løvik and Hovem 1979); and the resonance frequency seems to be elevated above the theoretical level at the depth of adaptation. There is little evidence, however, for anything but the most modest excess pressure within the swimbladder (Sundnes and Gytre 1972, Sand

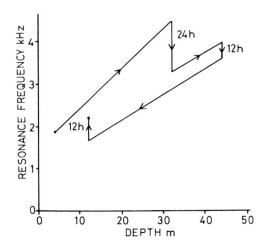

Figure 3-4. Change in resonance frequency of a 7.7 cm coalfish after various changes of depth with time allowed for adaptation (redrawn from Løvik and Hovem 1979).

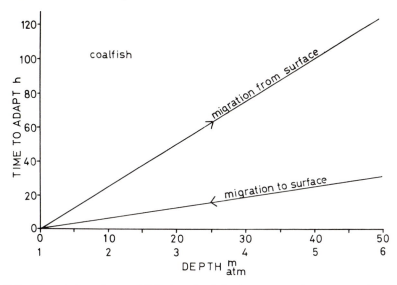

Figure 3-5. Time taken for coalfish to adapt to increases of pressure when moving from the surface to different depths and for decreases of pressure moving from different depths to the surface.

and Hawkins 1974) which could explain this although it may be that the stiffness of the body wall plays some role. If cod are allowed to adapt to a new depth for a while, the resonance frequency on a return toward the surface suggests gas secretion has occurred, the swimbladder is larger and the resonance frequency lower (Fig. 3-6B).

The swimbladder is probably also the site of pressure sensitivity. Tytler and Blaxter (1973, 1979) found that pressure sensitivity was lost immediately after rapid pressure changes were imposed on coalfish. Yet sensitivity was subsequently restored within a few minutes after incremental increases of pressure within the range 1 to 17 atm (Fig. 3-7), much more quickly than could be explained by gas secretion returning the swimbladder to its original volume. It seems that the volume of the swimbladder had been reduced in such a way that tonus is retained in at least some parts of the wall that has stretch receptors.

6 The Nature of Swimbladder Compression

It is possible to examine the way in which the swimbladder contracts under pressure in the apparatus shown in Fig. 3-8. A newly dead fish such as coalfish is prepared by removing part of the body wall and the viscera and is then pinned on to a plate and placed horizontally in a plexiglass cylinder filled with water. After bolting on the end plate and bleeding off air, pressure increases of 1 atm or more can be applied by a syringe. The swimbladder can be viewed through a mirror below the cylinder which is surrounded by a water-filled box to improve the optics.

Under pressure the coalfish swimbladder contracts mainly at the posterior end on account of the slant of the swimbladder within the body cavity. X-Radiographs were

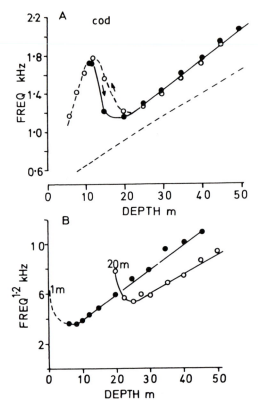

Figure 3-6. (A). Changes in resonance frequency of the swimbladder of a 16 cm cod adapted to 11 m, lowered to 50 m (filled symbols) and then immediately raised to 6 m (open symbols). The dashed line shows the expected change in frequency for a free spherical bubble containing the same mass of gas as the swimbladder.
(B) Changes in resonance frequency of a 16 cm cod after being adapted to the surface, lowered rapidly to 50 m (filled symbols) and then kept at 20 m for 20 hr and rapidly lowered to 50 m again (open symbols). Both figures redrawn from Sand and Hawkins (1973).

taken of intact fish and the angle measured between the axis of the body and the axis of the swimbladder (Fig. 3-9). In ten coalfish (Table 3-1) the mean angle was $7.0° \pm 0.8$.

The apparatus shown in Fig. 3-8 can be rotated through different angles to the horizontal which can be measured by a protractor and spirit level. If the head of the fish is lowered to simulate a shallow dive a critical angle is reached where gas ceases to be located preferentially at the anterior end of the swimbladder and the swimbladder flattens along its full length; at a steeper angle gas collects at the posterior end. This critical angle in five saithe was $6.1° \pm 0.8$. After a simulated dive in the apparatus, gas collects posteriorly but moves anteriorly once the fish becomes horizontal again.

During a descent in natural life it appears that coalfish would retain gas near the anterior end of the swimbladder in a shallow dive; after a steep dive gas would return to the anterior end once the fish became horizontal. It may be no coincidence that gas

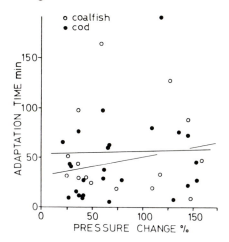

Figure 3-7. The time taken for cod and coalfish to restore their pressure sensitivity after different pressure increases. Neither regression (upper, coalfish; lower, cod) is significant (redrawn from Tytler and Blaxter 1973).

tends to be located at the anterior end of the swimbladder nearest to the ear, where it is best placed to amplify sound. Gas secretion tends later to cause refilling of the more posterior parts of the swimbladder, if the fish remains at the new depth.

7 The Special Case of the Clupeoids

Blaxter, Denton, and Gray (Chapter 2) have described the acousticolateralis system of clupeoids. The clupeoids have a special problem which is to ensure that gas is available to flow from the swimbladder into the otic bulla system *during* a dive; otherwise there is a danger of the bulla membrane bursting under the increased pressure. In the herring the swimbladder has a toughened front end and a compliant center section. On com-

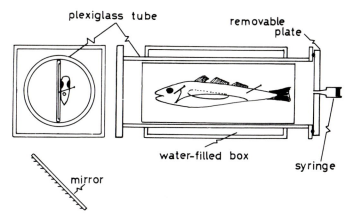

Figure 3-8. Plexiglass apparatus (left cross-section, right elevation), for applying pressure changes to swimbladders (see Blaxter et al. 1979). The apparatus can be rotated vertically through known angles to the fore-aft axis.

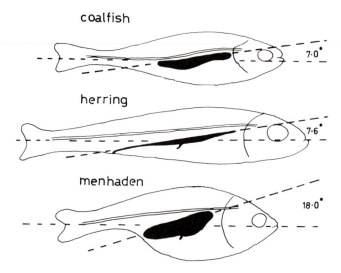

Figure 3-9. Diagrams drawn from X-radiographs showing the angle of the swimbladder in relation to the horizontal axis in coalfish, herring, and menhaden.

pression, gas tends to collect at the front end (Blaxter, Denton, and Gray 1979), partly because of this difference in the swimbladder wall and partly because the swimbladder is angled upwards at about 7° (Fig. 3-9, Table 3-1). During a simulated dive in the apparatus shown in Fig. 3-8, the critical angle is difficult to judge; it is possible that the center section seals off some gas in the anterior section during a steep dive.

An interesting adaptation is seen in the related clupeid, *Brevoortia tyrannus* (menhaden) which has a very baggy swimbladder with no toughened anterior end. The swimbladder is very steeply angled (Fig. 3-9, Table 3-1) at about 18°. This makes it more certain that gas collects anteriorly during a dive, the critical angle being 13° to 14°. Should the menhaden dive more steeply it is possible that the bulla might be starved of gas with consequent rupturing of the membrane if the amplitude of the dive was great.

Table 3-1. Measurements on swimbladders

Species	Length (cm)	Angle of swimbladder to horizontal	Critical angle (see text)
Coalfish	15-17	$7.0°$ ± 0.8 (n=10)	$6.1°$ ± 0.8 (n=5)
Herring	16-17	$7.6°$ ± 1.5 (n=13)	?
Menhaden	13-18	$18.0°$ ± 1.5 (n=10)	$13.3°$ ± 3.5 (n=5)

8 Conclusions

At least one function of an upwardly slanted swimbladder seems to be to locate gas (especially if limited) as near the ear as possible. After a downward migration the reduction in swimbladder volume causes an increase in the resonance frequency. In fish like cod the resonance frequency thus becomes more offset from the hearing frequency. Sand and Hawkins (1973) considered that the cod swimbladder was heavily damped so that any advantage gained at the resonance frequency was much reduced. A resonating swimbladder should still transduce sound pressures to displacements at frequencies well below the resonance frequency. At very low frequencies, however, free-field displacements become more important. The offset resonance system of fish like gadoids may confer slightly less sensitive hearing but hearing which is more stable since it is less likely to change in sensitivity with fish size or depth.

The swimbladder may have many functions, in buoyancy, hearing, sound production, as a gas reservoir in clupeoids, or perhaps as a source of respiratory oxygen. In an evolutionary sense some compromise may have been adopted to allow the swimbladder to fulfill a number of roles.

References

Blaxter, J. H. S., Denton, E. J., Gray, J. A. B.: The herring swimbladder as a gas reservoir for the acoustico-lateralis system. J. Mar. Biol. Assoc. U.K. 59, 1-10 (1979).

Chapman, C. J., Hawkins, A. D.: A field study of hearing in the cod. J. Comp. Physiol. 85, 147-167 (1973).

Chapman, C. J., Sand, O.: Field studies of hearing in two species of flatfish *Pleuronectes platessa* (L.) and *Limanda limanda* (L.). Comp. Biochem. Physiol. 47A, 371-385 (1974).

Coombs, S., Popper, A. N.: Hearing differences among Hawaiian squirrel fish (family Holocentridae) related to differences in the peripheral auditory system. J. Comp. Physiol. 132, 203-207 (1979).

Dunn, W. I.: The depth and species dependence of the target strength of gadoids. Proc. Conf. Acoustics in Fisheries. Inst. Acoustics, Underwater Acoustics Group, Hull 26/27 September, 1978 Paper No. 1.5. 24 pp. (mimeo) (1978).

Fay, R. R., Popper, A. N.: Modes of stimulation of the teleost ear. J. Exp. Biol. 62, 379-387 (1975).

Løvik, A., Hovem, J. M.: An experimental investigation of swimbladder resonance in fishes. J. Acoust. Soc. Am. 66, 850-854 (1979).

Popper, A. N., Fay, R. R.: Sound detection and processing by teleost fish: A critical review. J. Acoust. Soc. Am. 53, 1515-1529 (1973).

Sand, O., Enger, P. S.: Evidence for an auditory function of the swimbladder in the cod. J. Exp. Biol. 59, 405-414 (1973).

Sand, O., Hawkins, A. D.: Acoustic properties of the cod swimbladder. J. Exp. Biol. 58, 797-820 (1973).

Sand, O., Hawkins, A. D.: Measurements of swimbladder volume and pressure in the cod. Norw. J. Zool. 22, 31-34 (1974).

Schmidt, U.: Beiträge zur Biologie des Köhlers (*Gadus virens* L.) in den isländischen Gewässern. Ber. Deut. Wiss. Kommn. Meeresforsch. 14, 46-82 (1955).

Sundnes, G., Gytre, T.: Swimbladder gas pressure of cod in relation to hydrostatic pressure. J. Cons. Perma. Int. Explor. Mer. 34, 529-532 (1972).

Tytler, P., Blaxter, J. H. S.: Adaptation by cod and saithe to pressure changes. Neth. J. Sea Res. 7, 31-45 (1973).

Tytler, P., Blaxter, J. H. S.: The effect of swimbladder deflation on pressure sensitivity in the saithe *Pollachius virens*. J. Mar. Biol. Assoc. U.K. 57, 1057-1064 (1979).

Discussion

TAVOLGA: Concerning the acoustic mismatch of the swimbladder, it is even more obvious when you compare the frequency most fish use in sound production (about 100 Hz) with resonant frequency. They may not be gaining much from resonance, but, I suggest, they are not losing much either. This is because, by and large, swimbladders have a lousy acoustic Q.

SCHUIJF: Yes, the resonance peak is very shallow, but that is important for the phase comparison that I suggested in my model, where phase changes can take place at much lower frequencies.

Chapter 4

The Utricle in *Ictalurus punctatus*

DAVID B. JENKINS*

1 Introduction

The inferior division of the inner ear (saccule and lagena) is considered the primary area of audition in most fishes and has received most of the attention in anatomical investigations of the labyrinth. Recent ultrastructural studies have provided information on surface features of the sensory areas in numerous species, defining various polarization patterns and several types of hair cell bundles. Speculation on the function of particular types of hair cells has been possible to some extent based on physiological data available on the inferior division. The ultrastructure of the superior division of the labyrinth has received less attention; this may be due, at least in part, to indications that anatomically this region displays less interspecific diversity than the inferior division, and functionally the superior division is considered primarily a gravistatic organ in most fishes (Lowenstein 1971). Differences in the types of hair cell bundles have been reported in the utricular maculae of a few osteichthyan species (Dale 1976, Platt 1977, Popper 1978). Attempts to interpret the functional significance of these bundles has often been based upon their presence in sensory areas other than the utricular macula, sometimes in different species, where the function of the region is better known. If bundle type indicates specific function, the presence of cells in the utricle that are similar to those in areas of the inner ear not considered as gravistatic organs suggests that the utricular macula in osteichthyans may have more than one function.

This report is part of a study undertaken to provide gross morphological, histological, and ultrastructural information on the utricle in the channel catfish, *Ictalurus punctatus,* to determine how conditions in the utricle compare to those in the inferior division of the labyrinth and to compare data reported in other species.

*Department of Anatomy, University of North Carolina, Med. Res. Building "D" 331H, Chapel Hill, North Carolina 27514.

2 Materials and Methods

Specimens used in this study were albino channel catfish ranging in length from 6 to
16 cm. Methods for light microscopy and scanning electron microscopy were those
utilized in previous studies of the inner ears in catfish (Jenkins 1977, 1979).

3 Anatomy of Utricle

The gross morphology of the utricle in fishes has been described in preceding accounts.
In this report a few of the significant findings on the utricle in *I. punctatus* will be discussed.

3.1 Otolithic Membrane

The entire utricular macula is covered by an otolithic membrane. Anteriorly the membrane is thickened, follows the curve of the anterior part of the macula, and extends
up the lateral wall to cover the lacinia. Its anterolateral edge has a trabeculated appearance. Posteriorly, the membrane is much thinner and appears more granular. The
undersurface of the entire membrane exhibits small cavities into which the apical processes of the hair cells apparently extend.

Relatively little has been reported on the otolithic membrane in the utricle of osteichthyan species. In the cod Dale (1976) has described a "knotted" appearing membrane which resembles that seen in the posterior region of *I. punctatus*. No differences
in thickness or presence of a trabeculated membrane in the utricle were noted, but the
saccular otolithic membrane was described as trabeculated.

3.2 Relation of the Lacinia/Otolithic Membrane/Lapillus

In *I. punctatus* the lacinia, the lateral process of the utricular macula, extends up the
lateral wall, and continues a short distance onto the utricular roof; thus, this part of
the lacinia is in a horizontal position. The thickened region of the otolithic membrane
(see Section 3.1) covers the lacinia and fills the gap between the sensory epithelium
(on the lateral wall) and the lapillus; dorsally, the membrane enlarges and spans between the horizontal part of the lacinia and dorsal surface of the otolith. Therefore,
although the membrane increases in thickness to fill the gap between the lacinia and
otolith, the sensory epithelium still maintains an otolithic membrane-otolith association. This condition is different from that in species where the lacinia is considered
to have no association with the otolith (Platt 1977).

3.3 Utricular Macula

Until recently little information was available on the surface features of the utricular
macula in osteichthyan fishes. Flock (1964) described the polarization pattern of hair
cells and several hair cell bundle types in the burbot based on transmission electron
microscopic studies. Similar polarization patterns have subsequently been reported in

the cod (Dale 1976), flatfish (Jørgensen 1976), goldfish (Platt 1977), and bichir (Popper 1978). In addition, variations in hair cell bundles in the cod, goldfish, and bichir were noted by these authors.

In *I. punctatus* the polarization of the hair cells of the utricular macula is similar to that usually defined as the "typical" vertebrate pattern with opposing orientation along a line parallel to the anterolateral border of the macula. Variation in bundle types, based upon the relative lengths of the kinocilium and stereocilia are evident, particularly in the anterior and lateral regions of the macula (Fig. 4-1). For descriptive purposes the distribution of the bundle types will be discussed as a series of "bands" beginning with the anterior edge of the macula; these bands are not exclusive and mixing of types is apparent.

The most peripheral cells exhibit short stereocilia; the kinocilia vary in length but are longer than the stereocilia (Fig. 4-1). In a second band the cells have longer stereocilia and kinocilia; the rows of stereocilia in each bundle are often arranged in two groups, a shorter group in graded sizes and a taller group graded but with the tallest stereocilia nearly as long as the kinocilium (Fig. 4-2). The diameter of the bundles is noticeably greater in a third band and the stereocilia exhibit a more uniform steplike arrangement (Fig. 4-2). The fourth band varies in width; the tallest stereocilia are shorter than those in the preceding band and the kinocilium is usually no longer than adjacent stereocilia (Fig. 4-3). The bundles in the fifth and widest band, when compared to those in the fourth band, have longer processes but a kinocilium usually no longer than the tallest stereocilia (Fig. 4-3). The change in polarization of the hair cells occurs within the fourth and fifth bands. A sixth band has a mixed population with bundles varying in diameter and length. Posteriorly the cells are more difficult to examine due to the matted appearance produced by remnants of the otolithic membrane. Preliminary data show graded stereocilia and a kinocilium as long as or slightly longer than the tallest stereocilia. In the lacinia the bundle types are mixed. The fifth band extends into the lower part of the lacinia, but most of the rest of the cells have bundles with a long kinocilium.

The most peripheral cells are probably developing cells, but there is so much modulation in the length of the processes and bundle diameter in an anteroposterior direction that the bundle types probably are not merely different stages in the morphogenesis of the macula. The bundle characteristics may be related to function. For example, the bundles in the region of opposing polarization in *I. punctatus* (where the kinocilium is usually no longer than the tallest stereocilia) are similar to those of the utricular striola in the cod (Dale 1976), the thick bundles in the goldfish (Platt 1977) and type F1 of Popper (Popper 1977). Both the thick bundles of Platt and F1 of Popper are found widely distributed in the saccular maculae in the species they studied; in *I. punctatus* bundles similar to those in bands four and five are the predominant type in the saccular macula (work in progress). Based in part on the presence of similar cells in auditory areas of other vertebrates, Dale (1976, 1977) has postulated that such cells with shorter bundles may be sensitive to higher frequencies. Such correlation of structure and function is only speculative and other interpretations have been reported. In recent work on the herring, Popper and Platt (1979) have suggested that the "tall" bundles of the utricular macula (a sensory area considered to have an auditory function in that species) may be sensitive to higher frequencies than are the shorter bundles.

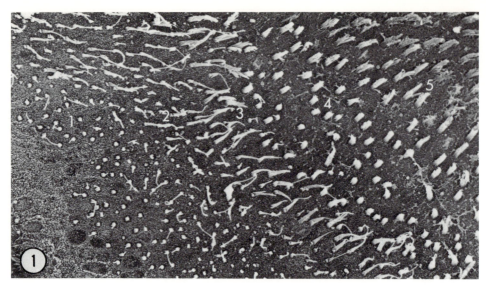

Figures 4-1–4-3. Anterior region of the utricular macula. Numbers indicate the "bands" described in the text. (1) Anterior to lower left. ✕ 780. (2) Anterior to bottom of micrograph. Note three peripheral cells in lower right corner. ✕ 1300. (3) More posterior than (2). Band four in the middle of micrograph and band five in upper left. ✕ 1300.

4 Summary

The results obtained in this study show regional variation in the thickness of the otolithic membrane, extension of the lacinia onto the utricular roof, relation of the lacinia to both the otolithic membrane and lapillus, and numerous hair cell bundle types. These data, and recent work on the utricle in other osteichythan species, suggest that the utricle shows more anatomical diversity than previously suspected. In addition, the presence of hair cell bundle types similar to those in other sensory areas may indicate functional diversity. Dual function in a macula, as in the utricle in the ray (Lowenstein and Roberts 1951), and interspecific functional differences of a particular region of the inner ear, as proposed in the utricle of the clupeids (for anatomical information see Popper and Platt 1979), are not uncommon. More information is needed, however, to accurately correlate function with particular regions and cell types.

Acknowledgments. The author is grateful for the technical assistance provided by Ms. Connie R. Guild. This study was supported by NIH Grant NS14961.

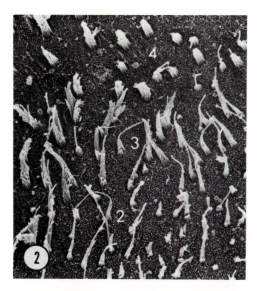

Figure 4-2.

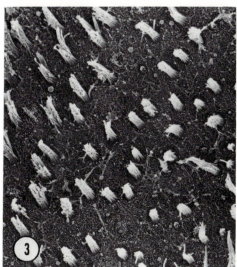

Figure 4-3.

References

Dale, T.: The labyrinthine mechanoreceptor organs of the cod *Gadus morhua* L. (Teleostei: Gadidae). Norw. J. Zool. 24, 85-128 (1976).

Dale, T.: Functional-morphological correlations in acoustico-lateralis sensory organs. SEM/II, IIT Research Institute, Chicago, Illinois 60616, 445-452 (1977).

Flock, Å.: Structure of the macula utriculi with special reference to directional interplay of sensory responses as revealed by morphological polarization. J. Cell Biol. 22, 413-431 (1964).

Jenkins, D.: A light microscopic study of the saccule and lagena in certain catfishes. Am. J. Anat. 150, 605-630 (1977).

Jenkins, D.: A transmission and scanning electron microscopic study of the saccule in five species of catfishes. Am. J. Anat. 154, 81-102 (1979).

Jørgensen, J. M.: Hair cell polarization in the flatfish inner ear. Acta Zool. (Stockholm) 57, 37-39 (1976).

Lowenstein, O.: The labyrinth. In: Fish Physiology, Vol. V. Hoar, W. S., Randall, D. J. New York: Academic Press, 1971, pp. 207-240.

Lowenstein, O., Roberts, T. D. M.: The localization and analysis of the responses to vibration from the isolated elasmobranch labyrinth. J. Physiol. 114, 471-489 (1951).

Platt, C.: Hair cell distribution and orientation in goldfish otolith organs. J. Comp. Neur. 172, 283-298 (1977).

Popper, A. N.: A scanning electron microscopic study of the sacculus and lagena in the ears of fifteen species of teleost fishes. J. Morphol. 153, 397-418 (1977).

Popper, A. N.: Scanning electron microscopic study of the otolithic organs in the bichir (*Polypterus bichir*) and shovelnose sturgeon (*Scaphirhynchus platorynchus*). J. Comp. Neur. 181, 117-128 (1978).

Popper, A. N., Platt, C.: The herring ear has a unique receptor pattern. Nature 280, 832-833 (1979).

Discussion

COOMBS: Have you looked under the light microscope at the patterns of the very small branches? Do you see any significance in the way they spread out?

JENKINS: I have only looked at a gross preparation. There is a separate band from the rest of the utricular nerve that fans out. There is a different offshoot that goes to the roof of the utricle.

COOMBS: Isn't that utricle extremely large?

JENKINS: It's fairly large, but it varies from species to species among the catfishes. Considering that seen in *Arius,* it is not so large, however.

HAWKINS: Earlier, you showed a membrane or series of fibers coming from underneath the otolithic membrane which you said was attached to the otolith. We have seen something similar in the cod.

JENKINS: I've seen this in the saccule also. It seems to be attached on the inferior process of the sagitta.

HAWKINS: Do you know if there is a fine membrane investing the whole of the otolith? Another question: How does the otolith grow?

JENKINS: I have not looked at the surface of the otolith in detail to determine if it is completely invested by a fine membrane.
   On otolith growth, I know that it does show growth rings. In transmission E.M., you can see these rings clearly, like the rings of a tree trunk.

BLAXTER: This is quite well known. You can actually see what seasonal problems the fish has had from looking at the growth rings of the otoliths.

CORWIN: Could you comment on the possible age changes in the receptor structures? There are some who have suggested that age changes can account for most of the variations, although morphogenetic changes can be accompanied by functional changes.

JENKINS: I can't rule out age changes, but the function of the otolithic membrane may be different when the hair cells are longer than when they are shorter.

CORWIN: They are not mutually exclusive, so you can have both an age difference and a functional difference.

Chapter 5

Audition in Elasmobranchs

JEFFREY T. CORWIN*

1 Introduction

This article presents a current assessment of elasmobranch auditory biology and an evaluation of future directions for research. Excellent review articles have recently detailed the pre-1975 development of this field (Popper and Fay 1977, Myrberg 1978), so I will not emphasize that work. Instead, I will focus on more recent investigations, including some new data.

I will begin by addressing two questions; "What sounds do elasmobranchs hear?" and "How do they detect them?" Those questions are simple, but due to the nature of the elasmobranchs they will probably not have simple answers.

Along with the holocephalans, the elasmobranchs are our closest links to the early jawed vertebrates that first inhabited the earth over 400 million years ago, 200 million years before the first mammals. Yet, extant elasmobranchs are modern. At least two divergent evolutionary lines have existed in this group for 150 million years or more, and many species and structures have become specialized in that time (Schaeffer 1967). Therefore, the modern elasmobranchs possess mixtures of characters. Some of their characters are primitive, never changed during evolution and thus similar to ancient forms. Others are derived, new adaptations shaped by recent selective pressures. Yet, there is no clear way that we can examine characters in a single species and confidently identify them as more or less primitive. In order to make those identifications, to determine the primitive characters that are the basic forms within a group, we must look for characters that are the most common in all the divisions of that group.

Ideally, in discussing elasmobranch audition I should point out some features that are thought to represent the basic primitive traits in this subclass. In that way I might provide a framework for simple answers to the questions I have posed. However, that is not presently possible since nearly all of our knowledge of elasmobranch audition

*Department of Neurosciences, School of Medicine and Neurobiology Unit, Scripps Institution of Oceanography, University of California, San Diego, La Jolla, California 92093.
Present address: Department of Zoology, University of Hawaii, Honolulu, Hawaii 96822.

stems from studies focused on one group, the galeomorph sharks. Four major divisions, the squalomorph sharks (spiny dogfish and relatives), the squatinomorph sharks (angel sharks), the batoids (skates and rays), and the galeomorph sharks (requiem sharks and relatives) are currently recognized in the elasmobranchs (Compagno 1977), but few studies have investigated hearing in members of the nongaleomorph divisions. Therefore, in the first sections of this article I will concentrate on galeomorph audition. Readers should be cautious, however, in generalizing from this account to the other elasmobranch groups. Even in evolutionary terms, 150 million years is a long time, especially when we are considering a highly adaptive sense. Therefore, we should expect the limited and incomplete answers of this article to be replaced by a more complete account of elasmobranch hearing as the nongaleomorphs are investigated.

2 What Sounds Do Sharks Hear?

2.1 Measurements in the Laboratory

As theory predicted, measurements have shown that sharks are sensitive to the so-called displacement or kinetic component of sound (Banner 1967, Kelly and Nelson 1975). Since sharks lack any known pressure-to-displacement transducers, such as the gas bladders of some bony fish or the sealed middle ears of some tetrapods, they presumably must rely directly on the displacement sensitivity of their mechanoreceptive cells (Fig. 5-1). Thus, sharks detect a form of acoustic stimulus that is basically different from the acoustic pressure wave that is most familiar to humans. Unlike acoustic pressure, the kinetic stimulus is inherently directional, but its magnitude rapidly decreases, relative to the pressure decrement, as it propagates outward from the sound source in the near field. In the frequency range where they are most sensitive carcharhinid sharks physiologically respond to displacements as low as 5×10^{-8} cm (Banner 1967, Corwin 1981b).

Measurements have also shown that sharks are low frequency sensitive. The upper range of behavioral sensitivity is 600 to 800 Hz in both the scalloped hammerhead shark, *Sphyrna lewini* (Olla 1962), and the lemon shark, *Negaprion brevirostris* (Nelson 1967). At least 40 dB better sensitivity is found between 40 Hz and 300 Hz (Nelson 1967). Also, Nelson (1967) clearly demonstrated that at least in that more sensitive lower part of the spectrum sharks are able to discriminate between sounds on the basis of frequency alone.

2.2 Natural Responses to Sound

2.2.1 Attraction

As originally demonstrated by Nelson and Gruber (1963) and illustrated in Figure 5-2 some sharks show urgent attraction to sounds that resemble those produced by struggling wounded fish. At least eighteen species, all galeomorphs, are known to show this response, in which they are not only aroused by the sound, but also can localize

A. With Pressure-to-Displacement Transducer

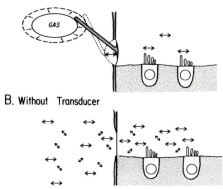

B. Without Transducer

Figure 5-1. A diagram of the two basic types of vertebrate sound detectors; (A) those with a pressure-to-displacement transducer, and (B) those without a transducer. In many tetrapods and many bony fishes the inner ear is coupled to a flexible gas-filled chamber analogous to that drawn in (A). The gas changes volume in response to sound pressure, thereby transforming the acoustic pressure stimulus into a displacement of the chamber's surface. The displacement can then be transmitted to the hair cells of the ear, so that they indirectly respond to the pressure component of the sound, a stimulus that they cannot sense directly. No elasmobranchs possess gasbladders, however, so the ear in these animals must function as in (B). There the hair cells of the ear directly respond to molecular displacements caused by acoustic pressure gradients in the medium. These gradients are relatively large near most sound sources, but decrease rapidly with distance, so that this type of detector may be confronted with low signal to noise ratios when far from a source.

source direction (Nelson and Johnson 1972, Myrberg 1978). This localizing ability is remarkable in that it extends well into the acoustic far field, so that Myrberg and co-workers (1972) were able to observe oriented responses in silky sharks, *Carcharhinus falciformis,* at 250 meters from a sound source.

The sounds which elicit these responses are characterized by energy in the 10 to 800 Hz range and low frequency amplitude modulation (Myrberg 1978). Thus, the upper frequency limit corresponds with the upper limit of hearing measured in conditioning studies. The most effective spectral range also roughly corresponds to the conditioned shark's best sensitivity range, 40 Hz to 300 Hz (Myrberg 1978).

Apparently these sharks use remote sound detection to identify and then to locate stressed or injured prey. Since the attracted species of sharks are predators that expend substantial energy in pursuing prey to capture, they would undoubtedly receive selective advantages from the ability to locate stressed and easily captured prey.

The sounds which could provide cues for this might be those generated by the stressed prey itself or in some cases sounds generated by active attackers. In fishing for sharks one occasionally notices an increase in new arrivals during or just after the struggling of a hooked shark (Nelson and Johnson 1976, pers. comm.). Since many of the attractable species have dentition suitable for piecemeal feeding on large prey, responding to sounds generated by other feeding sharks might allow an individual to

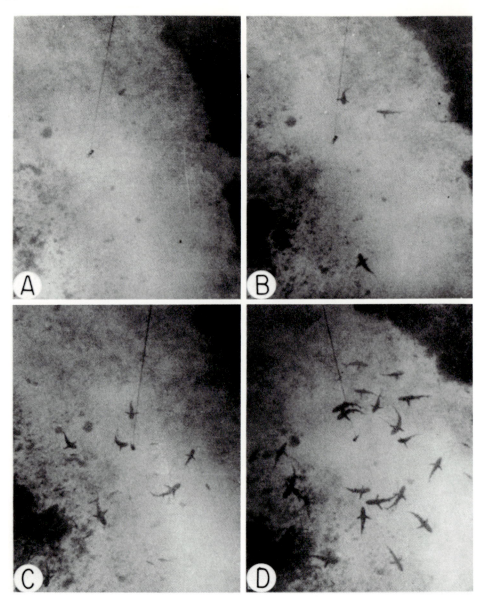

Figure 5-2. The urgent attraction of gray reef sharks to a suspended sound projector during playback of pulsed low frequency noise: (A) At onset of sound. (B) 10 sec after onset. (C) 30 sec after onset. (D) 2 min after onset. Water depth, 20 m; speaker depth, 15 m. (From Nelson and Johnson 1972).

share in a large kill. The situation faced by these sharks might in this way be similar to that affecting some birds and mammals that appear to gain foraging information by attending to the behavior of other foraging individuals (Ward and Zahari 1973, Bertram 1978). However, we still need to know the magnitude and composition of the sounds feeding sharks generate, and whether they are in fact attractive.

2.2.2 Withdrawal Responses

Recently, the repellant effects of abruptly increased sound intensity have been documented in three of the same shark species which show attraction responses (Klimley 1976, Myrberg, Gordon, and Klimley 1978, Klimley and Myrberg 1979). The effective stimuli for these responses are again sounds with at least a large portion of their energy below 800 Hz. In fact the same pulsed, low frequency sound used to attract sharks can drive them away when its intensity is abruptly increased.

In first noting this type of withdrawal response in small lemon sharks Banner (1972) proposed that it might serve an antipredatory function and as Klimley and Myrberg have noted this might also be true for the larger sharks.

Another natural situation where abrupt sound intensity increase might signal danger and elicit an appropriate withdrawal response might be found near shore. Breaking waves could provide powerful sounds in the spectral range audible to sharks (cf. Wenz 1962) and might be useful orienting cues for many elasmobranchs. Also, these sounds should have particularly far reaching kinetic detectability since they are generated by large wavefronts that resemble line sources rather than point sources. As a result of that source geometry, the amplitude of the sonic displacements that they generate should decrease with distance at a slower rate than it would for comparable sounds generated at a point source (Ford 1970). Furthermore, as pointed out earlier it is this kinetic component of sound that sharks detect. Still, measurements are needed to determine the specific sensitivity of elasmobranchs to these natural sound sources and also the effects that breaking wave sources may have on all kinetically dependent, i.e., not gas bladder mediated, sound detection by fishes.

3 How Do Sharks Hear?

3.1 The Role of the Ear

Early in this century, before behavioral studies had delimited what sounds sharks could hear, Parker (1909) conducted ablation experiments to identify the organs of sound and vibration detection in the smooth dogfish, *Mustelus canis*. By transecting cranial nerves and observing responses to a pendulum used to strike the wall of a shark's aquarium, he showed that the labyrinths were far more sensitive to that stimulus than either the lateral line organs or the trigeminal pathways. Years later Dijkgraaf (1963) came to the same conclusion in using cat sharks, *Scyliorhinus cannicula*. He conditioned the sharks to 180 Hz tones generated by an underwater sound projector and was able to show that the threshold at that frequency increased by at least 13 dB when the two eighth nerves were severed.

Recently, my own work has provided further evidence that suggests the labyrinths are primarily responsible for the most sensitive sound detection in sharks (Corwin 1979, 1981b). Using lemon sharks, *Negaprion brevirostris,* the same species studied behaviorally by Nelson (1967), I have measured frequency-threshold properties of the eighth nerve compound action potential response. For this a blunt stainless steel electrode is placed on the intracranial trunk of the eighth nerve in a restrained and curarized shark. The animal is held partially submerged in a vibration isolated seawater tank with shaped tone bursts presented from an airborne loudspeaker (Fig. 5-3). Then at each tone frequency the threshold for a just noticeable response at the eighth nerve is measured. Figure 5-4 shows the action potential threshold curve constructed from seven ears tested in this way. All seven individual curves have similar shapes, as do curves measured with the experimenter "blind" to the stimulus level.

The dashed curve in that same figure shows Nelson's (1967) threshold curve for operantly conditioned lemon sharks. It is not possible to compare the two curves directly, since Nelson's behavioral measures have been plotted against a filtered noise level specific to each frequency, but the overall similarity of the curves is still evident. Both the animal's behavior and its eighth nerve electrical responses show best sensitivity to sound at low frequencies, from at least 40 Hz at the low end to 100 or 160 Hz at the higher end. Above those frequencies both the behavioral sensitivity and the sensitivity of the ear fall off rapidly until at 600 Hz they can only be stimulated by sounds that are at least 20 dB louder than those effective at the low frequencies. Thus, in terms of best spectral sensitivity the response of the ear closely resembles the animal's behavioral response.

As a whole then this evidence has confirmed that the labyrinths are involved in the acoustic sensitivity of sharks, but the labyrinths are complex organs with equilibrium

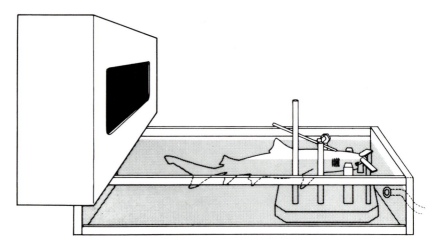

Figure 5-3. A diagrammatic view of a lemon shark positioned in the experimental tank in preparation for recording from the eighth cranial nerve. Notice that the dorsal surface of the head is in air. The large loudspeaker suspended from the ceiling posterior to the animal provided acoustic stimuli.

as well as acoustic sensitivity. It is therefore important to examine the responses of individual detectors within the ear. Lowenstein and Roberts (1950, 1951) pioneered in this area using isolated labyrinths from the skate, *Raja clavata,* and though these preparations never gave responses to sound, their work on tilt and vibration responses has guided subsequent studies in audition. Recording from individual branches of the eighth nerve they found two regions of otolithic maculae that were sensitive to vibration, the anterior sacculus and the lacinia of the utriculus. They also found particularly sensitive vibration responses in the eighth nerve branch running from the nonotolithic macula neglecta. The posterior portion of the sacculus, the main body of the utriculus, and the lagena showed tilt sensitivity, but no responses to vibration.

3.2 The Nonotolithic Channel

In reporting the notable vibration sensitivity of the seldom studied macula neglecta, Lowenstein and Roberts (1951) suggested that its proximity to the fenestra ovalis, a dorsal opening in the otic capsule cartilage, might allow stimulation by sound selective-

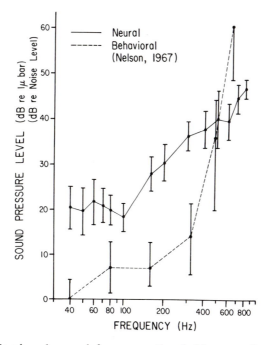

Figure 5-4. Behavioral and neural frequency threshold curves for the lemon shark, *Negaprion brevirostris.* The solid curve is a mean threshold for seven lemon shark ears in dB re 1 μbar. The dashed curve is from Nelson's (1967) operant conditioning study of sound thresholds for the same species. Direct comparisons between these curves are not possible since the values for the dashed curve are referenced to filtered noise level around each frequency. However, a low frequency plateau of best sensitivity below 200 Hz and a gradual threshold increase above that frequency are evident in both curves.

ly transmitted through that opening. Interestingly, a similar function for the fenestra ovalis had been suspected many years before (Howes 1883).

After behavioral studies had reported the urgent sound attraction response of *Carcharhinus* and some other galeomorph sharks, Tester, Kendall, and Milisen (1972) provided an account of the labyrinth's structure in that genus, particularly focusing on the macula neglecta. Finding that its two patches of sensory epithelium lined a duct that was directly apposed to the fenestra ovalis in *Carcharhinus,* they amplified the earlier suggestion that a select sound path was directed at the neglecta. This suggestion received experimental support when Fay et al. (1974) showed that microphonic potentials recorded near the neglecta were most responsive to a local vibrator applied to skin over the fenestra ovalis. When the vibrator was applied to other regions of the head the response was much smaller.

More recently ultrastructural studies of the macula neglecta in *Carcharhinus* have shown that its sensory epithelia contain remarkably large, well-aligned populations of hair cells (Fig. 5-5) (Corwin 1977a). In a typical adult gray reef shark, *Carcharhinus amblyrhynchos* = *menisorrah,* the posterior sensory epithelium contains over 220,000 hair cells, all aligned with their kinocilia ends directed posteroventrolaterally along the

Figure 5-5. A scanning electron micrograph of hair cell and supporting cell surfaces in the macula neglecta of a carcharhinid shark. The cilia complexes projecting from the surfaces of these hair cells contain approximately 50 stereocilia and a single kinocilium located at the end of the cell next to the longest stereocilia. In carcharhinids nearly all of these ultrastructurally polarized cilia complexes are aligned within each sensory epithelium; however, that is not the case for all elasmobranches. Bar equals 3 μm.

longitudinal axis of the posterior canal duct (Fig. 5-6). The smaller anterior patch contained over 40,000 hair cells which were also well aligned, but polarized in the opposite direction from those in the posterior patch. Together the two epithelia of the macula neglecta contained the largest hair cell populations that had been described in any single detector within a vertebrate ear.

Since all of these cells were directed along the axis of the posterior canal duct, i.e., nearly perpendicular to the plane of the fenestra ovalis, their great number and the internal alignment within the patches suggested that they might provide enhanced detectability for signals directed along that axis if their outputs were summed. Counts of the myelinated nerve fibers in the ramulus neglectus indicate that this is probably the case. Fewer than 5,000 axons reach the population of 260,000 hair cells in the adult's macula neglecta, so that the minimum average convergence would be approximately 60 hair cells to each nerve fiber. Since this mechanism should essentially act like a signal averaging circuit, the ratio of the outputs for an appropriately directed signal and a randomly directed noise should be improved by approximately the square of the number of inputs converged, for the macula neglecta with 60:1 convergence this would be an improvement of more than a factor of seven over the output of elements in a circuit with a 1:1 hair cell to nerve fiber ratio.

On the basis of this predicted directionality enhancement and consideration of the acoustic impedance characteristics of the tissues associated with the ear I added to the earlier proposals the suggestion that the macula neglecta might provide *Carcharhinus* with an instantaneous mechanism for directional hearing in both the near field and the far field (Corwin 1977a). As illustrated in Fig. 5-7 that mechanism would depend on preferential sound transmission through loose connective tissue filling the parietal fossa, then through the fenestra ovalis to the cupula over the hair cells of the macula neglecta contralateral to the sound source. The macula neglecta's geometry, its convergent afferent innervation, its proximity to the fenestra ovalis, and its association with a light cupula are all consistent with this proposal. Yet at the time when this was

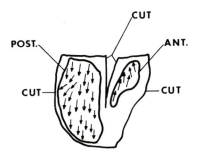

Figure 5-6. A diagrammatic view of the lumenal surface in the posterior canal duct from the right ear of *Carcharhinus*. The duct has been cut longitudinally as indicated and reflected open so that it lies flat. Arrows indicate the general directions of the kinocilium end in the hair cell cilia complexes throughout the anterior (ANT.) and posterior (POST.) sensory patches of the macula neglecta. Since the physiological polarity of these cells corresponds to this ultrastructural polarity, the arrows also represent the direction of the displacements that depolarize these cells. (From Corwin 1977a).

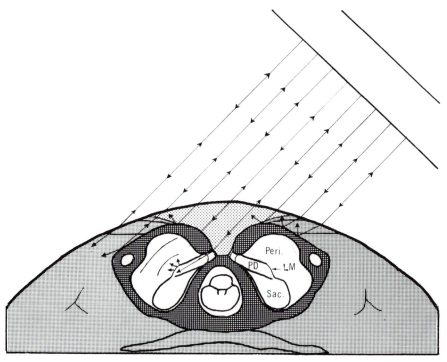

Figure 5-7. A diagrammatic cross-sectional view of the carcharhinid shark head presenting the proposed transmission path for sound reaching the head from above. The diagonal lines at the upper right represent wavefronts. The thinner lines running perpendicular to the wavefronts are rays representing the path of propagation and the direction of molecular displacements caused by the sound. This diagram is an oversimplification in that it omits displacements of the skull that the sound would produce, but according to this hypothesis the skull's displacements should be lagging and smaller than those of the seawater medium and the structures in the parietal sound transmission path. Loose connective tissue, muscle, and cartilage are respectively represented by light, medium, and dark shading. L.M., lateral release membrane; Peri., perilymph; P.D., posterior canal duct; Sac., sacculus. (From Corwin 1977a).

advanced no study had even verified that the macula neglecta was sensitive to sound. All previous physiological studies of the macula neglecta had employed vibratory stimuli, not sound.

Recently, through recordings of compound action potentials at the intracranial eighth nerve trunk and through unit recordings in the ramulus neglectus itself, the macula neglecta's sound sensitivity has been confirmed (Corwin 1979, 1981b). As Figs. 5-8 and 5-9 illustrate individual peaks in click-evoked compound action potentials can be reversibly decreased by shielding the parietal fossa or occluding the fenestra ovalis, two elements in the proposed parietal sound channel. Of course this does not specify the macula neglecta's role, but it supports the suggested sound transmitting function of those parietal structures, adding to the evidence from microphonic potentials (Fay et al. 1974) and brain evoked potentials (Bullock and Corwin 1979).

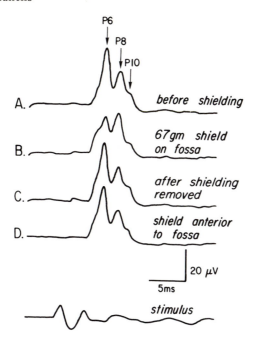

Figure 5-8. Eighth nerve compound action potentials evoked by airborne click stimuli. (A) The response in the undisturbed preparation. (B) The response while a 67 gm brass plate is resting on the skin over the parietal fossa, a proposed portal for sound transmission to the macula neglecta. The effect of placing the brass shield on the skin was specific to its location over the parietal fossa. Each trace is the average of 64 sweeps. The lower trace shows the microphone output monitoring the airborne click stimulus at +36 dB re 1 μbar peak to peak pressure.

Unit recordings specifically confirm the sound sensitivity proposed for the macula neglecta. As the spike train in Fig. 5-10 illustrates, ramulus neglectus units respond to tone bursts with frequency following responses typical of acoustic units recorded in the eighth nerve trunk. The few neglecta units that have been recorded also have thresholds within the normal range for units in the main trunk of the eighth nerve. Yet they may be tuned to different frequencies than the conglomerate population of all eighth nerve units (Fig. 5-11). That difference in tuning and the persistence of significant compound potentials in the parietal occlusion experiments suggested that at least one other inner ear detector besides the macula neglecta must be responding to sound. Further, since the macula neglecta and the semicircular canal cristae are the only non-otolithic detectors in this ear the other detector is likely to be otolithic.

3.3 The Otolithic Channel

Unit recordings from the intracranial portion of the eighth nerve in preparations where the ramulus neglectus has been lesioned confirmed that a second channel is functioning in sound detection. These units also showed a tendency toward tuning to lower fre-

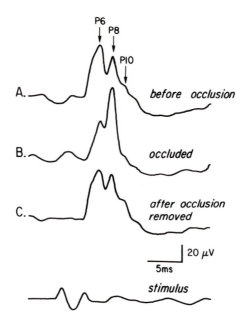

Figure 5-9. Click-evoked eighth nerve compound action potentials in a lemon shark where the skin and connective tissue have been surgically removed from the parietal fossa. (A) The response before occlusion of the fenestra ovalis. (B) The response during occlusion of the fenestra ovalis, an opening in the otic capsule cartilage opposite the end of the posterior canal duct which contains the macula neglecta. (C) The response after removal of the occluder. Along with the results in Figure 5-8 these confirm the sound transmitting role of these parietal structures and suggest that the eighth nerve responses to sound may originate at more than one detector within the ear. Each trace is the average of 64 sweeps. The lower trace shows the airborne microphone output for the +36 dB re 1 μbar click stimulus.

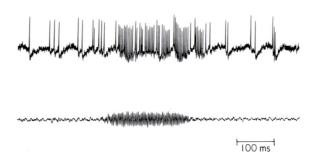

Figure 5-10. A single unit tone burst response recorded extracellularly in the ramulus neglectus of the lemon shark, *Negaprion brevirostris*. Notice that the unit spikes generally follow the frequency of this 250 Hz stimulus. The rms sound pressure of the stimulus was +29 dB re 1 μbar as monitored by an LC-32 hydrophone in the lower trace.

quencies than most neglecta units, but again the number of units recorded was small (Fig. 5-11).

Compound action potential recordings from the eighth nerve trunk combined with selective lesioning of individual detectors in the labyrinth further indicated that the sacculus is the primary site of otolithic sound detection. Figure 5-12 shows the great decrement in an averaged click evoked compound action potential caused by lesioning the sacculus. In this species lesions of the lagena and the utriculus did not cause such an effect. It therefore appears that the sacculus and the macula neglecta are the primary acoustic detectors in the lemon shark. However, there are some indications from studies of isolated ears from rays and guitarfish that the utriculus may play some role in sound detection in the batoids (Lowenstein and Roberts 1951, Budelli and Macadar 1979).

Scanning electron microscope examination of the lemon shark sacculus revealed an epithelium with hair cells that resemble those of the macula neglecta (Fig. 5-13). However this epithelium forms a single patch that covers an even larger area than the epithelia of the macula neglecta.

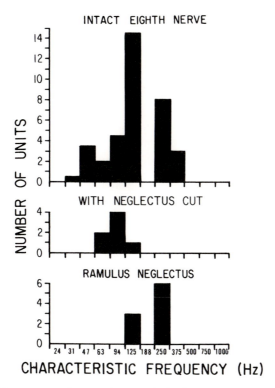

Figure 5-11. Three histograms of unit characteristic frequencies in the lemon shark eighth nerve. The upper histogram contains units recorded in the intracranial portion of intact eighth nerves. The middle histogram contains units recorded in that trunk following cutting of the ramulus neglectus. The lower histogram contains units recorded in the ramulus neglectus, a branch of the eighth nerve that serves only the macula neglecta.

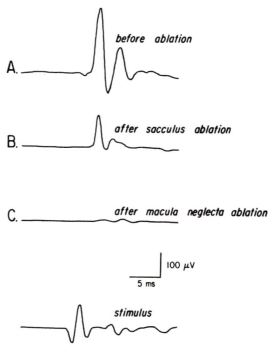

Figure 5-12. Click-evoked eighth nerve compound action potentials in the lemon shark. (A) The response from the ear after the lateral wall of the otic capsule was removed and the endolymph-perilymph barrier breached by a large cut in the sacculus wall. (B) The response after the saccular macula was lesioned with an electrocautery. (C) The response of the ear following lesioning of both the sacculus and the macula neglecta. These results strongly suggest that the sacculus and the macula neglecta are the primary auditory receptors in the lemon shark ear. Each trace is the average of 16 sweeps. The lower trace is the output of an airborne microphone monitoring the +49 dB re 1 μbar clicks just above the shark's head.

The saccular macula forms an S-shaped band extending along the anterior wall of the sacculus, down along its ventral floor, and then along the medial wall above the lagena (Fig. 5-14). A line of hair cell polarity reversal runs the entire length of this epithelium so that cells on either side are oriented with their kinocilia facing away from that line (Fig. 5-14). This simple outward orientation of hair cells has also been observed in the sacculus in rays (Lowenstein, Osborne, and Wersäll 1964), sturgeons (Popper 1978), and many ostariophysine teleosts (Platt 1977, Jenkins 1979a, 1979b), but in the lemon shark the complex three-dimensional structure of the macula gives the hair cells a spectrum of directional orientations. This should presumably allow responses to many different directions of head vibration and thus to differently directed sonic displacements.

Figure 5-14 also illustrates the relative shapes and sizes of both the saccular macula and the macula neglecta. The saccular macula is in fact six times larger in surface area than the macula neglecta of the same ear. Counts show that the spatial density of hair

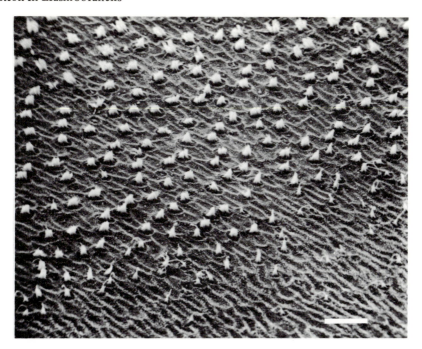

Figure 5-13. A scanning electron micrograph of hair cells at the edge of the saccular macula in the lemon shark, *Negaprion brevirostris*. These cells closely resemble those of the macula neglecta, but the saccular population is six times larger. Bar equals 10 μm.

cells is the same in both detectors, so that this sixfold area difference reflects a sixfold difference in the hair cell population sizes. In a 1.5-kg juvenile lemon shark the macula neglecta contains approximately 50,000 hair cells, while the macula of the sacculus already contains about 300,000 hair cells. If postembryonic production of hair cells and growth of the sensory epithelium proceed in this macula to the extent that has been documented in the macula neglecta of *Carcharhinus,* then there might be an eightfold increase in the size of this epithelium as the juvenile grows (Corwin 1977b, 1981a). Thus, it seems reasonable to expect that the saccular macula of an adult lemon shark might contain over two million hair cells.

In summary then carcharhinid sharks appear to detect sounds using two labyrinthine detectors functioning in parallel. One, the macula neglecta, is a nonotolithic type of sensor. A light gelatinous cupula lies above its sensory hair cells, and here stimulation is thought to depend on the relatively good acoustic coupling of the light cupula and the loose connective tissue of the parietal fossa to the surrounding seawater. A suitably directed portion of the kinetic component of the sound transmitted through the parietal fossa should cause displacements of this cupula along the long axis of the posterior canal duct, so that this macula might function like the type of mechanical detector diagrammed in Fig. 5-15A. The surrounding duct should vibrate relatively little and with a phase lag, while the light tectorial structure should vibrate with nearly the same amplitude and phase as the surrounding medium. It should be inherently directional

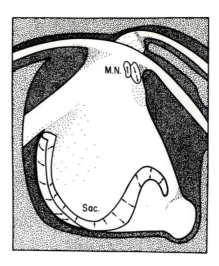

Figure 5-14. A diagrammatic lateral view of the left labyrinth in the lemon shark, *Negaprion brevirostris*. This shows the labyrinth as it would appear when the lateral wall of the otic capsule has been removed by a parasagittal cut. The macula neglecta (M.N.) and the saccular macula (SAC.) are drawn approximately to scale in their normal locations. The small arrows in those maculae represent the directions of hair cell cilia polarities. Arrow heads point in the direction of an excitatory shearing force, i.e., from the stereocilia toward the kinocilium end of the cells. Dorsal is up; anterior is to the left.

since the tectorial structure is in a duct where its motion is constrained except along the longitudinal axis (Fig. 5-16).

The other detector, the sacculus, is otolithic. A dense mass of calcium carbonate otoconia lies above its sensory hair cells, and here stimulation is thought to depend on the relatively inertial behavior of this otolithic mass and its poor acoustic coupling to the surrounding seawater (Dijkgraaf 1960). Theoretically this detector should function in a manner similar to that diagrammed in Fig. 5-15C. In response to the molecular displacements in the external medium the lighter surrounding structure should move with greater amplitude and less phase lag than the dense central mass. Also, in this case the motion in the detector is not necessarily contrained to any one axis. A force from any direction should move the outer structure equally well, so that with differently directed shear detectors incorporated into the walls it should be possible to detect sonic displacements from a number of directions. In the same manner, the sacculus of the lemon shark with its spectrum of hair cell orientations should respond to many displacement directions.

The possession of parallel otolithic and nonotolithic sound channels may therefore be expected to give the lemon shark more information concerning directional features of sounds. The dissimilar modes of stimulus detection at these detectors may also provide mechanisms for discriminating frequency and sound intensity, if the detectors have different frequency response functions. However, further physiological investigations of these possibilities are needed.

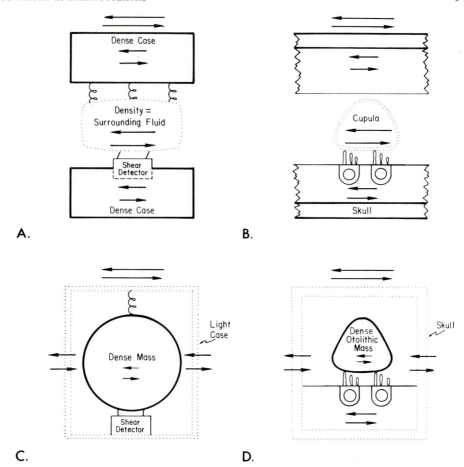

Figure 5-15. Diagrammatic representations of mechanisms for detecting displacements in a surrounding fluid medium. (A) A mechanical detector that has a light central element exposed to the medium and a dense surrounding case. This type of detector senses displacements as resultant shears between the central element, which is well coupled to the medium's movement, and the dense case, which has poorer coupling. (B) A model of a nonotolithic inner ear detector such as the shark macula neglecta. This detector is thought to function in a manner similar to the mechanical sensor diagrammed in (A). A light cupula exposed to the medium via the parietal sound path rests above the ciliated surfaces of hair cells anchored to the much denser skull. (C) A mechanical detector that has a dense central element and a light outer case. This type of detector, which is commonly employed in seismic recording, also senses displacements as resultant shears, but here the case has better coupling to the medium than the central mass has. (D) A model of an otolithic detector such as the shark sacculus. This is thought to function in a manner similar to (C). Here a dense otolithic mass resting above the hair cell cilia should have even poorer coupling to the medium than the skull to which they are attached.

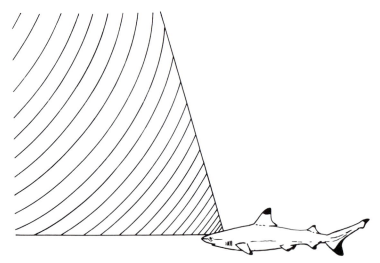

Figure 5-16. A representation of the zone predicted as most effective for stimulation of the macula neglecta via the parietal sound path in Carcharhinus. The hypothesis, based on the anatomical descriptions of these structures, is that they provide a forward biased sensitivity. The sacculus on the other hand is expected to function with essentially symmetrical anterior and posterior sensitivities.

4 The Auditory Brainstem Response

A promising approach to future questions in auditory biology of elasmobranchs and other fishes stems from recent experiments which have demonstrated that an Auditory Brainstem Response (ABR) analogous to the mammalian ABR can be recorded from elasmobranchs and other nonmammalian vertebrates (Corwin, Bullock, and Schweitzer 1981). This approach which employs evoked response averaging in conjunction with the use of large metal electrodes placed near, but not in the brain tissue, appears to be suitable for studies requiring that the subjects remain relatively intact or located in naturalistic acoustic environments.

The ABR is a series of waves generated as auditory information progresses from the eighth nerve on through higher levels in the nervous system. As surgical transections of the brainstem demonstrate, the ABR in the elasmobranch *Platyrhinoidis* originates as sequentially generated waves in the eighth nerves, the medulla, and the midbrain (Fig. 5-17). ABR recording provides a number of advantages over conventional techniques in permitting measurement of persistent whole animal responses, without the need of tedious training. It also leaves the animal nearly intact and does not require precise placement of the electrodes. Electrodes do not contact brain tissue so that restraint and laboratory conditions are not required. Finally, although this technique has only recently been employed in studies of fish audition it is hoped that it will have wide applicability. Through use of a differential masking procedure (Dallos and Cheatham 1976, Bullock and Corwin 1979) this response could provide a frequency response curve for otherwise difficult species. In the elasmobranchs it might

AUDITORY BRAINSTEM RESPONSES

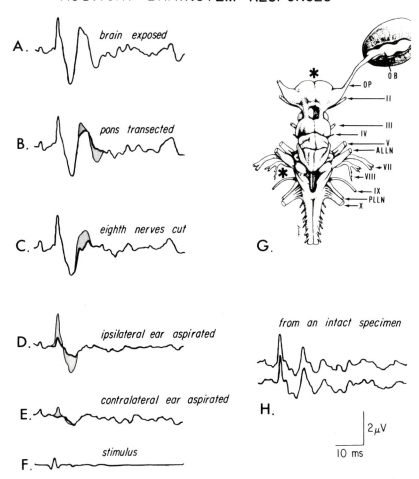

Figure 5-17. Auditory brainstem responses (ABRs) recorded in the thornback ray, *Platyrhinoidis triseriata*. (A)-(E) are a series of ABRs recorded from an exposed brain during surgical transections. Stimulus shown in (F). The shaded areas in these traces indicate the waveform changes caused by the manipulations described above each trace. As these traces demonstrate the complex waveform of this ABR arises as a series of individual volume conducted waves originating in the auditory pathway as neural excitation passes from the eighth nerve to the medulla and the midbrain. (G) A drawing of the brain in *Platyrhinoidis triseriata* (used by kind permission of Dr. R. G. Northcutt). (H) Two ABR traces recorded from a nearly intact ray with electrodes inserted into the cerebrospinal fluid through small holes in the chondrocranium. All traces are averages of 256 sweeps.

allow the first evaluations of auditory sensitivity in nongaleomorph species. As stated earlier, we know very little about the hearing sense in nongaleomorphs. Most of our knowledge stems from comparative anatomical descriptions that point out some major differences in the structure of the labyrinth, its sensory epithelia, and the surrounding structures. On the basis of these comparisons some functional correlates have been proposed (Corwin 1978), but solid behavioral or physiological measures are needed both to test those proposals and to provide general data. In dealing with the auditory system of elasmobranchs we may be able to gain insight into the more than 400 million year history of this sense. Yet, we will need information on a much wider range of elasmobranch groups before we can make full use of even the present data.

Acknowledgments. The original studies reported in this article were supported by NSF and NIH grants and a private donation from Mr. Milton Shedd to Dr. Theodore H. Bullock. The author also gratefully acknowledges the advice and comments of Drs. Bullock, W. F. Heiligenberg, R. G. Northcutt, R. H. Rosenblatt, and A. A. Myrberg.

References

Banner, A.: Evidence of sensitivity to acoustic displacements in the lemon shark. *Negaprion brevirostris* (Poey). In: Lateral Line Detectors. Cahn, P. H. (ed.). Bloomington, Indiana: Indiana Univ. Press, 1967, pp. 265-273.

Banner, A.: Use of sound in predation by young lemon sharks, *Negaprion brevirostris* (Poey). Bull. Mar. Sci. 22, 251-283 (1972).

Bertram, B. C. R.: Living in groups: predators and prey. In: Behavioural Ecology an Evolutionary Approach. Krebs, J. R., Davies, N. B. (eds.). Oxford: Blackwell, 1978, pp. 64-96.

Budelli, R., Macadar, O.: Statoacoustic properties of utricular afferents. J. Neurophysiol. 42, 1479-1493 (1979).

Bullock, T. H., Corwin, J. T.: Acoustic evoked activity in the brain in sharks. J. Comp. Physiol. 129, 223-234 (1979).

Compagno, L. V.: Phyletic relationships of living sharks and rays. Am. Zool. 17, 303-322 (1977).

Corwin, J. T.: Morphology of the macula neglecta in sharks of the genus *Carcharhinus*. J. Morphol. 152, 341-362 (1977a).

Corwin, J. T.: Ongoing hair cell production, maturation, and degeneration in the shark ear. Soc. Neurosci. Abstr. 3, 4 (1977b).

Corwin, J. T.: The relation of inner ear structure to feeding behavior in sharks and rays. In: Scanning Electron Microscopy. Johari, O. (ed.). Chicago: S.E.M. Inc., 1978, pp. 1105-1112.

Corwin, J. T.: Parallel channels for sound detection in the fish ear. Soc. Neurosci. Abstr. 5, 18 (1979).

Corwin, J. T.: Postembryonic production and aging of inner ear hair cells in sharks. (1981a). (submitted)

Corwin, J. T.: Peripheral auditory physiology in the lemon shark: Evidence of parallel otolithic and non-otolithic sound detection. J. Comp. Physiol. (1981b) (in press).

Corwin, J. T., Bullock, T. H., Schweitzer, J.: The non-mammalian auditory brainstem response: a physiological measure of brain evolution (1981) (in prep.).

Dallos, P., Cheatham, M. A.: Compound action potential (AP) tuning curves. J. Acoust. Soc. Am. 59, 591-597 (1976).

Dijkgraaf, S.: Hearing in bony fishes. Proc. R. Soc. London Ser. B 152, 51-54 (1960).

Dijkgraaf, S.: Sound reception in the dogfish. Nature 197, 93-94 (1963).

Fay, R. R., Kendall, J. I., Popper, A. N., Tester, A. L.: Vibration detection by the macula neglecta of sharks. Comp. Biochem. Physiol. 47A, 1235-1240 (1974).

Ford, R. D.: Introduction to Acoustics. New York: Elsevier, 1970.

Howes, G. B.: The presence of a tympanum in the genus *Raia*. J. Anat. Physiol. XVII, 188-190 (1883).

Jenkins, D. B.: A transmission and scanning electron microscopic study of the saccule in five species of catfishes. Am. J. Anat. 154, 81-102 (1979a).

Jenkins, D. B.: Anatomical investigation of the saccule in *Clarius batrachus*. Scanning Electron Microsc. 1979, III, 949-954.

Kelly, J. C., Nelson, D. R.: Hearing thresholds of the horn shark, *Heterodontus francisci*. J. Acoust. Soc. Am. 58, 905-909 (1975).

Klimley, A. P.: Analysis of acoustic stimulus properties underlying withdrawal in the lemon shark, *Negaprion brevirostris* (Poey). University of Miami: M.S. Thesis (1976). (1976).

Klimley, A. P., Myrberg, Jr., A. A.: Acoustic stimuli underlying withdrawal from a sound source by adult lemon sharks, *Negaprion brevirostris* (Poey). Bull. Mar. Sci. 29, 447-458 (1979).

Lowenstein, O., Osborne, M. P., Wersäll, J.: Structure and innervation of the sensory epithelia of the labyrinth in the thornback ray (*Raja clavata*). Proc. R. Soc. London Ser. B 160, 1-12 (1964).

Lowenstein, O., Roberts, T. D. M.: The equilibrium function of the otolith organs of the thornback ray (*Raja clavata*). J. Physiol. 110, 392-415 (1950).

Lowenstein, O., Roberts, T. D. M.: The localization and analysis of the responses to vibration from the isolated elasmobranch labyrinth. A contribution to the problem of the evolution of hearing in vertebrates. J. Physiol. 114, 471-489 (1951).

Myrberg, Jr., A. A.: Underwater sound—Its effect on the behavior of sharks. In: Sensory Biology of Sharks, Skates and Rays. Hodgson, E. S., Mathewson, R. R. (eds.). Arlington, Virginia: Office of Naval Research, 1978, pp. 391-417.

Myrberg, Jr., A. A., Ja, S. J., Walewski, S., Banbury, J. C.: Effectiveness of acoustic signals in attracting epipelagic sharks to an underwater sound source. Bull. Mar. Sci. 22, 926-949 (1972).

Myrberg, Jr., A. A., Gordon, C. R., Klimley, A. P.: Rapid withdrawal from a sound source by open-ocean sharks. J. Acoust. Soc. Am. 64, 1289-1297 (1978).

Nelson, D. R.: Hearing thresholds, frequency discrimination and acoustic orientation in the lemon shark, *Negaprion brevirostris* (Poey). Bull. Mar. Sci. 17, 741-768 (1967).

Nelson, D. R., Gruber, S. H.: Sharks: Attraction by low-frequency sounds. Science 142, 975-977 (1963).

Nelson, D. R., Johnson, R. H.: Acoustic attraction of Pacific reef sharks: Effect of pulse intermittency and variability. Comp. Biochem. Physiol. 42A, 85-89 (1972).

Nelson, D. R., Johnson, R. H.: Some recent observations on acoustic attraction of Pacific reef sharks. In: Sound Reception in Fish. Schuijf, A., Hawkins, A. D. (eds.). Amsterdam: Elsevier, 1976, pp. 229-239.

Olla, B.: The Perception of Sound in Small Hammerhead Sharks, *Sphryna lewini*. Univ. of Hawaii, M. S. Thesis (1962).

Parker, G. H.: The sense of hearing in the dogfish. Science 29, 428 (1909).

Platt, C. J.: Hair cell distribution and orientation in goldfish otolith organs. J. Comp. Neurol. 172, 283-298 (1977).

Popper, A. N.: Scanning electron microscopic study of the otolithic organs in the bichir (*Polypterus bichir*) and shovel-nose sturgeon (*Scaphirhynchus platorynchus*). J. Comp. Neurol. 181, 117-128 (1978).

Popper, A. N., Fay, R. R.: Structure and function of the elasmobranch auditory system. Am. Zool. 17, 443-452 (1977).

Schaeffer, B.: Comments on elasmobranch evolution. In: Sharks, Skates and Rays. Gilbert, P. W. (ed.). Baltimore: Johns Hopkins Press, 1967, pp. 3-35.

Tester, A. L., Kendall, J. I., Milisen, W. B.: Morphology of the ear of the shark genus *Carcharhinus* with particular reference to the macula neglecta. Pacif. Sci. 26, 264-274 (1972).

Ward, P., Zahari, A.: The importance of certain assemblages of birds as 'information-centers' for food-finding. Ibis 115, 517-534 (1973).

Wenz, G. M.: Acoustic ambient noise in the ocean: Spectra and sources. J. Acoust. Soc. Am. 34, 1936-1956 (1962).

Discussion

BELL: I note the lack of evoked potentials after you removed the sacculus and the macula neglecta. Do you think that this implies that the other elements are not involved in hearing?

CORWIN: To me it implies that they do not make a major contribution. It has been reported that the utriculus has some sound sensitivity in *Rhinobatos*, and, at least, if not sound then it was a response to vibrations that we might consider analogous to sound in animals that detect only particle motion. This study [Budelli and Macadar, J. Neurophysiol. 42, 1479-1493 (1979)] used an unusual technique. The animal's head was pointed upward, and they may have obtained a response that you might not normally obtain. Lowenstein and Roberts [J. Physiol. 114, 471-489 (1951)] also reported some sensitivity from the lacinia, a small portion of the utriculus, not weighted by otoconia. Looking across species, and comparing Lowenstein's work on *Raja* with mine, I find that the detectors are quite different. The macula neglecta in rays, for example, is very small, with a single sensory epithelium and without the alignment of hair cells as in the macula neglecta in carcharhinids. In *Carcharhinus,* it is a huge organ with two sensory epithelia. It may be that when we get the full picture, we shall find the different detectors to have different roles in various elasmobranchs. The utriculus may have a greater or lesser role. In the species I worked with, it seems to have a minor part.

PLATT: The rays that have been studied do not have a large macula neglecta and they may, in fact, have a large lacinia. It would be interesting in this context to try a recording from the lacinia of one of the carcharhinids.

CORWIN: I agree, and such recordings should not be difficult given the favorable position of the utriculus.

TAVOLGA: You showed the shark ear as being essentially a pure near-field detector, yet you presented your threshold data in terms of pressure values, i.e., in dB re one microbar. It seems to me that at the low frequencies you used, the intensities of the stimuli, and with the size of the tank, your system would have an overall impedance approaching that of air. There is clearly a substantial near field, i.e., displacement, component in your stimulus. Perhaps you should consider that the near-field energy is crucial to this entire receptor system.

CORWIN: I agree that this is a displacement detector, although I don't like to use the near field/far field distinction. I don't think the distinction exists. The absolute threshold, and where the near field/far field boundary ends up is only determined by the frequency. It is sensitive to displacement regardless of whether the displacement component is of near- or far-field origin.

TAVOLGA: I didn't want to get into a semantic argument here, but when I said "near field" I meant displacement, i.e., the actual particle movement or velocity. In the shark ear, there appears to be no effective pressure transducer as a gas bladder, for instance. How did you measure the sound pressure, and where, in relation to the inner ear?

CORWIN: I measured the sound pressure in the air above the animal, directly above the parietal fossa. I should measure velocity with a velocity or displacement hydrophone, but that is something I have yet to do.

FAY: You mentioned acoustic impedance of different shark tissues as opposed to that of water. How large an impedance difference to you think there is?

CORWIN: This is an unknown, and the possible contribution of impedance differences is subject to some disagreement. The problem involves refraction and diffraction that may allow sound to either pass through or pass around tissues and structures that are small relative to the wavelengths. Some have claimed that the animal oscillates as a whole and has, essentially, one acoustic impedance. If this is really true, then these animals should not be able to hear at all, and yet they clearly respond to sound. Nobody has yet made any local measurements of oscillations or impedances, except for some on the saccular otoliths of teleosts.

SCHUIJF: It all depends on how the volume displacements are coupled to the hair cell cupula. There are a number of possible ways that displacements could activate the sensors.

CORWIN: Could you explain that?

SCHUIJF: The parietal fossa has a shape such that compression from all sides would produce distortions. These, in turn, would produce the shearing forces necessary to stimulate hair cells.

CORWIN: Are you suggesting actual pressure detection?

SCHUIJF: Yes, but in a different sense from what we usually think. The type of coupling involved would, in effect, produce a pressure detector.

CORWIN: I guess that we have to do now is to determine which of these detector mechanisms are actually at work.
    I should point out that both Platt and Jenkins have found macula neglecta structures that are quite similar to those I found in the shark. Two distinct epithelia are present, with oppositely polarized populations of hair cells, in the Ostariophysi. There is a possibility that we may have an otolithic pathway and a nonotolithic pathway in both teleosts and elasmobranchs.

ENGER: Blaxter, Denton, and Gray (Chapter 2) have described a small macula neglecta in the clupeid ear. We know very little of that.

CORWIN: Every major division of the teleosts has been shown to have a macula neglecta with a bipartite structure, i.e., two sensory epithelia. Retzius described this long ago.

BULLOCK: Corwin has indicated that there are populations of hair cells of different ages. It is hard for me to believe that these cells of different ages would have the same physiology and sensitivity. Changes with time and turnover of sensory cells could lead to significant changes in sensory properties.

CORWIN: This has been supported in part by the work of E. R. Lewis (University of California, Berkeley), and may account for some of the functional differences among different detectors.

PART TWO

Auditory Capabilities and Psychophysics

Morphological (Part One) and physiological (Part Three) investigations provide data on subunits of the auditory system. Behavioral sound detection and analysis, on the other hand, involve the integrated use of all portions of the auditory system. These psychophysical capabilities of a fish may thus reflect far more extensive and complex questions than seen when studying only single portions of the system. Investigations of the hearing ability of a species is accomplished using a variety of behavioral techniques through which it has been possible to demonstrate capabilities for detecting a wide range of signals and for reasonably sophisticated analyses of sound. Hawkins, in Chapter 6, reviews the various behavioral and psychophysical techniques used and the data obtained from them. Buwalda (Chapter 7) uses one of these techniques to demonstrate the degree to which directional and nondirectional aspects of the sound stimulus are independently detected and analyzed in the cod auditory system and provides important new data relevant to the theoretical work of Schuijf (Chapter 14) on sound source localization. Coombs (Chapter 8) demonstrates that fish may show marked species differences in auditory capabilities for signal analysis, a finding that reflects the discussion in Chapter 1 by Platt and Popper. Finally, Hall, Patricoski and Fay (Chapter 9) describe peripheral and central neurophysiological correlates of intensity discrimination behavior which may explain a significant divergence from Weber's law seen in the goldfish and in a variety of vertebrate taxa.

Chapter 6

The Hearing Abilities of Fish

A. D. Hawkins*

1 Introduction

Earlier this century, von Frisch and his pupils paved the way to the objective measurement of the hearing abilities of fish. They trained fish to respond in an unambiguous way to sounds, presenting them with tones of differing frequency and amplitude to explore the limits of their sensitivity. Such techniques form the basis for most modern studies of the hearing capacities of fish.

The ability of the fish to detect a sound is usually expressed in terms of some minimal detectable level or threshold, often defined as the sound level to which the fish will respond on a specified proportion of presentations. The absolute threshold for a fish measured under quiet background conditions, is not necessarily fixed and unvarying but may change with time and physiological state. A formal theory, signal detection theory, has developed around the concept of the threshold, a full exposition being given by Green and Swets (1966). The theory predicts that the operation of the sensory receiver is based on statistical operation rather than abrupt on-off functions, and it follows from this that the method for defining any threshold by experiment is crucial, and must be carefully chosen. So far, signal detection theory has largely been developed only for human subjects but it has great practical relevance to research into the sensory capacities of animals.

The absolute threshold is dependent on various features of the applied stimulus; these include the stimulus duration, its direction, and above all its frequency. A plot of absolute thresholds for a range of frequencies is termed an audiogram. Audiograms have been constructed now for over 25 different species of fish (see Popper and Fay 1973, Fay and Popper 1980). Since the majority of these audiograms have been measured in the last decade one might at first suppose that they constitute a clear and comprehensive body of evidence on the hearing capacities of fish, but this is not so. The validity of many published audiograms, including some of the most recent, can seriously be questioned, mainly because of deficiencies in the methods adopted.

*Marine Laboratory, Victoria Road, Aberdeen, Scotland.

2 Methods of Investigating the Hearing Abilities of Fish

2.1 Conditioning Experiments

In the simplest experiments fish are trained to approach a particular area, or to show a particular response on hearing a sound, by means of a food reward offered each time the sound is presented (von Frisch 1938, Poggendorf 1952, Kritzler and Wood 1961, Enger 1966, Schuijf, Baretta, and Wildschut 1972). The response of the fish is often ambiguous and erratic, especially near threshold. In addition, only a rather limited number of trials can be performed in a given time because feeding motivation must be kept high. Such methods are therefore most useful for difficult species, which cannot be trained by other methods.

An instrumental avoidance conditioning method has been practiced by Tavolga and Wodinsky (1963), Weiss (1966), Cahn, Siler, and Wodinsky (1969), and Popper (1970), based on a method first developed by Wodinsky, Behrend, and Bitterman (1962). With this technique, a fish is placed in a shuttlebox of two compartments separated by a barrier. The animal is presented with a sound, and a short while later (10 sec or so) an electric shock is administered. The sequence is repeated until the fish learns to cross the barrier to avoid the electric shock. Trials can be repeated at a rate of only about 25 per day, and it may take several days to produce a response on 90% of the presentations. Some species are more reluctant to cross barriers and are difficult to condition by this method. Also, the technique relies on the fish moving, and special care must therefore be taken to ensure that the sound field is uniform, or that any spatial variation in stimulus level is known.

Perhaps the most popular current method is the classical conditioning of heartbeat or gill ventilation, described by Otis, Cerf, and Thomas (1957). Conditioning of the heartbeat has been applied by Buerkle (1967), Offutt (1968), Chapman and Hawkins (1973), and Sawa (1976). Conditioning of the ventilatory rhythm has been exploited by Fay (1969), and Popper, Chan, and Clarke (1973). Both the heart rate and ventilatory rate of an undisturbed fish are relatively constant, and presentation of any strong stimulus may suppress either of the two rhythms. Usually the inhibition of heartbeat or ventilation wanes with repeated stimulation without reinforcement, but if the sound is coupled with a delayed electric shock the fish associates the sound with the shock, and shows cardiac and ventilatory suppression on every stimulus presentation. Full conditioning is often complete within 10 trials, and can be achieved within an hour or so, so that several thresholds can be determined in a day. This method is therefore much quicker to apply than instrumental avoidance conditioning, and more elaborate experiments are possible. As the fish does not have to move to show a response it can be positioned at a particular point in the sound field, where the sound level is known.

Cardiac conditioning does not appear to work well with some species of shark [e.g., the lemon shark, *Negaprion brevirostris* (Nelson 1967a); the dogfish, *Scyllium canicula* (Hawkins unpublished)]. Though these fish may show a pronounced slowing of the heart, or bradycardia, in response to sounds the response does not become conditioned and soon wanes. Nevertheless, the technique has worked successfully for the horn shark, *Heterodontus francisci* (Kelly and Nelson 1975), and conditioning of the ventilatory rhythm may also prove satisfactory for other species.

2.2 Threshold Determination

Having trained a fish to respond unambiguously to a sound the next step is to determine a threshold. Two procedures are common. The first is the von Békésy staircase or tracking method (Cornsweet 1962), applied in conjunction with training to food, or avoidance conditioning, but also practiced with cardiac conditioning. The stimulus level is progressively reduced in fixed increments with every positive conditioned response of the fish, or is increased by the same increment when the fish does not show a response. In this way, the threshold is alternately bracketed by positive and negative responses and an average can be calculated to represent the threshold, i.e., the level to which the fish responds on 50% of presentations. With avoidance conditioning, or training to food, a positive response is readily defined, i.e., the animal either responds or does not. With classical conditioning of the heart rate the response is a graded slowing of the heart rate, which may diminish close to threshold. The inhibition of the heartbeat must therefore be defined statistically by comparing the heartbeat intervals during stimulus presentation with a sample taken between trials.

An alternative way of threshold determination is by the method of limits, where a series of stimuli are applied at levels above and below the expected threshold level. In practice a short descending series of sound levels can be applied, and then repeated several times. The percentage of positive responses occurring at each sound level can then be determined, leading to a threshold. Fay (1969, 1974a) has described an application of the methods of limits to classical conditioning of the ventilatory rhythm. Fundamentally, the method of limits is not very different from the staircase method and the thresholds are comparable. Popper et al. (1973) have obtained thresholds for the goldfish by classically conditioned ventilatory suppression and the method of limits and have compared the results with those obtained by avoidance conditioning and the staircase method. The thresholds obtained did not differ significantly. There is no strong evidence therefore that the particular criteria adopted in defining a threshold make a major contribution to any of the variations observed between results.

2.3 Unconditioned Responses

Much of our knowledge of the acoustical behavior of fishes has been derived not from controlled laboratory experiments but from long periods of direct observation upon free-ranging and captive animals. Thus, field and laboratory observations by such workers as Moulton (1956), Tavolga (1958), Brawn (1961), Winn (1967), and Myrberg and Spires (1972) on the sound producing behavior of fish have helped to clarify the role played by sounds in the lives of vocal fish. The experiments by Myrberg and Spires (1972) have shown clearly that the bicolor damselfish, *Eupomacentrus partitus*, can discriminate between various sounds, based on some factor of the structural patterning in time. Myrberg (1972) was also able to demonstrate a facilitating effect of the courtship sounds on the initiation of courtship in the male damselfish.

Hawkins and Rasmussen (1978) have drawn attention to the fact that the calls of various vocal gadoid fish differ in their temporal patterning, and that within a particular species, the haddock *Melanogrammus aeglefinus*, the distinctive calls produced in various contexts also differ in terms of their structure in the time domain.

Manipulation of the structure of sounds has shown that the temporal characteristics are very important in attracting predatory fish, and especially sharks. Studies by Nelson, Myrberg, Banner, Johnson, Richards, and their co-workers (reviewed by Myrberg, Gordon, and Klimley 1976) has been particularly important in establishing the ability of fish to orientate toward underwater sound sources. Indeed, the abundant information on the rapid and directed orientation of free-ranging sharks to a distant sound source was at one time the only evidence against the view of van Bergeijk (1964) that directional hearing for fish was only possible close to a sound source.

2.4 Presentation and Measurement of the Sound Stimulus

Several workers have stressed the need for experiments on the hearing of fish to be carried in a very special environment, where sounds can be accurately reproduced and measured (Griffin 1955, Parvulescu 1964, 1967, Weiss 1969, Popper 1970, Chapman and Hawkins 1973). Laboratory aquarium tanks are commonly used for this purpose but are usually quite unsuitable because the fish and sound measuring instruments are close to an interface between air and water. Because of the different acoustical properties of the two media any sound wave is reflected, and the relationship between sound pressure and particle motion in the applied stimulus changes. In a free sound field the particle motion can be calculated from the sound pressure, whereas in a perturbed sound field this cannot be done with any confidence. Many fish are sensitive primarily to particle motion, but because hydrophones which measure particle motion are difficult to obtain and to calibrate, most underwater acoustic measurements are made with hydrophones sensitive to sound pressure on the assumption that free field conditions prevail or that the fish is pressure sensitive. A small thin-walled aquarium tank is a particularly poor acoustic environment because the fish is effectively surrounded by "soft" reflecting boundaries, and any sound pressure inside the tank is accompanied by relatively massive particle motion. This particle motion may well stimulate the fish ear directly, and the conventional description of the sound in terms of measured sound pressures may therefore be highly misleading.

The problems of reflection are especially important in experiments with directional sound stimuli. Particle motion is a vector quantity, and in a small tank, the particle motion vectors may be greatly affected by reflecting boundaries, giving a distorted directional sound field.

Attempts have been made to reduce the effects of reflection inside aquarium tanks by lining them with sound absorbent materials, but it is clear from theoretical considerations that to be effective the absorbent layer must be very thick, approaching a significant fraction of the sound stimulus wavelength. This is often impractical (at 100 Hz the wavelength is 15 m).

There are several ways of providing a more satisfactory sound field, where sound stimuli can be properly specified (Fig. 6-1). Some workers have described acoustic ranges in the sea, where the fish, hydrophones, and sound projectors are positioned well away from reflecting boundaries. Under these conditions, sound pressure measuring hydrophones can be used to describe the sound stimuli with some confidence. This solution has been adopted with success by Schuijf, Baretta, and Wildschut (1972), Chapman and Hawkins (1973), and their collaborators. Working in the open field is not entirely satisfactory. Many technical difficulties are encountered, and in addition the technique is difficult to apply with very low frequency sounds. The lower the frequency the greater is the working depth required to eliminate effects due to reflection of sound by the sea surface.

One great advantage of working in the open field is that it is possible to vary the ratio of sound pressure to particle motion in the applied sound field, making use of the near-field effect, described by Harris and van Bergeijk (1962). Thus, by bringing the fish closer to the sound projector, a greater magnitude of particle motion can be applied for a given sound pressure. By this means, Chapman and Hawkins (1973) established that the cod, *Gadus morhua,* is sensitive to sound pressure, while Chapman and Sand (1974) and Hawkins and Johnstone (1978) showed that the dab, *Limanda limanda,* and salmon, *Salmo salar,* respectively, were sensitive to particle motion.

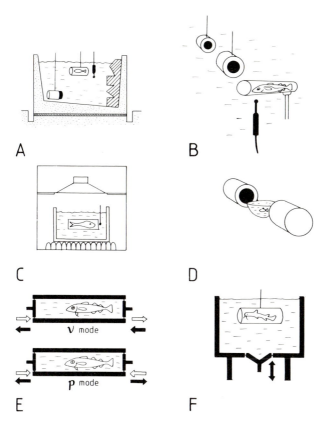

Figure 6-1. Methods of presenting sound stimuli to fish. (A) Open tank with submerged projector (Buerkle 1967). (B) Acoustic range in the sea, two projectors at different distances from the fish (Chapman and Hawkins 1973). (C) Small tank within an airfilled box, sound applied by an air loudspeaker (Fay 1974a). (D) Version of (C) with a small tank between two opposing air loudspeakers (Sawa 1976). (E) Water-filled standing wave tube. By varying the phase of the projectors the internal acoustic impedance can be altered (Hawkins and MacLennan 1976). (F) Open tank with submerged projector in the floor (Poggendorf 1952).

An alternative method of obtaining a satisfactory sound field much favored by investigators of freshwater species, and especially applied to those fish believed to be sensitive to sound pressure, is to generate the sounds in air, around a very small thin-walled water-filled tank (Fig. 6-1C). This arrangement was first suggested by Parvulescu (1964) as a means of applying measured sound pressures without the large attendant particle motions generated by the use of an immersed sound projector in a small body of water. The principle on which the method is based is that if the wavelength of the sound in air is large compared with the tank dimensions then the boundary surfaces of the water in the tank will be uniformly compressed by the airborne sound field. The sound pressure in the water will vary with the air pressure. Moreover, since the pressure acts toward the center of the water mass, the particle motion at the center of the tank will be small enough to be ignored. The main disadvantage of this method is that it is really only suitable for fish sensitive to sound pressure. In addition, an upper frequency limit is set by the need for the tank to be smaller in dimensions than one wavelength of the sound in air. If the tank is any larger, different parts of the tank experience pressure waves of differing phase, or amplitude, and large particle motions may be generated. A tank 1 m^3 in volume is only satisfactory at frequencies below 50 Hz. Despite this stricture, such tanks are very popular. A special case of a tank providing stimulation by sound pressure in the absence of particle motion was provided by Sawa (1976), who suspended a thin-walled tank between two opposing air loudspeakers.

A completely different kind of tank was designed by Poggendorf (1952) (Fig. 6-1F). A sound projector was mounted in the floor of an open topped water filled tank. In such a tank the ratio of pressure to displacement is highest near the floor, close to the the projector, and lowest near the water surface. Such an arrangement therefore offers a means of varying the ratio of sound pressure to particle velocity in the applied stimulus, though generally particle velocity predominates in a shallow tank.

Some rather elaborate tubular tanks have been constructed for applying sound stimuli to fish. Banner (1967) described a long tank, with a sound projector at one end, and sound absorbent material at the other end. Weiss (1967), Cahn et al. (1969), and Hawkins and MacLennan (1976) have all built short water-filled tanks intended to operate as standing wave tubes for sounds of long wavelength. The tube has a sound projector at each end, and by driving the two projectors at different phases it is possible to effectively vary the acoustic impedance at the center of the tube within wide limits. Thus, with the two projectors driven out of phase, the water at the center of the tube is oscillated from side to side with a sound pressure null at the center. With the two projectors in phase the water at the center of the tube is alternately compressed and relaxed, with a null in particle motion at the center. The great disadvantage of such a tank is that it is very inconvenient to use, since the fish is not very accessible to the experimenter. Some of these problems are alleviated if the same kind of sound field is set up inside an open body of water, or a large tank as described by Buwalda and van der Steen (1979). However, large particle motions may be generated normal to any adjacent air/water interface if only two projectors are used, though in other respects the sound field may be satisfactory. With additional pairs of sound projectors the vertical field can also be adjusted (see Buwalda, Chapter 7). A disadvantage common to all standing wave devices is that the nulls set up between the sound projectors can be very sharp, and strong spatial gradients of the cancelled parameter may therefore exist. The experimental animal must therefore be confined within rather narrow limits.

Some of the acoustical problems encountered in the laboratory can be overcome by monitoring particle motion directly with calibrated pressure gradient or velocity hydrophones. Bauer (1967) has commented on the difficulties of measuring particle velocity and of calibrating particle velocity hydrophones. Schuijf et al. (1972), Banner (1973), and Hawkins and MacLennan (1976) describe particle velocity hydrophones constructed from seismic accelerometers. It is important to note that in most sound fields an array of three velocity hydrophones, arranged at mutual right angles, is necessary to describe the sound field adequately.

2.5 The Effects of Background Noise

The limit to the detection of sounds may be set not by any limit in sensitivity but by the level of background noise in the environment. The thresholds determined for fish under noisy conditions may be determined by the need for the signal to be above the background noise. The masking of a presented signal by background noise can be established by showing that the threshold changes with any change in the background noise. Only if the thresholds are independent of the noise level can they be regarded as absolute.

The underwater environment is rarely quiet and ambient noise levels, particularly in the sea, are often high and rather variable. Chapman and Hawkins (1973) showed that for cod in the sea unmasked thresholds were only obtained under the calmest of sea conditions. Laboratory tanks offer the best opportunity for obtaining absolute thresholds, but often such tanks are noisy because of their coupling to the floor and laboratory bench, or because of the water flow through the tank. In these circumstances it is wise to assume that any thresholds are masked, unless it can be demonstrated otherwise.

3 The Hearing Capacities of Fish

3.1 The Audiogram

The audiogram has become the standard way of expressing the auditory sensitivity of fish. It shows at a glance the band of frequencies audible to fish, and is of special value since it represents the response of the whole intact animal, rather than the electrophysiological response of a particular receptor organ.

The audiogram must be expressed in the most appropriate units. That is, it should be expressed in sound pressure if the fish is sensitive to sound pressure, and in particle velocity or displacement if the fish is particle motion sensitive. The thresholds of the audiogram should not be affected by the background noise, or if they are then the masked frequencies must be indicated. Dual thresholds have been reported by Tavolga and Wodinsky (1963) and by Cahn et al. (1969), and one interpretation is that some fish may have separate channels sensitive to sound pressure and particle motion, and they may sometimes respond to one parameter, and sometimes the other.

There is great variability in the audiograms presented for given species of fish (Hawkins 1973, Popper and Fay 1973, Popper, Salmon, and Parvulescu 1973, Sawa

1976, Fay 1978). As examples we can take two species, the cod studied by Buerkle (1967), Chapman and Hawkins (1973), and Offutt (1974); and the goldfish, studied by Enger (1966), Weiss (1966), Jacobs and Tavolga (1967), Offutt (1968), Fay (1969), Popper (1971), Popper et al. (1973), and Sawa (1976).

Three audiograms for the cod are given in Fig. 6-2, indicating the differences obtained in threshold by different authors. In practice, the discrepancies are readily explained. The lowest thresholds, those of Offutt, were obtained with the fish in a quiet vibration-isolated tank, stimulated by a loudspeaker in air outside the tank. The low thresholds can be attributed to low prevailing ambient noise levels. By contrast, Buerkle obtained much higher thresholds under conditions of high ambient noise in the laboratory. Buerkle (1968) confirmed the masking of the thresholds, showing that as the ambient noise increased, so the thresholds were raised. Finally, the intermediate threshold values reported by Chapman and Hawkins were obtained under conditions of natural sea noise and varied as sea noise varied, again indicating masking.

By placing cod at different distances from the sound source, exposing them to the near field effect, Chapman and Hawkins showed that over a wide range of frequencies the cod was sensitive to acoustic pressure, rather than particle motion. Though the acoustic conditions employed by Offutt and Buerkle were rather poor, their threshold levels were presented in terms of the measured sound pressures and are therefore directly comparable with those of Chapman and Hawkins. Thus, the audiogram of the cod is well established, showing a steep cut off at frequencies above about 250 Hz, and

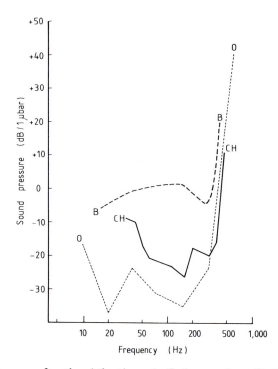

Figure 6-2. Audiograms for the Atlantic cod, *Gadus morhua*. B, Buerkle 1967; CH, Chapman and Hawkins 1973; O, Offutt 1974.

showing some variability in the thresholds at lower frequencies, even under natural sea conditions.

The results for the goldfish are much more variable even than those for the cod (Fig. 6-3). Differences of over 60 dB are apparent in the thresholds at some frequencies. The differences may have resulted from the various conditioning techniques employed [though experiments by Popper et al. (1973) suggested that this was not so], or by innate differences between European and American goldfish as suggested by Offutt (1968). However, it is more likely that the differences arise from the various acoustic conditions under which the experiments were performed, and by a greater or lesser degree of masking by the background noise. Taken together, the differences in these thresholds illustrate the care that must be taken in preparing an audiogram for a species.

If we look at the audiograms published for fish it is probable to draw several general conclusions. First, some species can be grouped together because they respond only to particle motion, and yield audiograms of rather narrow bandwidth. These species include the two flatfish, the dab, *Limanda limanda* (Fig. 6-4), and plaice, *Pleuronectes platessa,* studied by Chapman and Sand (1974), and the atlantic salmon, *Salmo salar,* studied by Hawkins and Johnstone (1978). The thresholds for these particle motion sensitive fish rise steeply above about 100-160 Hz, and even at their most sensitive frequencies, the thresholds do not fall below –10 dB re 1 μvar (i.e., 1 μvar = 6.49 × 10^{-6} cm sec^{-1}, equivalent to the particle velocity for a sound pressure of 1 μbar in the far field). Other species showing a similarly restricted frequency range, and which are probably particle motion sensitive, include the goby, *Gobius niger* (Dijkgraaf 1952), the perch, *Perca fluviatilis* (Wolff 1967), the ruff, *Acerina cernua,* the pike perch, *Lucioperca sandra* (Wolff 1968), the toadfish, *Opsanus tau* (Fish and

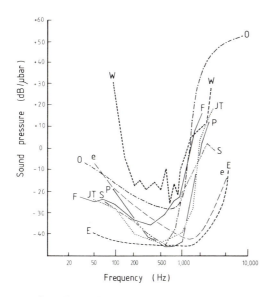

Figure 6-3. Audiograms for the goldfish, *Carassius auratus.* E, Enger 1966 (air loudspeakers); e, Enger 1966 (water loudspeakers); F, Fay 1969; JT, Jacobs and Tavolga 1967; O, Offutt 1968; P, Popper 1971; S, Sawa 1976; W, Weiss 1966.

Offutt 1972), and the tautog, *Tautoga onitis* (Offutt 1971). Taxonomically, many of these fish are unrelated.

A second group of telost fish, representing perhaps the opposite extreme, includes species with a much wider frequency range. Perhaps the most carefully researched example is the freshwater catfish *Ictalurus* (= *Amiurus*) *nebulosus* studied first by von Frisch (1923), and later by Poggendorf (1952) and Weiss, Strother, and Hartig (1969). Poggendorf obtained thresholds for this species (Fig. 6-4) under conditions where the ratio of sound pressure to particle motion could be varied, and clearly established that the catfish was sound pressure sensitive. This species not only shows a remarkably extended frequency range in comparison with the foregoing species, but in the far field, where sound pressure and particle velocity thresholds are readily compared, it is much more sensitive to sounds even at low frequencies. Thresholds as low as -40 dB re 1 μbar were obtained, the fish showing an acute sensitivity to sound frequencies even above 2 kHz.

A range of other species also belonging to the Ostariophysi have also shown an extended frequency range, and are probably sound pressure sensitive. As well as the goldfish they include the chub, *Semotilus atromaculatus* (Kleerekoper and Chagnon 1954), the Japanese carp, *Cyprinus carpio* (Popper 1972b), several species of the genus

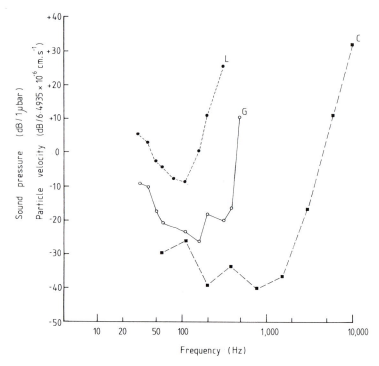

Figure 6-4. Audiograms for three teleosts. Thresholds for the cod, *Gadus morhua,* and the catfish, *Amiurus (Ictalurus) nebulosus,* are expressed in terms of sound pressure. Thresholds for the dab, *Limanda limanda,* are expressed in terms of particle velocity. The two kinds of threshold can only be compared in the far field of a sound source. C, *A. nebulosus,* Poggendorf 1952; G, *G. morhua,* Chapman and Hawkins 1973; L, *L. limanda,* Chapman and Sand 1974.

Astyanax (Popper 1970), the minnow, *Leucaspius delineatus* (Schade 1971), and the golder carp *Carassius carassius* (Siegmund and Wolff 1973). There are also fish belonging to other unrelated groups which have an extended frequency range. These species also seem to be sound pressure sensitive from anatomical evidence and include in particular the holocentrids *Myripristis* (Coombs and Popper 1979), and *Holocentrus ascensionis* (Tavolga and Wodinsky 1963). *Myripristis* gave thresholds of ~50 dB re 1 μbar for frequencies between 300 Hz and 2 kHz (Coombs and Popper 1979).

Many species, like the cod, the sea bream, *Sargus annularis* (Dijkgraaf 1952), a number of species studied by Tavolga and Wodinsky (1963) and the squirrelfishes, *Adioryx xantherythrus* (Coombs and Popper 1979) fall somewhere between these two extremes. Some may prove to be sound pressure sensitive though they lack an elaborate connection between the swimbladder and the ear.

Sharks have only been investigated under rather poor acoustic conditions and the published thresholds must be treated with caution. An early study by Kritzler and Wood (1961) on the bull shark, *Carcharhinus leucas,* indicated that this species had quite a wide hearing range, with greatest sensitivity at 400-600 Hz. However, Kritzler and Wood pointed out that their thresholds were influenced by ambient noise. In addition sound pressure measurements were made with a fish that may have been particle motion sensitive. Another study by Banner (1967) indicated that the lemon shark, *Negaprion brevirostris,* was sensitive to particle motion, while more recently Kelly and Nelson (1975) have shown that the horn shark *Heterodontus francisci* is also sensitive to this parameter, with its lowest threshold at 80 Hz. Though the horn shark showed greatest sensitivity to approximately the same frequencies as the particle motion sensitive teleosts, the threshold levels were much higher. Rather high thresholds were also reported by Nelson (1967b) for the lemon shark. These high thresholds measured for sharks may be associated with high background noise levels, but the data also suggest that these fish are inherently less sensitive to sounds than teleosts.

It is now known that on the whole the frequency range of fish is rather more restricted than for terrestrial vertebrates. It is also clear that both the absolute sensitivity and the hearing bandwidth shown by individual species depend very much on the anatomy of the ear, and in particular on any connection between the otolith organs of the ear and any gas-filled spaces. There can be great differences in the hearing abilities of species even within a family, depending on the structure of the auditory apparatus (Coombs and Popper 1979).

At least one marine species and probably many others both in the sea and freshwater are so sensitive to low frequency sounds that they are limited in the natural environment only by their ability to distinguish sounds from the prevailing background noise. It is very probable that many fish show hearing abilities adapted to their own acoustic environment, though such adaptation has not so far been subjected to close examination.

3.2 Auditory Masking

Tavolga (1967) first described the masking effect that background noise can have on the hearing of fish. In a series of experiments with squirrelfish, *Holocentrus rufus,* and blue striped grunt, *Haemulon sciurus,* thresholds to tones were measured at different ambient noise levels. As the level of the white noise background changed, so the threshold for the stimulus altered. Buerkle (1968) later showed that the hearing thresholds of cod were also influenced by changes in the level of noise in the aquarium tank, while Chapman and Hawkins (1973) showed that the hearing of this same species could be masked even under quiet conditions in the sea. Masking of auditory thresholds for goldfish in the laboratory has been demonstrated by Fay (1974a). Recently, Popper and Clarke (1978) have demonstrated that the noise does not need to occur at the same time as the signal for masking to take place. Nonsimultaneous backward and forward masking is possible.

For human subjects it has long been known that not all frequency components of a masking noise are equally effective in promoting the masking. A pure tone stimulus is masked most effectively by noise at the same, and immediately adjacent frequencies, the frequency span of noise which is effective being termed the critical band (Fletcher 1940). A useful analogy is that the ear behaves as a filter tuned to the frequency of the stimulus, eliminating remote frequencies.

There are several ways of demonstrating that masking is frequency selective and for determining the width of the critical masking band. One method is to show that as the bandwidth of the noise masking a pure tone is progressively narrowed, the threshold remains the same until the critical bandwidth is reached when the threshold changes significantly for narrower bands. Hawkins and Chapman (1975) adopted this technique with cod, and confirmed the existence of critical bands at several frequencies. A more precise method for determining the bandwidth is to measure thresholds for a pure tone, or very narrow band noise stimulus, in the presence of narrow bands of noise centered at different frequencies. Alternatively, the frequency and level of a narrow band of masking noise is kept constant and the frequency of the stimulus varied. Such methods have been successfully applied by Buerkle (1969) and Hawkins and Chapman (1975) for the cod, by Tavolga (1974) and Fay, Ahroon, and Orawski (1978) for the goldfish, and by Hawkins and Johnstone (1978) for the salmon. With these techniques, the stimulus is masked most effectively by noise centered at the same frequency, and masking diminishes as the frequency separation between stimulus and masker increases.

There is a rapid way of obtaining an approximate value for the critical band, based on the assumption that when white noise just masks a pure tone the noise in the relatively narrow band performing the masking is equal in power to the tone itself. Thus, the value of the masked threshold for a pure tone is determined in the presence of white noise and then the ratio of the threshold to the spectrum level of background noise (T_b) calculated. Since, for white noise, the level within a band of width Δf Hz is equal to the spectrum level plus $10 \log_{10} \Delta f_c$ then: $10 \log_{10} \Delta f_c = T_b$, where Δf_c is the width of the band responsible for the masking. Values for the bandwidth measured in this way (termed critical ratios to distinguish them from empirically measured critical bands) have been determined by Cahn et al. (1969), Fay (1974a), and Hawkins and Chapman (1975). The latter found that the critical ratios for cod were rather smaller than the critical bands.

The masking function obtained in critical band experiments by plotting the relative masking effect of noise at different frequencies is generally V-shaped, the apex corresponding to the frequency at which greatest masking is obtained. Tavolga (1974) reported very steep sided masking functions for the goldfish; those for the cod are broader (Hawkins and Chapman 1975); while those for the salmon even broader still (Hawkins and Johnstone 1978). Thus, there is evidence that narrow critical bands are associated with a wider hearing range, and more acute hearing. Hawkins and Chapman (1975), pointed to the difficulty in assigning a particular bandwidth to the masking function and suggested that the best way of describing the auditory filter was in terms of the bandwidth of an equivalent rectangular filter. Fay et al. (1978) have also reported V-shaped masking functions for signal frequencies below 350 Hz in the goldfish, but for signals above 350 Hz described peaks both at 350 Hz and at the frequency of the signal. Such complex functions cannot be described in terms of a simple filter of fixed bandwith (see also Coombs, Chapter 8).

The demonstration from psychophysical experiments of a filter-like process within the fish auditory system has led to considerable speculation about the physiological nature of the frequency filtering mechanism. Fay et al. (1978) attempted to equate the so-called psychophysical tuning curves with the tuning curves measured physiologically from primary afferent fibers. They have pointed out that the psychophysical tuning curves for goldfish are much narrow than the tuning curves of the primary fibers. The fish is more highly frequency selective than can be accounted for by the tuning of peripheral neural channels. Comparison of the neural tuning curves for the cod (Horner, Hawkins, and Fraser, Chapter 11) with critical bands for the cod (Fig. 6-5) (Hawkins and Chapman 1975) point to a similar conclusion.

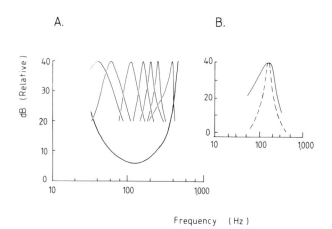

Figure 6-5. Critical bands in fish. (A) Critical bands for the cod, *Gadus morhua*, obtained by masking pure tone stimuli with 10 Hz wide noise bands centered at different frequencies. An audiogram is provided for comparison. (B) Critical bands for two species at 160 Hz, obtained with 10 Hz wide noise bands. The salmon, *Salmo salar* (-), and the cod *Gadus morhua* (---). Data from Hawkins and Chapman (1975) and Hawkins and Johnstone (1978).

3.3 The Relevant Stimulus

A question of some importance is whether the fish ear is predominantly sensitive to sound pressure, or to the oscillatory particle motion that also accompanies passage of a sound wave. Several psychophysical studies and accompanying physiological researches have shown that the otolith organs of the labyrinth are essentially particle motion detectors, but that in different teleosts the organs may be coupled to gas-filled accessory structures which render the fish sound pressure sensitive. Reviews of the experiments on this topic are provided by Hawkins (1973), Popper and Fay (1973), Fay (1978), and Fay and Popper (1980). In general the relevant stimulus has been investigated by changing the ratio of sound pressure to particle velocity in the applied sound field. The manipulation of the stimulus is achieved either by placing the fish at different distances from a source in a free sound field, or by placing the fish in a standing wave tube. Thus, by means of the former technique Chapman and Hawkins (1973) confirmed that over a broad frequency range the cod was sensitive to sound pressure. Chapman and Sand (1974) later showed that two species of flatfish, the plaice and dab, were particle motion sensitive while Hawkins and Johnstone (1978) made a similar finding for the salmon. In a kind of standing wave tank Poggendorf (1952) showed that the dwarf catfish was sound pressure sensitive while Banner (1967) showed that the lemon shark was sensitive to particle motion. Cahn et al. (1969) working with a standing wave tube found that two species of grunt could shift from velocity (or displacement) sensitivity to pressure sensitivity as frequency increased.

It has been suggested that sound localization in many fish may involve a comparison of particle motion and sound pressure inputs to the ears (Schuijf and Buwalda 1980), and that there may well be a segregation of the two forms of stimulation within the fish auditory system. Tavolga and Wodinsky (1963) found early evidence of this by obtaining dual thresholds from the blue-striped grunt, *Haemulon sciurus.* In a series of conditioning experiments with *Haemulon parrai,* Cahn, Siler, and Auwarter (1970) investigated the masking of one parameter by the other, employing a standing wave tube to apply both the stimulus and masking noise. It was discovered that while sensitivity to sound pressure was not affected by particle motion noise, sensitivity to particle motion was influenced by sound pressure noise. More recently, Schuijf and Buwalda (1980) have described very similar unpublished experiments by Buwalda on the cod, and reported that it was impossible within the limits of their apparatus to mask one sound variable (e.g., pressure) with another (e.g., velocity). The latter result certainly points to a separate processing of sound pressure and particle motion within the auditory system (see Buwalda, Chapter 7).

3.4 Hearing and Sound Duration

In most experiments on the hearing of fish, the animals are presented with sounds of relatively long duration. Generally, auditory thresholds are determined for single pulses of sound, lasting up to 10 sec, usually with a gradual rise to and from the maximum amplitude. However, many of the natural sounds heard by fish, including the calls produced by various species, and either relatively short, or can be divided into a series of repeated pulses (Hawkins and Rasmussen 1978). The temporal characteristics of the fish ear are therefore of special interest.

In man and other mammals, a reduction in the length of a single sound pulse below about 200 msec results in an increase in the threshold for detection. The shorter a sound pulse the higher it must be in amplitude to be detected (Plomp and Bouman 1959, Ehret 1976).

Offutt (1967) investigated the detection of repeated tone pulses of varying length by the goldfish. He found that the threshold essentially depended upon the sound energy, that is, the percentage of time occupied by the sound signal, or duty cycle. In a later study of the same species, Popper (1972a) reported quite different results. In experiments with single tone pulses Popper found no difference between the thresholds for short pulses and continuous tones. In addition, in experiments with repeated tone pulses the thresholds remained the same regardless of pulse length or duty cycle. Popper concluded that temporal summation did not occur within the goldfish auditory system, a finding which was not supported by the results of a later study (Popper and Clarke 1978), reporting that when short tone pulses were presented simultaneously with a longer noise pulse, the threshold was dependent upon the tone pulse length.

Hawkins and Horner have recently performed a series of experiments on cod, where auditory thresholds were determined against a white noise background both for single pulses of differing duration, and for sequences of pulses of differing duty cycle. Their results show that the shorter the duration of a tone pulse the higher the sound level need be for the signal to be detected against a noise background. In experiments with repeated pulses, it was clear that the lower the duty cycle, the higher the masked thresholds, i.e., the higher the amplitude necessary for detection (Fig. 6-6). Thus the auditory system of the cod, like that of mammals, performs a temporal summation or integration of the sounds presented to it. One important feature of this temporal integration is that it renders the auditory system relatively insensitive to randomly occurring transients. Temporal integration may also account for the nonsimultaneous masking phenomena described by Popper and Clarke (1978) for the goldfish.

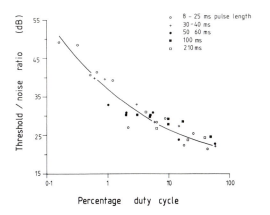

Figure 6-6. Effects of duty cycle upon detection of pulsed tone signals for the cod. The duty cycle is the percentage of a total 10 sec stimulus time occupied by the sound pulses. Relative sensitivity is expressed in terms of the threshold to noise ratio, against a standard white noise background.

3.5 Discrimination of Sound Quality

A knowledge of the auditory capacities of fish is not complete without data on their ability to distinguish between sounds of differing quality, or to detect a change in any sound stimulus. Those features of a sound which are particularly important in this respect are its frequency spectrum, amplitude, temporal characteristics, and direction.

The threshold for detection of a difference between two stimuli is termed a difference limen. Limens are often determined by presenting a repetitive series of sound pulses to the fish, and changing the quality of sound from one pulse to the next without altering the regular pattern of repetition. The fish is conditioned to the change in sound quality occurring in the interval between two pulses, the trial being followed by a reversion to the original stimulus state. The advantage in presenting a sequence of pulses is that switching from one sound quality to another in the silent interval avoids the introduction of transients or stimulus artifacts. Inherent in this technique is that the animal learns to ignore the repeated presentation of sound and is conditioned only to a change in quality. The switch from one stimulus quality may occur once, at the start of the trial, or it may continue as an alternation of one stimulus quality with another throughout the trial. With the latter, the fish has several opportunities to sense the change in quality, and the demands on its memory are minimal. Hence, it may yield a lower limen.

Another way of registering a discrimination between two stimulus states by the fish is to train the fish to do one thing when one stimulus is presented and a different thing when another is presented. One of these responses can simply be for the animal to ignore one of the stimuli. This method, which is sometimes called the absolute method, places a more severe demand on the animal and requires stringent definition of the two different responses.

3.5.1 Frequency Discrimination

Several early experimenters showed that teleosts could be trained to distinguish between tones separated in frequency. A good review of the earlier literature is provided by Lowenstein (1957). In many of the earlier experiments, however, no allowance was made for the fish detecting subjective intensity changes. Dijkgraaf and Verheijen (1950) therefore included in their study of frequency discrimination a series of controls to rule out intensity cues. They demonstrated that the minnow *Phoxinus laevis* could distinguish between tones differing by as little as 5%.

Dijkgraaf (1952) later extended his interest to several nonostariophysine fish including the sea bream *Sargus annularis* and the black goby, *Gobius niger,* and showed that these species could distinguish a 9% frequency change. Since these carefully performed early experiments Jacobs and Tavolga (1968) and Fay (1970a) have determined frequency limens for another ostariophysine, the goldfish, of less than 5%. In comparison with mammals and birds frequency discrimination by most fish is relatively poor, the limens being greater than those for man by about tenfold. However, Fay (1979b) points out that though the difference limens for the goldfish are higher than those for many mammals and birds they are not exceptionally so unless compared with those for man.

The only elasmobranch which has been investigated is the lemon shark, *Negaprion brevirostris* (Nelson 1967b). Tested by the absolute method the sharks discriminated tones separated by half an octave (i.e., by 50%). This single poor result from a shark may simply reflect the relatively crude method of investigation.

Fay (1970b) has reported a remarkable dependence of the conditioned respiratory response of goldfish on stimulus frequency. After conditioning the response to one frequency, the magnitude of the response to different frequencies decreased greatly with the frequency separation. That is, the fish did not generalize its response to other frequencies. This finding indicates the importance of frequency information to the goldfish.

3.5.2   Amplitude Discrimination

There have been few studies of amplitude discrimination by fish. Jacobs and Tavolga (1967) obtained intensity limens of between 3 and 6 dB for the goldfish, while Chapman and Johnstone (1974) showed that cod could discriminate changes of between 1.3 and 9.5 dB at different frequencies (see Hall, Patricoski, and Fay, Chapter 9). The limens for both species were higher at higher frequencies. In a series of stimulus generalization experiments Fay (1972) presented goldfish with amplitude modulated tones, where the depth of modulation was varied. The fish did not generalize well, which indicates that amplitude changes, or perhaps more complex qualities of the amplitude modulation, are important to the goldfish.

3.5.3   Discrimination of Phase

Piddington (1972) has shown by absolute discrimination conditioning that goldfish can distinguish between a given click and the same click inverted. That is, with the compression and rarefaction phases reversed.

3.5.4   Discrimination of Sound Direction

Early experiments on directional hearing indicated that fish only gave directional responses when they were close to the sound source (von Frisch and Dijkgraaf 1935, Reinhardt 1935). Where directional responses were obtained the studies were often criticized on the grounds of the poor acoustic conditions under which they were performed (Kleerekoper and Chagnon 1954, Moulton 1963b, Moulton and Dixon 1967, Nelson 1967b). However, despite this rather equivocal state of knowledge anecdotal reports indicated that predatory fish, and especially sharks, oriented toward sound sources in the field, often from considerable distances (Myrberg et al. 1976) though Richard (1968) emphasized that these fish may not have oriented directly to the sound source, but may have approached it through exploratory swimming.

The first definite evidence that fish could discriminate between sound coming from either of two spatially separated sources was provided by Schuijf et al. (1972). Wrasse, *Labrus berggytta,* were successfully trained to detect a change in the direction of a sound signal. Later experiments by Chapman and Johnstone (1974) similarly showed that cod could be conditioned to a change in the direction of a pulsed tone, switched between two equidistant sources. In addition, Chapman (1973) and Chapman and Johnstone (1974) demonstrated clearly that the masking effect of noise on detection of a pure tone was much less when the noise was transmitted from a different projector, spatially separated from the signal projector. This latter experiment indicated that a directional hearing sense not only enables a fish to determine the direction of a particular source but improves the ability of the fish to detect the presence of a localized sound source in a noisy environment.

Subsequently, Schuijf and Siemelink (1974) and Schuijf (1975) trained cod to indicate the active one of two alternative sound projectors by swimming to either of two opposing corners of a cage in return for a food reward. This experiment provided proof that cod are capable not only of discriminating between sounds from two different directions but can actually orientate toward a source. Furthermore, it has now been confirmed that cod can discriminate spatially separated sound sources in the median vertical, as well as the horizontal plane (Hawkins and Sand 1977) and that cod can distinguish between frontally incident and caudally incident sound waves (Schuijf and Buwalda 1975) (see Fig. 6-7).

Directional hearing does not appear to be confined to the wrasse and the cod. Popper et al. (1973) have shown that playbacks of alarm calls of squirrelfish, *Myripristis* sp., release an oriented movement (also see Salmon 1967), while Schuijf, Visser, Willers, and Buwalda (1977) succeeded in showing that the ide, *Leuciscus idus* (an ostariophysine), can discriminate sounds coming from the head from those coming from the tail. Moreover, there is a great deal of evidence from field studies that predatory sharks can locate a source of sound (Myrberg et al. 1976).

There is now no longer any reason to suppose, as van Bergeijk did in his influential 1964 paper, that the directional hearing abilities of fish are poor.

4 Conclusions

As we have seen, in the last two decades there has been extensive application of various psychophysical methods to the study of fish hearing. Though much of the work has been impaired by methodological and technical difficulties, there is now a substantial body of data on the hearing abilities of a range of species. Furthermore, the differences which have been established between different groups or types of fish have prompted more intensive physiological examination of the hearing mechanisms of fish, and have led to a great improvement in our knowledge of the fish auditory system. Is there still scope then, for further behavioral and psychophysical studies on fish hearing?

Certainly there is a need to investigate the hearing abilities of still more species. Next to nothing is known of hearing in Cephalochordates, such as *Amphioxus,* and Agnathans like the lamprey *Petromyzon* and hagfish *Myxine.* Detailed information on the hearing abilities of sharks and rays is still largely lacking, and it is not clear whether elasmobranchs are less sensitive to sounds than teleosts like the pleuronectid flatfish, and salmonids, or how well they can discriminate between sounds of differing frequency, amplitude, and timing. Nor is it known how well they can distinguish between spatially separated sound sources. Certain teleost groups also need closer investigation.

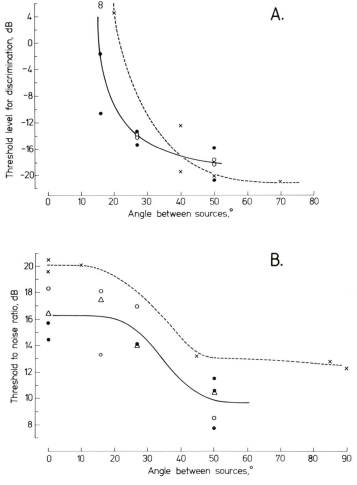

Figure 6-7. Directional hearing in the cod, *Gadus morhua*. (A) Minimum sound pressure thresholds for detection of an angular change in direction for a 110 Hz tone as a function of angle. Circles represent changes in the median vertical plane for two fish. The X and dashed line indicate values for changes in the horizontal plane. (B) Threshold to noise ratio as a function of one projector transmitting a pure tone, and the other a masking noise. Circles and open triangle represent data at 110 Hz from 3 fish with the projectors in the vertical plane. The X indicates mean values with the projectors in the horizontal plane for several tone frequencies. Data taken from Chapman and Johnstone (1974) and Hawkins and Sand (1977).

They include the herrings (Clupeidae), anchovies (Engraulidae), and the mackerels (Scombridae), groups that are of great commercial importance, and in the case of the herring and anchovy which are also of special physiological interest.

More information is required on the ontogeny of sound reception in fish. The majority of behavioral and psychophysical studies have been of adult fish, and the development of hearing has hardly been touched upon.

The degree of refinement shown by the fish auditory system deserves further investigation. For example, though repetitive pulsing of sounds is known to be important for species recognition, and for the attraction of predatory fish, there has been no psychophysical study of the discrimination by fish of sounds which differ in their temporal characteristics. There is also a need for behavioral data to augment the results of recent physiological investigations. Experiments are required to determine the degree of functional separation between the auditory channels detecting sound pressure and particle velocity. Work is needed on the precision of directional hearing at different angles of azimuth and elevation, especially combined with the selective elimination of otolith organs to ascertain the separate roles played by the utriculus, sacculus, and lagena in directional hearing. Discrimination between sound sources placed at different distances has not been explored.

How do fish distinguish between complex sounds? Is the distinction made purely on the basis of the spectral differences between sounds or are other factors like periodicity or the phase relationships between the various spectral components important?

Perhaps the greatest gap exists in our knowledge of those features of sounds which hold greatest interest for fish and which are especially important in call recognition. Special characteristics of the calls of anuran amphibians have been investigated by comparing the performance of certain relatively stereotyped behavior patterns during playback of the calls, and modified versions of the calls (for example, as in Gerhardt 1978). Though the calls produced by fish are relatively stereotyped, the behavior shown to their playback is often variable, and the fish may habituate to repeated playback, or modify their behavior as a result of experience. There are relatively few reports of clear-cut repeatable and unambiguous behavior from fish in response to sounds (see Myrberg, Chapter 20). The exceptions are movement toward a speaker emitting calls by a goby *Bathygobius soporator* (Tavolga 1958) and similar responses by the squirrelfish *Myripristis berndti* (Popper et al. 1973); the facilitatory effects of courtship sounds on courtship initiation in the bicolor damselfish, *Euopmacentrus partitus* (Myrberg 1972); and the calling of toadfish, *Opsanus tau,* in response to calls from conspecifics (Winn 1967). It is far from clear to what extent these responses are under strict genetic control, or whether auditory responsiveness can be modified. Nevertheless, playback studies of this kind offer promise for the future.

Acknowledgements. I am grateful to Kathleen Horner for her comments on this paper, and to Audrey Rae for her help in preparing the draft.

References

Banner, A.: Evidence of sensitivity to acoustic displacement in the lemon shark, *Negaprion brevirostris*. Bull. Mar. Sci. 17, 741-768 (1967).

Banner, A.: Simple velocity hydrophone for bioacoustic application. J. Acoust. Soc. Am. 53, 1134-1136 (1973).

Bauer, B. B.: Measurement of particle velocity in underwater sound. In: Lateral Line Detectors. Cahn, P. (ed.). Bloomington: Indiana Univ. Press, 1967.

van Bergeijk, W. A.: Directional and non-directional hearing in fish. In: Marine Bioacoustics. Tavolga, W. N. (ed.). Oxford: Pergamon Press, 1964, pp. 281-299.

Brawn, V. M.: Sound production by the cod *Gadus callarias* L. Behaviour 18, 239-255 (1961).

Buerkle, U.: An audiogram of the Atlantic cod, *Gadus morhua* L. J. Fish. Res. Bd. Can. 25, 1155-1160 (1968).

Buerkle, U.: Auditory masking and the critical band in Atlantic cod, *Gadus morhua*. J. Fish. Res. Bd. Can. 26, 1113-1119 (1969).

Buwalda, R. J. A., van der Steen, J.: The sensitivity of the cod sacculus to directional and non-directional sound stimuli. Comp. Biochem. Physiol. 64A, 467-471 (1979).

Cahn, P., Siler, W., Auwarter, A.: Acoustico-lateralis system of fishes: Cross-modal coupling of signal and noise in the grunt, *Haemulon parrai*. J. Acoust. Soc. Am. 49, 591-594 (1970).

Cahn, P., Siler, W., Wodinsky, J.: Acoustico-lateralis system of fishes: Tests of pressure and particle-velocity sensitivity in grunts, *Haemulon sciurus* and *Haemulon parrai*. J. Acoust. Soc. Am. 46, 1572-1578 (1969).

Chapman, C. J.: Field studies of hearing in teleost fish. Helgolander Wiss Meeresunters. 24, 371-390 (1973).

Chapman, C. J., Hawkins, A. D.: A field study of hearing in the cod, *Gadus morhua*. L. J. Comp. Physiol. 85, 147-167 (1973).

Chapman, C. J., Johnstone, A. D. F.: Some auditory discrimination experiments on marine fish. J. Exp. Biol. 61, 521-528 (1974).

Chapman, C. J., Sand, O.: Field studies of hearing in two species of flatfish, *Pleuronectes platessa* and *Limanda limanda*. Comp. Biochem. Physiol. 47A, 371-385 (1974).

Coombs, S., Popper, A. N.: Hearing differences among Hawaiian squirrelfish (Holocentridae) related to differences in the peripheral auditory system. J. Comp. Physiol. 132, 203-207 (1979).

Cornsweet, T. M.: The staircase method in psychophyics. Am. J. Psychol. 75, 485-491 (1962).

Dijkgraaf, S.: Uber die Schallwahrnehmung bei Meeresfischen. Z. Vergl. Physiol. 34, 104-122 (1952).

Dijkgraaf, S., Berheijen, F.: Neue Versuche über das Tonunterscheidungsvermogen der Elritze. Z. Vergl. Physiol. 32, 248-256 (1950).

Ehret, G.: Temporal auditory summation for pure tones and white noise in the house mouse *Mus musculatus*. J. Acoust. Soc. Am. 59, 1421-1427 (1976).

Enger, P. S.: Acoustic thresholds in goldfish and its relation to the sound source distance. Comp. Biochem. Physiol. 18, 859-868 (1966).

Fay, R. R.: Behavioural audiogram for the goldfish. J. Aud. Res. 9, 112-121 (1969).

Fay, R. R.: Auditory frequency discrimination in the goldfish (*Carassius auratus*). J. Comp. Physiol. Psychol. 73, 175-180 (1970a).

Fay, R. R.: Auditory frequency generalization in the goldfish (*Carassius auratus*). J. Expl. Anal. Behav. 14, 353-360 (1970b).

Fay, R. R.: Perception of amplitude modulated signals in the goldfish. J. Acoust. Soc. Am. 52, 660-666 (1972).

Fay, R. R.: Masking of tones by noise for the goldfish (*Carassius auratus*). J. Comp. Physiol. Psychol. 87, 708-716 (1974a).

Fay, R. R.: Auditory frequency discrimination in vertebrates. J. Acoust. Soc. Am. 56, 206-209 (1974b).

Fay, R. R.: Sound detection and sensory coding by the auditory systems of fishes. In: The Behaviour of Fish and Other Aquatic Animals. Mostofsky, D. (ed.). New York: Academic Press, 1978, pp. 197-231.

Fay, R. R., Ahroon, W. A., Orawski, A. A.: Auditory masking patterns in the goldfish (*Carassius auratus*): Psychophysical tuning curves. J. Exp. Biol. 74, 83-100 (1978).

Fay, R. R., Popper, A. N.: Structure and function in teleost auditory systems. In: Comparative Studies of Hearing in Vertebrates. Popper, A. N., Fay, R. R. (eds.). New York: Springer-Verlag, 1980, pp. 1-42.

Fish, J. F., Offutt, G. C.: Hearing thresholds from toadfish, *Opsanus tau,* measured in the laboratory and field. J. Acoust. Soc. Am. 51, 1318-1321 (1972).

Fletcher, H.: Auditory patterns, Rev. Mod. Phys. 12, 47-65 (1940).

Gerhardt, H. C.: Mating call recognition in the green tree frog. J. Exp. Biol. 74, 59-73 (1978).

Green, D. M., Swets, J. A.: Signal detection theory and psychophysics. New York: Wiley, 1966.

Griffin, D. R.: Hearing and acoustic orientation in marine animals. Papers on marine biology and oceanography. Deep Sea Res. 3 Suppl. 406-417 (1955).

Harris, G. C., van Bergeijk, W. A.: Evidence that the lateral-line organ responds to near field displacements of sound sources in water. J. Acoust. Soc. Am. 34, 1831-1841 (1962).

Hawkins, A. D.: The sensitivity of fish to sounds. Oceanogr. Mar. Biol. Annu. Rev. 11, 291-340 (1973).

Hawkins, A. D., Chapman, C. J.: Masked auditory thresholds in the cod *Gadus morhua* L. J. Comp. Physiol. 103A, 209-226 (1975).

Hawkins, A. D., Johnstone, A. D. F.: The hearing of the Atlantic salmon, *Salmo salar.* J. Fish. Biol. 13, 655-673 (1978).

Hawkins, A. D., MacLennan, D. N.: An acoustic tank for hearing studies on fish. In Sound Reception in Fish. Schuijf, A., Hawkins, A. D. (eds.). Amsterdam: Elsevier, 1976, pp. 149-169.

Hawkins, A. D., Rasmussen, K.: The calls of gadoid fish. J. Mar. Biol. Assoc. U.K. 58, 891-911 (1978).

Hawkins, A. D., Sand, O.: Directional hearing in the median vertical plane by the cod. J. Comp. Physiol. 122, 1-8 (1977).

Jacobs, D. W., Tavolga, W. N.: Acoustic intensity limens in the goldfish. Anim. Behav. 15, 324-335 (1967).

Jacobs, D. W., Tavolga, W. N.: Acoustic frequency discrimination in the goldfish. Anim. Behav. 16, 67-71 (1968).

Kelly, J. C., Nelson, D. R.: Hearing thresholds of the horn shark, *Heterodontus francisci.* J. Acoust. Soc. Am. 58, 905-909 (1975).

Kleerekoper, H., Chagnon, E. C.: Hearing in fish with special reference to *Semotilus a. atromaculatus.* J. Fish. Res. Bd. Can. 11, 130-152 (1954).

Kritzler, H., Wood, L.: Provisional audiogram for the shark, *Carcharhinus leucas.* Science 133, 1480-1482 (1961).

Lowenstein, O.: The sense organ: The acousticolateralis system. In: The Physiology of Fishes, Vol. 2. Brown, M. E. (ed.). New York: Academic Press, 1957, pp. 155-186.

Moulton, J. M.: Influencing the calling of sea robins *Prionotus* spp. with sound. Biol. Bull. 114, 357-374 (1956).

Moulton, J. M.: Acoustic behavior of fishes. In: Acoustic Behavior of Animals. Busnel, R. G. (ed.). Amsterdam: Elsevier, 1963a, pp. 655-693.

Moulton, J. M.: Acoustic orientation of marine fishes and invertebrates. Ergebnisse der Biologie 26, 27-29 (1963b).

Moulton, J. M., Dixon, R. H.: Directional hearing in fishes. In: Marine Bio-acoustics, Vol. 2. Tavolga, W. N. (ed.). Pergamon Press, 1967, pp. 187-228.

Myrberg, A. A.: Using sound to influence the behaviour of free-ranging marine animals. In: Behaviour of Marine Animals, Vol. 2. Winn, H. E., Olla, B. L. (eds.). New York: Plenum Press, 1972, pp. 435-468.

Myrberg, A. A., Gordon, C. R., Klimley, P.: Attraction of free-ranging sharks by low frequency sound, with comments on its biological significance. In: Sound Reception in Fish. Schuijf, A., Hawkins, A. D. (eds.). Amsterdam: Elsevier, 1976, pp. 205-228.

Myrberg, A. A., Spires, J.: Sound discrimination by the bicolor damselfish, *Eupomacentrus partitus*. J. Exp. Biol. 57, 727-735 (1972).

Nelson, D. R.: Cardiac responses to sounds in the lemon shark, *Negaprion brevirostris*. In: Sharks, Skates and Rays. Gilbert, P. W., Mathewson, R. F., Rall, D. P. (eds.). Baltimore: Johns Hopkins Press, 1967a, pp. 533-544.

Nelson, D. R.: Hearing thresholds, frequency discrimination and acoustic orientation in the lemon shark, *Negaprion brevirostris*. Bull. Mar. Sci. 17, 741-768 (1967b).

Offutt, G. C.: Integration of the energy in repeated tone pulses by man and the goldfish. J. Acoust. Soc. Am. 41, 13-19 (1967).

Offutt, G. C.: Auditory response in the goldfish. J. Aud. Res. 8, 391-400 (1968).

Offutt, G. C.: Response of the tautog, *Tautoga onitis*, to acoustic stimuli measured by classically conditioning the heart rate. Conditional Reflex. 6, 205-214 (1971).

Offutt, G. C.: Structures for the detection of acoustic stimuli in the Atlantic codfish *Gadus morhua*. J. Acoust. Soc. Am. 56, 665-671 (1974).

Otis, L. S., Cerf, J. A., Thomas, G. J.: Conditioned inhibition of respiration and heart rate in the goldfish. Science 126, 263-264 (1957).

Parvulescu, A.: Problems in propagation and processing. In: Marine Bioacoustics. Tavolga, W. N. (ed.). Oxford: Pergamon Press, 1964, pp. 87-100.

Parvulescu, A.: The acoustics of small tanks. In: Marine Bioacoustics, Vol. 2. Tavolga, W. N. (ed.). Oxford: Pergamon Press, 1967, pp. 7-14.

Piddington, R. W.: Auditory discrimination between compressions and rarefactions by goldfish. J. Exp. Biol. 56, 403-419 (1972).

Plomp, R., Bouman, M. A.: Relation between hearing threshold and duration for tone pulses. J. Acoust. Soc. Am. 31, 749-758 (1959).

Poggendorf, D.: Die absoluten Hörschwellen des Zwergwelses (*Amiurus nebulosus*) und Beiträge zur Physik des Weberschen Apparatur der Ostariophysen. Z. Vergl. Physiol. 34, 222-257 (1952).

Popper, A. N.: Auditory capacities of the Mexican blind cavefish *Astyanax jordani* and its eyed ancestor *Astyanax mexicanus*. Anim. Behav. 18, 552-562 (1970).

Popper, A. N.: The effects of size on auditory capacities of goldfish. J. Aud. Res. 11, 239-247 (1971).

Popper, A. N.: Auditory threshold in the goldfish as a function of signal duration. J. Acoust. Soc. Am. 52, 596-602 (1972a).

Popper, A. N.: Pure tone thresholds for the carp *Cyprinus carpio*. J. Acoust. Soc. Am. 52, 1714-1717 (1972b).

Popper, A. N., Chan, A. T. H., Clarke, N. L.: An evaluation of methods of behavioural investigations of teleost audition. Behav. Res. Meth. Instr. 5, 470-472 (1973).

Popper, A. N., Clarke, N. L.: Non-simultaneous auditory masking in the goldfish, *Carassius auratus*. J. Exp. Biol. 83, 145-158 (1978).

Popper, A. N., Fay, R. R.: Sound detection and processing by teleost fishes: a critical review. J. Acoust. Soc. Am. 53, 1515-1529 (1973).

Popper, A. N., Salmon, M., Parvulescu, A.: Sound localization by the Hawaiian squirrelfishes. Anim. Behav. 21, 86-97 (1973).

Reinhardt, F.: Über Richtungswahrnehmung bei Fischen, besonders bei der Elritze (*Phoxinus laevis* L.) und beim Zwergwels (*Amiurus nebulosus* Raf.). Z. Vergl. Physiol. 22, 570-603 (1935).

Richard, J. D.: Fish attraction with pulsed low frequency sound. J. Fish. Res. Bd. Can. 25, 1441-1452 (1968).

Salmon, M.: Acoustical behaviour of the menpachi, *Myripristis berndti,* in Hawaii. Pac. Sci. 21, 364-381 (1967).

Sawa, M.: Auditory responses from single neurons of the medulla oblongata in the goldfish. Bull. Jpn. Soc. Sci. Fish. 42, 141-152 (1976).

Schade, R.: Experimentelle Untersuchungen zur Hörvermogen an *Leucaspius delineatus.* Biol. Zbl. 90, 337-356 (1971).

Schuijf, A.: Directional hearing of cod under approximate free-field conditions. J. Comp. Physiol. 98, 307-332 (1975).

Schuijf, A., Baretta, J. W., Wildschut, J. T.: A field investigation on the discrimination of sound direction in *Labrus berggylta.* Neth. J. Zool. 22, 81-104 (1972).

Schuijf, A., Buwalda, R.: On the mechanism of directional hearing in cod (*Gadus morhua* L.). J. Comp. Physiol. 98, 333-343 (1975).

Schuijf, A., Buwalda, R.: Underwater localization—A major problem in fish acoustics. (1980).

Schuijf, A., Siemelink, M. E.: The ability of cod (*Gadus morhua*) to orient towards a sound source. Experientia 30, 773-774 (1974).

Schuijf, A., Visser, C., Willers, A. F. M., Buwalda, R.: Acoustic localization in an ostariophysan fish. Experientia 33, 1062-1063 (1977).

Siegmund, R., Wolff, D. L.: Experimentelle Untersuchungen zur Bestimmung des Horvermogens der Karausche (*Carassius carassius* L.). Fisch. Forsch. 11, 117-124 (1973).

Tavolga, W. N.: The significance of underwater sounds produced by males of the gobiid fish, *Bathygobius soporator.* Physiol. Zool. 31, 259-271 (1958).

Tavolga, W. N.: Masked auditory thresholds in teleost fishes. In: Marine Bioacoustics, Vol. 2. Tavolga, W. N. (ed.). Oxford: Pergamon Press, 1967, pp. 233-245.

Tavolga, W. N.: Signal/noise ratio and the critical band in fishes. J. Acoust. Soc. Am. 55, 1323-1333 (1974).

Tavolga, W. N., Wodinsky, J.: Auditory capacities in fishes. Pure tone thresholds in nine species of marine teleosts. Bull. Am. Mus. Nat. Hist. 126, 177-240 (1963).

von Frisch, K.: Ein Zwergwels der kommt wenn Man Ihm pfeift. Biol. Zbl. 43, 439-446 (1923).

von Frisch, K.: The sense of hearing in fish. Nature 141, 8-11 (1938).

von Frisch, K., Dijkgraaf, S.: Können Fische die Schallrichtung wahrnehmen? Z. Vergl. Physiol. 22, 641-644 (1935).

Weiss, B. A.: Auditory sensitivity in the goldfish. J. Aud. Res. 6, 321-335 (1966).

Weiss, B. A.: Sonic sensitivity in the goldfish (*Carassius auratus*). In: Lateral Line Detectors. Cahn, P. (ed.). Bloomington: Indiana Univ. Press, 1967, pp. 249-264.

Weiss, B. A.: Lateral-line sensitivity in the goldfish (*Carassius auratus*). J. Aud. Res. 9, 71-75 (1969).

Weiss, B. A., Strother, W. F., Hartig, G. M.: Auditory sensitivity in the bullhead catfish (*Ictalurus nebulosus*). Proc. Nat. Acad. Sci. USA 64, 552-556 (1969).

Winn, H. E.: Vocal facilitation and biological significance of toadfish sounds. In: Marine Bio-Acoustics, II. Tavolga, W. N. (ed.). Oxford: Pergamon Press, 1967, pp. 283-303.

Wodinsky, J., Behrend, E. R., Bitterman, M. E.: Avoidance conditioning in two species of fish. Anim. Behav. 10, 75-78 (1962).

Wolff, D. L.: Das Hörvermögen des Flussbarsches, *Perca fluviatilis* L. Biol. Zentr. 86, 449-460 (1967).

Wolff, D. L.: Das Hörvermögen des Kaulbarsches (*Acerina cernua* L.) und des Zanders (*Lucioperca sandra* Cuv. und Val.). Z. Vergl. Physiol. 60, 14-33 (1968).

Discussion

BLAXTER: As a behaviorist, I take issue with your disparaging remarks about spontaneous behavior. If you use, say, avoidance conditioning, you are imposing a regime on the fish which may mean it will respond in a way that may not be true under natural conditions.

HAWKINS: I didn't mean to belittle that kind of study. It is simply that it is not particularly useful to a physiologist. To an ecologist or a behaviorist, I agree. The work on sharks, for instances, has been very useful.

TAVOLGA: I should like to comment on your use of the critical band idea. I have used the term sometimes loosely, and been jumped on justifiably. When you talk about a critical band in fishes, you automatically include all the notions of critical bands as derived from mammalian studies. Since we are not even sure in fishes whether the place or the volley theory is applicable, it might be wiser to avoid the term critical band because of all the connotations it includes.

HAWKINS: You should remember that the term critical bands is also a misnomer when applied to human hearing, since there is nothing really critical about them. The alternative term is "psychophysical tuning curve," and I find that equally objectionable.

TAVOLGA: I don't care for it either.

HAWKINS: I think that "critical band" is quite suitable, because everybody knows what you are talking about. We measure it in exactly the same way that Fletcher defined it, and I see no objection to it.

TAVOLGA: It still carries with it certain assumptions about the physiology of the system. You mentioned that the narrow critical band seems to be present in species that have a broad frequency spectrum. I prefer to think that a narrow critical band is associated with species that have a high degree of frequency discrimination. Indeed, those that have a broad or, effectively, no critical band, may be virtually incapable of frequency discrimination.

HAWKINS: The only two I have compared are the salmon and the cod. I guess that they would also have distinct differences in frequency discrimination. There is a strong need for more data on frequency discrimination in a wider range of species.

FAY: I also think that the term "critical band" is not a good one to use here, since it means so much in mammalian and human psychophysics. The term is applied in very specific ways that have to do with the procedures used. Perhaps a better term would be "masking function" or "filter function," and I would vote for the latter. A psychophysical tuning curve is obtained with a different, specific procedure, and should not be used in general except to identify a particular technique in masking experiments. All masking experiments provide filter functions, and I think we should avoid other implications.

HAWKINS: Fine. You call them "filter functions" and I'll call them "critical bands."

BULLOCK: That's the trouble. The meaning of a word to you when you use it may not be the same to the hearer who hears it. Most people will understand the term based upon their experience with the literature on the subject, and may not be what you want them to understand. You really cannot use this term "critical band" if you are a physiologist looking for unit activity. It is a term peculiar to the behavioral level, and if you transfer it to a physiological experiment, somebody will jump on you— justifiably.

HAWKINS: I disagree. I don't think I should amend my term to take care of other people's confusion.

BULLOCK: You want to communicate. You would expect that other people understand by your words what you understand when you deliver them. Therefore, you have to communicate in the hearer's language, not yours.

CORWIN: In physiological experiments, we often find that there is masking by the water flow that we use to artificially ventilate the animal's gills. In my shark experiments, I always turn off the flow when I take measurements. I wonder if you have any idea if the same sort of hydrodynamic flow could be causing masking in a fish under natural conditions while particle notion is being detected?

HAWKINS: I don't agree with your use of the term "masking," but yes, we have observed similar conditions while detecting unit activity in the lateralis system. If there is a lot of water going over the gills, spontaneous activity is increased, and if you reduce flow, the activity drops.

CORWIN: Do you think that in the natural situation, when the animal is not curarized and immobilized, if its threshold might be elevated when it ventilates its gills?

HAWKINS: I find it difficult to conceive of an experiment which might demonstrate that.

MYRBERG: The one case I know of was the work done by Banner [Bull. Mar. Sci. 22, 251-283 (1972)]. He initially used a respiratory response in the laboratory to determine a threshold masked by ambient noise. A few years later, he went into the field and studied the same species, and found appropriately the same threshold at about the same noise level.

COOMBS: In regard to the anterior projection of the swimbladder, is there any connection between the swimbladder and the inner ear?

HAWKINS: There is no contact between the air-filled forward projections from the swimbladder and the auditory system. In juveniles, these projections run forward, but in older animals they are reflected back along the sides of the bladder. I do not know if these have any acoustical function. In the cod, the swimbladder is close enough to the inner ear to account for its sensitivity.

BUWALDA: Some ablation studies on these extensions have been done, and the results were negative.

BULLOCK: I have a suggestion regarding the wide variation in goldfish hearing sensitivity that you alluded to. Some workers have recently found a puzzling variability in insect neurons. Usually in insects, one expects to find the same neurons in the same position in every specimen of the species, but in laboratory species one finds more variation than expected. The explanation is that this is, in effect, a domesticated animal. It has been bred for many generations in such a way that it was not restricted by natural limits. As a result, there is a greater variation. When the same investigators looked at wild types, they found much less variability.

HAWKINS: It has been suggested that there are differences between Japanese, European, and American goldfish strains.

POPPER: We had the opportunity in Hawaii to work with wild strains of goldfish [J. Acoust. Soc. Am. 52, 1714-1717 (1972)]. These are quite different animals in that the results were much more stable from individual to individual, but thresholds were essentially the same as in pet-store strains, as reported separately by Fay [J. Aud. Res. 9, 112-121 (1969)] and Jacobs and Tavolga [Anim. Behav. 15, 324-335 (1967)].

In regard to the data you included from Weiss's work [J. Aud. Res. 6, 321-335 (1966)], Tavolga [Fish Physiology, Vol. 5, 135-205 (1971)] pointed out that this was really an extinction curve. It was Wodinsky (unpublished) who trained an animal in an avoidance task, then, fearing he might be causing trauma by the electrick shock, he discontinued the shock when the animal failed to respond to the sound. The thresholds rose dramatically.

HAWKINS: Weiss's acoustical methods were also curious. He used a tubular tank operating entirely in the velocity mode to measure response to pressure!

FINE: To return to the comment by Coombs, do you have any ideas about the possible functions of these swimbladder extensions that are found in so many species.

HAWKINS: I think that in the cod it has to do with growth and morphological changes with age. The adult cod has a much heavier head, and I believe that the swimbladder shape is simply an accommodation to the change in position of the center of gravity.

In deep sea cod, there is a distinct connection between the anterior end of the swimbladder and the auditory system, but this may be associated with the accommodation to depth changes rather than with acoustics.

MOK: Most fish are more sensitive to low frequency sounds, but some may be sensitive to higher frequencies that are associated with feeding sounds. I wonder if there is any correlation in high frequency hearing with feeding habit, say, in a comparison of a plankton vs. a coral reef feeder. The audiograms may represent ecological adaptations associated with feeding behavior.

HAWKINS: The acoustic anatomy is so different in different groups of fishes, that many may be limited in frequency range by structure. Furthermore, the habitats are widely different and impose different requirement. I doubt if we can lay down any simple rules for fish.

I had an idea once that pelagic fish like clupeoids that are preyed upon by aerial or surface predators might have been subjected to a selective pressure toward sensitivity to sound pressure rather than particle displacement. But I now think it would be difficult or impossible to generalize.

BUWALDA: In your reference to new areas of research, you spoke to distance perception and angular discrimination. I have done research in both these areas, and I can say that the study of distance perception looks promising, while that of angular discrimination does not.

HAWKINS: Both of these require some rather difficult and tedious experiments.

Chapter 7

Segregation of Directional and Nondirectional Acoustic Information in the Cod

ROBBERT J. A. BUWALDA*

1 Introduction

It is well known that the remarkable versatility of the acoustico-lateralis system is due to the variety of accessory structures that render the basically uniform receptor units, the hair cells, sensitive to different aspects of mechanical energy. For a system in which the hair cells are coupled to an inertial mass, exemplified by the otolith organs of fishes, the physically adequate stimulus is linear acceleration. The output of such a system may, depending on the latter's mechanical phase and frequency characteristics, behave differently from the physically adequate input, however. Therefore, it may be convenient (though incorrect from a physicist's point of view) to consider the parts of the otolith organs generally thought to be involved in reception of sound energy in fishes, i.e., the partes inferiores labyrinthi, acoustic particle velocity or displacement detectors rather than proper acceleration detectors.

If designating the adequate motional variable for the otolithic fish ear is then already somewhat of a problem (though mainly one of definition), the matter is even further complicated by the presence of a swimbladder or comparable structure in many teleosts, transforming the acoustic pressure to "indirect" motion superimposed upon and interfering with the original "direct" incident particle motion. Such interactions have been demonstrated by recording from the fish ear while controlling the ratio of the direct to the indirect stimulus by manipulation of either the swimbladder (Sand and Enger 1973) or the sound field's pressure to motion ratio (Fay and Olsho 1979, Buwalda and van der Steen 1979) or both (Fay and Popper 1974, 1975) (cf. Fay, Chapter 10).

We may resort to a widely used definition stating that the adequate input is that physical stimulus quantity to which a receptor is most sensitive in terms of input energy requirements. A transition from particle motion to pressure as the adequate stimulus depends, then, on the relative efficiencies of the direct and indirect stimu-

*Laboratory of Comparative Physiology, State University Utrecht, 40 Jan van Galenstraat, 3572 LA Utrecht, The Netherlands.

lation modes and on the amplitude ratio of the two variables. The pressure to motion ratio is (apart from manipulations) a function of acoustic conditions such as source distance, etc. (Siler 1969). The theoretical gain provided by the swimbladder's transforming action varies linearly with frequency and inversely with ambient pressure (depth) at frequencies below the swimbladder's natural frequency (cf. Sand and Hawkins 1973), and is of course also dependent on such characteristics as its size and its functional proximity to the labyrinths.

The obvious test of this dual-input concept, i.e., the behavioral or physiological demonstration of one input variable taking over from the other as the adequate stimulus, has so far been tried only in a few species exhibiting "normal" (i.e., adequate, but nonspecialized) coupling between swimbladder and labyrinths. The grunts *Haemulon sciurus* and *H. parrai* have been shown to respond to particle motion below 200 Hz, and to pressure above this frequency (Cahn, Siler, and Wodinsky 1969)[1]; the same holds for *Gadus morhua* (cod) below and above 50 Hz, respectively (Chapman and Hawkins 1973). Electrophysiological studies of the cod sacculus indicate crossover frequencies from appreciably below 122 Hz [since at that frequency the anterior sacculus proved already 10-15 dB more sensitive to pressure than to motion (Buwalda and van der Steen 1979)] up to ca. 100 Hz (Sand and Enger 1973). On the other hand, such demonstrations may well fail for fishes having very efficient or particularly inefficient coupling, since in those cases the pressure-motion crossover frequency will be below or above the fishes' hearing range under normal conditions (cf. Fay and Popper 1975).

The definition of the adequate stimulus figuring in the above is useful for predicting whether and when the scalar pressure input or the vectorial, and hence inherently directional, particle motion input will determine a fish's absolute sensitivity. It provides no starting point, however, for considering any interactions of the two inputs. If we are to assume that the fish ear simply integrates the two superimposed stimuli without any separate processing, then the increase in absolute sensitivity provided by the indirect stimulation mode becoming dominant should go at the cost of a loss of directional sensitivity. The reason, of course, is that the scalar pressure, acting via a single nondirectional transducer, should severely mask the directional information conveyed by the direct input and should even eliminate, by rendering the two labyrinths effectively a single receptor, any (physically and physiologically already highly improbable) mechanism based on binaural differences analogous to human directional hearing.

Van Bergeijk (1964) was the first to realize these implications and to provide a solution for them by proposing two separate sensory input channels for the directional and the nondirectional stimuli: the lateral line system and the swimbladder-labyrinth complex, respectively. His model was based on the electrophysiologically observed directional sensitivity of the lateral line to particle motion close to a source, in the acoustic near field (Harris and van Bergeijk 1962), and it has dominated concept building in fish hearing to such an extent that the possibility of the fish receiving direct motion input via its otolith organs was all but overlooked for a time.

---

[1]*Note added in proof*: Very similar results have recently been obtained by Myrberg and Spires (J. Comp. Physiol. 140A, 135-144, 1980) for the Damselfish *Eupomacentrus dorsopunicans*.

It has become increasingly clear, however, that van Bergeijk's model suffers from logical flaws and that, moreover, some of his basic assumptions are invalidated by new data:

By restricting motional inputs to the lateral line and thus directional hearing to the near field, van Bergeijk failed to appreciate which requirements the hearing system has to meet to fully qualify as the fish's foremost, or indeed only, detector of information on distant stimulus sources. The evidence endorses the validity of this argument: directional hearing, by now firmly established for several teleost species with swimbladder, appears not subject to a fundamental distance limit (cf. Schuijf and Buwalda 1980).

Van Bergeijk evidently assumed that the effectiveness of the directional hearing system is determined by its pressure immunity. It appears, however, that acoustic localization requires both pressure and particle motion input for eliminating the inherent $180°$ ambiguity of mechanisms based on detection of particle oscillation direction, by means of an analysis of the phase of pressure relative to motion (Schuijf 1975, 1976a, 1976b, Schuijf and Buwalda 1975; see Chapter 14). It is difficult to reconcile the cooperation between the two input channels that is needed for this phase analysis with the rather rigorous segregation between the ascending pathways or central nuclei of the VIIIth nerve and those of the lateralis system (cf. Northcutt, Chapter 16). Reception and processing of both inputs within one of either system seems more probable, but is not implicit in van Bergeijk's model.

Though apparently proposed by van Bergeijk as an input channel receiving directional information free from contamination by the nondirectional pressure input, the lateral line might very well respond to the steep motion gradients set up by the vibrating swimbladder, which functions, after all, as a secondary source with its own near field. That this effect proves rather minimal (see Sand, Chapter 23) appears to be more a consequence of the relatively low sensitivity of the lateral line than of the argument being faulty. Sand's data (Chapter 23) now minimize the probability of the lateral line being an important sound detector at all. Its involvement in directional hearing was already doubtful since it has been demonstrated that a cod with intact lateral line system but with unilaterally severed saccular and lagenar nerves is no longer capable of acoustic orientation (Schuijf 1975).

Rejecting van Bergeijk's model on these grounds means that some circumstantial evidence for separate perception of particle motion and pressure can no longer be explained by invoking the lateral line as an independent input channel. Consequently, the double threshold phenomenon, first observed by Tavolga and Wodinsky (1963) and characterized by identical experimental conditions resulting in low-frequency signal detection thresholds separated by a rather wider gap than can be accounted for by statistical variation in sensitivity, may point to the existence of a mechanism of segregating the direct and indirect inputs to the ears. Such a mechanism could also explain the fact that at 200 Hz *Haemulon* may respond to either pressure or motion, irrespective of the acoustic conditions governing the pressure-motion crossover frequency but apparently depending on training conditions (Cahn et al. 1969). On the other hand, Cahn's later results are less readily compatible with input segregation. Still taking van Bergeijk's model for granted, Cahn, Siler, and Auwarter (1970) undertook to investigate a possible coupling between the lateral line and the ear by masking (successfully) a particle motion signal by a pressure noise, and a pressure signal by particle motion noise (unsuccessfully). These results were explained by the authors in terms of one-way coupling between the two sensory systems. However, since in a subsequent report (Cahn, Siler, and Fujiya 1973) the lateral line was shown to be insensitive to the particle motion stimulus employed in previous studies, the ear and its direct and indirect inputs must have been the receptor and stimuli involved. The results of Cahn et al. (1970) are then readily explained by simply assuming that the indirect stimulation mode dominates over the direct mode; the partial success in these so-called cross-modality masking experiments even argues against any segregation of direct and indirect inputs.

On the one hand, then, the issue of segregation of the (directional) direct and the (nondirectional) indirect inputs is far from settled. On the other hand, there is some evidence that directional hearing is largely independent of factors that govern the relative efficiencies of the direct and indirect modes (cf. Schuijf and Buwalda 1980, pp. 51-53). Gaining insight into the mechanisms responsible for segregation of directional and nondirectional information clearly requires that more extensive and pertinent data be available. In the next sections this paper will present the results of some behavioral experiments with the cod, a standard species in studies of (directional) hearing, employing methods that are adapatations of and improvements on the cross-modality masking and standing-wave manipulation techniques pioneered by Cahn et al. (1969, 1970).

2 Materials and Methods

2.1 The Acoustic Setup

2.1.1 Sound Field Synthesis: The Superposition Principle

The idea of employing multiple sound sources for manipulating certain spatial properties of a sound field is well known among acousticians (cf. Urick 1967). Several students of fish hearing have resorted to this method in order to circumvent to some extent the problem of complex and uncontrollable acoustic conditions in small tanks (cf. Parvulescu 1964). Their solutions are mostly variants of the "standing wave tube" (Weiss 1967, Cahn et al. 1969, Fay and Popper 1974, Hawkins and MacLennan 1976). In such a device two opposite sound projectors, usually mounted at each end of a more or less rigid-walled tube, are driven with a specific amplitude and phase relationship with respect to each other, producing ideally a well-defined sound field. Under some assumptions (complete rigidity of the tube's walls, for instance) the sound field resulting from a given output of the projectors can be calculated (cf. Hawkins and MacLennan 1976). The underlying notion is that the sound fields produced by the separate projectors are completely additive. In a standing wave tube this *superposition principle* allows a good control, especially midway between the sound projectors, of either the amplitude ratio of the acoustic pressure to the particle motion, or the phase between these variables, along one axis only.

Schuijf and Buwalda (1975) were apparently the first to use the superposition principle in controlling, simultaneously, the amplitude and phase of the adequate input variables to the fish ear. Using four underwater sound projectors, they succeeded in inverting locally the polarity of the pressure in a horizontal traveling wave by means of superposition of a vertical standing wave with a high pressure/motion ratio. More important than this technical result is their finding that such a hybrid sound field, or even a fully synthesized sound field consisting solely of standing waves, elicits directional responses from fishes that are indistinguishable from those exhibited in an ideal traveling sound wave. In fact, the standing wave and the traveling wave situations are not even discriminated by cod, if they are sufficiently similar as to direction, phase, and amplitude of pressure and motion at the place of the fish (Buwalda, Schuijf, and Hawkins in press).

Such evidence warrants the use of sound field synthesis based on the superposition principle in behavioral studies of fish hearing, even if these be concerned with the aspects of fish hearing ostensibly most affected by the spatial characteristics of a sound field, i.e., directional hearing.

2.1.2 The Experimental Tank

The acoustic setup to be described next has been designed to combine the inherent benefits of the method of sound field synthesis, allowing on the one hand extensive control of the acoustic variables and permitting on the other hand the use of a relatively small volume of water. Still, the quality of the sound field depends on the dimensions and the symmetry of the system. A compromise between the ideal configuration (a large, spherical volume of water enclosed by a wall of uniform reflection characteristics) and practical operatability, our cylindrical polyethylene tank (Fig. 7-1) has a radius of 0.8 m and a height of 2 m, and is open at the top. The six identical, locally built underwater sound projectors, carefully positioned equidistant (0.5 m) from the tank center, have been conceived as dipole sources. They have a usable frequency range from well below 10 Hz to about 200 Hz, at which frequency the first major flexural vibrating mode of the diaphragm results in output distortion. The usefulness of the projectors above 200 Hz is therefore limited to selected frequencies.

Sound field synthesis is performed in this setup by superposition of standing waves produced by mutually perpendicular pairs of opposingly orientated sound projectors. By driving the two units of a pair in the "push-push" or in the "push-pull" mode, the pressure or the particle motion directed along the projectors' axes/hr, respectively, is maximized (or, vice versa, minimized). In this way a pair controls the amplitude and phase of a single sound variable: horizontal particle motion or pressure for the horizontal sets, vertical particle motion or pressure for the vertical projectors. The three mutually perpendicular motion components are controlled by three independent systems, permitting the production of any specified motion direction, or even of any elliptical motion pattern. If, in addition, sound pressure control is required, then the motion patterns are confined to the plane defined by the two projector pairs remaining for motion control.

The sound variables can be measured by an assembly of a pressure hydrophone (Bruel and Kjaer type 8103) and three velocity hydrophones (modified after Schuijf

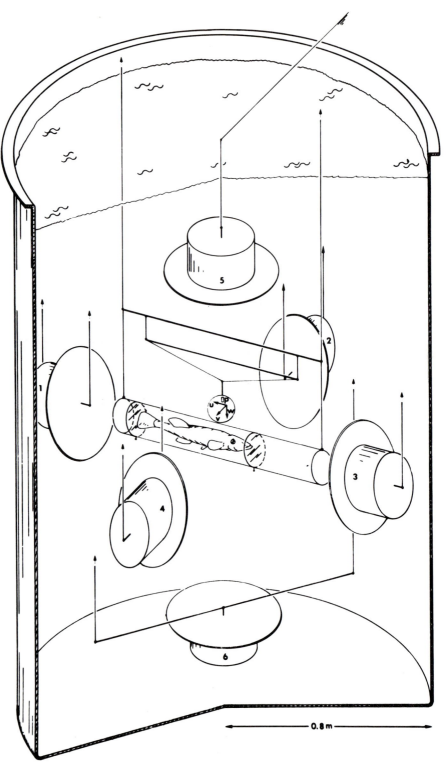

0.8 m

1974), mutually perpendicularly orientated and housed together in a perspex sphere of 9 cm diameter. The hydrophone assembly has a fixed orientation with respect to the tank and thus defines the tank's right-handed orthogonal coordinate system x, y, z with the corresponding horizontal and vertical components u, v, and w of the particle velocity $\vec{v}$. (N.B. The v in "p/v ratio" may denote either the magnitude of the y-component or, incorrectly, but conveniently, that of $\vec{v}$ itself). The assembly can be lowered or raised to bring the cage with the fish (always aligned along the x-axis) instead of the hydrophones at the tank center, the coordinate system's origin.

2.1.3 Instrumentation

The layout of the stimulus generating circuits (schematized in Fig. 7-2, middle part) is characterized by a hierarchy of phase and amplitude controls balancing the inputs to the sound projectors.

The output of an oscillator is fed through a transient-free main gate (pulsing the signal with a 50% duty cycle, rise and fall times of ca. 150 msec and a period of ca. 2 sec) and a 3 dB-step attenuator (constituting together the "gated oscillator") into a control unit ("phase/amplitude balance") with three outputs, each connected to a 1 dB-step attenuator and a phase-shifting analog delay line with polarity control amplifiers. In this way the inputs to each of the three standing-wave generating channels are independently adjustable as to phase and amplitude. In each channel another "phase/amplitude balance" control unit (consisting of an operational amplifier with two parallel outputs, each with vernier control of gain, and one of which is in series with an analog delay line) matches the two sound projectors of a channel to within 1° and 0.1 dB. The matched pair produces a very stable pure-tone standing wave with a high pressure/motion ratio in the tank center, or with an extremely low ratio if the polarity of the input to one of the two projectors is reversed.

In an entirely similar way narrow-band noise can be used in the production of quasi-steady standing waves. In our setup the output of a random-noise generator (one for each channel) is band-pass filtered, routed through a signal splitter, an amplitude balancer, and a polarity control amplifier ("phase/amplitude balance"); and mixed with the (gated) pure tone prior to a final amplification in one half of a stereo power amplifier ("summation/amplification"). Because of the independent polarity controls of the pure tone channel and the noise channel, one pair of sound projectors can be operated in the push-push mode for the pure tone, and at the same time in the push-pull mode for noise, or in any other combination. Logic control circuits, either manually operated or slaved to a KIM micro-processor, select which ones of the channel gates will open to produce more or less complex synthesized sound fields.

◄ Figure 7-1. Simplified perspective view of the experimental tank. Note the six underwater sound projectors, all equidistant from the tank center and arranged in three mutually perpendicular pairs of opposingly oriented units. The two pairs of horizontal sound projectors (1,3 and 2,4) are suspended from a common frame, which can be rotated around the vertical axis of the tank. A system of pulleys and ropes allows a remote control of the vertical position of the cylindrical mesh cage (containing the fish) and the hydrophones assembly [containing in a single housing of 9 cm ϕ a pressure (p) hydrophone and three velocity hydrophones sensitive to the horizontal and vertical particle velocity components u, v, and w].

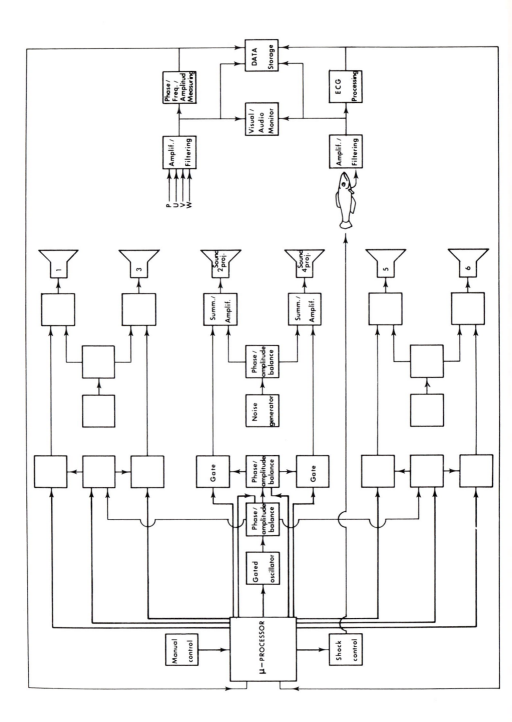

For sound field monitoring and measuring purposes either the acoustic pressure p, the horizontal particle velocity components u or v or the vertical particle velocity component w can be selected via a switchbox. Signal analysis is performed by measurement of amplitude, of frequency or of spectral content, and of phase relative to the oscillator's output. Since the pathway leading to the measuring apparatus is the same for each of the sound variables p, u, v, and w (see Fig. 7-2, top right), the data are directly comparable and the relative phase of, for instance, p to v is readily found (with an accuracy of ca. $2°$).

2.1.4 Some Properties of the System

Because of the difficulties in knowing the boundary conditions existing in the experimental tank, calculating the sound fields, though mathematically feasible, is all but impracticable (Schuijf pers. comm.). Therefore, only some empirical data on the system's behavior will be reported.

The sound field of a single, horizontal projector is characterized by a p/v amplitude ratio of -22 dB re far-field value and a phase difference $\psi_p - \psi_v$ of $93°$ as measured in the tank center, at a frequency of ca. 105 Hz. These results are remarkably close to the theoretical p/v amplitude and phase ratios existing at an acoustical distance $2\pi r/\lambda$ = 0.2 under free field conditions, which amount to ca. -20 dB re far-field value and $90°$, respectively, if a dipole source is assumed (Siler 1969). This good approximation of the theoretical near-field conditions holds throughout the frequency range relevant for the experiments.

The p/v amplitude ratio of a pure-tone sound field produced by a balanced pair of projectors can vary from an extreme -60 dB to $+20$ dB (re far-field value), since it proved entirely possible to reduce pressure or particle motion to below 1% in a standing wave. Slightly less extreme p/v ratios of -50 to $+10$ dB are observed for narrow band noise produced by the horizontal projectors. The situation is less auspicious for the vertical projector pair, because of the inherent vertical acoustical asymmetry of the tank with its free water surface. Even so, the most extreme p/w ratios for narrow-band noise are ca. -45 dB and $+5$ dB in a vertical standing wave.

Owing to the excellent linearity and stability of the entire system, the p/v amplitude ratio of a sound field, once adjusted, should be independent of the intensity level. In practice, however, the high background noise levels (ca. -6 dB re 1 μbar and 0 dB re 1 μvar for pressure and any particle velocity component, respectively, as measured in a frequency band from 80 to 150 Hz) and the projectors' maximum undistorted output (ca. $+42$ dB re 1 μbar or $+64$ dB re 1 μvar = ca. 10^{-2} cm sec^{-1}, at 105 Hz, for a balanced pair of projectors) limit the dynamic range of control of a sound field to ca. 10 dB, 50 dB, and 20 dB at p/v ratios of -60 dB, -20 dB, and $+20$ dB, respectively.

◄ Figure 7-2. Highly simplified schematic diagram of the employed electronic circuitry. Note the process control unit at the left-hand side of the figure, the physical and physiological data processing equipment at the right, the sound generating circuitry in the middle. The latter consists of three identical subcircuits, each driving a pair of opposingly oriented sound projectors. The heavy lines connecting the micro-processor to the sound generating circuitry represent logic or analog control lines. See text.

A more serious limitation is due to the fact that, while the maximized sound component remains relatively constant along the standing wave's axis, the pressure or velocity null is highly localized. This phenomenon, always present in a standing wave, is aggravated by the steep gradients resulting from the near-field conditions prevailing in the tank. Even at less than extreme p/v ratios the minimized components will interfere with both phase and amplitude control of complex synthesized sound fields, and this effect increases rapidly with distance away from the tank's center. The phenomenon is illustrated in Table 7-1, in which the off-center phase and amplitude behavior of one particular synthesized sound field (or rather three, differing only in pressure/velocity amplitude ratio) is presented, expressed as the range of deviations from the "Sollwerte" that are measured within a volume of water in the tank center with radius of 2.5, 5, or 7.5 cm. It can be seen that at the more extreme p/v ratios the control of amplitude and phase rapidly deteriorates with distance. It should be noted that the example, though typical, is only one kind of sound field, and that other sound fields will behave differently.

Any attempt, then, to produce sound field conditions differing from the "normal" (i.e., a p/v phase difference of $\pm 90°$ and an amplitude ratio of ca. -20 dB), must be a compromise between factors such as phase or amplitude control, extensiveness of the field, and dynamic range of control.

2.2 Behavioral Procedures and Experimental Control

Behavioral experiments designed to reveal possible differences in masked signal detection or directional discrimination capabilities correlated with differences in signal and noise composition were conducted with 15 juvenile cods (*Gadus morhua* L.) 28-48 cm long, caught in shallow water. Fishes of apparently normal behavior and health were selected for cardiac conditioning experiments and fitted with differential ECG electrodes. These consisted of two 0.1 mm trimel insulated stainless steel wires of unequal length and with the bare tips bent 180°. They were implanted by means of a blunt injection needle which carried the bent tips to the pericard and was then retracted, leaving the electrode in place. The electrode leads were anchored to the skin and to the lower jaw by means of sutures. After recovery from MS 222 anesthesia, the fish was placed in a cylindrical cage and positioned in such a way as to bring its swimbladder and labyrinths close to the tank's acoustical center, and allowed to settle down for 24-48 hr. Depending on its condition and response behavior, a fish was subjected to experimenting for 4-8 days at a time.

The ability of a cod to detect the presentation of a pulsed tone signal in a continuous noise or of a repeated change in direction of a continuously present pulsed tone was studied by means of classically conditioned bradycardia (see Fig. 7-3). The cod was considered to have responded positively in such a task if the maximum heartbeat interval measured during the 10 sec trial period exceeded the maximum interval of the immediately preceding 10 sec pretrial period by more than a specified percentage (usually 15-30%). The latter was chosen such as to keep the probability of the fish exceeding this floating criterion during blank trials below 5%, and it was kept constant for a given animal. The stimulus to be detected or discriminated was presented three times as a 1 sec pulse during a trial. This was followed immediately by a 0.2 sec 50 Hz

Table 7-1. Typical Amplitude Variation and Phase Variation as a Function of the Measuring Position in a Sound Field Produced by Means of a Superposition of Standing Waves[a]

Conditions of measurements			Measured data						
$p/u,v$ in dB re far-field value	Displacement from center of tank (cm)	Along x-, y-, or z-axis	Amplitude variation (dB)			Phase variation (degrees of arc)			w maximum (dB re u,v)
			p	u	v	p	u	v	
−40	±2.5	x	1.7	.1	.1	58	1	0	−22.9
		y	1.5	.0	.0	64	0	1	−22.4
		z	.2	.5	.0	5	1	0	−22.8
	±5	x	4.3	.7	.2	99	1	0	−22.9
		y	4.3	.0	.3	102	0	2	−24.4
		z	.5	.5	.1	17	1	0	−22.5
	±7.5	x	6.8	.8	.5	121	2	0	−22.9
		y	7.1	.2	.8	124	0	3	−24.4
		z	.7	.7	.3	27	1	1	−21.0
−20	±2.5	x	.2	.2	.0	6	4	0	−22.2
		y	.2	.1	.1	8	0	7	−23.9
		z	.0	.4	.0	1	1	0	−15.6
	±5	x	.2	.6	.2	14	7	0	−22.2
		y	.2	.1	1.1	16	0	17	−23.9
		z	.2	.5	.2	2	1	1	−10.7
	±7.5	x	.2	1.0	.6	22	11	0	−21.9
		y	.2	.2	1.3	24	1	31	−23.6
		z	.5	.8	.4	2	1	1	− 7.0
0	±2.5	x	.2	3.8	.7	1	60	2	−16.5
		y	.2	.4	2.4	1	7	60	−18.8
		z	.0	.1	.4	1	3	3	+ 2.9
	±5	x	.2	6.5	1.4	1	100	7	−13.0
		y	.2	.4	5.8	1	7	105	−15.3
		z	.2	.4	.5	1	5	5	+ 8.0
	±7.5	x	.2	8.5	2.0	2	121	12	− 9.1
		y	.2	.4	8.4	2	10	124	−10.9
		z	.5	1.0	.7	1	10	8	+12.6

[a] The u- and the v-component in this sound field were controlled by the two horizontal sound projector pairs, the p-component was controlled by the vertical pair. The "Sollwerte" of this particular sound field (f = 105 Hz) were for p: 5 dB re 1 μbar or 25 dB re 1 μbar; for u and v: 25 dB re 1 μvar or 45 dB re 1 μvar; for the phase difference $\psi_p - \psi_{u,v}$: 180°; for w: a minimum value. Combination of the different values for p and u,v resulted in three different pressure-to-velocity ratios (−40 dB, −20 dB, 0 dB re far field value).

The measured data indicate for the three sound components (p, u, and v) the range of phase and amplitude values encountered when the hydrophones assembly was moved 2.5, 5, or 7.5 cm either way from the acoustic center of the tank along the three mutually perpendicular x-, y- and z-axes, at three different $p/u,v$ ratios. The data for any given sound variable are arranged in a column, the data obtained in any one of the 3 x 3 x 3 = 27 possible measuring conditions are arranged in a horizontal row. The right-hand columns shows the measured maximum values of w relative to u,v. See text.

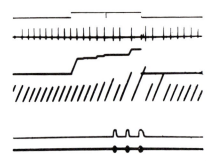

Figure 7-3. Oscillogram of a signal detection trial with cod. The traces represent (from top to bottom): pretrial- and trial-period indicator (ca. 10 sec each; on = high); the ECG of the cod; the output of the ECG peak interval reader; the output of the ECG interval reader; the output of the phase meter (only during sound presentation); the measured signal pulses (i.e., the horizontal particle velocity u at a high S/N ratio). Note the shock artifact at the end of the trial period in the ECG, and the clear brady-cardia in response to signal presentation. Compare trace 3 and 4 and note, that since sensitivity of peak reader is reduced automatically by, e.g., 20% during trial, only trial intervals exceeding pretrial peaks by more than 20% will be detected.

a.c. shock delivered through two silver strip electrodes flanking the cod's tail, at a peak current density of ca. 5-10 mA. cm^{-2}.

The limits of masked signal detection and directional discrimination under various conditions of signal and noise p/v composition were tested by means of the staircase method of threshold approximation. Starting with a high S/N ratio ($\geqslant$ 20 dB) or a maximal (90°) directional change in order to establish consistent conditioned brady-cardia, the experiment proceeded by decreasing and increasing (in steps of 3 dB or 12.8°) the signal level or the magnitude of the directional change for the next trial (spaced 2-6 min) after a positive and negative response, respectively, until a minimum of about 10 response reversals had occurred, or until one level or angular difference had been presented 6-8 times (starting=training conditions excepted). The statistical method described by Dixon and Massey (1957) was used to calculate the resulting threshold. The reproducibility of thresholds thus obtained was, for a given situation, usually better than 5 dB or 8 dB difference for an individual fish or different animals, respectively.

The 105 Hz signal frequency used exclusively in this study fits well into the range used in previous studies of directional hearing in cod, and is of interest as it might approximately coincide with a transition from particle motion sensitivity to pressure sensitivity in this species. Masking noise was limited in bandwidth to 35 Hz, bracketing the signal frequency, and well within the corresponding critical bandwidth of masking as determined by Hawkins and Chapman (1975) for the cod's hearing system as a whole. (For measurement of the effective masking noise level, however, the filter in the receiving circuit was tuned to match this critical bandwidth.)

The experimental control procedures, decisions and criteria described above were compatible with manual control as well as with micro-processor-supervised automatic operation. In the latter mode, up to eight threshold determinations, each with its own programmed conditions, could be executed consecutively in one automated session of 36-72 hr, including rests.

3 Experiments and Results

3.1 Masked Signal Detection with Varying Signal and Noise Composition

3.1.1 The Cod's Hearing System: Responding to Pressure or to Particle Motion?

The first experiment, aiming at settling the basic question of pressure or motion sensitivity of the cod at 105 Hz, was designed on the principle that only the adequate stimulus components of a complex signal or masker can affect the corresponding masked threshold of detection.

By increasing and decreasing by 15 dB the pressure and the particle velocity (component u), respectively, or vice versa, relative to a situation in which the direct and indirect stimulation modes should be about equally effective [i.e., at a p/u ratio of ca. -10 to -15 dB (Buwalda and van der Steen 1979)], a total of four different conditions were created for observing masked detection threshold shifts. These conditions, characterized by low or high p_s/u_s (i.e., p/u ratio in the signal) and/or p_n/u_n (i.e., noise p/u ratio), were: p_s/u_s =ca. -30 dB (low) and p_n/u_n =ca. 0 dB (high); low p_s/u_s and low p_n/u_n; high p_s/u_s and high p_n/u_n; high p_s/u_s and low p_n/u_n. The exact amplitude values for the sound variables are given in Table 7-2. The phase difference between p_s and u_s was here (as in the other experiments, unless stated otherwise) $0°$; p_n and u_n, being produced by independent noise generators, were uncorrelated.

Looking at the results (nine thresholds obtained from one cod) shown in the 2×2 matrix of Table 7-2, one notices that in the second column (at a p_n/u_n of -30 dB) a ca. 15 dB increase in p_s results in a virtually equal increase in signal attenuation at threshold, in spite of a decrease in u_s. This indicates a perfect responding to p, as might be expected for the input to the indirect channel prevailing over the direct input as to magnitude. However, at the p_n/u_n of 0 dB (1st column) the same increase and decrease in p_s and u_s, respectively, result in a positive threshold shift of about the same magnitude, indicating control of sensitivity by u! Similarly, an increase in u_n shows the hearing system responding to p at a high p_s/u_s, but to u at a low p_s/u_s. Moreover, a transition to the threshold condition characterized by a high p/u for both signal and noise from a condition in the same column follows the rules of u-control, whereas the "horizontal" transition (from the "low p_n/u_n; high p_s/u_s" condition) indicates p-control. Only in transitions to the conditions "low p_s/u_s; high p_n/u_n" and "high p_s/u_s; low p_n/u_n" does the hearing system behave consistently as a u- or a p-detector, respectively. Since these two conditions are characterized by a relatively high signal/noise ratio for u and p, respectively, masked signal detection appears to be ruled by the component with the highest S/N ratio.

3.1.2 Perceptual Segregation of Pressure and Motion?

Basically similar to the previous design, this experiment tests a possible cross-modality masking for three sound variables: the acoustic pressure p, the horizontal particle velocity component u along the fish body, and the vertical velocity component w. The

Table 7-2. Masked Detection Threshold Data Obtained in Four Different Acoustic Conditions: High or Low p/u Ratio in the Signal (105 Hz pure tone), and High or Low p/u in the Masking Noise Band ($90 < f < 125$ Hz)[a]

Noise band level							
p_n (dB re 1 μbar)				14.9		0.3	
u_n (dB re 1 μvar)				15.2		29.7	
Signal start level							
p_s (dB re 1 μbar)	u_s (dB re 1 μvar)	p/u (dB re far-field ratio)		ca. 0		ca. -30	
23.9	23.2	ca. 0	8.4	1.7	$\overleftrightarrow{p}$	20.7	1.6
			Mean and *range* of signal $\downarrow u \uparrow$ attenuation (dB) at $\downarrow p \uparrow$ detection thresholds				
9.0	38.2	ca. -30	21.2	0.4	$\overleftrightarrow{u}$	5.9	1.0

[a]The differences in the attainable levels of attenuation (mean of two or more threshold determinations for each condition) shown in the 2 x 2 matrix indicate that signal detection is controlled by either p or u (as indicated by the characters near the arrows symbolizing transition from one condition to another), depending on the p/u ratio in signal and noise. See text.

design is illustrated in Fig. 7-4 (see also legend): masked signal detection was studied for three different compositions of the signal (three columns of Fig. 7-4) and of the noise masker (three horizontal rows of Fig. 7-4), each characterized by a 6-10 dB increase of one component, relative to a reference condition in which the ratio of p to u and to w was ca. -20 dB. Accordingly, nine different conditions (plus the reference condition) were tested.

One way of analyzing the results (21 thresholds obtained from three cods) is by measuring the threshold shift with respect to the reference condition. It then appears that the data can be pooled into two groups. One is characterized by very small threshold shifts (mean -1.8 dB) and contains the thresholds obtained in the conditions "p_s; p_n", "u_s; u_n", and "w_s; w_n", i.e., the conditions in which the same component in signal and noise was increased. The other group is characterized by relatively large negative threshold shifts (mean -7.1 dB) of about the same absolute magnitude as the average increase in the signal components; this group contains all thresholds obtained in conditions in which the increased signal and noise components differed from each other. The implication is clear: the increase in strength (and hence in detectability) of a signal can only be cancelled by increasing the *same* component in the noise masker.

Another type of analysis follows from Fig. 7-4, and is based on the notion that the critical ratio (i.e., the ratio of signal and masker levels at detection threshold) should be about 0 dB if the masker energy is contained within a bandwidth smaller than the critical bandwidth of masking. The graphical representation of the threshold determination procedures and results in Fig. 7-4 shows that the threshold is not reached until the "last" (i.e., relatively strongest) signal component has about reached its corresponding noise level.

Both analyses endorse the conclusion drawn in Section 3.1.1: signal detection is ruled by the component with the highest S/N ratio. Moreover, they indicate that there is little or no cross-modality masking between pressure and motion, nor reciprocal (or even one-way) influencing of the horizontal and vertical particle velocity components, in the conditions used.

3.1.3 Masked Signal Detection at Extreme p/v Ratios in Signal and Noise

Driving the equipment at the limits fixed by the background noise levels on the one hand, and overstimulation levels for the fish on the other hand, this experiment tests possible cross-modality masking effects at ratios of direct to indirect stimulation appreciably more extreme than the ca. ± 15 dB range employed in Sections 3.1.1 and 3.1.2, and should thus reveal more clearly any remaining cross-talk between the inputs. Three extreme signal compositions were produced, in which either p, u, or w was maximized, while the remaining two variables were minimized, resulting in p/u and p/w ratios of ca. +20 dB and +15 dB; -60 and -40 dB; -40 and -60 dB, respectively, for the unattenuated signal. Masking noise p/u and p/w ratios varied from ca. +12 dB to -45 dB and from ca. +5 dB to -30 dB, respectively.

Table 7-3 displays 12 thresholds (one cod) obtained in such extreme conditions, expressed as critical ratios of the maximized signal component to each of the three masker components p_n, u_n, and w_n. The corresponding noise compositions can be calculated by comparing the three critical ratios, which describe, after all, the threshold conditions for each single threshold determination. As the signal composition was always more extreme than the noise composition, the minimized signal components were all far below the corresponding noise levels at threshold, and can therefore be omitted in this analysis.

Critical ratios of around 0 dB appear systematically only in the columns labeled "p_s/p_n", "u_s/u_n", and "w_s/w_n." Again, it can be concluded that the component with the highest S/N ratio governs masked signal detection. This principle appears to hold even when the noise levels of the other components are very much higher than the signal component governing detection: this is evidenced by the critical ratios of down to ca. −40 dB, if expressed in μbars and/or μvars (see Table 7-3), or down to ca. −25 dB if expressed in terms of direct and/or indirect input strength (assuming a 15 dB gain of the indirect stimulation mode; see also Section 3.1.1).

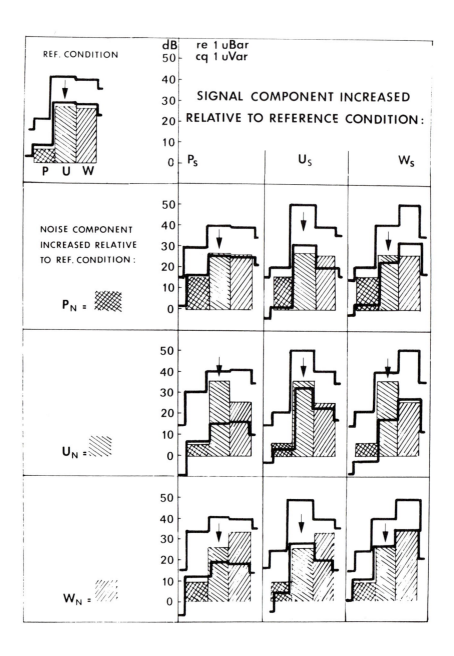

3.2 Directional Discrimination

It has been shown in Section 3.1 that the cod is capable of detecting pressure in an excess of velocity noise and vice versa, which indicates that there is, for signal detection purposes at least, perceptual segregation of these sound components. However, it remains entirely possible that the nondirectional indirect input interferes (negatively) with the extraction of the directional information conveyed by the direct inputs. The experiments described next have been designed to reveal such interactions, if any, by measuring angular resolution of the cod's hearing system as a function of pressure/motion ratio.

3.2.1 Angular Resolution as a Function of
Signal Composition

Directional discrimination thresholds were determined in two types of sound field, differing as to p/v ratio as well as in the mode of production. In both cases the direction of particle motion (simulating the direction of sound incidence) was determined by the vector sum of the two perpendicular velocity components u and v. Starting from a situation in which these variables should be equal as to phase and amplitude, a change of direction without any corresponding change in amplitude of this resultant vector can be accomplished by multiplying u and v by cos Φ and sin Φ, respectively, where Φ denotes the desired angle of incidence relative to the fish (in the x,y plane).

   In the first type of sound field, u and v were produced by two mutually perpendicularly oriented horizontal sound projectors (No. 3 and No. 2, respectively, see Fig. 7-1). In the sector $0° \leqslant \Phi \leqslant 90°$ to which testing of directional discrimination was restricted in these experiments, the p/v ratio is then about normal, i.e., ca. -20 dB. In such conditions the indirect stimulus is already strong enough to distort the polar plot

◀ Figure 7-4. Masked signal detection threshold data obtained in ten systematically varied conditions of pressure/velocity ratio in the signal (105 Hz pure tone) and in the masking noise band ($90 < f < 125$ Hz). Nine conditions are characterized by a ca. 6-10 dB increase in one component of the signal (either pressure p_s, horizontal particle velocity u_s, or vertical velocity w_s) and in one masking noise component (either p_n, u_n or w_n) relative to a reference condition (top left), in which the p/u and the p/w ratios were ca. -20 dB for both signal and noise. In each of the ten figures the heights of differently shaded columns represent the respective levels of (from left to right) the noise components p_n, u_n, and w_n. The upper heavy line contour represents the levels of the three signal components p_s, u_s, and w_s at the start of threshold determination, whereas the lower contour indicates the situation at signal detection threshold, after simultaneous, stepwise attenuation (downward arrow) of the signal components. Note that in the reference condition, in the $p_s;p_n$ condition, in the $u_s;u_n$ condition, and in the $w_s;w_n$ condition the lower signal contour virtually coincides with the noise contours, indicating a critical ratio at threshold of ca. 0 dB for each component. In each of the other six conditions a critical ratio of around 0 dB is found for one component only: the one with the highest S/N ratio. The remaining critical ratios are all negative: the involved components apparently did not contribute to signal detection at threshold. See text.

Table 7-3. Masked Signal Detection Threshold Data Obtained in Various Extreme Pressure/Velocity Ratio Conditions for Both Signal and Masker are Here Arranged to Display the Ratio of the Prevailing (maximized) Signal Component [either pressure (p_s), horizontal (u_s) or vertical (w_s) particle velocity] to Each of the Components p_n, u_n, and w_n Present in the Masking Noise[a]

Maximized signal Component	Ratio of maximized signal component p_s, u_s, or w_s to each of the three noise components p_n, u_n, and w_n at masked signal detection threshold (expressed in dB re 1 μbar or re 1 μvar)		
p_s	p_s/p_n	p_s/u_n	p_s/w_n
	− 0.9 dB	+11.5 dB	+ 4.3 dB
	+ 1.2	−38.9	−17.3
	− 4.5	−33.6	−12.5
	− 3.0	−14.6	−32.7
Range	5.7 dB	50.4 dB	37 dB
Average	− 1.8 dB	−18.9 dB	−14.5 dB
u_s	u_s/p_n	u_s/u_n	u_s/w_n
	− 7.7 dB	+ 4.0 dB	− 4.8 dB
	−10.4	− 4.7	− 7.4
	+45.9	+ 6.3	+27.7
Range	56.3 dB	11 dB	35.1 dB
Average	+ 9.3 dB	+ 0.9 dB	+ 5.2 dB
w_s	w_s/p_n	w_s/u_n	w_s/w_n
	− 2.0 dB	+ 9.7 dB	+ 1.0 dB
	− 6.9	− 1.2	− 3.9
	+ 4.2	− −	+ 1.8
	+ 9.1	0	0
	+15.6	−29.0	− 4.9
Range	22.5 dB	38.7 dB	6.7 dB
Average	+ 4 dB	− 5.1 dB	− 1.2 dB

[a] It can be seen that the variation in critical ratios thus arranged is small only for the columns labeled p_s/p_n, u_s/u_n, and w_s/w_n, indicating that these critical ratios govern signal detection. Moreover, they average close to zero. As the 35 Hz bandwidth of the noise masker was smaller than the critical bandwidth around the signal frequency (105 Hz), the critical noise should in fact be around 0 dB for the effective masking component.

Note also the large negative critical ratios appearing in most rows (a row representing a single threshold determination) indicating that, before threshold was reached, detection of the signal must have been occurring at a level about 10-40 dB *lower* than the prevailing noise component. A given acoustic variable can apparently be masked only by the corresponding component in the noise. See text.

of saccular directional sensitivity (Buwalda and van der Steen 1979). [N.B. Since u and v add vectorially, as opposed to the scalar p-contributions of No. 3 and No. 2, the pressure component varies slightly: 3 dB at most, for a change from $\Phi = 0°$ to $\Phi = 45°$. This extra contribution to the detectability of a directional change is negligible, however, since it is below the amplitude discrimination limen as determined by Chapman and Johnstone (1974) and later confirmed by us for synthesized sound fields.] The second type of sound field, on the other hand, was produced by superposition of horizontal standing waves, resulting in p/v and p/u ratios well below -50 dB. Only background noise levels were present during the experiments.

For each condition, five thresholds were obtained (totaling 10 thresholds from one cod). The frontal angular resolution in the horizontal plane, expressed as the just noticeable change $\Delta\Phi$ in particle motion direction (with starting point $0°$), proved virtually equal in the two conditions.

for $p/u,v < -50$ dB: $\Delta\Phi = 16.3°$ (range 7.9° to 30.7°)
for $p/u,v = -20$ dB: $\Delta\Phi = 16.0°$ (range 12.6° to 19.1°)

Directional discrimination then appears to be independent of pressure for at least a moderate range of p/v ratios in the directional stimulus. Acoustic localization (i.e., determining the source direction), on the other hand, requires the presence of sound pressure in a particular phase relationship with respect to the motion (Schuijf 1975, 1976a, 1976b, Chapter 14, Schuijf and Buwalda 1975). It has been found in a similar acoustic setup that sound fields of the first type here employed elicit normal acoustic orientation responses in a cod for the sector $0° \leqslant \Phi \leqslant 90°$, but erratic behavior for the sector $0° > \Phi > -90°$ where the p-contributions add negatively, resulting in a pressure null for $\Phi = -45°$ (Buwalda and Ehrich pers. comm.).

3.2.2 Masked Directional Discrimination

Optimal angular resolution as described in the preceding section is only possible at high stimulus levels (Chapman and Johnstone 1974, Hawkins and Sand 1977), or rather, at high S/N ratios (Schuijf and Buwalda 1980). In order to determine whether the noise motion component or the noise pressure component is most effective in deteriorating angular resolution, the just noticeable change in particle motion direction was studied at various S/N ratios and at various p/v ratios in the masking noise.

Again, the directional stimulus was produced by the two horizontal standing waves, at a $p/u,v$ ratio well below -50 dB. The velocity components of the noise masker were produced by the two horizontal systems as well. In order to prevent any directional masking effect (see Section 3.3; not to be confused with masking of directional discrimination, the effect studied in the present section), the inputs for these systems were generated by independent noise generators, so that the u_n and v_n components were uncorrelated, resulting (via the vector addition principle) in random directions simulating all-sided random noise incidence. Uncorrelated pressure noise p_n was added by the vertical sound projector pair to attain $p_n/u_n,v_n$ ratios of -40 dB to +5 dB.

Eighteen discrimination thresholds from two cods were obtained, which were plotted both as a function of the ratio of the signal velocity component to the noise

velocity component (Fig. 7-5, top) and as a function of the signal velocity component
to the noise pressure component (Fig. 7-5, bottom). A comparison of both plots will
make it immediately clear, that directional discrimination is not masked by the pressure noise, whereas angular resolution is increasing with the velocity S/N ratio from a
poor 50° or worse at an S/N ratio of ca. +10 dB, to a best performance of ca. 13° at
S/N ratios exceeding +30 dB, with most of the improvement occurring in the S/N
range from +10 dB to +20 dB.

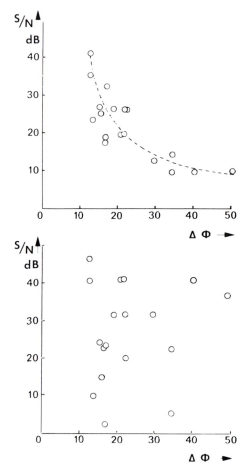

Figure 7-5. Angular discrimination threshold data plotted as a function of the ratio
(in dB) of the prevailing signal component (horizontal particle velocity) to the noise
horizontal particle velocity component (top), and replotted as a function of the ratio
of the prevailing signal component (in dB re 1 μvar) to the noise pressure component
(in dB re 1 μbar) (bottom). The pressure/velocity ratio in the signal was minimal (ca.
−60 dB) but was varied between ca. −40 and +5 dB (re far field value) in the noise on
successive discrimination threshold determinations. Note that only in the upper plot a
relation is apparent between the just noticeable angular difference $\Delta\Phi$ and the ordinate variable. This indicates that directional acuity depends on the velocity S/N ratio.
See text.

3.3 Directional Sensitivity and Directional Masking

Directional effects in masking and unmasking, well known from human directional hearing (the "cocktail party effect"), have first been observed in fish by Chapman (1973), Chapman and Johnstone (1974), and Hawkins and Sand (1977). These investigators of cod hearing noted that angular separations between signal and noise sources in excess of about $10°$ to $20°$ of an arc result in a negative threshold shift (i.e., unmasking) of ca. 6-10 dB. Directional unmasking is evident as a special case in the present study (Sections 3.1.2 and 3.1.3), where it was found that the two mutually perpendicular velocity components u and w do not mask each other at S/N ratios down to at least -10 dB, or even -30 dB.

   Such effects may also have been responsible for the lack of cross-modality masking between pressure and motion evidenced by all of the hitherto presented results of this study. Pressure, though in itself a nondirectional scalar quantity, is transformed by the swimbladder into ("indirect") motion which may well, according to the coupling mechanism involved, arrive at the labyrinths with a definite direction of incidence differing sufficiently from the longitudinal or vertical axis of the fish to show considerable unmasking with respect to u and w. If such be the case, then it should be possible to find a direction of incidence of the direct stimulus for which a cross-modality masking effect can be demonstrated.

   In order to investigate this potential mechanism of segregating direct and indirect stimuli, the directional sensitivity of cod, defined as the masked threshold of detecting a pure particle motion signal as a function of particle motion direction, was tested in a number of conditions. These differed as to the composition of the masking noise or, as can be seen in Fig. 7-6, by having a characteristic angular distribution of velocity noise. This distribution was not determined by rotating a velocity hydrophone (this obvious method being impracticable owing to the fixed orientation of the hydrophones' assembly) but by calculating the noise leve for each direction Φ from the measured perpendicular, noncorrelated velocity noise components u_n and v_n (expressed in μvars) by means of the formula

$$\text{Noise level } (u_n, v_n, \Phi) \text{ in dB re 1 } \mu\text{var} = 10 \log[(u_n \cos \Phi)^2 + (v_n \sin \Phi)^2] \qquad (1)$$

The two procedures are equivalent, as has been checked by rotating the sound projector carrying frame (and by that, the synthesized sound field) with respect to the hydrophones and comparing the measured data with the calculated values.

   In considering the data of Fig. 7-6, it should be kept in mind that the ordinate variable (degrees azimuth or elevation) does not denote "direction" or "direction of incidence" (as such terms are meaningless in standing-wave sound fields as employed here, except when relating to the instantaneous particle motion), but rather one end of an alternating orientation axis. For the sake of clarity of arrangement, however, we have refrained from indications like $30°/210°$ or $30°/-150°$ for the "direction," a term that we will continue to use for convenience. Therefore, the representation of results in a sector from $+90°$ to $-90°$ actually describes the situation completely.

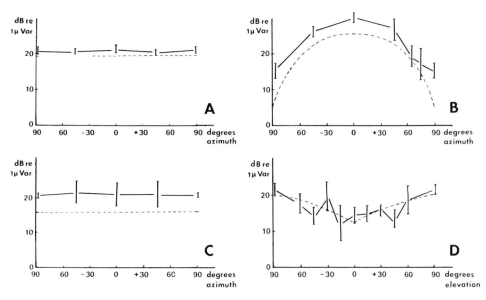

Figure 7-6. Masked signal detection threshold data as a function of signal "direction" and noise "direction" at low (A, B) and at high (C, D) pressure/velocity ratios in the masking noise (but always at low signal p/v). The vertical bars indicate the range of signal detection thresholds obtained at a given signal "direction"; the solid lines connect the respective means. The dashed lines represent the output level of a (hypothetical, rotatable: see text) velocity-noise measuring hydrophone as a function of its orientation and demonstrate thus the directional characteristics of the various sound fields employed for producing the masking noise. Note that the thresholds follow these angular distribution contours of the kinetic noise components quite closely, irrespective of the pressure/velocity ratio in the noise. See text.

3.3.1 Overall Directional Sensitivity of the Cod's Hearing System

First, the directional sensitivity of cod was studied for simulated all-sided incidence (see Section 3.2.2) of a masking noise characterized by a very low pressure/velocity ratio. As directional masking effects should then be absent (or rather remain constant) for different "directions" of the signal, this experiment provides a reference for subsequent experiments in which such effects may be expected to appear. Moreover, this experiment is interesting in itself since it might reveal the existence of one or more sensitivity axes of the hearing system such as have been demonstrated electrophysiologically for single (unilateral) parts of the labyrinth (Enger, Hawkins, Sand, and Chapman 1973, Sand 1974, Fay and Olsho 1979, Buwalda and van der Steen 1979, Hawkins and Horner, Chapter 15), and might thus lead to conclusions concerning the organization of the (directional) hearing system. The rationale of this experiment is, of course, that while in these conditions the noise input to a directionally sensitive system is constant and even independent of the system's orientation, the effectiveness of the signal should vary according to the signal's "direction" (theoretically decreasing to zero for a "direction" perpendicular to the system's axis), resulting in varying "effective" S/N ratios, with correspondingly varying thresholds. The more acute the directional characteristics of the system, the more pronounced this threshold shift effect (the behavioral analogon of the electrophysiological polar sensitivity plot) should be.

The results of 10 threshold determinations (one cod) in the horizontal plane are shown in Fig. 7-6A. It is clear that the threshold S/N ratio is quite constant and independent of signal "direction." There is no evidence for the cod's hearing system as a whole having any remaining major sensitivity axis in the horizontal plane. As the results obtained in the median and transverse vertical planes (not shown; 18 thresholds, one cod) are entirely similar, it is doubtful whether there is any major departure from a uniform spatial distribution of sensitivity.

3.3.2 Directional Masking

In an attempt to determine the magnitude of the directional masking effect, if any, the directional sensitivity of the cod in the horizontal plane was measured in a sound field in which the angular distribution of noise energy was altered relative to the preceding experiment by changing the u_n/v_n ratio to +20 dB (i.e., the velocity noise was predominantly parallel to the fish). It appears from the result (20 thresholds from two cods) in Fig. 7-6B that a directional masking effect occurs in this condition which, on the one hand, sets in more gradually than is the case in the experiments of Chapman and Johnstone (1974) and Hawkins and Sand (1977), but which produces, on the other hand, more pronounced unmasking at an angular separation of 90°, of about the same magnitude as found for u with respect to w and vice versa (see Sections 3.1.2 and 3.1.3).

It is probable that the differences in results between our and others' studies originate in the different sound fields employed. The absence of pressure in our experiments may well render the directional masking effect presented here the purer, or at least less complex, manifestation of the phenomenon.

3.3.3 Directional Sensitivity at High p/v Ratios
in the Noise

It is clear from the preceding section that in our conditions the directional masking effect is rather extensive with maximum unmasking occurring only for mutually perpendicular directions of signal and noise. For a situation in which the pressure noise is to provide the directional masker, this should mean that directional masking should be demonstrable in the form of a cross-modality masking effect (or vice versa) even if the "direction" of the noise component reaching the labyrinths via the indirect stimulation mode and the "direction" of the velocity signal are not precisely parallel, provided, of course, that the ratio of the indirect to the direct noise contribution is sufficiently high to produce near the labyrinths an appreciably anisotropic angular distribution of noise motion components.

Accordingly, the directional sensitivity of cod was tested in the median plane at extreme p_n/w_n and p_n/u_n ratios of +6 dB and +13 dB, respectively (i.e., at indirect/direct motion ratios of at least +20 dB), as well as in the horizontal plane, at p_n/v_n and p_n/u_n ratios of ca. -2 dB to +4 dB (i.e., at indirect/direct motion ratios of ca. +10 to +15 dB). The results, 19 thresholds from three cods and 13 thresholds from two cods, respectively, are presented in Fig. 7-6D and Fig. 7-6C. In spite of rather elevated thresholds in Fig. 7-6C (but no more so than in Fig. 7-6B) and rather irregular threshold distribution contours in Fig. 7-6D, the general pattern is much the same as for the results obtained in the two experiments at low p/v ratios in the noise: the threshold distribution is congruent with the angular distribution of velocity noise. The threshold levels, then, seem to be determined by the quantity of noise that would be received by a velocity hydrophone (or by any particle motion detector exhibiting cosine-law directional sensitivity) oriented in the same direction as the signal. No influence of cross-modality masking is evident, although the direction of the indirect motion cannot have been missed by more than 45°, at which angular separation between signal "direction" and noise "direction" only ca. 3 dB unmasking is to be expected.

4 Discussion

4.1 Perceptual Segregation of Inputs to the Cod's Hearing System: Pressure- and Motion-Selective Systems?

The pattern of results emerging from the preceding sections is remarkably consistent, which seems to exclude the possibility that the observed perceptual segregation of acoustic inputs by the cod is a chance effect. Nor is this phenomenon likely to be an artifact evoked by experimental design, since the conducted experiments, even if based on the same behavioral paradigm, should be considered sufficiently different in many other respects to ensure that the cause of the uniform results is not a snag common to all designs.

Of course, one might be suspicious of the wholly artificial acoustic conditions employed. Such objections, however, can be raised against virtually all behavioral and physiological studies of fish hearing thus far, acoustic free-field studies excepted. This study, at least, employs underwater stimulation of which the relevant characteristics were measured in detail. Anyway, it should be realized that an experimental problem like the present dictates an approach in which either the stimulus composition is manipulated and controlled, or the contribution of the involved sensory systems is interfered with directly (e.g., by surgery). While the latter method may be instrumental and even necessary in clarifying the underlying mechanisms, the former method can be used to demonstrate unequivocally the sensory ability in question without any prior assumptions as to the possibly involved sensory systems, and should therefore be used first.

On the other hand, while it has been argued in Section 2.1.1 that synthesized sound fields are not per se inadequate, or even perceptibly different from "natural" sound fields of propagating waves, the conditions for this statement to be valid may not always have been satisfied in these experiments. Particularly at extreme p/v ratios in signal or noise the volume in which the sound field composition could be controlled was small relative to the size of a fish like the cod, and quite steep gradients in amplitude or phase of the acoustic variables could occur along the fish's body. Even if of no consequence to a lumped sensory system (i.e., the ears) such abnormal gradients might well influence the output of a distributed system (i.e., the lateral line), or they might even evoke from such a system a contribution to segregation of acoustic variables which is normally absent. However, as such gradients were not always present, they cannot be the sole explanation for the consistent perceptual segregation.

If it is then accepted that the results reflect a real sensory ability of cod, the assumption of (sub)systems selective to pressure or to the kinetic inputs provides a convenient way of expressing the evident lack of any cross-masking effects between these sound components. For the moment, such p-, u-, v-, and w-systems are understood to be abstract constructs useful in discussing system characteristics rather than actual, physically separate parts of the (peripheral) hearing system. Whether such separate subsystems may have an adequate physiological basis will be discussed further on. However, from the outset it seems reasonable to suppose that, whatever the physiological correlate, the material parts making up any one such system are not distributed over different sensory input channels. This assumption, if valid, would enable us to decide upon the role of the lateral line system which, theoretically, might constitute, or contribute to, any one or all of the subsystems. The adequate stimulus for a $\vec{v}$-system is, by definition, particle motion, and according to our results, independent of the mode of production at that. Since the lateral line system is not stimulated in the motion antinode of a standing wave (Cahn et al. 1973, Sand, Chapter 23), though it may well respond to the steep gradients accompanying a particle motion null, it can constitute at best only a part of a compound $\vec{v}$-system, and this is incompatible with the above assumption. The inertial ear, on the other hand, responds to any supraliminal particle motion and exhibits thus the required characteristics of a $\vec{v}$-system. The relatively low sensitivity of the lateral line renders its functioning as the p-system doubtful, too (Sand, Chapter 23).

A further point of discussion concerns the critical bandwidths of the systems. It has been found in Section 3 that all signal detection threshold situations are characterized by a critical ratio of about 0 dB for at least one of the participating systems. In general, the magnitude of the critical ratio for any system at threshold depends on the bandwidth of the filter that is used for measuring the noise band level, on the bandwidth of the physiological filter (i.e., the system's critical bandwidth of masking), and on the width of the masking noise band itself. The interdependency of the effects of these three variables is such, that with the spectral composition of the masking noise varying from that of the background noise (for a minimized component) to a concentration in a 35 Hz narrow band, a constant critical ratio can only result if the physiological and physical filters have equal bandwidths. And since the tuning of the receiving circuitry remained the same throughout the experiments, the different systems must all have approximately equal critical bandwidths. In fact, the reason for trying to limit the noise masker to a 35 Hz bandwidth was the lack of a priori knowledge concerning the filtering functions of the possibly little related systems which might participate in the detection of different sound components. That these filtering characteristics appear not to differ significantly after all may indicate that the systems are all part of the same supersystem: the ear and associated central nervous apparatus.

4.2 The Organization of the Motion-Sensitive Systems

Since the study of the discrimination of changes in particle motion direction described in Section 3.2 results in optimum values for angular resolution which are quite similar to the data of previous studies of directional acuity in cod conducted with "real" sound sources in free-field conditions (Chapman and Johnstone 1974, Hawkins and Sand 1977), our results seem to confirm once more that the vectorial quantity particle motion constitutes the adequate stimulus for directional discrimination and, as a consequence, for directional hearing. Any discussion of the motion-sensitive systems' properties may then be useful for gaining insight into the organization of the directional hearing system.

The experiments have disclosed a striking perceptual segregation of the mutually perpendicular particle velocity components u, v, and w. In view of the directional sensitivity characteristics of the hair cell, such a result could be explained by assuming that the hypothetical u-, v-, and w-systems actually represent three receptor units, each consisting of a group of hair cells that are oriented parallel to the longitudinal, transversal, and vertical axis of the fish, respectively. That this simple explanation is inadequate, however, follows from a consideration of the results of Section 3.3.2: If the directional masking effect were dependent upon the individual masked detection performance of two u- and v-selective receptor units which, according to the above model, should be the only ones involved in this experiment, then the curve relating the masked detectability threshold of the particle motion signal to its "direction" should exhibit a much steeper path on either side of the $0°$-point, since the magnitude of the v-component of the signal changes rapidly there. If, for instance, the direction of a signal at detection threshold level (criticial ratio x dB, say) is changed from $0°$ to $+45°$ or $-45°$, then the S/N ratio of the u-component drops to $x - 3$ dB, whereas the S/N ratio for v increases to $x + 17$ dB, so that the detectability of a signal at $±45°$ incidence should improve by about 17 dB, relative to the $0°$-incidence situation. In point of fact, however, the unmasking at $±45°$ is a mere 3 dB, and it is only at $±90°$ that the maximum unmasking effect (of ca. 15 dB) is reached.

Actually, the results of Section 3.3 on directional masking suggest an alternative organization of the motion-sensitive systems. It has been demonstrated there that the distribution of signal detection thresholds follows the angular distribution contours of the noise energy quite closely. It seems reasonable to interpret this result in terms of directional signal detectability being determined by the quantity of noise received by a particle motion detector directed toward the signal source (cf. Section 3.3.3). Certainly the motion-sensitive systems do behave as if for every single signal direction a separate input channel is available. Such system behavior suggests a physiological basis of many differently oriented receptor units [whether arranged in an orderly (e.g., fanwise) pattern or not] rather than of just two or three perpendicular groups.

An organization of the directional hearing system in which the distribution of the sensitivity axes of hair cell groups covers (the better part of) a circle, or a sphere in the three-dimensional case, constitutes, in some respects, a departure from Dijkgraaf's (1960) classic model based on two bilaterally symmetrical (essentially monaxial) detectors. On the other hand, it does explain in a simple way the uniform distribution of (directional) sensitivity apparent from Section 3.3.1, as well as the lack of any sectors of especially high or poor directional acuity (Buwalda pers. comm.) that are to be expected for Dijkgraaf's model (Sand 1974, see also Schuijf and Buwalda 1980, Hawkins, Chapter 6). Moreover, this model could account for the special relations shown to exist between u, v, and w, since it includes the simple model proposed in the second paragraph of this section. In fact, it predicts that such lack of reciprocal masking should exist between any mutually perpendicular motion components. A further implication is that a central neural representation of auditory space, such as found for owls and frogs (Knudsen and Konishi 1978, Pettigrew, Chung, and Anson 1978) might, in fishes, be based on the peripheral distribution of sensitivity axes, via a principle of topographic organization.

The available physiological and anatomical evidence on this matter is inconclusive so far. While distinct sensitivity axes for the sacculus have been demonstrated favoring either Dijkgraaf's model (Sand 1974, in perch) or the notion of orthogonal u- and v-detectors (Enger et al. 1973, in haddock), several authors have reported the existence of a rather wider spread of directional sensitivity axes in sacculus, lagena, and/or utriculus (Fay and Olsho 1979, in goldfish; Buwalda and van der Steen 1979, Hawkins and Horner, Chapter 15, in cod). As the underlying patterns of hair cell polarization seem to be tied to the gross organization of the labyrinths (cf. Platt and Popper, Chapter 10), it may well be necessary to establish which otolith organs partake in directional hearing before the proposed organization of the motion-sensitive systems can be provided with a sound basis. In view of the results of Dijkgraaf's experiments in unilateral elimination of the pars inferior (Schuijf 1975), the sacculi and/or the lagenae are, in cod, certain to be involved.

4.3 Mechanisms for Segregating Direct and Indirect Stimulus Components

Since a model of segregation of pressure and motion inputs based on the participation of the lateral line is doubtful to begin with and, moreover, unlikely to provide new insights, the discussion in this section will be restricted to the auditory system, and, therefore, to mechanisms which may reduce or obviate the effects of the interference between direct and indirect stimulus components.

A quite simple mechanism is suggested by the bilateral symmetry properties of the labyrinths and of the indirect stimulus, on the one hand, and by the recently discovered presence in the cod's torus semicircularis of auditory neurons with ipsilateral excitatory and contralateral inhibitory inputs (Horner, Sand, and Enger 1980) on the other hand: a perfect elimination of the indirect contribution would theoretically result if the responses of the left and right labyrinth were subtracted from each other centrally. The resulting difference is a function of sound direction (though also of intensity) (Sand 1974), and as such a variable potentially useful for the CNS in reconstructing auditory space in a manner completely differing from the proposal of Section 4.2. Central subtraction of bilateral information does provide a means for segregating the directional component, but similar simple schemes for obtaining the pure indirect information are less obvious.

On the other hand, the very complexity of the peripheral auditory apparatus (Dale 1976, for cod; Platt and Popper, Chapter 1) suggests a regional functional differentiation which may well be connected with, among other things, a segregation of direct and indirect stimuli. Potential peripheral mechanisms can be divided in two classes. The common principle in the first class is structural differentiation affecting the efficiency ratio of the direct to the indirect stimulation mode, thereby promoting a (local) selectivity for one of the stimulus components. For instance, whereas rigid skull structures (essentially transparent to propagated sound waves, but opaque to near-field oscillations) will not affect the mass-loaded hair cells' sensitivity to an external sound source (near or far) because the acoustically small skull will be carried along as a whole with the particle motions, such structures can function as a screen to the swimbladder's near field, because the skull is elastically constrained in its motion relative to the swimbladder's center (cf. Schuijf 1976b). The resulting specificity for direct stimulus components may be reversed, however, for parts of the labyrinths which are closely associated with near-field transparent "acoustic windows" in the skull, or with other structures locally improving a coupling to the swimbladder.

Specificity for indirect stimulus components may also be found among the hair cells which are not, or very loosely, coupled to an inertial mass. Such receptors will, like the lateral line neuromasts, be insensitive to the particle motions in a nondivergent sound field, but they may respond quite well in the strong spatial gradients in the swimbladder's near field, and may thus indirectly be selective to the indirect stimulus component in all but the most extreme near-field conditions.

Mechanisms of the second class rely on directional sensitivity as a means to reduce cross-talk between stimulus components. For instance, hair cells oriented perpendicularly with respect to the (body-fixed) direction of the swimbladder-induced vibrations will respond only to the direct particle motion components. A number of objections can, however, be raised against this mechanism or similar ones. First, perfectly selective particle motion detectors would seem to be confined to essentially a single set of planes, perpendicular to the indirect stimulus. Cod, however, as opposed to, e.g., man, exhibits directional detection in three-dimensional space (Hawkins and Sand 1977, Buwalda, Schuijf, and Hawkins in press). Second, the fact that sound can come from all directions also excludes an indirect-stimulus selectivity operating on this same principle of orthogonality. Finally, an objection against directional sensitivity-based mechanisms in general is that the specificity for direct or indirect stimulus components should break down for situations of more or less parallel incidence of these variables. This, indeed, was the rationale for the experiments reported in Section 3.3.3. The failure to demonstrate any cross-modality masking effect indicative of this breakdown seems to argue against mechanisms of this class. However, the objections mentioned above have a common ground in the assumption that the indirect stimuli have (nearly) the same direction for all parts of the labyrinth. This assumption would seem valid for most mechanisms of coupling between swimbladder and otolith organs (cf. Schuijf 1976a, 1976b) but not for the swimbladder's near field reaching the labyrinths via a bilaterally symmetrical arrangement of "acoustic windows" or functionally analogous structures which will act, according to Huijgens' principle, as sources in their own right. The resulting strong inhomogeneity of the indirect stimulus field would then virtually guarantee that no situation can develop in which direct and indirect stimulus have the same direction for all parts of the labyrinths, and this eliminates, in principle, the third objection. The invalidation of the other objections is equally evident.

It then appears that special conduction paths for the swimbladder-induced vibrations are almost a prerequisite for segregation of direct and indirect stimuli, either as a means for enhancing locally the relative efficiency of the indirect stimulation mode or as a source of (extra) divergence. Pertinent physiological data, however, are scarce. Only for the goldfish is there evidence that the obvious structural differentiation between sacculus (connected to the swimbladder via a specialized mechanohydraulic link) and lagena is paralleled by a functional differentiation resulting in pressure specificity for part of the sacculus and possibly motion specificity for the lagena, respectively (Fay and Olsho 1979). The situation in cod is less clear. Thin spots in the opisthoticum near the macular isthmus of the sacculus may constitute acoustic windows (Schuijf 1976b). These parts of the sacculus prove indeed 10-15 dB more sensitive (at 120 Hz) to the indirect than to the direct stimulus components. The posterior macula sacculi, on the other hand, is closer to the swimbladder and seems, with its rostrocaudad sensitivity axis, optimally situated to receive the swimbladder's near field. Its pressure selectivity, however, is some 10 dB lower than that of middle and anterior sacculus (Buwalda and van der Steen 1979). The screening of the posterior sacculus apparent from this result might even be more effective for the nearby lagena and render this structure essentially motion specific. Indeed, Offutt's (1974) surgical elimination experiments have disclosed that the lagena's contribution to absolute sensitivity as expressed in sound-pressure thresholds is much lower than that of the sacculus. His sound fields are insufficiently specified, however, for us to check if his lagenar thresholds may have been determined by particle motion instead of pressure.

4.4 Concluding Remarks

The preceding sections have made it amply clear that the acute perceptual segregation of directional and nondirectional acoustic inputs by the cod can adequately be explained by mechanisms that do not require the participation of the lateral line system. Most theoretical objections to a directional hearing system based on particle motion detection by the otolith organs in fishes with a swimbladder (cf. van Bergeijk 1964) appear then to be invalidated. The solution of the remaining problem, i.e., the inherent 180° ambiguity in determining wave propagation direction from particle motion, requires that pressure and motion be compared in the hearing system as to phase (Schuijf 1975, 1976a, 1976b, Chapter 14; Schuijf and Buwalda 1975). The phase analysis involved may well occur at the CNS level, and a model of directional hearing based on more or less absolute peripheral segregation (cf. Section 4.3), on peripheral particle motion vector detection (cf. Section 4.2) and on unilateral and bilateral convergence and processing of phase-locked afferent information makes up an adequate synthesis of our pertinent knowledge so far. On the other hand, the model synthesis proposed by Schuijf in Chapter 14 makes use of the time structure of the stimulus components as a means to perform both the phase analysis and the detection of incident particle motion direction on a peripheral level. Perceptual segregation of inputs, however, is a result of central processing. The model is, in principle, of the type based on directional sensitivity and orthogonality, and would seem to require a peripheral organization as discussed in Sections 4.2 and 4.3. It should be stressed that the views presented in this chapter and Chapter 14 are not mutually exclusive, but even complementary in many respects.

Decisive data on the underlying mechanisms can only be obtained through physiological or surgical methods. What is needed now is an electrophysiological determination of the level of first occurrence of segregation or selectivity, if any. Peripheral selectivity based on gross functional differentiation may be demonstrated by elimination techniques in behavioral experiments. In this respect the anode-blocking technique designed by Sand (Horner et al. 1980) could prove particularly useful, if it can be adapted for use in the relatively intact, unanesthetized, unrestrained fish. This method permits evoking and reversing the effect at will, as opposed to surgical elimination, e.g., by severance of cranial nerve roots (Dijkgraaf 1973). First attempts may be directed to unilateral elimination; mechanisms based on structural differentiation resulting in peripheral segregation are clearly less likely to be affected by such an operation than mechanisms based on binaural interaction.

The physiological data needed in cod are available in goldfish (Fay and Olsho 1979). For this latter species, however, no firm behavioral evidence of input segregation is available. On the contrary, Cahn, Siler, and Auwarter (1971, cited in Cahn et al. 1973) have shown that severe cross-modality masking occurs in goldfish, which, while not explicitly arguing against segregation of direct and indirect stimuli, does not tell strongly in its favor, either. In a species equipped so well in this respect, this seems certainly a surprising result. On the other hand, the goldfish's cyprinid relative *Leuciscus idus* (the ide) exhibits phase analysis of p to v just like cod (Buwalda, Schuijf, Visser, and Willers in press) but this can be considered only extremely circumstantial evidence for input segregation.

In general, more comparative data are needed. In view of the variety of mechanisms imaginable and the pronounced interspecific structural variation (cf. Platt and Popper, Chapter 1), it may well turn out that generalization is only possible to a limited extent. Indeed, the results of the few directly comparable studies thus far (Cahn et al. 1970, 1973) indicate that the acute perceptual segregation of acoustic inputs found in cod may not be representative for all, or even most, fish.

Segregation of pressure and motion implies that, in principle, any stimulus dimension characterized by a particular relation between these inputs is amenable to analysis in the auditory system. One of these is frequency, because of the indirect stimulation mode's relative efficiency being frequency dependent. A peripheral mechanism of segregation could even provide a basis for a place principle of frequency selectivity (cf. Enger, Chapter 12), which is, however, very unreliable since the p/v amplitude ratio providing the frequency cue is even more dependent on other acoustic parameters, in particular the source distance. This, again, suggests the feasibility of acoustic distance perception. Not only the amplitude ratio, but also the phase difference between pressure and motion is relevant then (cf. Fig. 7 in Schuijf and Buwalda 1980). Indeed, it has been found that (simulated) distance is discriminated by cod on phase cues alone (Buwalda, Wester, and Schuijf in prep.).

Acknowledgments. The author wishes to acknowledge the help received from the graduate students in biology Mr. A. Portier and Mr. D. J. Wester in performing some of the experiments in Sections 3.1.3 and 3.3.3. He is much indebted to his colleague Dr. Ir. A. Schuijf, who has provided a theoretical starting point for this study with his ideas on directional hearing, as well as a material starting point by designing important equipment, and who has continued to be involved through stimulating discussions and timely advice, and through direct contributions in related studies. Prof. Dr. F. J. Verheijen's unabating interest has proved most encouraging. Thanks are further due to Mr. R. van Weerden and Mr. R. Loots for developing and building some of the electronics and for helping in micro-processor programming; to Mr. W. Maasse, Mr. G. Stooker, and Mr. E. van der Veen for constructing the sound projectors and the velocity hydrophones; to Miss Selma van Cornewal, Mr. C. Versteeg, and the O.M.I. photographic department for preparing the figures, and to Miss Renée Enklaar for typing the manuscript.

The journey to Florida for presenting the lecture on which this chapter is based has been made possible by a travel grant from the U.S. National Science Foundation.

References

van Bergeijk, W. A.: Directional and Non-directional Hearing in Fish. In: Marine Bioacoustics, Vol. 1. Tavolga, W. N. (ed.). Oxford: Pergamon Press, 1964, pp. 281-299.

Buwalda, R. J. A., Steen, J. van der: Sensitivity of the cod sacculus to directional and non-directional stimuli. Comp. Biochem. Physiol. 64A, 467-471 (1979).

Buwalda, R. J. A., Schuijf, A., Hawkins, A. D.: On the discrimination, in three dimensions, of sound waves from opposing directions by the cod. (in press)

Buwalda, R. J. A., Schuijf, A., Visser, C., Willers, A. F. M.: Experiments on acoustic orientation in an Ostariophysian fish. (in press)

Buwalda, R. J. A., Wester, D. J., Schuijf, A.: Discrimination of phase between sound pressure and particle motion: a basis for acoustic distance perception in the cod. (in prep.).

Cahn, P. H., Siler, W., Wodinsky, J.: Acoustico-lateralis system of fishes: Tests of pressure and particle-velocity sensitivity in grunts, *Haemulon sciurus* and *Haemulon parrai*. J. Acoust. Soc. Am. 46, 1572-1578 (1969).

Cahn, P. H., Siler, W., Auwarter, A.: Acoustico-lateralis system of fishes: Cross-modal coupling of signal and noise in the grunt *Haemulon parrai*. J. Acoust. Soc. Am. 49, 591 (1970).

Cahn, P. H., Siler, W., Fujiya, M.: Sensory detection of environmental changes by fish. In: Responses of Fish to Environmental Changes. Chavin, W. (ed.). Springfield, Illinois: Charles C Thomas, 1973, pp. 363-388.

Chapman, C. J.: Field studies of hearing in teleost fish. Helgoländer Wiss. Meeresunters. 24, 371-390 (1973).

Chapman, C. J., Hawkins, A. D.: A field study of hearing in the cod, *Gadus morhua* L. J. Comp. Physiol. 85, 147-167 (1973).

Chapman, C. J., Johnstone, A. D. F.: Some auditory discrimination experiments on marine fish. J. Exp. Biol. 61, 521-528 (1974).

Dale, T.: The labyrinthine mechanoreceptor organs of the cod *Gadus morhua* L. (Teleostei: Gadidae). Norw. J. Zool. 24, 85-128 (1976).

Dijkgraaf, S.: Hearing in bony fishes. Proc. R. Soc. London Ser. B 152, 51-54 (1960).

Dijkgraaf, S.: A method for complete and selective surgical elimination of the lateral-line system in the codfish, *Gadus morhua*. Experientia 29, 737-738 (1973).

Dixon, W. J., Massey, F. J.: Introduction to statistical analysis. 2nd ed. New York: McGraw-Hill, 1957, pp. 318-328.

Enger, P. S., Hawkins, A. D., Sand, O., Chapman, C. J.: Directional sensitivity of saccular microphonic potentials in the haddock. J. Exp. Biol. 59, 425-434 (1973).

Fay, R. R., Olsho, L. W.: Discharge patterns of lagenar and saccular neurons of the goldfish eighth nerve: Displacement sensitivity and directional characteristics. Comp. Biochem. Physiol. 62A, 377-387 (1979).

Fay, R. R., Popper, A. N.: Acoustic stimulation of the ear of the goldfish (*Carassius auratus*). J. Exp. Biol. 61, 243 (1974).

Fay, R. R., Popper, A. N.: Modes of stimulation of the teleost ear. J. Exp. Biol. 62, 379-387 (1975).

Harris, G. G., van Bergeijk, W. A.: Evidence that the lateral-line organ responds to near-field displacements of sound sources in water. J. Acoust. Soc. Am. 34, 1831-1841 (1962).

Hawkins, A. D., Chapman, C. J.: Masked auditory thresholds in the cod *Gadus morhua* L. J. Comp. Physiol. 103 A, 209-226 (1975).

Hawkins, A. D., MacLennan, D. N.: An acoustic tank for hearing studies on fish. In: Sound Reception in Fish. Schuijf, A., Hawkins, A. D. (eds.). Amsterdam: Elsevier, 1976, pp. 149-169.

Hawkins, A. D., Sand, O.: Directional hearing in the median vertical plane by the cod. J. Comp. Physiol. 122, 1-8 (1977).

Horner, K., Sand, O., Enger, P. S.: Binaural interaction in the cod. J. Exp. Biol. 85, 323-332 (1980).

Knudsen, E. I., Konishi, M.: A neural map of auditory space in the owl. Science 200, 795-797 (1978).

Offutt, G. C.: Structures for the detection of acoustic stimuli in the Atlantic codfish *Gadus morhua*. J. Acoust. Soc. Am. 56, 665-671 (1974).

Parvulescu, A.: Problems of propagation and processing. In: Marine Bio-Acoustics, Vol. 1. Tavolga, W. N. (ed.). Oxford: Pergamon Press, 1964, pp. 87-100.

Pettigrew, A., Chung, S. H., Anson, M.: Neurophysiological basis of directional hearing in amphibia. Nature 272, 138-142 (1978).

Sand, O.: Directional sensitivity of microphonic potentials from the perch ear. J. Exp. Biol. 60, 881-899 (1974).

Sand, O., Enger, P. S.: Evidence for an auditory function of the swimbladder in the cod. J. Exp. Biol. 59, 405-414 (1973).

Sand, O., Hawkins, A. D.: Acoustic properties of the cod swimbladder. J. Exp. Biol. 58, 797-820 (1973).

Schuijf, A.: Field studies of directional hearing in marine teleosts. Thesis, University of Utrecht, Utrecht: Elinkwijk, 1974, 119 pp.

Schuijf, A.: Directional hearing of cod (*Gadus morhua*) under approximate free field conditions. J. Comp. Physiol. 98, 307-332 (1975).

Schuijf, A.: The phase model of directional hearing in fish. In: Sound Reception in fish. Schuijf, A., Hawkins, A. D. (eds.). Amsterdam: Elsevier, 1976a, pp. 63-86.

Schuijf, A.: Timing analysis and directional hearing in fish. In: Sound Reception in Fish. Schuijf, A., Hawkins, A. D. (eds.). Amsterdam: Elsevier, 1976b, pp. 87-112.

Schuijf, A., Buwalda, R. J. A.: On the mechanism of directional hearing in cod (*Gadus morhua* L.). J. Comp. Physiol. 98, 333-344 (1975).

Schuijf, A., Buwalda, R. J. A.: Underwater localization—A major problem in fish acoustics. In: Comparative Studies of Hearing in Vertebrates. Popper, A. N., Fay, R. R. (eds.). New York: Springer-Verlag, 1980, pp. 43-77.

Siler, W.: Near- and far-fields in a marine environment. J. Acoust. Soc. Am. 46, 483-484 (1969).

Tavolga, W. N., Wodinsky, J.: Auditory capacities in fishes. Pure tone thresholds in nine species of marine teleosts. Bull. Amer. Mus. Nat. Hist. 126, 177-240 (1963).

Urick, R. J.: Principles of underwater sounds for engineers. New York: McGraw-Hill, 1967, 342 + x pp.

Weiss, B. A.: Sonic sensitivity in the goldfish (*Carassius auratus*). In: Lateral line detectors. Cahn, P. (ed.). Bloomington: Indiana Univ. Press, 1967, pp. 249-264.

Chapter 8

Interspecific Differences in Hearing Capabilities for Select Teleost Species

Sheryl Coombs*

Recent anatomical studies have demonstrated considerable interspecific variability in both gross and ultrastructural features of the auditory system in fish (Platt and Popper, Chapter 1) and electrophysiological investigations have pointed to interspecific differences in the relative pressure and motion sensitivity of the fish ear (see Fay and Popper 1980 for review). It is reasonable to expect that the auditory capabilities of different fish species may reflect some of these anatomical and neurophysiological differences.

The most frequently cited example of such a structural-functional relationship involves the postulated contribution of a pressure-sensitive cavity, usually the swimbladder, to enhanced auditory sensitivity. Since a number of taxonomically unrelated groups of fish seem to have evolved some form of coupling system between air-filled cavities and the ear (Blaxter and Tyler 1978), it is likely that this morphological arrangement has some significant advantages for these groups of fish. Yet with the presently available data, it is not at all clear what advantages increased hearing sensitivity brings to these animals. It is possible that hearing sensitivity observed to be enhanced by a swimbladder-inner ear connection is far less relevant to these animals than other more complex analytical capabilities yet to be measured, such as the ability to sound localize or the ability to discriminate signals from ambient noise. Unfortunately, very few comparatiave data are available to evaluate these possibilities. The data reported here have been chosen to demonstrate that (1) interspecific differences can be found in complex analytical capabilities as well as in basic hearing sensitivity and that (2) some of these differences can be tentatively correlated with anatomical differences in the peripheral auditory system, particularly those concerned with the relationship between the swimbladder and inner ear.

Data have been obtained from two marine fish from the family Holocentridae, *Adioryx xantherythrus* and *Myripristis kuntee,* plus one freshwater species from the family Notopteridae, *Notopterus chitala.* Operant conditioning techniques were used in training fish to make behavioral responses in the presence of a pulsed pure tone

*Department of Zoology, University of Hawaii, and Department of Anatomy, Georgetown University Schools of Medicine and Dentistry, Washington, D.C. 20007.

signal for determining: (1) auditory sensitivity and frequency range of hearing and (2) discriminability of a 500 Hz signal from simultaneously presented pure tone maskers of different frequencies. The sound system and techniques for measuring auditory sensitivity are described elsewhere (Coombs and Popper 1979). For signal discriminability tests, the fish were trained to report the occurrence of a pulsed 500 Hz signal, fixed in intensity at between 10 and 15 dB above the animal's measured threshold, in the presence of a continuously pulsed pure tone masker (see also Coombs and Popper in prep.). The intensity of the masker was manipulated until a measure of the animal's ability to just detect the signal in the presence of the masker was obtained. By plotting the intensity of the masker at which threshold detection of the signal occurs for maskers of different frequencies, a psychophysical tuning curve (PTC) can be generated. This function describes how well sounds of different frequencies interfere with or "mask" the detection of any given signal and as such, provides a measure of the frequency resolution of the auditory system.

Figures 8-1 and 8-2 show the results of these tests along with results from other investigators (Popper 1972, Fay, Ahroon, and Orawski 1978) on the goldfish, *Carassius auratus* for comparison. Figure 8-1 shows that while all species can apparently hear as low as 100 Hz (the lowest frequency tested), the upper end of the frequency range is highest for *Myripristis* at 3000 Hz, intermediate for *Carassius* at 2000 Hz, and lowest

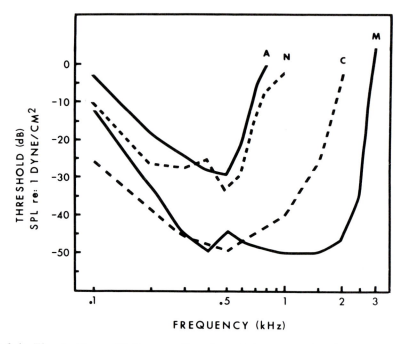

Figure 8-1. Threshold sensitivity as a function of frequency for four different fish species: *Adioryx xantherythrus* (A) and *Myripristis kuntee* (M) from Coombs and Popper (1979); *Notopterus chitala* (N) from Coombs and Popper (in prep.); and from *Carassius auratus* (C) from Popper (1972). Curves connect the mean of pooled threshold data from several individuals of each species.

for *Adioryx* and *Notopterus* at between 800 and 1000 Hz. Best sensitivity is similar for both *Carassius* and *Myripristis,* averaging about -45 dB re 1 μbar for frequencies between 300 and 800 Hz. Best sensitivity for *Adioryx* and *Notopterus* peaks at 500 Hz at about -30 dB. However, although the averaged data in Fig. 8-1 do not show this, the range of thresholds for frequencies between 300 and 700 Hz for *Notopterus* is much broader than it is for either *Adioryx* or *Myripristis,* with a number of individual threshold determinations which approach sensitivity measurements for *Myripristis* and *Carassius* (see Coombs and Popper in prep. for further detail).

Although there appear to be differences in the sensitivity data between *Notopterus* and *Carassius,* the psychophysical tuning curve data for these species are more similar, with peak masking occurring in two separate frequency regions: at the same frequency as the signal, 500 Hz, and also at 300 Hz (Fig. 8-2). These sharply tuned, multipeaked curves seen for *Notopterus* and *Carassius* are quite different from the rather broadly tuned, unipeaked curve generated for *Adioryx*. While data for *Adioryx* show that masking of a 500 Hz signal is a relatively simple inverse function of the frequency separation between masker and signal, masking in *Notopterus* appears more complex.

While all three species tested differ in both gross and ultrastructural features of the auditory system (Popper 1977, Coombs and Popper in prep.), both *Myripristis* and *Notopterus* have in common a nonhomologous specialization of the anterior end of the swimbladder which results in a physical coupling between the swimbladder and saccular end-organ of the inner ear (Nelson 1955, Dehadrai 1957). This condition is analogous to what occurs in *Carassius,* where the swimbladder and saccule are connected via the Weberian ossicles, but quite different from that found in *Adioryx,* where the anterior end of the swimbladder is situated some distance away from the ear (Nelson 1955).

In addition, the saccular macula in *Notopterus* is quite unusual in that it is subdivided into three distinct areas which can be distinguished from one another by a num-

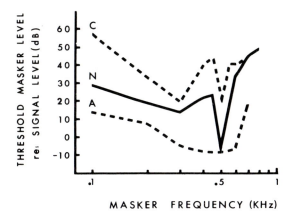

Figure 8-2. Masker intensity needed to just mask a 500 Hz signal as a function of masker frequency for three different species: *Adioryx* (A) and *Notopterus* (N) from Coombs and Popper (in press); and *Carassius* (C) from Fay et al. (1978). Tuning curves connect median data points for *Adioryx* and *Notopterus* and the means for *Carassius* from a single individual in each case.

ber of features, including the position of the macula regions relative to the swimbladder extension (Coombs and Popper in prep.). Only the two most posterior regions of the saccular macula in *Notopterus* appear to be associated with the portion of the otolith closest to the swimbladder, whereas the entire macula appears to be involved in *Myripristis*. On the other hand, the anterior region of the saccular macula in *Notopterus* is covered by a fingerlike extension of the otolith which is lodged in a bony groove some distance away from the swimbladder. Moreover, the hair cell orientation pattern shared by the two posterior regions in *Notopterus* is different from that of the anterior end.

Although there are other peripheral differences in the auditory system of the species examined here and perhaps central differences as well which may be relevant to the observed differences in auditory capabilities, the following correlations between auditory functions and structures are suggested:

(1) The frequency range of hearing seems to be unrelated to the swimbladder/inner ear arrangement, with the upper end of the frequency range varying from as low as 1000 Hz to as high as 3000 Hz in species which are all characterized by some type of connection between the ear and swimbladder.

(2) Hearing sensitivity seems to be somewhat related to the swimbladder/inner ear arrangement, with *Carassius* and *Myripristis* having significantly better sensitivity than *Adioryx,* the only fish tested without any specialized relationship between the swimbladder and inner ear. Although sensitivity data for *Notopterus* appear to be inconsistent with this trend, it has been suggested that this may be due to the unusual configuration of the swimbladder-inner ear relationship in this fish (Coombs and Popper in prep.). The anatomical differentiation of the saccular macula into three separate regions suggests that there may be some functional differentiation of the macula as well, with perhaps some portions of the macula more responsive to swimbladder input than others. Thus, the wide range in thresholds observed for this fish may actually represent two populations of thresholds, only one of which reflects sensitivity to sound channeled through the swimbladder.

(3) Multipeaked, sharply tuned PTCs have so far only been observed for species with swimbladder/inner ear connections, whereas the broadly tuned, unipeaked curve has only been observed for a species without such an adaptation. It is possible that the irregularities apparent in the multipeaked curves (particularly the apparent ease with which *Carassius* detects 500 Hz signals in the presence of 600 and 400 Hz maskers) is a function of the way in which masker and signal combine when presented simultaneously. That is, the amplitude modulations (beats) produced by presenting signal and masker at the same time may form the basis for signal detection at particular signal-masker frequency differences. Indeed, recent work with the goldfish by Fay (1980) shows that best sensitivity for amplitude modulation detection occurs when the frequency separation between two tones is between 100 and 200 Hz. Since the frequency separation between a 500 Hz signal and a 400 or 600 Hz masker falls within this range, it has been suggested that the apparent ineffectiveness of these maskers is not a function of reduced masking, but rather a result of beat detection (Fay and Popper 1980). This is further supported by recent evidence from *Notopterus* showing that when masker and signal are not presented simultaneously, the irregularities, or regions of reduced masking formerly produced by simultaneous masking techniques, do not

appear (Coombs and Popper in press). If detection of amplitude modulations is responsible for the irregular peaks in the multipeaked tuning curve function, then one might hypothesize, based on the PTC data shown in Fig. 8-2, that fish without tightly coupled swimbladder-inner ear systems, such as *Adioryx,* are relatively poorer at amplitude modulation detection than fish with such adaptations.

The data presented here demonstrate that there are qualitative and quantitative differences in both basic and more complex auditory capabilities of different fish species. While these data cannot specify the mechanisms responsible for the observed differences in auditory capabilities, some tentative correlations between these differences and gross anatomical differences in the peripheral auditory system of the species tested provide some clues as to what these mechanisms might involve. Furthermore, they suggest that increased hearing sensitivity is not the only or even major advantage accrued by fishes with specialized connections between the swimbladder and inner ear (see also Coombs and Popper in prep.) and that there may be other ways in which the swimbladder is involved in acoustic processing in fish, as has been suggested by others (Schuijf, Chapter 14, Buwalda, Chapter 7, and Blaxter et al., Chapter 3). Clearly more comparative data on a wide variety of auditory capabilities for carefully selected species are needed in the future.

References

Blaxter, J. H. S., Tyler, P.: Physiology and function of the swimbladder. In: Advances in Comparative Physiology and Biochemistry, Vol. 7. New York: Academic Press, 1978, pp. 311-367.

Coombs, S., Popper, A. N.: Hearing differences among Hawaiian squirrelfish (Family Holocentridae) related to differences in the peripheral auditory system. J. Comp. Physiol. 132, 203-207 (1979).

Coombs, S., Popper, A. N.: Comparative frequency selectivity in fishes: simultaneously and forward-masked psychophysical tuning curves. J. Acoust. Soc. Am., in press.

Coombs, S., Popper, A. N.: Structure and function of the auditory system of the clown knifefish, *Notopterus chitala.* (in prep.)

Dehadrai, P. V.: On the swimbladder and its relation with the internal ear in genus *Notopterus* (Lacepede). J. Zool. Soc. India 9, 50-61 (1957).

Fay, R. R.: Psychophysics and neurophysiology of temporal factors in hearing by the goldfish: Amplitude modulation detection. J. Neurophysiol. 44, 312-332 (1980).

Fay, R. R., Popper, A. N.: Structure and function in teleost auditory systems. In: Comparative Studies of Hearing in Vertebrates. Popper, A. N., Fay, R. R. (eds.). New York: Springer-Verlag, 1980, pp. 3-42.

Fay, R. R., Ahroon, W. A., Orawski, A. A.: Auditory masking patterns in the goldfish (*Carassius auratus*): Psychophysical tuning curves. J. Exp. Biol. 74, 83-100 (1978).

Nelson, E. M.: The morphology of the swimbladder and auditory bulla in the Holocentridae. Fieldiana. Zool. 37, 121-130 (1955).

Popper, A. N.: Auditory threshold in the goldfish (*Carassius auratus*) as a function of signal duration. J. Acoust. Soc. Am. 52(2), 596-602 (1972).

Popper, A. N.: A scanning electron microscopic study of the sacculus and lagena in the ears of fifteen species of teleost fishes. J. Morphol. 153(3), 397-418 (1977).

Discussion

HAWKINS: Is it not true that critical bands in mammals when measured using pure tone maskers also show pronounced notches of the kind you describe?

COOMBS: Yes, when the frequency separation (between masker and signal) is very small, about 10 Hz.

BULLOCK: The typical mammalian single unit tuning curve has a single high peak, but if you use single tone masking, you get a notch, as in the goldfish.

HAWKINS: Could the fish be hearing amplitude modulations?

COOMBS: Exactly. But my point is that under the same conditions, the species give different results when you use simultaneous masking. When you use forward masking, the curves all look the same. In *Adioryx,* the simultaneous masking curve looks just like the forward masking curve in *Notopterus.*

TAVOLGA: I was particularly interested in the plateau effect shown in your testing tracks. We got this sort of thing in several species, but not with the squirrelfish. The one we tested was *Holocentrus ascensionis.* We saw our plateau effect as the analog of the rod-cone dark adaptation curve in vision studies. We assumed at that time [Bull. Am. Mus. Nat. Hist. 126, 177-240 (1963)] that the animals were switching from one modality to another. In your case, the plateau seemed to be reversible. Do you have any thoughts on that?

COOMBS: I think a significant difference between our data and yours was that you used underwater sound sources, while ours was in air. With an underwater speaker, there would be a higher velocity to pressure ratio. Furthermore, your plateau effects occurred at low frequencies only. Ours were in the range of 300 to 700 Hz.

Chapter 9

Neurophysiological Mechanisms of Intensity Discrimination in the Goldfish

L. HALL, M. PATRICOSKI and R. R. FAY*

1 Introduction

The detection of changes in sound intensity (ΔI) is one of the fundamental auditory capacities for which auditory systems have presumably been "designed" throughout evolution. Questions of the neural representations of sound intensity, and the ways these are processed by the brain are basic to an analysis of sensory coding. The psychophysical literature contains two major observations: (1) The ratio $\Delta I/I$ is approximately constant for noise signals within a wide range of I values (i.e., Weber's law holds) (Miller 1947, Rodenburg 1972). (2) For tone signals, $\Delta I/I$ declines somewhat at larger I values (i.e., there is a "near miss" to Weber's law) (Reisz 1928, McGill and Goldberg 1968, Jesteadt, Wier, and Green 1977, Steigel 1977).

Pure tone intensity discrimination limens (DLs) for a variety of vertebrate species also show the "near miss" (see Fig. 9-1). Questions which have arisen from these observations are: (1) What aspect of neural activity accounts for Weber's law behavior? and (2) Why is there a "miss" to this law for tones?

Three models have been proposed to explain the "near miss" for tonal stimuli. Zwicker (1975) suggested that as intensity of a tonal stimulus increases the high-frequency side of the excitation pattern along the basilar membrane grows more quickly than the rest of the pattern. Thus, at higher levels it takes a smaller increment to produce the same amount of neural excitation. Viemeister (1972) argued that Weber's law holds at moderate intensities but not at high intensities where observers use information from harmonics produced because of nonlinearities in the ear. McGill and Goldberg (1968) suggested the existence of a neural counting mechanism which generates a Poisson-like distribution of spike counts. When energy is added to or subtracted from a signal the mean and standard deviation of the Poisson distribution changes and some other neural mechanisms detects this change. In this conception, whether Weber's law holds or not is simply a function of the relation between stimulus intensity and spike count statistics.

*Parmly Hearing Institute, and Department of Psychology, Loyola University of Chicago.

Given the fundamental differences in structural and functional organization between the mammalian cochlea and the goldfish ear, it was thought that a combined study of the psychophysical capacities and the neurophysiological coding of intensity differences in the goldfish could shed some light on the processes operating in man and other species.

2 Methods

Psychophysical intensity discrimination thresholds were obtained for four, 15 cm goldfish using classical respiratory conditioning, a modified method of limits, and the methods and apparatus described by Fay, Ahroon, and Orawski (1978). Animals were exposed to a continuous carrier signal (800 Hz pure tone or white noise) and were conditioned to respond to the onset of 2.5 Hz sinusoidal amplitude modulation (SAM) impressed upon the carrier. The degree of amplitude modulation was reduced from 1 (100% modulation) in small steps until the modulation was no longer detectable. A modulation threshold was defined as the modulation depth producing a median respiratory suppression ratio of 0.4. For each animal, thresholds were determined twice at a number of overall sound pressure levels (SPLs) for both tone and noise.

The activity of single neurons of the saccular nerve was recorded in two animals using an acoustic field and stimulus conditions similar to the psychophysical experiment and using the recording and stimulation apparatus described in Fay (1978).

Activity of 13 single neurons of the saccular nerve was recorded in one animal using glass electrodes. In one experiment, PST histograms were obtained to the onset of a 2.5 Hz SAM impressed upon a continuous noise or tone carrier at a variety of m values and SPLs. In a second experiment, modulation period histograms were obtained and analyzed for steady state SAM stimulation, again at a number of m values and SPLs.

Gross averaged evoked potentials were recorded from a midbrain region in one animal. In this experiment, the response evoked by the onset of 2.5 Hz SAM impressed upon a continuous tone or noise carrier was measured as a function of m and overall SPL. A microelectrode with an exposed 25 μm tip was used.

3 Results and Discussion

Behavioral intensity difference limens obtained for the goldfish using SAM noise and tone are presented in Fig. 9-1 along with similar data for a number of other species. Intensity DLs in dB [20 log (peak/trough) sound pressure] are plotted as a function of sensation level in dB. For AM noise the resulting function is flat with just detectable ΔI remaining the same (approximately 2.6 dB) for the sensation levels tested, indicating that Weber's law holds. For the 800 Hz tone there is a steady decrease in threshold ΔI as sensation level increases from approximately 3.7 dB at 5 dB SL to 0.2 dB at 55 dB SL. Thus, there is a "clear miss" to Weber's law for the tonal stimulus.

The goldfish ΔIs in this study are significantly smaller than those reported previously for goldfish (Jacobs and Tavolga 1967) and for all other species tested except man. This remarkable sensitivity is most likely due to the method used in which the animal is trained to detect modulation added to a continuously presented tone. In all other nonhuman experiments illustrated, observers were trained to detect an intensity difference between successively presented tone bursts. The human data (Stiegel 1977) are somewhat different in that observers detected the difference between continuous tone bursts and those which were amplitude modulated. Stiegel pointed out that the ΔI values determined using this method differed little from those obtained using the more traditional tone burst procedure (Jesteadt, Wier, and Green 1977). In any case, it

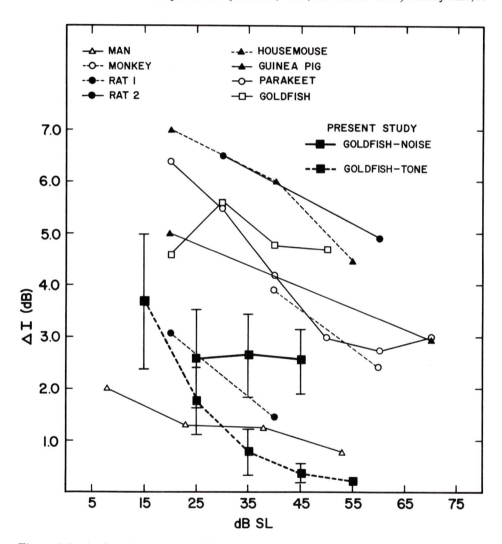

Figure 9-1. A plot of intensity difference limens in dB as a function of sensation level in dB for the present results (vertical lines indicate ±1 SD) and for 7 vertebrate species. All species exhibit the "near miss" to Weber's law when tones are used as stimuli: man (Stiegel 1977); monkey (Stebbins, Pearson, and Moody 1969); rat 1 (Hack 1971); rat 2 (Terman 1970); housemouse (Ehret 1977); guinea pig (Prosen et al. 1980); parakeet (Dooling and Saunders 1975); goldfish (Jacobs and Tavolga 1967).

appears that the miss to Weber's law is a very general characteristic of vertebrate auditory systems responding to tones.

Figure 9-2 shows PST histograms for two representative saccular units in response to SAM added to a continuous tone (upper and lower panels) and to continuous noise (center panel). Unit 304 responded to amplitude fluctuations with sharply peaked bursts of spikes, and with a particularly highly synchronized burst at the beginning of modulation. Note that this unit shows significantly more long term adaptation to the tonal carrier than to the noise carrier, and that the peaks of the PST histogram lead the tone modulation envelope in phase, indicating further short term adaptation. The major point is that the degree of spike rate modulation caused by the stimulus amplitude fluctuation declines gradually with sound pressure level over a 30 dB range (from left to right). Since the modulation depth in dB is constant for a given row, this neuron's response to tones clearly does not code stimulus intensity in accord with Weber's law. Note, however, that in response to the noise carrier (center panel), coding of amplitude is in accord with Weber's law, except very near threshold. In other words, the degree of spike rate modulation depends on stimulus modulation depth (the peak-to-trough intensity fluctuation in dB) and not on overall SPL. These noise data are representative of all neurons studied.

Unit 305 shows less adaptation to the tone carrier, and codes intensity fluctuations more in accord with Weber's law, except near threshold. Note that this unit is more sensitive than 304 in terms of overall SPL, but not in terms of modulation depth. Compare, for example, the histograms corresponding to 25% modulation and note that the response of 304 is 100% modulated while that of 305 is not.

Figure 9-3 presents the results of experiments in which the amount of spike rate modulation, R_1, was measured (see Fay, Chapter 10, for a definition of R_1) as a function of m and SPL in saccular neurons. Each point shows the stimulus m value which produces R_1 = 50 spikes per second. These modulation sensitivity thresholds generally decline with SPL for the tone carrier but not for noise. The open circles replot the behavioral intensity DLs of Fig. 9-1 and illustrate the clear relations between these and the ensemble of neural sensitivities for both carrier types. Not all saccular neurons code tone intensity similarly. Neuron 13 (similar in properties to 304) "misses" Weber's law behavior, while neuron 16 (similar to 305) does not. These results show that aspects of the psychophysical intensity discrimination data, including the Weber's law behavior for noise, the "miss" for an 800 Hz tone, and the relative sensitivities for tonal and noise modulation find clear correlates in individual saccular neurons. Models of intensity discrimination in the goldfish, therefore, need not include hypotheses about the growth or shape of excitation patterns on the saccular macula (a la Zwicker 1975), or about the generation of harmonic distortion (a la Viemeister 1972).

Figure 9-4 shows averaged evoked potentials (AEP) from the midbrain elicited by a sinusoidally shaped increment in a continuous 800 Hz tone presented at different SPLs but always with the same percentage modulation. Clearly, the amplitude and latency of the response depends on overall SPL in a way that is not in accord with Weber's law (no AEPs could be recorded for increments of noise). Figure 9-5 shows the peak-peak voltage of the AEPs as functions of m, with SPL as a parameter. Voltage depends upon SPL as well as m. The values of m producing a 20 μV response are plotted in Fig. 9-3.

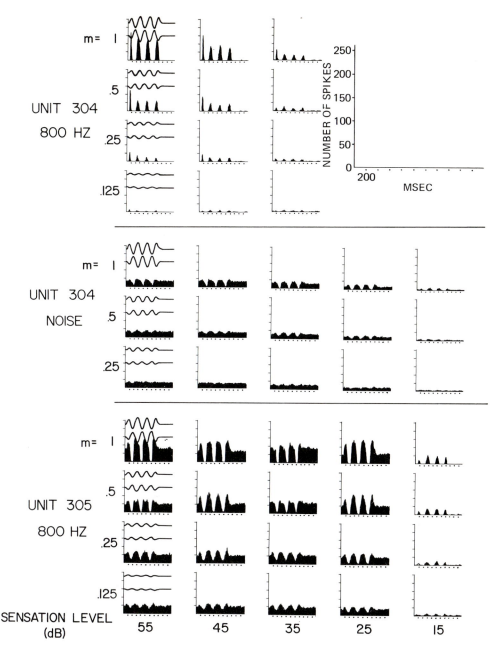

Figure 9-2. PST histograms showing the response of Units 304 and 305 to the onset of 2.5 Hz SAM impressed upon a continuous noise or tonal carriers. Columns consist of five different sensation levels: for tones: 55, 45, 25, and 15 dB SL; for noise: 45, 25, 15, 10, and 5 dB SL. Rows indicate different modulation depths: 100%, 50%, 25%, and 12.5% modulation. Note that percent modulation divided by 100 equals the modulation index, m. The waveform above each histogram in the left column shows the envelope of the SAM stimulus. Note that these envelopes, and also that in Fig. 9-4, show that the first modulation peak is lower in amplitude and reaches a maximum more quickly (about 55 msec) than successive peaks. This is due to a nonlinearity in the electronic system producing the modulation.

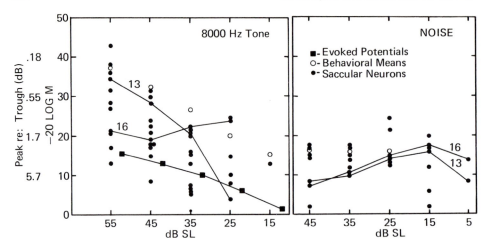

Figure 9-3. Modulation depth necessary to produce a criterion response plotted as a function of sensation level for 800 Hz tone and noise carriers. Modulation depth is expressed both as a ΔI in dB and as $-20 \log m$, where m is defined as in Fig. 9-2. The open circles are the behavioral results of Fig. 9-1. The solid squares show the modulation depths necessary to produce a 20 μV evoked potential from the goldfish midbrain (see text for details). The filled circles represent the modulation depths necessary to produce a spike rate modulation in individual saccular neurons of 50 spikes/second. The points for two units are connected to illustrate behavior consistent with Weber's law (unit 16) and behavior which "misses" (unit 13). The ensemble of units shows best sensitivity which "misses" Weber's law, and is consistent with the behavioral sensitivity.

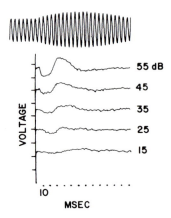

Figure 9-4. Midbrain AEPs evoked by the stimulus envelope illustrated at the top. The continuous carrier frequency was 800 Hz presented at the sensation levels shown, and the modulation was a 2.5 Hz sinusoid gated on at a positive-going zero crossing.

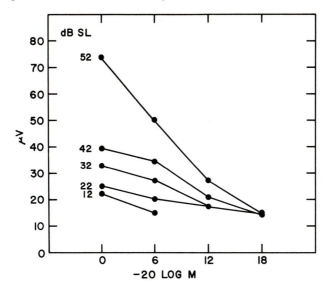

Figure 9-5. The amplitude of midbrain AEPs in response to the onset of modulation. The potentials are plotted as a function of modulation depth where $m \times 100$ is the percentage modulation.

While the sensitivity of this measure is not as great as that for single peripheral neurons, it is clear that the behavioral "miss" to Weber's law finds electrophysiological correlates at least to midbrain levels.

Acknowledgments. This research was supported by NSF and NIH grants to R. R. Fay. Great appreciation is expressed to Susan Guszcza for her assistance in producing this manuscript.

References

Dooling, R. J., Saunders, J. C.: Auditory intensity discrimination in the parakeet (*Melopsittacus undulatus*). J. Acoust. Soc. Am. 58, 1308-1310 (1975).

Ehret, G.: Comparative psychoacoustics: Perspectives of peripheral sound analysis in mammals. Naturwissenschaften 64, 461-470 (1977).

Fay, R. R.: Coding of information in single auditory-nerve fibers of the goldfish. J. Acoust. Soc. Am. 63, 136-146 (1978).

Fay, R. R., Ahroon, W., Orawski, A.: Auditory masking patterns in the goldfish (*Carassius auratus*): Psychophysical tuning curves. J. Exp. Biol. 74, 83-100 (1978).

Hack, M. H.: Auditory intensity discrimination in the rat. J. Comp. Physiol. Psychol. 74, 315-318 (1971).

Jacobs, D. W., Tavolga, W. N.: Acoustic intensity limens in the goldfish. Anim. Behav. 15, 324-335 (1967).

Jesteadt, W., Weir, C., Green, D.: Intensity discrimination as a function of frequency and sensation level. J. Acoust. Soc. Am. 61, 169-177 (1977).

McGill, W. J., Goldberg, J.: Pure-tone intensity discrimination and energy detection. J. Acoust. Soc. Am. 44, 576-581 (1968).

Miller, G. A.: Sensitivity to changes in the intensity of white noise and its relation to masking and loudness. J. Acoust. Soc. Am. 19, 609-619 (1947).

Prosen, C. A., Moody, D. B., Stebbins, W. C., Hawkins, J. E., Jr.: Absolute and intensity difference thresholds in the guinea pig after cochlear injury by kanamycin and amikacin. Abstracts of the third midwinter research meeting of the Association for Research in Otolaryngology. St. Petersburg Beach, Florida, 1980, p. 36.

Reisz, R. R.: Differential sensitivity of the ear for pure tone. Phys. Rev. 31, 867-875 (1928).

Rodenburg, M.: Sensitivity of the auditory system to differences in intensity. Ph.D. Thesis, Medical Faculty, Rotterdam, 1972.

Stebbins, W. C., Pearson, R. D., Moody, D. B.: Hearing in the monkey (*Macaca*): Absolute and differential sensitivity. Paper presented at the meeting of the Acoustical Society of America, San Diego, 1969.

Stiegel, M. S.: Intensity discrimination with three signal ensembles. Unpublished Ph.D. Thesis, University of Pittsburg, 1977.

Terman, M.: Discrimination of auditory intensities by rats. J. Exp. Anal. Behav. 13, 145-160 (1970).

Viemeister, N. F.: Intensity of discrimination of pulsed sinusoids: The effects of filtered noise. J. Acoust. Soc. Am. 51, 1265-1269 (1972).

Zwicker, E.: Scaling. In: Handbook of Perception, Vol. 5/2. Keidel, W., Neff, W. (eds.). Berlin: Springer-Verlag, 1975, p. 441.

Physiological Mechanisms of the Ear

Investigations of the different pattern in auditory nerve fibers have provided new data leading to a more complete understanding of the ear's function and of the neural representation of acoustic information. In Part Three, Fay (Chapter 10) reviews what is known about the auditory nerve physiology in fishes and provides new data on neural coding by the goldfish ear. Perhaps most significant is the observation of a faithful representation of the acoustic waveforms in temporal patterns of nerve activity, a feature that will be of significance in later discussions of sound detection and analysis by fishes (e.g., Fine, Chapter 13; Schuijf, Chapter 14; Myrberg, Chapter 19). Horner, Hawkins, and Fraser (Chapter 11) describe the frequency response and temporal characteristics of the ear of another well-studied teleost, the cod. Enger (Chapter 12) provides the first direct evidence for topographic frequency analysis within the ear using ultrastructural techniques, and relates this to behavioral capacities for frequency discrimination. Fine (Chapter 13) has studied the relations between vocalization and the discharge patterns of auditory neurons in one of the most soniferous of teleost species, the toadfish.

Chapter 10

Coding of Acoustic Information in the Eighth Nerve

Richard R. Fay*

1 Introduction

The saccule and lagena are thought to be the major acoustic receptor organs in most fishes (however, see Blaxter, Denton, and Gray, Chapter 2). Recent ultrastructural studies of these organs in a wide variety of teleosts (Platt and Popper, Chapter 1; Popper 1977) have shown significant intraspecific variability in a number of features such as size, shape, and curvature of the macula; size, sculpturing and positioning of the otolith and its otolithic membrane over the macula; and types of sound pathways to the otolith organs. The functional correlates and adaptive significance of much of this structural variation are not yet clear.

A large and growing literature exists on the hearing sensitivity and discriminative capacities of fishes (Fay and Popper 1980; see Hawkins, Chapter 6) and we are now well aware of many of the similarities and differences between these functions in several teleost species, especially the goldfish and the cod, and those characteristics of birds (Dooling 1980) and mammals (see Popper and Fay 1980 for reviews). However, correspondence between interspecific variation in inner ear structure and the variation in behavioral hearing capacities among fishes has yet to emerge clearly. One of the barriers to this understanding is that we know little about what stimulus dimensions are transduced, transformed, and encoded in the auditory periphery, and how this neural code process is related to variation in underlying structural organization. Progress toward this understanding should include the development of a paradigm to be used in analyzing the neurally coded output of the ear, and then the accumulation of systematic and quantitative descriptions of this neurally represented information in species carefully chosen for systematic structural variation along selected dimensions.

This chapter reviews what is presently known about the neurally coded output of the ear of fishes, and the goldfish in particular. While we are far from a truly comparative analysis of auditory neurophysiology in fishes, some of the results for the goldfish

*Parmly Hearing Institute, and Department of Psychology, Loyola University of Chicago.

will be helpful in establishing a quantitative experimental paradigm, and will begin to illustrate the range of physiological variation that is associated with a common and simple, yet specialized form of structural organization characteristic of the Ostariophysi.

2 Overview of the Goldfish Auditory System

The following describes some of the important features of the goldfish auditory system which will help in understanding and interpreting the neurophysiological data. For a description of the systems of other teleost species see recent reviews by Fay and Popper (1980) and Platt and Popper (Chapter 1).

The goldfish, along with carp, catfish, minnows, and a number of other freshwater groups are referred to as Ostariophysi because a chain of modified vertebrae link the anterior swimbladder with the fluid systems of the labyrinth. These Weberian ossicles [named for E. H. Weber (1820), an early comparative physiologist and one of the founders of psychophysics] conduct movements of the swimbladder walls to the ear. The Ostariophysi are sound pressure sensitive to the extent that the swimbladder expands and contracts with the sound pressure waveform. The motions of the swimbladder are conducted first to the unpaired perilymphatic sinus, and then equally to the two sacculi via the endolymphatic transverse canal. Note that the transverse canal opens directly into the sacculus and may affect the lagena only indirectly, if at all. The common conception (e.g., van Bergeijk 1967) is that volume displacements of the endolymph bring the saccular otolith into motion over the underlying hair cells, thereby causing predominantly dorsal-ventral shearing motion across the sensory hair cell cilia. A relative motion between the otoliths and underlying tissue may also occur as the fish's body moves with the water particles transmitting the sound while the more dense otoliths (predominantly calcium carbonate) may lag in motion due to their greater inertia (Wever 1971). This accelerometer principle is thought to operate similarly for all of the otolithic organs, but to play a dominant role in stimulating the lagena and utricle which lack direct input from the swimbladder.

Platt (1977) has described in detail features of the sensory maculae of the goldfish otolithic organs. Figure 10-1 shows the spatial relations among the organs, and maps of the hair cell orientation patterns characteristic of each. The maculae are composed of support and hair cells, the latter varying in density between about 200 and 500 per 10^4 μm^2, and numbering about 7,500 and 11,000 in the saccule and lagena, respectively, for a fish 51 mm long. Each hair cell has an eccentrically placed kinocilium and 40 or more stereocilia. The ciliary bundles have been described as thick (1.42 μm) and thin (0.93 μm). The thick bundles are generally found over the central areas of the macula while the thin are generally found at the macula's periphery. The length of the thick bundles underlying the saccular otolith grows somewhat from anterior to posterior. The saccular macula also "twists" somewhat so that the anterior portion is oriented nearly vertically while the posterior surface lies in a more dorso-lateral plane. Thick hair cell bundles are found in two distinct patches under the lagenar otolith with a bridging zone of cells having very long kinocilia.

The hair cell orientation pattern of the goldfish saccule is apparently characteristic of the Ostariophysi which are nearly unique among fishes for having a simple dorsal-ventral, bidirectional pattern. This orientation is consistent with the notion that the

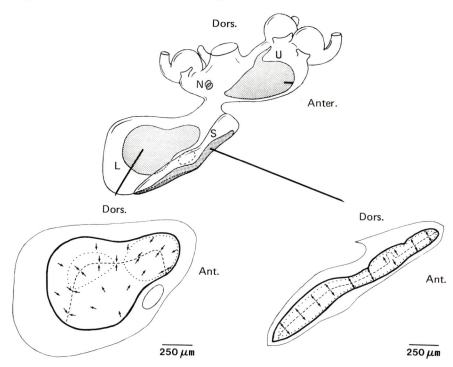

Figure 10-1. Otolith organs of the goldfish. Above: General organization of the left ear from a medial view showing the maculae (stippled) of the saccule (S), utricle (U), lagena (L) and macula neglecta (N). Below: Maps of the hair cell orientation patterns are shown for the right lagena (left) and the right saccule (right). The heavy line indicates the macula border and the dashed line indicates the region where opposed hair cell groups abut. Adapted from Platt (1977).

saccular otolith moves dorsally and ventrally in response to endolymphatic displacements through the transverse canal. The dorsally placed hair cells are oriented so that an upward movement of the cilia is excitatory, while the ventrally placed cells are oriented oppositely. The orientation maps for the goldfish lagena and utricle are generally similar to those of all species which have been studied.

Hama (1969) and Nakajima and Wang (1974) have described the ultrastructure of the hair cells and their synapses in the goldfish saccule. Three types of synaptic connections have been observed in the sensory epithelium. One presumably excitatory afferent type is found between the hair cell and nerve fibers with vesicles on the hair cell side of the junction. A second presumably efferent type is found between the hair cell and nerve ending with vesicles concentrated on the nerve side of the junction. A third presumably efferent inhibitory synapse is found between two types of nerve endings (see Fig. 1-2 in Chapter 1).

Rosenbluth and Palay (1961), Furukawa and Ishii (1967), and Furukawa (1978) have described anatomical features of the goldfish saccular nerve. Cross sections show that fiber diameters range between about 5 and 20 μm, with the large diameter fibers lying ventrolaterally in the nerve trunk. The afferent fibers are bipolar types with cell bodies either scattered throughout the trunk (large diameter fibers) or concentrated at the point where the saccular nerve passes under the transverse canal (small diameter fibers). The cell bodies of large diameter fibers are long and fusiform, appearing to be elongated thickenings in the nerve fiber. The cell bodies of the small fibers are more compact and globular. These cell bodies are generally myelinated, but the myelin can be either "loose" or "compact" (Rosenbluth and Palay 1961). Furukawa (1978) has shown that one group of large diameter saccular fibers terminates either among the group of dorsally oriented hair cells, in which case they respond to the compression phase of the acoustic waveform, or among the ventrally oriented cells and having an excitatory response to the rarefaction phase. Remarkably, some of the large diameter fibers were shown to bifurcate, and even trifurcate some distance from the macula and end on both dorsally and ventrally oriented hair cells. These fibers were shown to respond to both phases of the acoustic waveform. Small diameter fibers responded either to the compression or the rarefaction phase, but could not be successfully traced to their site of termination. Generally, the small diameter fibers end in the posterior portion of the macula while the large diameter fibers end anteriorly.

3 Physiological Response of the 8th Nerve

Enger (1963), Furukawa and Ishii (1967), Furukawa, Ishii, and Matsuura (1972), Furukawa (1978), Moeng (1978), Fay (1978a,b), Fay and Olsho (1979), Fay (1980), and Fay and Patricoski (1980) have presented systematic studies of various aspects of the physiological response of single saccular and/or lagenar fibers of the teleost 8th nerve (see also Hawkins and Horner, Chapter 15; Horner, Hawkins, and Fraser, Chapter 11; and Fine, Chapter 13). In the sections that follow, these data will be reviewed, and previously unpublished data from my laboratory will be presented and discussed in the context of previously published results.

3.1 Spontaneous Activity

Descriptions of spontaneous activity patterns have been made for *Cottus scorpius* (sculpin), a nonostariophysine (Enger 1963), and goldfish (Furukawa and Ishii 1967, Fay 1978a) 8th nerve fibers. Generally, four broad classes of spontaneous patterns are observed; silent (no activity), regular (periodic activity), irregular (Poisson-like distributions of interspike intervals) and burst (bimodal interspike interval distributions). The latter two types are illustrated in Fig. 10-2. All patterns occur in the saccular and lagenar nerves of the goldfish and in the saccular lagenar and utricular nerves of the sculpin (Enger 1963). Regular units are generally, but not always, unresponsive to sound and/or vibration in sculpin, but were never observed to respond in the goldfish.

Spontaneous rates vary between 0 and over 300 impulses per second (ips) in the goldfish saccular nerve, with irregular units showing the highest rates. Burst units generally show intermediate long-term rates, but rather high instantaneous rates within a burst. Regular units have rates generally below 100 ips in all three species. Figure 10-3 shows the distribution of rates for 102 goldfish saccular units selected for response to

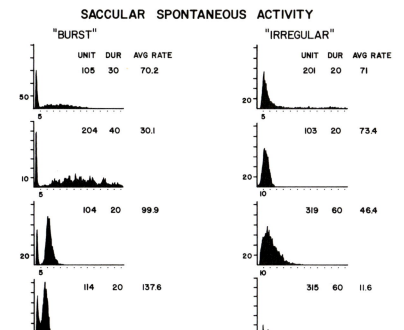

SACCULAR SPONTANEOUS ACTIVITY

Figure 10-2: Spontaneous interspike interval historgrams for eight saccular neurons illustrating the "burst" and "irregular" types and some of the variation in rate and pattern that may be seen. For each histogram the unit is identified and the duration of the record analyzed (sec) is shown along with the average spontaneous rate in impulses per second.

an 800 Hz tone. Note that this distribution has a greater mean and greater variability than distributions for mammalian and avian auditory neurons (Sachs, Wolff, and Sinnott 1980). An unbiased sampling of the spontaneous rate distribution in the goldfish 8th nerve has yet to be carried out.

The functional significance of spontaneous activity is not known. While it is clear that regular (periodic) activity can serve as a carrier for ultra low frequency information in vestibulrar units (e.g., position and tilt coding neurons), it is not clear to what extent modulation of spontaneous patterns or rates carries useful information in auditory neurons. For example, most spontaneous saccular and lagenar units in the goldfish exhibit phase-locking to sinusoidal signals at sound levels as much as 20 dB below those which cause a change in the long term average rate (Fay 1978a). Is this "temporal rearrangement" of impulses detectable, and if so, is the periodicity or temporal pattern discriminable from others producing the same average impulse rate?

The origin of spontaneous activity is also not clearly understood. Intracellular recordings from goldfish saccular neurons (Furukawa, Ishii, and Matsuura 1972, Ishii, Matsuura, and Furukawa 1971) provide strong evidence that synaptic activity between hair cell and nerve fiber is chemically mediated and suggest that the physiology of

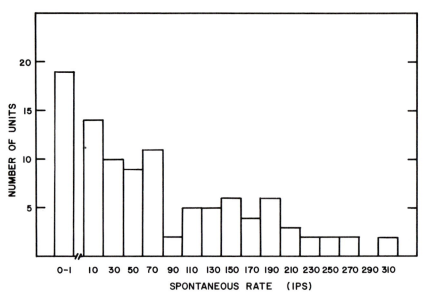

Figure 10-3. Frequency distribution of spontaneous rates from 102 goldfish saccular neurons selected for a response to an 800 Hz tone at 10 dB re 1 dyne/cm^2.

transmitter release may determine some aspects of spontaneous patterns. Given the quite distinct spontaneous patterns observed in the goldfish 8th nerve, a search for ultrastructural correlates in both pre- and postsynaptic mechanisms may be fruitful.

Spontaneous activity of some goldfish saccular fibers can be suppressed during sound stimulation (Fig. 10-4). Figure 10-4 also illustrates that the degree of adaptation and suppression grows with increasing frequency (unit PST41). This effect may be related to the observation of Furukawa and Ishii (1967) that intracellularly recorded generator potentials tend to fuse and decline in overall amplitude, or even disappear in some fibers, during maintained high frequency sound stimulation. Note that this example of neural suppression is unlike the two tone rate suppression (TTRS) characteristic of other vertebrates (Capranica and Moffat 1980) in that: (1) spontaneous as well as driven activity is suppressed in goldfish, and (2) the suppression simply grows with the frequency of the suppressor tone in the goldfish, regardless of the neuron's best frequency. While this suppression effect has not been systematically studied in the goldfish, there is at present no evidence for the type of TTRS seen in other vertebrates.

Figure 10-5 illustrates another unusual effect of stimulation on the spontaneous activity of a small number of burst type units. Here, an intense stimulus which may or may not drive the impulse rate above spontaneous levels produces a rhythmic afterdischarge which may last over 500 msec. The periodicity of the afterdischarge is independent of the stimulus characteristics and suggests the convergence of a number of oscillating inputs which are "reset" synchronously by external stimulation. A roughly similar phenomenon has been described for certain utricular neurons of the guitarfish (Budelli and Macadar 1979).

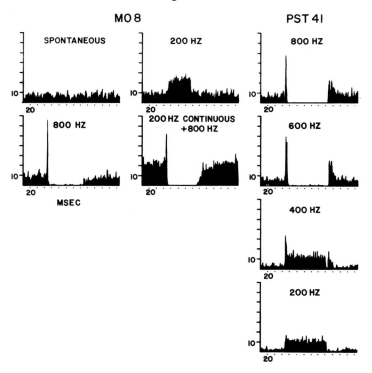

Figure 10-4. PST histograms illustrating the suppression of both evoked and spontaneous activity in neuron M08 and the effect of frequency on the suppression of spontaneous activity in neuron PST41. In all cases, tones were gated on with 5 msec rise-decay times at 10 dB re 1 dyne/cm^2 50 msec after the time axis origin.

3.2 Classification of Saccular Units

Furukawa and Ishii (1967) have used the presence or absence of spontaneous activity in a classification scheme for saccular neurons. Nonspontaneous units with best response at high frequencies (600-800 Hz) were called S1, and spontaneous (burst and regular) units showing a best response at 200-400 Hz called S2. The S1 neurons responded to either or both phases of the pressure waveform, and showed varying degrees of adaptation to steady state tones, particularly at the higher frequencies. The S2 units respond only to one stimulus phase or the other, and show less adaptation to tones.

Observations from my laboratory on the goldfish saccular nerve are generally in accord with those of Furukawa and his colleagues. However, I have observed an additional class of units having no spontaneous activity which are tuned to frequencies below 200 Hz and have relatively high thresholds (Fay 1978a). In addition, an analysis of tuning curves for a population of spontaneously active units suggest a continuous distribution of best frequencies ranging from below 100 to about 800 Hz. Questions of tuning, sensitivity and frequency response area of saccular neurons are treated in more detail in a later section. For now, it is enough to point out that the S1-S2 functional

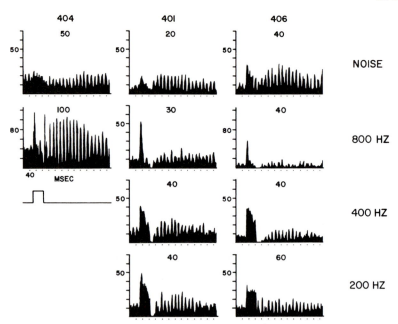

Figure 10-5. PST histograms for three saccular neurons showing a persistent periodic afterdischarge following noise and tonal stimulation at 10 dB re 1 dyne/cm². These neurons were classified as having a "burst" type spontaneous activity, and the afterdischarge appears to represent a regular oscillation of bursts comprised of from 5 to 10 spikes each. Frequency dependent adaptation effects are also evident. Note that in some cases regular bursts appear to occur *prior* to stimulus onset. This is due to the persistence of the periodic bursting for the entire period of time between stimulus presentations (1 sec).

classification may be incomplete, and that a substantial overlap occurs between spontaneous and nonspontaneous units in tuning characteristics.

In classifying units of the sculpin auditory nerve on the basis of functional characteristics, Enger (1963) identified four types of spontaneous patterns (silent, regular, irregular, burst) and described correlated differences in adaptation patterns, thresholds, and frequency response areas. Generally, silent units showed low-pass frequency response areas with a rapid decline in sensitivity above 100 Hz. Irregular units showed low-pass characteristics but with greater high frequency sensitivity than the silent units. Bursting units showed band-pass frequency response characteristics with best frequencies between 150 and 200 Hz. Regular units were generally unresponsive to sound. Adaptation was greatest in the silent and irregular units and least in the burst units. While comparisons are difficult due to limited samples and methodological differences, it appears that the sculpin units correspond roughly to the silent low frequency units (Fay 1978a) and to the S2 units (Furukawa and Ishii 1967) of the goldfish. There are no units in the sculpin comparable to the goldfish S1 type.

3.3 Adaptation

Enger (1963), Furukawa and Ishii (1967), and Fay (1978a) have described aspects of adaptation in sculpin and goldfish 8th nerve. While a systematic study has yet to be done, it appears that patterns of adaptation in the fish differ considerably from those of mammals (Harris and Dallos 1979, Smith 1977) in two fundamental ways. First, there is wide variation among neurons in the occurrence and degree of adaptation. Some neurons adapt rapidly and completely while some show little adaptation at all. Generally, the nonspontaneous and irregular spontaneous neurons show the greatest adaptation while bursting spontaneous cells show the least. Second, adaptation is clearly frequency dependent in many neurons, with response to high frequencies adapting most and those to low frequencies the least (see Figs. 10-4 and 10-5). Another interesting aspect of adaptation behavior is that some neurons adapted to a high frequency (800 Hz) tone appear to have a heightened sensitivity to amplitude modulation (AM). This is illustrated in Fig. 10-6 where the response of a neuron to the onset of an AM tone is smaller and less synchronized than its response to the onset of modulation of a tone continuously present. This effect is not seen when continuous noise is modulated. In effect, certain temporal contours seem to be enhanced under adapted conditions, and this may be at least partially responsible for the goldfish's remarkable behavioral sensitivity to amplitude fluctuations under high intensity, high frequency adaptation (Hall, Patricoski, and Fay, Chapter 9; see also Section 6 of this chapter). The morphological and cellular physiological bases for these adaptation phenomena are not at all clear and remain significant areas for future study.

3.4 Tuning and Sensitivity

Questions about the tuning or frequency selectivity of fish 8th neurons are important for two major reasons. First, inasmuch as the mechanisms responsible for the frequency selectivity of terrestrial vertebrate auditory neurons are not completely understood, giving rise for example, to the hypothetical construct "second filter" (Evans and Wilson 1973), comparative data from species varying widely in physiological and morphological organization may provide new insights on the origin of tuning. For example, studies on electroreceptors (Bullock, Chapter 27) and cochlear hair cells of certain turtles (Crawford and Fettiplace 1978) have demonstrated that frequency selectivity in different fibers may arise from the tuning inherent in the physiological processes of the receptor cell. Second, the fact that peripheral auditory neurons are frequency selective in a given species is evidence for the "place principle" of frequency representation and leads to the conclusion that the central nervous system is adapted for processing the acoustic spectrum through an analysis of spatially coded input from the periphery (see Enger, Chapter 12). This would be an important conclusion to reach because it suggests that this type of frequency coding scheme is a very general organizational principle throughout the vertebrates, and probably evolved independently many times. However, this conclusion is difficult or impossible to reach on the basis of physiological data alone, and relies on a correspondence between the characteristics of

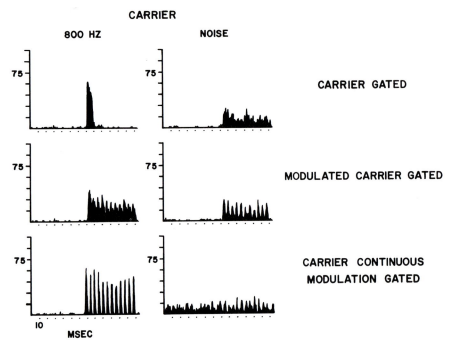

Figure 10-6. PST histograms for saccular neuron M07 in response to an 800 Hz tone and a noise carrier at 10 dB re dyne/cm^2. In the first row, the carriers are gated on at 50 msec for a duration of 50 msec. Note the rapid adaptation to the tone, and less to the noise. In the second row, the gated carriers are 100% amplitude modulated at 40 Hz. Note a reduction in adaptation and a periodic spike rate modulation at 40 Hz. In the third row, the carrier signals are on continuously, and the modulation is gated on at 50 msec. Note that there is greater steady state adaptation to the tone than to the noise, and that the response to envelope fluctuations is enhanced for the tone relative to the noise.

the putative neural code and the behavioral capacities of the animal to use the coded information in some way (see Section 7).

Enger (1963), Furukawa and Ishii (1967), and Fay (1978a) have described aspects of the tuning and frequency selectivity of 8th neurons in sculpin and goldfish. Figure 10-7 shows frequency threshold curves (FTC) published for the sculpin (Enger 1963) and those published for the goldfish (Furukawa and Ishii 1967, Fay 1978a). In the sculpin, there are two broad classes of FTCs; a wideband, low-pass type, and band-pass type with a best frequency (BF) between about 150 and 200 Hz. In the goldfish, Furukawa and Ishii's data show two band-pass types with different BFs (250-400 Hz for S2, 600-800 Hz for S1). The data from my lab show a broad class of FTCs most sensitive below 300 Hz generally showing low-pass characteristics, and another broad class with BFs above 300 Hz showing clear band-pass characteristics. As noted earlier, Furukawa and Ishii (1967) found S2 neurons to be spontaneously active while S1 were not. I have found both spontaneous and nonspontaneous neurons with best frequen-

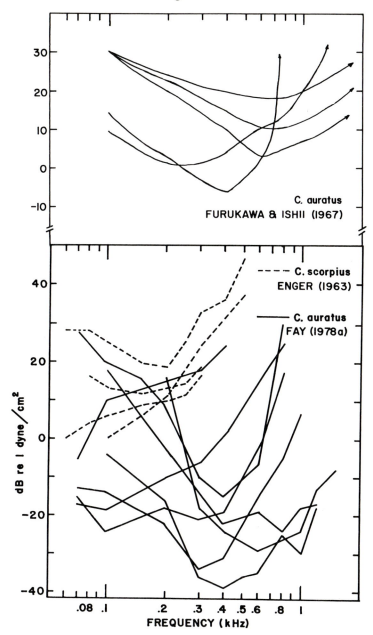

Figure 10-7. A comparison of published tuning curves for saccular neurons of the gold-fish (Fay 1978a, Furukawa and Ishii 1967) and the sculpin (Enger 1963). Note that Furukawa and Ishii's goldfish were stimulated in air while the sculpin and Fay's gold-fish were underwater. Threshold definitions for Fay's goldfish were based on the degree of phase-locking to continuous tones while the other curves probably were based on spike rate criteria.

cies ranging from below 100 Hz to 800 Hz (Fay 1978a). Enger (1963) noted that in the sculpin, low-pass neurons were either nonspontaneous or "irregularly" active while the band-pass neurons were burst types.

These kinds of data should provide a description of the frequency selectivity of neurons which may then be used to draw inferences about the more peripheral mechanisms responsible for tuning, to help evaluate hypotheses about the central neural coding of certain stimulus dimensions, and to begin a comparative structure-function analysis of the fish ear. Hindsight now allows us to criticize these data on several points, however. First, each author used different, and often vague, definitions for threshold. Certainly, meaningful comparative analysis requires a set of objective rules, consistently applied, to the definition of "sensitivity." Second, each laboratory used an arbitrary and often incompletely described system for acoustic stimulation, calibration, and ambient noise control. Physiological studies of terrestrial vertebrate auditory systems are now very close to the point of using truly standardized equipment and procedures so that comparison may be made with confidence across laboratories. Not only are we far from this sort of standardization in studies of underwater hearing, but any solutions at all to many of the complex problems of underwater stimulus specification are simply not yet at hand. Furukawa's lab, for example, has made good and creative use of a preparation stimulated in air, while both Enger's and my own studies have used waterborne sound. This fundamental methodological difference may account for some of the clear differences between Furukawa's and my tuning functions for goldfish saccular neurons (Fig. 10-7).

Related to this is the question of pressure versus vibrational (motional) sensitivity which has been important in the interpretation of behavioral measures of threshold. Different species, different organs within a species, and even different neurons from a single organ may have quite different sensitivities to pressure and particle movement. It is possible, for example, that the sound pressure thresholds for the sculpin and goldfish of Fig. 10-7 actually represent a response to a motion component, or that a neuron responds to one stimulus component in one frequency range and to another component at other frequencies. In this case, not only the neuron's sensitivity, but also the shape of its tuning curve, would be critically dependent on the impedance of the local sound field. Both of these problems (stimulus and response definitions) could lead to a misapprehension of the extent of species differences in sensitivity and tuning, and to erroneous conclusions regarding the origin of tuning and the neural codes for frequency.

While clear solutions to these problems are not at hand, the following section describes some recent studies of the sensitivity and tuning of goldfish 8th neurons which combine a set of promising approaches.

4 New Experiments on Goldfish 8th Nerve

For the experiments described below, attempts were made to effectively isolate the preparation from both ambient sound and vibration, to use appropriate objective criteria for a definition of sensitivity, and to determine both sound pressure and vertical vibration thresholds for saccular and lagenar neurons. Except as noted, materials and methods correspond to those published previously (Fay 1978a). Recording took place on a pneumatically vibration-isolated table in an IAC sound room. The fish was suspended underwater (except for the top of its head) in a 20 X 20 cm cylindrical water tank. Sound was generated into a closed air chamber below the water tank. (While there is no doubt that sound generated in this arrangement has a large motional component, due primarily to the proximity of the fish to the water's surface, this appears to be unavoidable in recording single unit responses of the 8th nerve.) The fish's head was clamped to a metal respirator tube which, in turn, was attached to a Goodman 5 lb electromechanical shaker by a rod projecting through the bottom of the tank. In this way, a relatively low-impedance stimulus could be produced by the shaker and a higher-impedance stimulus produced by the loudspeaker. While this stimulation system is simple to implement, it is limited to two fixed impedance values which may or may not allow the independent measurement of sound pressure and vibration thresholds from a given neuron.

Extracellular spikes from neurons of the saccular or lagenar nerve recorded with KCl-filled micropipettes. Responses were analyzed by accumulating 20-bin period histograms to long-duration (.5 to 10 sec) tones. The period histogram is a frequency distribution of the phase angles within individual stimulus cycles at which spikes occur. This is a most useful response measure since it provides an estimate of the average driven spike rate, the stimulus-response phase angle, and several measures of phase-locking. Recently, Kim and Molnar (1978) have described and compared these various measures in their representation of spatial and temporal characteristics of the mammalian cochlea's neurally coded output. They concluded that a comprehensive response measure of cochlear nerve fibers to tones should be based on measures of spontaneous rate, driven rate, synchronization (phase-locking), and the phase angle, and not simply upon driven rate. The most useful synchronization measure is essentially the first harmonic component of the discrete Fourier transform of the period histogram having units of impulses per second (R_1). The value R_1 depends both on driven rate and the degree of phase-locking between stimulus and response, and goes to zero if either of the others goes to zero. An intuitive way to view R_1 is that it is the amplitude of sinusoidal spike rate modulation occurring at the frequency of the stimulating tone. Note that this measure is more valuable than the coefficient of synchronization (Anderson 1973) which I have used in the past (Fay 1978a, 1978b, Fay and Olsho 1979, Fay and Patricoski 1980, Fay 1980) since it describes the sensitivity of spontaneous and nonspontaneously active neurons equally well. In the frequency threshold curves (FTC) presented in Figs. 10-8 and 10-9, period histograms were obtained for vibratory and sound stimulation and the interpolated stimulus level corresponding to R_1 = 50 ips was arbitrarily defined as threshold. It should be noted that once data are collected in this way, it is possible to express threshold or responsiveness in virtually all of the ways that have been used in the past. For each type of stimulation (loudspeaker and direct vibration) the sound pressure and the degree of vertical head vibration was measured so that a neuron's sensitivity could be expressed either in terms of sound pressure or particle movement.

Figure 10-8 shows the FTCs for six representative saccular neurons stimulated by the loudspeaker (and expressed in sound pressure units; left panel) and stimulated by the vibrator (and expressed in terms of vertical displacement; right panel). The sound pressure thresholds illustrate some of the most and least sensitive neurons encountered,

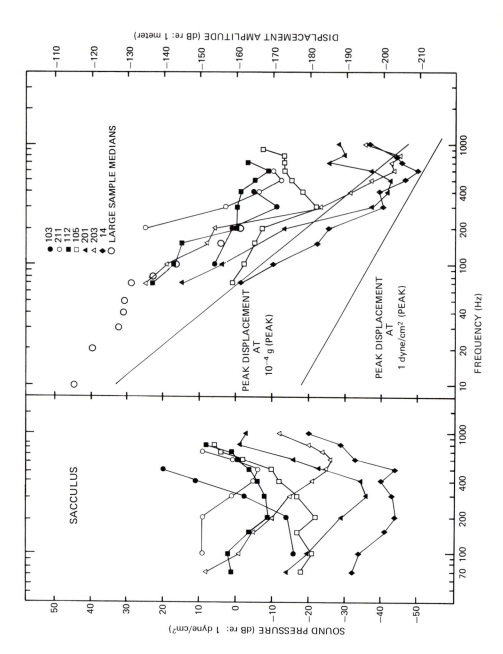

and those with the highest and lowest best frequencies. Note that the neuron with the highest BF (203) was spontaneously active, and that one broadly tuned toward the lower frequencies (112) was not spontaneous. These do not conform to Furukawa and Ishii's S1, S2 classification scheme. Unit 14 had the highest spontaneous rate encountered (438) and also the greatest sensitivity. Perhaps the best summary of the sound pressure FTCs is that there is significant diversity in best sensitivity, best frequency, bandwidth, and high and low frequency slopes.

For all saccular neurons studied in this way (28), the vertical displacement amplitudes produced at threshold for loudspeaker stimulation were found to be from 10 to 30 dB below those which existed at threshold for vibrator stimulation. This is an indication that the motional component of the loudspeaker stimulus did not determine the thresholds for loudspeaker stimulation (left panel of Fig. 10-8) in any of the neurons studied. In other words, it appears that none of the saccular neurons in this study responded significantly to the motional component of the loudspeaker stimulus. This finding is contrary to the earlier conclusions of Fay and Olsho (1979) in a study of the displacement thresholds at 100 Hz for two groups of low frequency saccular units; one group stimulated by the loudspeaker, and one by vertical head vibration. In that study, a correspondence between the distributions of displacement thresholds for the two types of stimulation suggested that vertical head vibration was the effective stimulus in each case. A reexamination of the data from that study shows that the conclusion may have been unwarranted since the distributions' medians differed by about 12 dB (Fig. 7 in Fay and Olsho 1979). In the present study, the distributions of displacement thresholds for loudspeaker and vibrator stimulation differ in the same direction by about 15 dB at 100 Hz, and by considerably greater amounts at higher frequencies. There is thus no clear evidence that any saccular neurons respond primarily to the motional component of the loudspeaker stimulus. The hypothesis (Fay 1978a, Fay and Olsho 1979) that the origin of tuning in saccular neurons lies in the degree to which a given neuron receives pressure-dependent input from the swimbladder (at the higher frequencies) is not supported by these results. The origin of tuning therefore remains a mystery and will continue to be a topic of experimental and theoretical interest.

The right panel of Fig. 10-8 shows displacement tuning curves for the same neurons. The ordinal relations among the neurons in sound pressure sensitivity, best frequency, and bandwidth are preserved for displacement sensitivity, and similar comparisons for 21 additional neurons studied in this way are consistent with this observation. It appears, therefore, that the ratio of pressure to diplacement sensitivity is approximately constant across saccular neurons having motion-detecting as opposed to pressure-

◀ Figure 10-8. Tuning curves for seven goldfish saccular neurons from 3 animals expressed in sound pressure units for loudspeaker stimulation (left) and in displacement units for vertical head vibration (right). Threshold was defined as the stimulus level corresponding to $R_1 = 50$ ips (see text for details). The open circles are medians for 50 neurons studied previously by Fay and Patricoski (1980) and are based on a coefficient of synchronization of 0.3. The two straight lines show the peak displacements in the far field necessary to produce an acceleration of 10^{-4} g, and that necessary to produce a sound pressure of 1 dyne/cm^2. Note that −200 dB re 1 m = 1 Å.

detecting functions. Note, however, that this conclusion relies on the assumption that the neurons responded to a motional component of the vibratory stimulus, and not to a concomitant pressure component. Measurements of the sound pressures generated during vibratory stimulation indicated that, in every case, the sound pressures at vibration threshold were well below (−10 to −40 dB) those necessary to reach sound pressure threshold (as defined in the left panel).

In response to vibration, all neurons show band-pass tuning with high-pass slopes averaging about 12 dB/octave. The most sensitive neurons show greatest sensitivity in the 500-800 Hz region, responding to peak displacements below 1 Å (−200 dB re 1 m). The neurons tend to show an approximately constant threshold in units of acceleration (5×10^{-4} to 5×10^{-5} g) up to the best frequency point, as would be expected under the view that the otolith organ functions as a mass loaded accelerometer for vibratory stimulation. Fay and Patricoski (1980) studied vertical vibration sensitivity for over 50 saccular and lagenar neurons at frequencies between 10 and 200 Hz. The median thresholds (based upon somewhat different criteria) are plotted as large unconnected open circles in Figs. 10-8 and 10-9, and show that the tendency for nearly constant acceleration thresholds extends down to 10 Hz. Note that at frequencies below 300 Hz or so, saccular and lagenar neurons are quite similar in sensitivity to vibration. Of additional interest here is that Fay and Patricoski (1980) also measured behavioral thresholds in goldfish to a vertical head vibration stimulus, and found a close correspondence (±5 dB) between these and the medians plotted in Figs. 10-8 and 10-9.

The lower solid line in the right panel (Fig. 10-8) shows the particle displacement amplitude in the far field accompanying a sound pressure level of 1 dyne/cm². Clearly, the vibration sensitivity of saccular neurons is such that pressure levels considerably above 1 dyne/cm² would be necessary to produce detectable particle motions, particularly at the lower frequencies. At the same time, the left panel of Fig. 10-8 shows that sound pressure thresholds may be as low as 43 dB below 1 dyne/cm². Therefore, far-field sound pressure levels would have to be from 40 to about 60 dB above pressure detection threshold in order to reach motion detection threshold. It is thus unlikely that the goldfish saccule directly codes significant information about the axis of particle motion. In order for the goldfish to localize sound using the mechanism proposed by Schuijf (Chapter 14), then, input from organs in addition to the saccule would appear to be necessary.

Figure 10-9 shows displacement FTCs for lagenar neurons in response to head vibration. Sound pressure thresholds are not shown since displacement FTCs were quite similar regardless of whether the vibrator or loudspeaker source was used, indicating that vibration amplitude determined the response in both cases. Lagenar neurons show a similar acceleration sensitivity down to 10 Hz, lower best frequencies (between 200 and 300 Hz), and less diversity in frequency selectivity compared with saccular neurons. While the ratio between displacement and pressure sensitivity is clearly higher for lagenar neurons, it nevertheless appears that rather large far-field sound pressure levels would be required for stimulation. Lagenar neurons are not clearly more sensitive to vibration than saccular neurons (at a given frequency), and retain an advantage over saccular neurons in being able to code particle motion directly *only* in the sense that they are less subject (to an unmeasured extent) to competing pressure-dependent input.

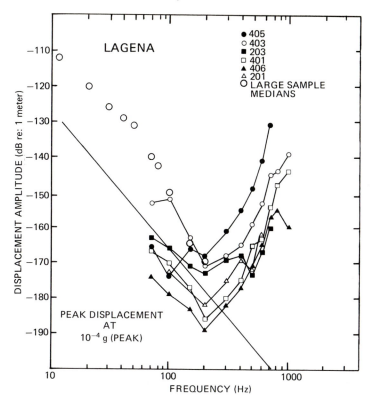

Figure 10-9. Tuning curves for 6 goldfish lagenar neurons from two animals expressed in displacement units for vertical head vibration. The large open circles show median sensitivity for 30 lagenar neurons from Fay and Patricoski (1980) defined as in Fig. 10-8.

These data can be summarized as follows:

1. Saccular neurons show great variation in sensitivity, best frequency, bandwidth, and tuning curve slopes.
2. Using objective response measures which are equally appropriate for spontaneous and nonspontaneous neurons, and using underwater rather than airborne sound, exceptions are found to Furukawa and Ishii's (1967) S1-S2 classification of saccular neurons.
3. All saccular neurons tested in response to loudspeaker stimulation appears to be pressure sensitive even in the small test tank's relatively large near field.
4. Tuning curves for the most sensitive saccular neurons combine to closely resemble the goldfish behavioral audiogram in sensitivity and shape. A similar correspondence exists between behaviorally measured sensitivity to vertical head vibration and the displacement sensitivity of saccular and lagenar neurons.

5. Thresholds to vertical displacement fall below 1 Å at the higher frequencies for the most sensitive neurons.
6. Saccular neurons show a relative constant threshold to acceleration (about 5×10^{-4} to 5×10^{-5} g) from 10 Hz up to a high frequency point which varies between neurons (from about 300 to 800 Hz), and which is correlated with the neuron's sound pressure FTC.
7. While saccular neurons are quite sensitive to head vibration, the far-field sound pressure levels necessary to bring displacement levels above threshold (from 0 to 40 dB above 1 dyne/cm^2) far exceed the neurons' pressure thresholds.
8. Pressure sensitivity for lagenar neurons is poor, and could not be measured. Lagenar displacement and acceleration sensitivity is quite similar to that for saccular neurons between 10 and 200-300 Hz, but declines thereafter. Lagenar neurons are thus better able to code particle motion parameters than saccular neurons, at least up to about 300 Hz.

5 Phase-Locking

All saccular and lagenar neurons responsive to sound show phase-locking or the tendency for spikes to occur during particular times within a stimulus cycle. Phase-locking can be quantitatively described by various measures derived from the period histogram (Kim and Molnar 1979), including the coefficient of synchronization or vector strength, a dimensionless number equal to the degree of spike rate modulation (R_1) divided by twice the average spike rate. This value can be used to estimate the dispersion of the period histogram (e.g., its standard deviation) in milliseconds, and is a measure of the temporal error with which a phase-locked response represents the period of the stimulus. Figure 10-10 shows these standard deviation values for saccular neurons in response to tones presented 35 dB above behavioral threshold (Fay 1978b). Throughout the range studied, the tendency is for the temporal error to be an approximately constant fraction of the period length (t-10%). The solid line in Fig. 10-10 shows the size of the temporal error (the jnd for period length) made by goldfish in a behavioral frequency discrimination task (Fay 1970). The fact that the behaviorally defined error bears a close and consistent relation to the neural error is an indication that frequency discrimination in the goldfish is based on estimates of period length (i.e., a temporal measurement).

Period histogram analysis also provides the stimulus-response phase angle. Furukawa and Ishii (1967) have observed qualitatively that spikes tend to be evoked during either the rarefaction or compression phase of stimulation (in S2 and some S1 fibers), or during both phases (a subset of S1 fibers). This is thought to arise from the existence of two oppositely oriented hair cell groups and their modes of innervation by saccular fibers (Furukawa 1978). At a quantitative level, this predicts a bimodal distribution of period histogram phase angles with the peaks separated by 180°. Fay and Olsho (1979) plotted these distributions for saccular and lagenar neurons stimulated by head vibration at 100 Hz and found them to be widely dispersed with only a weak tendency to form the predicted bimodal shape. This indicates that stimulus phase is not represented unambiguously by the population response. While the reasons for this variation

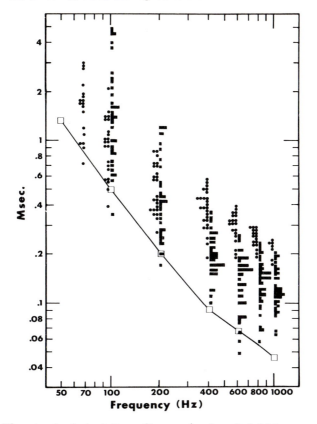

Figure 10-10. The standard deviations (in msec) of period histograms for 45 high-frequency (solid squares) are 22 low-frequency saccular neurons (solid circles) to sinusoidal stimulation at 35 dB above behavioral threshold (Fay 1969) from nine animals. The line connects the mean just noticeable differences in stimulus period (in msec) measured psychophysically for four animals (Fay 1970). (Taken from Fay 1978b).

are not known, Fay and Olsho (1979) suggested that it could be due to the relative degree to which a given neuron is stimulated by particle velocity or acceleration. These data also suggest that the motion of the otolith under these conditions of stimulation may be rather complex and nontranslatory (Fay and Olsho 1979, Sand and Michelson 1978, Hawkins and Horner, Chapter 15).

Figure 10-11 shows the distributions of phase angles for 20 saccular neurons responding well at high frequencies. Curiously, the predicted bimodal distribution is clear at 400 Hz but tends to flatten at lower and higher frequencies. This indicates that stimulus phase is unambiguously coded only in the 400 Hz region. For a given neuron, phase lag is an approximately linear function of frequency, indicating that the lag is simply due to a fixed delay between stimulus and response. However, each of these functions has a slightly different slope, and intersects the others at about 400 Hz. Note that 400 Hz is also the region of best auditory sensitivity (Fay 1969), best intensity discrimination (Jacobs and Tavolga 1967), and best relative frequency discrimination

(Fay 1970) for the goldfish. A similar frequency-dependent clustering of phase angles was not seen in lagenar responses to vibration. (See Hawkins and Horner, Chapter 15, for some phase-locking data for the cod 8th nerve.)

6 The Coding of Temporal Structure

The temporal structure of an acoustic signal may contain information both in the fine structure of the waveform and in its envelope. The mechanisms by which temporal structure is transduced and coded by the ear are not the same for waveform and envelope, and while there have been several descriptions of phase-locking to waveform fine structure (Furukawa and Ishii 1967, Fay 1978a, 1978b, Enger 1963, Horner, Hawkins, and Fraser, Chapter 11), there are comparatively fewer data on envelope coding. Fay (1980) has described the neural representation of envelope structure in the goldfish saccular nerve in response to sinusoidally amplitude modulated (SAM) stimuli which varied in carrier type (tones vs. noise), modulation (2 Hz to over 300 Hz), and modulation depth. The response of a typical saccular neuron to these SAM signals is shown in the form of modulation period histograms in Fig. 10-12. One of the remarkable features of envelope coding is that the degree of neural spike rate modulation often greatly exceeds the stimulus amplitude modulation. In other words, temporal contours of the stimulus envelope are enhanced in the temporal structure of the neural response. This is most clearly evident for the 100 and 200 Hz modulation rates (tonal carrier) where a highly synchronized response is evident even for 3% modulation. This temporal contour enhancement occurs over a wide range of overall sound pressure levels, but generally declines toward the lower modulation rates, and is not seen for noise carriers. This pattern of response is seen in the mammalian cochlear nucleus, but not in the 8th nerve (Møller 1976), suggesting that it results from neural interaction in the mammal but not in the goldfish.

A complete description of a neuron's modulation sensitivity consisted of measuring the minimum modulation depth causing a certain neural synchronization to the modulation envelope as a function of stimulus modulation rate. Figure 10-13 shows modulation thresholds for 28 saccular neurons, and complete tuning functions (or temporal modulation transfer functions, TMTFs) for five individual units in response to an AM tone (upper panel). The lower panel shows the response of the same neurons to AM noise. The large open circles in both panels are the mean detection thresholds for modulation measured behaviorally (Fay 1980). These data can be summarized as follows:

1. Neurons differ according to the modulation rate at which they are most sensitive. "Best modulation rates" vary from below 50 Hz to over 300 Hz for the tonal carrier. The origin of this "temporal tuning" is not clear but may arise from intracellular mechanisms.
2. The threshold modulation depths for the tone carrier are remarkably small at high modulation rates (a peak-to-trough amplitude ratio of less than 0.1 dB) for the most sensitive neurons.
3. Neurons are always less sensitive to the modulation of noise than of tones.
4. The ensemble of 80 neurons studied shows best sensitivity to the modulation of tones which grows at about 3 dB/octave for modulation rates between 2.5

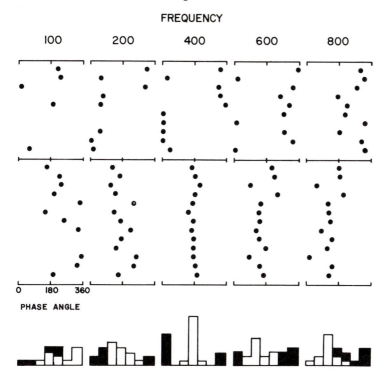

Figure 10-11. Distributions of phase angles for 23 high frequency saccular neurons in 2 animals. Phase angle is defined as the difference (in degrees) between the phase of the period histogram (see text for details) and an arbitrary point on the oscillator's output waveform. Thus, the phases shown have an arbitrary reference which is constant within a frequency. An analysis of these phases with a real time reference indicates that they result primarily from an approximate 5 msec delay between the oscillator output and time of stimulation. The neurons were classified in two groups (upper and lower panels) on the basis of a clustering of phase angles to 400 Hz stimulation. The histograms at the bottom show the overall distributions of the phase angles, and the relative contribution to the distribution of the two groups classified at 400 Hz.

and about 250 Hz and declines thereafter. Best sensitivity to the modulation of noise tends to be a flat function of modulation rate in the same range.

5. There is a close correspondence between the sensitivity of the neuron ensemble and the behaviorally measured detectability of the amplitude modulation of both tone and noise carriers (large open circles). These data have implications for questions of temporal versus spatial ("place") coding of frequency which are discussed in the following section.

Another aspect of the neural response to AM signals which is of interest is illustrated in Fig. 10-14 which shows the instantaneous spike rate evoked by an 800 Hz carrier AM modulated at 200 Hz in two saccular neurons as a function of the phase relation between the carrier and modulator waveforms. These waveforms are shown at the bottom of the figure. One type of neuron appears to respond whenever a fine structure

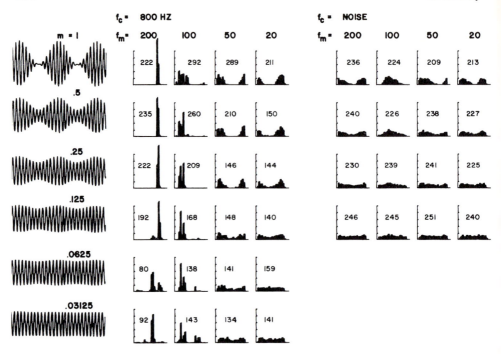

Figure 10-12. One saccular neuron's response to an 800 Hz tone or wide band noise sinusoidally modulated in amplitude at several modulation rates (columns), and at various modulation depths (rows). At left are illustrations of waveforms modulated at the different depths (m). The data are 20 bin modulation period histograms showing the relative distribution of spikes within individual cycles of modulation. The abscissa is time (the modulation period) and the ordinate is marked off in 10% increments. The number associated with each histogram is the long term average driven impulse rate in ips. The stimuli were presented 35 dB above behavioral detection threshold. Note that the 800 Hz tone carrier was phase-locked to the modulation sinusoid.

peak exceeds some threshold, and the spike rate doubles the modulation rate as two excitatory peaks per burst reach approximately equal amplitudes. Another type of neuron shows the opposite behavior as the spike rate declines to near zero for the "double peaked" bursts. Both effects could be observed over a 20 dB range (or greater) in overall sound pressure level, indicating that simple threshold effects are not the cause. These data indicate that a neuron's spike rate can be critically dependent on changes in the stimulating waveform's fine structure (phase spectrum) even when the amplitude spectrum remains unchanged.

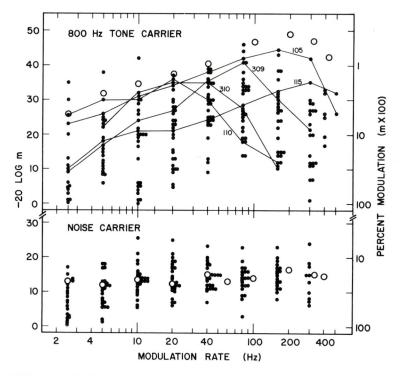

Figure 10-13. Amplitude modulation sensitivity for 31 saccular neurons in five animals in response to an 800 Hz tone carrier (top panel) and to a noise carrier (bottom panel) as a function of modulation frequency. Sensitivity is defined as the degree of amplitude modulation (modulation depth, m) necessary to produce a given degree of phase-locking synchronization to the sinusoidal envelope. Solid lines connect points from five individual neurons in order to illustrate some of the differences between neurons in modulation frequency response. The large open circles are mean behavioral detection thresholds for modulation impressed upon a continuous tone or noise carrier. (Taken from Fay 1980).

7 The Neural Codes for Waveform

The foregoing neural data show that the acoustic waveform is coded spatially to the extent that different neurons have different best tonal frequencies, and also best modulation frequencies. Enger (Chapter 12) presents evidence that this spatial or across-fiber code may also exist topographically in the cod saccule. The neural data also show that the acoustic waveform is coded temporally in responses phase-locked to both waveform fine structure and envelope. The critical question now is whether and to what extent these candidate codes are used by the nervous system in gaining information about sound.

A promising approach to these questions is to quantitatively compare behavioral performance in extracting information from certain signals with the amount and quali-

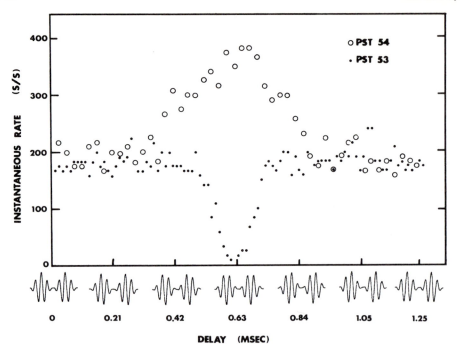

Figure 10-14. Instantaneous spike rate evoked in two neurons from one animal by an 800 Hz tone 100% amplitude modulated at 200 Hz as a function of the time delay between zero crossings of the carrier and modulator sinusoids. Progressing from left to right, the delay continously advances through one complete period of the 800 Hz signal (1.25 msec), so that the end points (0 and 1.25 msec) produce identical signals. Notice that the fine structure of the waveform begins with a single positive going peak, and is essentially inverted by a one-half period (.63 msec) delay. Both neurons responded to the upward-going components of the waveform.

ty of information transmitted to the CNS via one candidate code or another. An aspect of the temporal hypothesis has been evaluated in this way and presented earlier (Section 5). Evidence was found that in a pure tone frequency discrimination task, the jnd for stimulus period was approximately equal to the least variability (standard deviation) with which stimulus period is represented in the temporal intervals between spikes in saccular neurons. In an experiment designed to evaluate the spatial code hypothesis, Fay, Ahroon, and Orawski (1978) compared the characteristics of psychophysically defined "auditory filters" (psychophysical tuning curves) with the tuning curves for saccular neurons in goldfish (Fay 1978a). At frequencies below about 300 Hz, psychophysical "band-pass filters" were revealed which appeared not to exist in the neural data. Through a comparison of the high- and low-pass attenuation slopes of the psychophysical and neural tuning curves, however, the hypothesis was advanced that the animals based detection decisions on input from whichever neural channels had the best signal-to-noise ratio for a given combination of signal and masker. The psychophysical "band-pass filters" were thus thought to result from a combination of the high-pass characteristics of some high frequency neurons, and the low-pass characteristic of low frequency neurons. At signal frequencies above 300 Hz, psychophysical and neural curves showed a greater correspondence in overall bandwidth. However, the psychophysical curves showed a marked reduction in masking when the masker was within 100 to 200 Hz of the signal, resulting in very sharply "tuned" functions (see Coombs, Chapter 8). This effect was subsequently determined to be due to the animals' high sensitivity to beats (amplitude modulations; see Fig. 10-13) created when the masker and signal were separated by 100 to 200 Hz (Fay 1980). It appears then that while caution must be exercised in interpreting the phychophysical masking functions (the problems of "off-frequency listening" and beat detection) the correspondence between the behavioral and neural data show that the spatial neural code for frequency (peripheral tuning) has some behavioral relevance, at least in a signal detection task.

A third type of experiment having implications for questions of waveform coding involves the comparison between neural and behavioral measures of amplitude modulation detection. Figure 10-15 compares the sensitivity of the goldfish (also plotted in Figure 10-13) and man in detecting SAM impressed upon a continuous 800 Hz tone. The goldfish data have already been described and may be easily understood in terms of the sensitivity of saccular neurons to SAM. The human data may be understood in the following way: At low rates, modulation is detected as a loudness fluctuation, or a sensation of roughness. At rates above about 5 Hz, this cue is reduced, presumably because of temporal integration processes which tend to "smooth out" the neural representation of "fluctuation." However, at modulation rates between 25 and 40 Hz, this trend reverses and sensitivity to modulation grows steadily. In this range, the spectral side bands which are spaced above and below the carrier frequency by an amount equal to the modulation rate begin to be independently detected by human observers as tonal components (Hartmann 1979). The human curve thus illustrates two different but interrelated phenomena: temporal integration and the ability to analyze the acoustic spectrum in the frequency domain. The goldfish curve is quantitatively and qualitatively unlike that for man, showing a smooth growth in modulation sensitivity out to 200 Hz which parallels that of saccular neurons (Fig. 10-13). My interpretation of the differences between man and fish in this respect is that the fish temporally integrates (in the sense of a minimum integration time) far less than man, and does *not* switch to a frequency domain solution (side band detection) to this particular signal detection problem. This is an indication that the goldfish auditory system is not well adapted for frequency domain analysis, and is, instead, a superior analyzer in the time domain.

Through quantitative comparisons of psychophysical and neurophysiological results, these three types of experiments show that peripheral tuning has its consequences for behavioral signal detection, but that more complex processing such as frequency discrimination and the analysis of complex waveforms is more likely to be accomplished in the time domain.

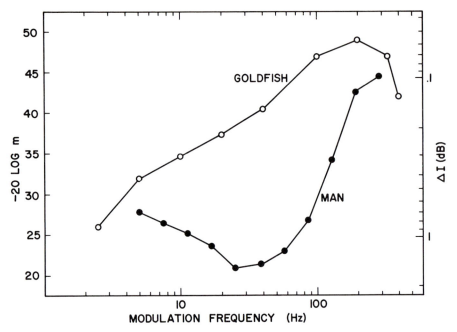

Figure 10-15. A comparison of the sensitivity of goldfish and human psychophysical observers in detecting sinusoidal AM impressed upon a continuous 800 Hz carrier at 35 dB SL (fish) and 55 dB SL (man), as a function of modulation rate. Note that the larger values of −20 log *m* correspond to small values of *m* and small percentages of modulation (see Fig. 10-13). The right ordinate shows the sound pressure level (in dB) at the peak of a modulation cycle relative to that at the trough for the corresponding values of −20 log *m*. (Data taken from Fay 1980).

8 Directional Sensitivity

Fay and Olsho (1979) measured the sensitivity of goldfish lagenar and saccular neurons to 100 Hz vibration in three orthogonal directions (vertical, transverse, rostrocaudal). Individual neurons differed in their sensitivity to vibration in different directions by as much as 30 dB, and these sensitivities were used to calculate "best" directions of stimulation in the saggital and horizontal planes. Figure 10-16 shows these data in the form of frequency distributions in polar coordinates. In the horizontal plane, these directions clustered about 15°-30° from the rostrocaudal axis, presumably reflecting the orientation of the saccular and lagenar maculae. In the saggital plane, the distribution of best directions of saccular neurons centered on the saccular hair cells' morphologically defined axis of greatest sensitivity (Platt 1957; see Fig. 10-1), but was more widely dispersed than might be predicted from the hair cell orientation maps. These results show that the axis of vibratory motion in the saggital plane may be coded spatially among both saccular and lagenar neurons. (See Hawkins and Horner, Chapter 15, for data on the directionality of single 8th neurons in the cod.)

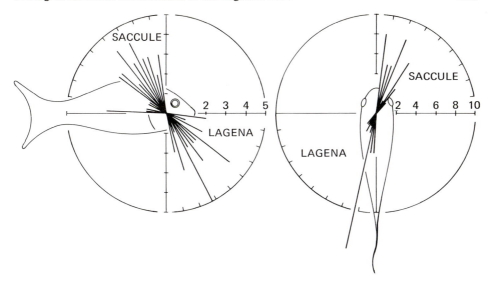

Figure 10-16. Polar coordinate frequency distributions of calculated best directions for 37 saccular and 29 lagenar neurons from the right ear of the goldfish. The length of each line corresponds to the number of neurons encountered falling in a given angular class interval ($5°$ increments). The saccular and lagenar distributions are projected into different quadrants for convenience of display. In fact, each class interval should be considered an axis passing through the origin, with a $180°$ directional ambiguity. It is also not known whether the distributions fall in the quadrants indicated ($\pm180°$), or whether they belong in adjacent quadrants, as if folded along the vertical axis of each plot. They are placed in the quadrants shown so as to correspond best to what is known about the orientation of the saccule in the saggital plane (left plot; see Fig. 10-1), and the orientation of the saccule and lagena in the horizontal plane (right plot). (Data taken from Fay and Olsho 1979).

9 Conclusion

The data presented lead to tentative conclusions on questions of the stimulus component effective in stimulating saccular and lagenar neurons, on questions of neural codes for temporal and spectral acoustic information, and on the dimensions of variation in response properties among 8th neurons.

9.1 Sensitivity to Pressure versus Particle Motion

It is clear that even in the rather large near fields produced in the experimental setups described, saccular neurons respond in proportion to sound pressure level, and not to the particle motion. The relative sensitivities of saccular neurons to pressure vs. particle motion are such that it is unlikely that they code particle motion direction except in the most extreme near field conditions. The dorsal-ventral bidirectional hair cell orientation patterns of the ostariophysan saccule appear to be special adaptations for responding to pressure-dependent input from the swimbladder.

The lagena is not more sensitive to particle motion than the saccule at low frequencies, and is, in fact, less sensitive at high frequencies (above 300 Hz). Since the lagena is far less sensitive to sound pressure than the saccule, however, it is more likely to code the direction of those particle motions which exceed threshold (accelerations of 10^{-4} g or so). In the far field, sound pressure levels above 20 dB re 1 dyne/cm^2 at an optimum frequency of 200 Hz appear to be necessary to excite lagenar neurons. Based on these data, then, rather large low frequency sound pressure levels would be required for sound localization by the goldfish in the far field.

9.2 Neural Codes for Spectral and Temporal Information

The goldfish auditory system appears to code aspects of both the acoustic spectrum and the temporal fine structure of the waveform. Spectral analysis through frequency selective properties of saccular neurons certainly exists, and a significant challenge remains to determine the origin of these effects. Mechanical inhomogeneities in the saccular macula (or more peripheral structures), differences in the micromechanics of the hair cell cilia or their immediate environment, or intracellular tuning mechanisms are some of the possible explanations that should be experimentally explored.

It appears doubtful, however, that the rather crude and variable frequency selectivity of saccular neurons provides a very effective "place" code for frequency in the goldfish auditory system. Rather, the system appears to be specially adapted for processing waveform details. First, the bidirectional hair cell orientation patterns double the density and detail with which the waveform is temporally coded in addition to providing a mechanism to discriminate stimulus polarity (Piddington 1972). Second, all saccular and lagenar neurons phase-lock with a precision characteristic of mammals and birds (Sachs, Wolff, and Sinnott 1980), and the phase-locking error correlates closely with errors made in behavioral frequency discrimination (Fay 1978b). Third, stimulus envelope is coded robustly in the temporal pattern of response, and even enhanced for tonal carriers. Fourth, the behavioral amplitude modulation detection data, along with the neural modulation sensitivity data, show that the goldfish auditory system has a minimum integration time which is significantly shorter than that for mammalian and avian systems (Fay 1980), and appears to be better adapted for processing the acoustic waveform in the time domain than in the frequency domain. Thus, temporal structure is preserved in more faithful detail in the goldfish than it is in terrestrial vertebrates whose auditory systems appear to be especially adapted for spectral processing.

9.3 Dimensions of Variation in Response Properties

One of the most striking aspects of the physiological description of the goldfish 8th nerve is the enormous variation in response properties found among saccular neurons. Spontaneous rates range more widely than those in mammals and birds (Sachs et al. 1980) and the patterns of spontaneous activity are more diverse. The appearance of adaptation also varies widely from one neuron to another and shows a frequency dependence not evident in mammalian systems. A neuron's best frequency, bandwidth, sensitivity, and tuning slopes vary significantly among neurons. Neurons differ also in their ability to represent amplitude modulation at different rates and modulation depths, with some showing clear tuning to different modulation rates. Some neurons show behavior consistent with Weber's law in coding intensity fluctuations, while others do not (Hall, Patricoski, and Fay, Chapter 9). Neurons of both the saccule and lagena vary in their directional sensitivities and in stimulus-response phase angle. Different neurons may code a complex signal's phase spectrum in quite different ways.

The cellular physiological and ultrastructural bases for much of this variation, as well as its functional significance, is not known. It is noteworthy that the goldfish saccule is, by some criteria, rather more simply organized than that for some non-ostariophysine species (see Platt and Popper, Chapter 1). At the same time, it is difficult to imagine a set of physiological responses which vary more than what we have already seen for the goldfish saccular nerve. Clearly, more experimental and theoretical work will have to be carried out on the goldfish before the causes and consequences of some of this variation can be made more understandable.

Acknowledgments. This research was supported by NSF and NIH grants and a NINCDS Research Career Development Award to R. R. Fay. Great appreciation is expressed to Susan Guszcza for her assistance in producing this manuscript.

References

Anderson, D.: Quantitative model for the effects of stimulus frequency upon synchronization of auditory nerve discharges. J. Acoust. Soc. Am. 54, 361-364 (1973).

Bergeijk, W. A.: The evolution of vertebrate hearing. In: Contributions to Sensory Physiology, Vol. 2. Neff, W. (ed.). New York: Academic Press, 1967, pp. 1-49.

Budelli, R., Macadar, O.: Statoacoustic properties of utricular afferents. J. Neurophys. 42, 1479-1493 (1979).

Capranica, R., Moffatt, A.: Nonlinear properties of the peripheral auditory systems of anurons. In: Comparative Studies of Hearing in Vertebrates. Popper, A., Fay, R. (eds.). New York: Springer-Verlag, 1980, pp. 139-165.

Crawford, A., Fettiplace, R.: Ringing responses in cochlear hair cells of the turtle. J. Physiol. (Lond.) 284, 135 (1978).

Dooling, R.: Behavior and psychophysics of hearing in birds. In: Comparative Studies of Hearing in Vertebrates. Popper, A., Fay, R. (eds.). New York: Springer-Verlag, 1980, pp. 261-288.

Enger, P. S.: Single unit activity in the peripheral auditory system of a teleost fish. Acta Physiol. Scand. 59, Suppl. 3, 9-48 (1963).

Evans, E., Wilson, J. P.: The frequency sensitivity of the cochlea. In: Basic Mechanisms in Hearing. Moller, A. (ed.). London: Academic Press, 1973, pp. 519-554.

Fay, R. R.: Behavioral audiograms for the goldfish. J. Aud. Res. 9, 112-121 (1969).

Fay, R. R.: Auditory frequency discrimination in the goldfish (*Carassius auratus*). J. Comp. Physiol. Psychol. 73, 175-180 (1970).

Fay. R. R.: Coding of information in single auditory-nerve fibers of the goldfish. J. Acoust. Soc. Am. 63, 136-146 (1978a).

Fay, R. R.: Phase-locking in goldfish saccular nerve fibers accounts for frequency discrimination capacities. Nature 275, 320-322 (1978b).

Fay, R. R.: Psychophysics and neurophysiology of temporal factors in hearing by the goldfish: Amplitude modulation detection. J. Neurophysiol. 44, 312-332 (1980).

Fay, R., Ahroon, W., Orawski, A.: Auditory masking patterns in the goldfish (*Carassius auratus*): Psychophysical tuning curves. J. Exp. Biol. 74, 83-100 (1978).

Fay, R. R., Olsho, L. W.: Discharge patterns of lagenar and saccular neurons of the goldfish eighth nerve: Displacement sensitivity and directional characteristics. Comp. Biochem. Physiol. 62A, 377-386 (1979).

Fay, R. R., Patricoski, M. L.: Sensory mechanisms of low frequency vibration detection in fishes. In: Abnormal Animal Behavior Proceeding Earthquakes. Buskirk, R. (ed.). Conference II. U.S. Geological Survey Open File Report, no. 80-453, pp. 63-91 (1980).

Fay, R. R., Popper, A.: Structure and function of teleost auditory systems. In: Comparative Studies of Hearing in Vertebrates. Popper, A., Fay, R. R. (eds.). New York: Springer-Verlag, 1980, pp. 3-42.

Furukawa, T.: Sites of termination on the saccular macula of auditory nerve fibers in the goldfish as determined by intracellular injection of procian yellow. J. Comp. Neurol. 180, 807-814 (1978).

Furukawa, T., Ishii, Y.: Neurophysiological studies on hearing in goldfish. J. Neurophysiol. 30, 1337-14403 (1967).

Furukawa, T., Ishii, Y., Matsuura, S.: Synaptic delay and time course of post synaptic potentials at the junction of hair cells and eighth nerve fibers of the goldfish. Jpn. J. Physiol. 22, 617-635 (1972).

Hama, K.: A study of the fine structure of the saccular macula of the goldfish. Z. Zellforsch. Mikrosh. Ant. 94, 115-171 (1969).

Harris, D., Dallos, P.: Forward masking of auditory nerve fiber responses. J. Neurophysiology 42, 1083-1107 (1979).

Hartmann, W.: Detection of amplitude modulation. J. Acoust. Soc. Am. (Abstr.) 65, S59 (1979).

Ishii, Y., Matsuura, A., Furukawa, T.: Quantal nature of transmission at the synapse between hair cells and eighth nerve fibers in the goldfish. Jpn. J. Physiol. 21, 78- 79 (1971).

Jacobs, D., Tavolga, W.: Acoustic intensity limens in the goldfish. Anim. Behav. 15, 324-335 (1967).

Kim, D., Molnar, C.: A population study of cochlear nerve fibers: Comparison of spatial distributions of average rate and phase-locking measures of responses to single tones. J. Neurophysiol. 52, 16-30 (1979).

Moeng, R.: Characterization of saccular nerve responses in the catfish. J. Acoust. Soc. Am. 64, S85 (Abstr.) (1978).

Møller, A.: Dynamic properties of primary auditory fibers compared with cells in the cochlear nucleus. Acta Physiol. Scand. 98, 157-167 (1976).

Nakajima, Y., Wang, D. W.: Morphology of afferent and efferent synapses in hearing organ of the goldfish. J. Comp. Neurol. 156, 403-416 (1974).

Piddington, R. W.: Auditory discrimination between compressions and rarefactions in goldfish. J. Exp. Biol. 56, 401-419 (1972).

Platt, C.: Hair cell distribution and orientation in goldfish otolith organs. J. Comp. Neurol. 172, 283-297 (1977).

Popper, A. N.: A scanning electron microscopic study of the sacculus and lagena in the ears of fifteen species of teleost fishes. J. Morphol. 153, 397-418 (1977).

Popper, A. N., Fay, R. R. (eds.): Comparative Studies of Hearing in Vertebrates. New York: Springer-Verlag, 1980.

Rosenbluth, J., Palay, S.: The fine structure of nerve cell bodies and their myelin sheaths in the eighth nerve ganglian of the goldfish. J. Biophys. Biochem. Cytol. 9, 853-877 (1961).

Sachs, M., Wolff, N., Sinnott, J.: Response properties of neurons in the avian auditory system: Comparison with mammalian homologues and consideration of the neural encoding of complex stimuli. In: Popper, A., Fay, R. (eds.). Comparative Studies of Hearing in Vertebrates. New York: Springer-Verlag, 1980, pp. 323-353.

Sand, O., Michelson, A.: Vibration measurements of the perch saccular otolith. J. Comp. Physiol. 123, 85-89 (1978).

Smith, R. L.: Short term adaptation in single auditory nerve fibers: Some poststimulatory effects. J. Neurophysiol. 40, 1098-1112 (1977).

Weber, E. H.: De aure et auditu hominis et animalium. Pars I. De aure animalium aquatilium. Lipsiae (1820).

Wever, E. G.: The mechanics of hair-cell stimulation. Trans. Am. Otol. Soc. 59, 89-107 (1971).

Discussion

PLATT: I was interested in the resonance after effects of the tone burst. I was not sure if the responses were all from one unit or from different units. The frequency of oscillation of the afterdischarge seems very constant. Do you think there might be some mechanical effect unrelated to the tone itself?

FAY: If you mean from the vibration of the head holder, I cannot see anything that correlates with that.

ROBERTS: Were your preparations curarized?

FAY: Yes.

ROBERTS: Where were you recording from?

FAY: From the saccular nerve itself, peripheral to the cell body. The cell bodies are scattered, and some recordings were from the cell body.

ROBERTS: The nerve itself was intact?

FAY: Correct. Nothing was cut.

ROBERTS: Could you have been getting any efferent activity?

FAY: That is a possibility. There were efferent fibers present.

ROBERTS: What would happen if the fish had not been curarized?

FAY: That is an interesting point. I don't know.

CORWIN: Did you follow the course of the fibers peripheral to where you were recording?

FAY: No. We did not do the kind of histology required to determine this. However, the branches are very clear, and you can follow one to the lagena, and one to the saccule.

CORWIN: Was there any contribution to your recording from the utricular macula?

FAY: No. That is far anterior to where I was recording.

SCHUIJF: How much contribution could hydrostatic changes make to your recordings?

FAY: I don't know.

SCHUIJF: I calculated that to be about one micron of displacement, about 40 dB of pressure below a microbar. There must be a mixture of displacement and pressure in your stimulus.

FAY: I am sure there is, however, I measured the sound pressure in the area of the fish, and the value is far below the animal's measured pressure threshold, by more than 20 dB.

HAWKINS: Is there any relationship between the phase of your synchronized response and the stimulus?

FAY: There is a phase angle between the stimulus and response, and this shows the lag between stimulus and response. There is no systematic relationship between phase of stimulus and of response. You might expect that there would be a bimodal distribution of phase angles between stimulus and response: one set of fibers innervating the hair cells polarized in one direction and another set to oppositely polarized hair cells. This does happen in recordings from the sacculus, but only at 400 Hz. If you go lower or higher in frequency, the distribution of phase angles tends to flatten out. I don't really understand why, but it could suggest that phase is represented only within a narrow frequency range.

FINE: In the toadfish, high frequency neurons showed normal PSCs and a high spontaneous activity. These neurons would not fire for perhaps 5 or 10 milliseconds, and then return to the spontaneous rate. Some fibers would fire from 2 to 30 times per second. After very high frequency stimulation they would be inhibited for a few milliseconds. This might represent a mechanism whereby the signal-to-noise ratio would be increased, enabling the animals to respond to courtship calls of nearby fish, while tuning out those less loud and farther away.

CORWIN: What were the conditions under which you measured spontaneous activity? What was the average level, and were these conditions of no water flow when the animal was completely still?

FAY: This was done in a soundproof room on a pneumatic vibration isolation table. The water flow over the gills was momentarily turned off during recording. I have noticed this effect on spontaneous activity only in the most sensitive neurons.

CORWIN: Do you have this effect more in the lagena?

FAY: That's a good question. I don't know.

CORWIN: What was the spontaneous rate?

FAY: Anywhere from zero to 400 per second.

POPPER: In the catfish, *Ictalurus,* there is a high sensitivity, behaviorally determined, up to 4,000 or 5,000 Hz. Measurement of unit activity by Moeng [J. Acoust. Soc. Am. 64, S85 (1978)], showed that some units respond best at 1,200 to 1,300 Hz.

TAVOLGA: The suppression effect you mention seems rather similar to what I obtained in recording from the lateral line in the marine catfish, *Arius.* I also saw the afterresponse that you described. In both cases, I assumed there was something wrong with my equipment.

FAY: I don't think there was anything wrong with the equipment.

Chapter 11

Frequency Characteristics of Primary Auditory Neurons from the Ear of the Cod, *Gadus morhua* L.

KATHLEEN HORNER, A. D. HAWKINS and P. J. FRASER*

1 Introduction

The auditory capacities of fish have recently been reviewed by Tavolga (1971), Popper and Fay (1973), Hawkins (1973) and Fay and Popper (1980), following early reviews by Griffin (1950), Kleerekoper and Chagnon (1954), and Moulton (1963). Not only can fish hear but they can distinguish between tones of differing frequency (Wohlfahrt 1939, Dijkgraaf and Verheijen 1950, Dijkgraaf 1952, Jacobs and Tavolga 1968, Fay 1970). For an ostariophysine fish, the goldfish, *Carassius auratus,* relative pitch discrimination lies between 3-6% (Jacobs and Tavolga 1968), and for several nonostariophysines it lies between 9 and 15% (Dijgraaf 1952).

Further evidence that the hearing of fish is frequency selective comes from experiments on the masking of pure tone thresholds by background noise. For humans, Fletcher (1940) observed the impairment of pure tone thresholds by noise and introduced the concept of the critical band, defined as that frequency band of noise which when presented to a subject simultaneously with a pure tone masks the presence of the tone. Buerkle (1968), Tavolga (1974), and Hawkins and Chapman (1975) have established that fish do show a critical band for masking and have estimated the width of the band for a range of pure tone frequencies.

Both pitch discrimination and the existence of critical bands in higher vertebrates may be at least partially explained by the presence of a peripheral frequency analyzer in the cochlea of the inner ear. Von Békésy (1960) demonstrated that a sound sets up a traveling wave along the basilar membrane of the cochlea and that each pure tone stimulus produces maximum movement at a single locus. There is some spread along the membrane from that site, however, and background noise with constituent frequencies centered around the same pure tone frequency will stimulate the same region of the membrane, masking the pure tone. Noise that stimulates other regions of the membrane has little or no masking effect. It is the neural integration over this limited

---

*Marine Laboratory, Victoria Road, Aberdeen, Scotland.

area of the basilar membrane which is believed to give rise to the critical band, and which defines the critical bandwidth (Zwicker, Flottorp, and Stevens 1957).

In fish there is no equivalent to the mammalia cochlea, but some form of peripheral frequency analysis based on a crude mechanical filter has been proposed by several workers. Van Bergeijk (1967) suggested that the macula of the otolith organ behaved as a bounded membrane with sufficient asymmetry to show a localized response to different frequencies. Sand (1974) on the other hand proposed that the vibration pattern of the otolith itself was frequency dependent, the part of the macula stimulated by the otolith changing with frequency. Fay (1974) and Hawkins and Chapman (1975) have tended to favor a central mechanism for frequency analysis which depends on differences in frequency being transmitted from the ear to the central nervous system without degradation, the afferent nerve fibers responding with a phase-locked response to all frequencies, rather than fibers having a response tuned to any one particular frequency. Fay (1978) quantitatively analyzed the degree of phase-locking in goldfish saccular nerve fibers, and concluded that the observed responses could account for the frequency discrimination capacities of this fish, working on the hypothesis that the period of the stimulus frequency was represented in 8th nerve fibers as the distribution of time intervals between phase-locked impulses.

Single unit recordings by other workers, both for the ostariophysine goldfish (Furukawa and Ishii 1967, Sawa 1975) and for a nonostariophysine species, the sculpin, *Cottus scorpius* (Enger 1963), have also shown phase-locked responses. There is a general agreement on the broad classification of types of primary afferent fiber, based on the patterns of spontaneous activity; they include irregular spontaneous, regular spontaneous, bursting spontaneous, and nonspontaneously active units. Of these, only the regular spontaneous units are not sound responsive. All the main classes of primary unit show phase-locking, but the different classes are reported as differing in their frequency response. Enger's nonspontaneous units responded over the lower frequency range, while the bursting and the irregular units responded to higher frequencies (see Fig. 10-7). All types adapted rapidly except the bursting units. Furukawa and Ishii (1967) associated high frequency units with large diameter fibers of the anterior sacculus and lower frequency sensitive units with small diameter fibers, found in both the anterior and posterior sacculus. The first type had no spontaneous activity and many adapted quickly, while the second fibers were normally spontaneously active and did not show adaptation. Fay (1978b) also reported high frequency responsive units associated with the anterior sacculus. Some of these units, however, were spontaneously active, while some low frequency sensitive units were not. All the nonspontaneously active and most spontaneously active units showed some adaptation during sound stimulation.

The present study examines the response of single units from the 8th nerve of the codfish to pure tone stimuli differing in frequency and amplitude.

2 Methods

2.1 Fish Handling

The experimental animals were juvenile cod of 100-500 g captured by trawl or baited hooks off the West Coast of Scotland. The fish were held in a recirculating seawater aquarium at 10°C for at least 2 weeks prior to the experiments.

Each fish was anaesthetized by immersion in a solution of 1:15000 MS 222, weighed, and intraperitoneally injected with Alphaxalone (Saffan, Glaxo) at 10-20 mg/kg live weight. *d*-Tubocurarine chloride (3 mg/kg) was then administered by injection into the lateral superficial muscle of the flank. The fish was clamped in a fish holder within a small perspex tank. A set of eye bars, and two sets of body bars secured the fish while the gills were irrigated with fresh seawater. The level of water in the tank was allowed to rise to the dorsal margin of the eye (Fig. 11-1).

The dorsal musculature overlying the labyrinth was removed on one side, and a small window cut into the cranium, just posterior to the exit of the accessory lateral or cutaneous nerve. The membranes overlying the brain and labyrinth were gently torn with a hooked needle.

To expose the auditory nerve, some perilymph was aspirated away, and the medulla displaced medially until the roots of nerve VIII were visible. The saccular nerve trunk runs ventrally to innervate the macula on the medio-lateral convex surface of the sacculus, and could be easily identified for selective recording from any ramus (Fig. 11-2). The dissection was routinely performed with no blood loss.

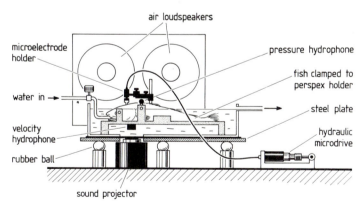

Figure 11-1. The fish clamp. The anesthetized fish was firmly held by eye bars and body clamps in a perspex holder, with the gills irrigated by fresh seawater at 10°C. A J9 sound projector was mounted under the head of the fish. Velocity transducers measured the stimulus amplitude in three orthogonal planes. The air loudspeakers were used in a particular experiment not reported on here.

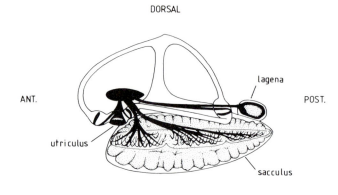

Figure 11-2. Lateral view of the left ear of the cod showing the principal rami of cranial nerve VIII. The sacculus is innervated by two main rami.

2.2 Sound Stimulus

Pure tone stimuli were generated by an oscillator (AIM Electronics), and the frequency monitored digitally. The sine wave was fed through a transient-free gate (Synthi VCS-3) to give a short sound pulse, 2 sec in duration with rise and fall times of 40 msec. The gated stimulus terminated in an amplifier and moving coil projector (Type J9). The diaphragm of the J9 was in close contact with a rubber diaphragm stretched over a circular hole in the floor of the fish tank (Fig. 11-1). The particle motion generated just below the head of the fish had a strong vertical component and minimal sound pressure. The maximum stimulus level was equivalent to a particle velocity of -70 dB re 1 cm sec^{-1} (equivalent to a free field sound pressure of +34 dB/μbar). The particle velocity was monitored by 3 seismic accelerometers at mutual right angles (Sensor Nederland, SM2), attached to the fish holder just ventral to the head of the fish.

All signals were recorded on a multitrack FM tape recorder (Bell and Howell, type 3600, bandwidth DC-2.5 kHz) for later analysis. During the search for acoustically sensitive units, low frequency sound pulses were presented via the J9 at low amplitude. Once a unit was picked up, the tape recorder was switched on, and the stimuli were presented in a series of frequencies (starting with highest) at a number of stimulus amplitudes.

2.3 Recording Unit Activity

The dissection to expose the nerves was always carried out from a dorsal aspect. A glass, fiber-filled microelectrode, filled with 3 M KCl and with an impedance of 20-50 $M\Omega$ was lowered into the particular nerve ramus by means of a hydraulic micro-drive until unit activity was registered. The units were 200 μV to 10 mV in amplitude.

The signal from the electrode was preamplified (Neurolog NL 100), band-pass filtered (80 Hz-2.5 kHz) and displayed on an oscilloscope. A spike trigger (NL 200) allowed a continual display of the number of spikes per second on a pen recorder using a scope-raster/stepper. An audio monitor of the spikes was essential during the search for units.

2.4 Analysis

The recorded signals were used to produce poststimulus time (PST) histograms and a statistical measure of the degree of phase-locking was obtained from these. The analysis was performed by feeding the triggered spikes into a Z80 based microprocessor, where the number of spikes per second, number of spikes per cycle, and the synchronization indices were calculated for 64 cycles of the stimulus waveform. Interspike interval histograms were displayed on a VDU.

3 Results

3.1 General

Results are presented for 22 fish, from which a total of 89 units were recorded. Once a unit was contacted it could be held for a few seconds to a few hours. Normally 45 minutes was adequate to record all the appropriate responses; but few units lasted that long.

The units differed both in their spontaneous activity and in their responses to pure tone stimuli. The character of each unit was assessed by noting the mean spontaneous rate of spike discharge (spikes/sec); by preparing an interspike interval histogram during spontaneous activity; and by investigating the response to pure tone stimuli at frequencies between 30 Hz and 400 Hz, recording the number of spikes/sec, the number of spikes/stimulus cycle and the degree of synchrony with the stimulus.

3.2 Spontaneous Activity

Spontaneous spike rates recorded from 44 units from the anterior ramus and 46 from the posterior ramus of the saccular nerve are presented in Fig. 11-3 in the form of a histogram for each ramus. A Mann-Whitney U test (0.001 significance level; Siegel 1956) showed that the distribution of spontaneous activity from the units in the anterior ramus (mean 49.3 spikes/sec) was significantly different from that in the posterior ramus (mean 94.0 spikes/sec).

Units could be broadly grouped on the basis of their interspike interval histograms. In Fig. 11-4 an oscillogram for each class of spontaneously active unit is shown together with its interval histogram.

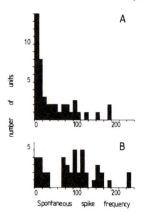

Figure 11-3. Histograms showing the frequency of occurrence of units with different rates of spontaneous spike activity. (A) Anterior ramus of the saccular nerve. (B) posterior ramus.

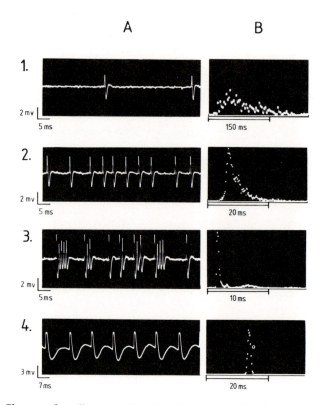

Figure 11-4. Classes of auditory unit, with their interspike interval distributions for spontaneous activity. (1 and 2) Slow and fast irregular types, both with a skewed distribution. (3) Bursting type, with a bimodal distribution. (4) Regular type with a symmetrical unimodal distribution.

Irregular units (Figs. 11-4.1 and 11-4.2) were sound responsive and showed a spontaneous rate varying between 1/sec and 250/sec. The spike intervals showed a strongly skewed distribution.

Bursting units were also sound responsive and showed a high spontaneous rate of activity (100-160 spikes/sec). The interval between successive spikes within a burst was always much shorter than the interval between bursts. This is reflected in the bimodal distribution of the interval histogram (Fig. 11-4.3). Bursting units were especially common in the posterior ramus of the saccular nerve (Table 11-1).

Regular units, with spontaneous rates of 10-100 spikes/sec, characteristically did not respond to sounds. The interval between spikes was almost constant, and the interval histogram showed a symmetrical unimodal distribution (Fig. 11-4.4). These units were probably from static or dynamic position receptors. The spontaneous rate might therefore have depended on the position or orientation of the fish but this was not tested.

Sound responsive units which were not spontaneously active were also recorded. Such units together with units which had a slow irregular activity were especially found in the anterior ramus of the saccular nerve (Table 11-1). Some of these units adapted very rapidly to sound stimuli.

3.3 Response to Pure Tone Stimulation

With a few exceptions, all sound responsive units responded to high amplitude particle velocity stimuli with a regular rate of discharge giving at least one spike per cycle.

The number of spikes produced per unit time, or per stimulus cycle depended on the stimulus amplitude (Fig. 11-5A) and stimulus frequency (Fig. 11-5B). At high stimulus amplitudes, the number of spikes per unit time often corresponded with the number of stimulus cycles. As the stimulus level was progressively reduced, the number of spikes per unit time became fewer, and the stimulus following response was lost. At low frequencies, several spikes were sometimes associated with each stimulus cycle.

Figure 11-6 illustrates the response of a unit with slow irregular spontaneous activity. During 160 Hz stimulation, spikes were produced at intervals equivalent to the period of the stimulus and at multiples of the period. A full stimulus following response would provide a unimodal interval histogram. Figure 11-7 shows that spikes tended to occur at a particular phase of each stimulus cycle. At low stimulus levels the distri-

Table 11-1. Number of Units Recorded from the Saccular Nerve

Spontaneous activity (spikes/sec)	Anterior	Posterior
Inactive	5 (4 adapting)	0
Slow irregular (1-20)	17 (3 adapting)	8
Fast irregular (21-250)	17	13
Burst (100-160)	5	15
Regular (10-100)	0	9
Total	44	45

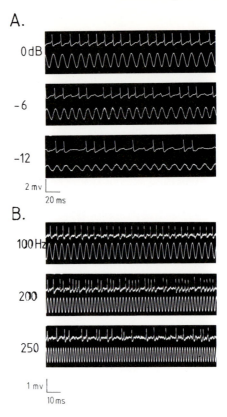

Figure 11-5. Typical responses of primary afferent fibers to pure tone stimuli. (A) For unit (W12.1), as the stimulus amplitude was reduced the total number of spikes decreased, and stimulus following was lost (125 Hz). (B) For unit (W12.2), as the stimulus frequency was increased at a fixed stimulus level, the total number of spikes decreased.

bution of spikes about the mean phase angle was more widespread than at high stimulus levels. The spread of spikes for a fixed number of stimulus cycles was described by an index of synchrony. The phase of each spike was considered a vector on a unit circle, and the index calculated as the length of the mean vector. [This is that statistic r defined by Goldberg and Brown (1969) and Mardia (1972).] From the data used to obtain r, the Rayleigh test (Mardia 1972) provides a basis for deciding whether the distribution of phase angles is significantly nonuniform and therefore determines whether the response is synchronized or not. For all units examined in this study the response was synchronized to only one half of the stimulus cycle for a given frequency of stimulation.

For units that were not spontaneously active, the synchrony index remained high at most stimulus levels, although the number of spikes diminished as the stimulus level decreased (Fig. 11-8). For irregular units, both the synchrony index and the number of spikes decreased as stimulus level was reduced (Fig. 11-9). Bursting units maintained a

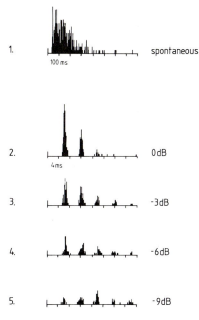

Figure 11-6. (1) The interval histogram for a slowly spontaneously active unit (W15.2). (2) During 160 Hz stimulation there is a multimodal distribution of spikes with peaks occurring at the stimulus period and multiples of it. (3, 4, and 5) The total number of spikes decreases as the stimulus amplitude is reduced.

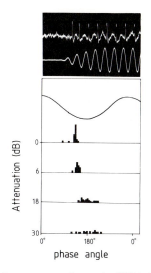

Figure 11-7. (A) Phase-locked response of a unit (W12.3). (B) The peak of the post-stimulus time histogram coincides with a particular stimulus phase angle. The distribution becomes more spread, and the total number of spikes decreases, as the stimulus level is reduced.

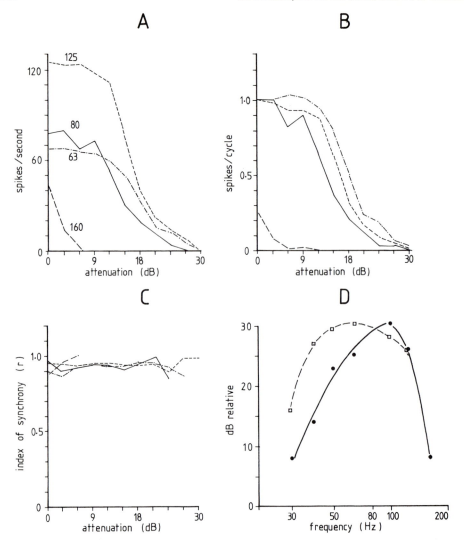

Figure 11-8. Frequency response characteristics of a unit showing no spontaneous activity (W13.2). (A) At high stimulus levels the number of spikes per second is equivalent to the number of stimulus cycles. As the stimulus amplitude is decreased, the number of spikes decreases. (B) At high stimulus levels there is about 1 spike per stimulus cycle. The number decreases as the stimulus amplitude is decreased. (C) The index of synchrony is high at all stimulus amplitudes. (D) Relative frequency sensitivity of the unit, derived from (i) spikes per cycle data (the level of stimulus required to give 0.3 spikes/cycle), dashed line; (ii) spikes per second data (the level of stimulus required to give 40 spikes/sec), full line.

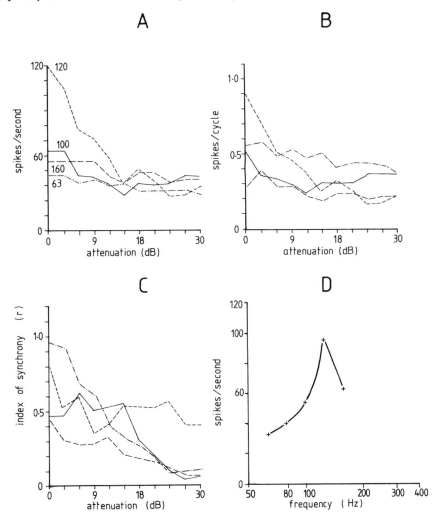

Figure 11-9. Frequency response characteristics of a slow irregular unit W10.5. (A) Spikes per second. (B) Spikes per cycle. (C) Index of synchrony decreases as the stimulus level is reduced. (D) Relative sensitivity curve for this unit is derived from isointensity data, i.e., the number of spikes per second occurring at a particular attenuation (−3 dB), for each frequency tested.

constant mean rate of spike production even during sound stimulation and responded only with a change in synchrony. For the bursting units, groups of spikes were allocated to each stimulus cycle (Fig. 11-10) so that the number of spikes/cycle multiplied by the frequency always gave a rate approximately equal to the spontaneous rate of discharge.

Some units did not produce spikes at the same rate throughout the entire duration of the stimulus, but adapted rapidly. These units (7 in total) were recorded from the

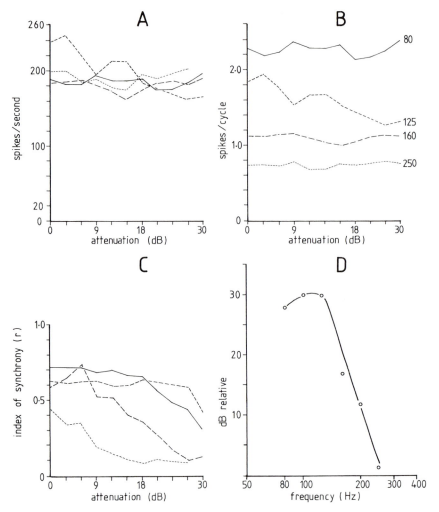

Figure 11-10. Frequency response characteristics of a bursting unit (16.2). (A) Shows the spikes per second remaining approximately the same at all stimulus frequencies and amplitudes. (B) Shows the spikes per cycle remaining approximately constant for any stimulus frequency as the amplitude of the stimulus is decreased. (C) The index of synchrony of the group of spikes decreases as stimulus amplitude is decreased. (D) Relative frequency sensitivity of the unit derived for a given index of synchrony (the level of stimulus attenuation required to give a synchrony index of 0.4).

anterior ramus of the saccular nerve (Table 11-1). The spontaneous activity of these adapting units was either zero or very low (5/sec). They were similar to spontaneously active units in their frequency response. Unit 13.2, shown in Fig. 11-11, was such a unit, responding at frequencies from 30 Hz up to 160 Hz but being completely insensitive on either side of this frequency range. Adaptation was least evident at 80-100 Hz and increased at stimulus frequencies at either side of this frequency. One exceptional case was an adapting unit which showed an even more restricted frequency range, only responding over the range 63-125 Hz and then only at high stimulus levels.

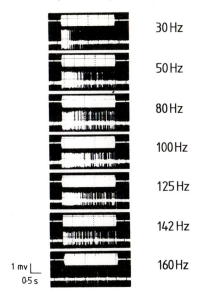

30 Hz

50 Hz

80 Hz

100 Hz

125 Hz

142 Hz

1 mv
0·5 s

160 Hz

Figure 11-11. An example of a unit (W13.2), with no spontaneous activity, but which responded over the whole frequency range, as shown in Fig. 11-8. Adaptation occurred at all stimulus frequencies, but was less prominent at 80 and 100 Hz.

3.4 The Frequency Response of Units

The response of the various classes of units to pure tone stimuli differed. To compare their frequency response characteristics several different criteria were adopted. For many units and especially the irregular and burst units, the level of stimulus required to produce a certain index of synchrony was used to draw up a relative sensitivity curve (Figs. 11-10 and 11-12). However, most units which were not spontaneously active showed a high index of synchrony (approx. 1.0) at most stimulus levels for all frequencies to which they responded. Therefore, relative sensitivity curves based on isorate data were drawn, that is, based on the stimulus level required to produce a given number of spikes. Isorate curves based on spikes/sec, and spikes/cycle were drawn. It should be noted that the spikes/sec data accentuate sensitivity at higher frequencies. With rapidly adapting units, the sampling period of 64 stimulus cycles included the adapted response, especially at low frequencies. For some units with slow spontaneous activity the response to increasing stimulus levels was not always an increase in number of spikes, i.e., it was not monotonic. For these units, and for some units whose spontaneous activity was higher than the stimulus frequency, iso-intensity curves were drawn. In this case the number of spikes/sec that occurred at a specific stimulus amplitude was determined. The ordinate of the sensitivity curve was therefore not in relative dB but was expressed in spikes/sec (Fig. 11-9).

The frequency ranges over which most units were sensitive were broad, especially relative to the critical bands for cod presented by Hawkins and Chapman (1975). There was generally a sharp cutoff in sensitivity above 80-160 Hz and a more gradual decline in sensitivity at lower frequencies.

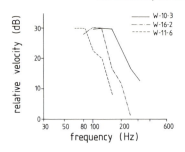

Figure 11-12. Relative frequency sensitivity of three units derived from synchrony index data. Bursting type (W16.2), with spontaneous rate of 170 spikes/sec. Irregular type (W11.6), with spontaneous rate of 50 spikes/sec. Irregular type (W10.2), with spontaneous rate of 115 spikes/sec.

For a given frequency of stimulation, the phase of the particle velocity cycle coinciding with spike production varied only slightly with stimulus level. The phase angle increased with frequency.

4 Discussion

In this first and preliminary account of activity within the cod auditory nerve our purpose has been, first, to examine the spontaneous activity of the primary afferent neurones and, second, to describe the response of the units to pure tones of differing frequency and amplitude. In a separate paper we consider the directional characteristics of the primary units (Hawkins and Horner, Chapter 15).

Many of the units contacted in the saccular nerve, indeed all the units in the posterior ramus, were spontaneously active even in the absence of sound stimulation. Spontaneous activity has been reported in previous studies of the fish auditory nerve (Löwenstein and Roberts 1951, Enger 1963, Furukawa and Ishii 1967, Fay 1978b), and is commonplace in the cochlear nerve of mammals (see, for example, Evans 1975). In common with other workers, we have classified the units into four types, based on their patterns of activity in the absence of direct stimulation; viz., silent, irregular, bursting, and regular.

Regular or rhythmic activity has been reported in units from the vestibular part of the labyrinths of fish and from tonic units of the otolith organs (Lowenstein and Roberts 1949, 1951, O'Leary, Dunn, and Honrubia 1976, Budelli and Macadar 1979). In our study these regular units were not sound or vibration sensitive, though Budelli and Macadar in their study of the utriculus of the guitar fish *Rhinobatis productus* concluded that the regular fibers were fit to transmit acoustical information. A regular maintained rate of discharge has been linked with the bidirectionality of response, as first noted by Lowenstein and Sand (1940), the discharge being excited and inhibited by motion in opposing directions. Such tonically active primary afferents are reported to be of narrow diameter and to have synaptic connection with many hair cells from the peripheral regions of the sensory epithelium (O'Leary et al. 1976). In addition, Highstein and Politoff (1978) have suggested that it is the simultaneous release of a chemical transmitter from many hair cells into profusely branching dendrites of the afferent fiber that results in a regular pattern of spike initiation.

Of the sound sensitive units, none of which were regularly active, the majority showed fast irregular trains of spikes, at rates of up to 250 spikes/sec. O'Leary et al. (1976) have pointed out that in the vestibular system irregular firing nerve fibers tend to be of wide diameter. Highstein and Politoff (1978) have associated irregular impulse initiation with neurons having restricted dendritic trees and associated with only one or two hair cells. We have seen that the interval histograms of irregular fibers are skewed distributions, as found for many cochlear nerve fibers in the cat by Kiang et al. (1965). Such a characteristic is an indication that the intervals between spikes are independent. Walsh et al. (1972) suggest that the irregular activity relates to a sto-chastic excitation associated with chemical synaptic transmission.

Sound sensitive bursting units have been reported from other species of fish (Lowen-stein and Roberts 1951, for the ray, Enger 1963, for the bullhead, Furukawa and Ishii 1967, and Fay 1978b, for the goldfish). Enger drew attention to these units and sug-gested that burst activity was associated with comparatively thick fiber diameters.

When stimulated by sound, phase-locking to the stimulus cycle was observed from all sound sensitive units. The phase-locking was sometimes maintained, even at low stimulus levels, when the number of spikes per unit time was greatly reduced. This type of response forms the basis for the well-known volley theory of Wever and Bray (1930) in which different fibers, firing individually at a low rate, are said to produce a composite discharge that faithfully reproduces the periodicity of the stimulus wave-form. A phase-locked response has also been reported for other fish species by Lowen-stein and Roberts (1949, 1951), Enger (1963), Furukawa and Ishii (1967), and Fay (1978b). In our case we found that the units responded only to one half of the stimu-lus cycle at any given frequency, unlike the goldfish in which a group of fibers from the sacculus respond to both phases (Furukawa 1978). Hawkins and Horner (Chapter 15) suggest that the locking to only one phase of stimulation results from an associ-ation of the unit with hair cells of a particular directional polarization; in the case of saccular units, situated in one half of the macula.

Though all sound responsive units showed phase-locked responses at high stimulus levels we have seen that the different classes of unit responded differently to sounds. A high degree of synchrony was shown by units which had no resting activity, even when the stimulus following was much reduced. This type of response adapted during a 2 sec stimulus period. The irregular spontaneous units were often very sensitive to sounds and a few could be influenced by background noise in the experimental appa-ratus. Their spontaneous discharge was increased or decreased during pure tone stimu-lation, depending on whether the stimulus frequency was higher or lower than the spontaneous frequency. The response normally did not adapt. Bursting units showed a unique characteristic, maintaining a rate of discharge equivalent to the spontaneous rate during stimulation. Their response was manifest only as a synchrony of groups of spikes with the stimulus waveform.

The precise phase of the cycle at which the spikes occurred differed between units and at different stimulus frequencies for each unit. The phase change was approxi-mately a linear function of frequency. Such a response indicates that the total auditory system including acoustical, mechanical and neural elements cannot be understood in terms of a simple model, for example, a tuned system, but must involve a series of time and phase delays between passage of the sound wave and production of the afferent spike.

The preservation of information on the phase of the stimulus in the response of the primary afferent fibers may be related to the general importance of phase to the fish. Piddington (1972) has shown that goldfish are capable of discriminating between a given click and its phase inversion, a discrimination that man cannot perform. Piddington went on to stress that phase analysis of the sounds generated by a swimming fish could be used to determine if the source was approaching or receding. Schuijf (1976) also stressed the importance of phase information to the cod for the unambiguous determination of the direction of propagation of a sound wave, a subject discussed by Hawkins and Horner (Chapter 15) and Schuijf (Chapter 14). A further aspect of this phase-locking is the preservation of information on the stimulus waveform in the form of a temporal code. If the central nervous sytem of the fish possesses some form of periodicity detector then the fish might discriminate between different sounds by the analysis of this temporal code.

Previous workers on the fish auditory system have been concerned with the analysis of frequency by the ear, and in particular with the presence of some form of peripheral mechanical frequency analyzer (van Bergeijk 1967, Sand 1974). Our study of the primary afferents of cod has shown, as has the work of most other workers on fish, that the units do not respond equally well at all frequencies. This restriction of the frequency response is clearly shown regardless of how the response is expressed and is particularly evident as a reduction in the response at higher frequencies. The frequency response curves for units from the cod sacculus were broad, with a gradual decrease in sensitivity at lower frequencies, but with a more abrupt loss of sensitivity at high frequencies than would be expected from comparison with the cod audiogram. This discrepancy at high frequencies is perhaps accounted for by the stimulation of the fish in our experiments by a source generating particle motion, with a minimum of sound pressure. Under free-field acoustic conditions, where the audiogram was obtained, the swimbladder of the fish serves as an accessory hearing organ, transforming incident sound pressures into reradiated particle motion. The transformation is believed to involve a progressively greater amplification of the particle motion received by the otolith organ as frequency increases (Chapman and Hawkins 1973, Sand and Enger 1973). The very low sound pressures generated by our sound source means that this stimulus amplification was absent in our experiments.

Though the units showed a restricted frequency response they were not highly tuned, and were much wider in bandwidth than the critical bands determined for cod from psychophysical experiments (Hawkins and Chapman 1975). Nor was there any evidence of a spatially organized arrangement of fibers tuned to different frequencies. There were, however, some qualitative differences between the anterior and posterior parts of the saccular macula. While units that were not spontaneously active were found only in the anterior ramus, bursting units predominated in the posterior ramus of the saccular nerve. Hawkins and Horner (Chapter 15) have described differences in the directionality of units from the two rami. These differences in the types of unit across the macula and the comparatively slight differences reported in microphonic potentials across the macula reported by other workers (Anderson and Enger 1968,

Sand 1974) are in no way comparable with the spatial organization of frequency found in the cochlea of mammals. We consider it very unlikely that the macula forms a mechanical frequency analyzer.

Acknowledgments. We thank our friends and colleagues Charles Robb and Gordon Smith for their advice and technical help, and Miss Ann-Marie Meconi for help in preparation of the manuscript. Kathleen Horner's research was financed by the Northern Ireland Education Department.

References

Anderson, R. A., Enger, P. S.: Microphonic potentials from the sacculus of a teleost fish, Comp. Biochem. Physiol. 27, 879-881 (1968).

van Bergeijk, W. A.: In Marine Bioacoustics, Vol. 2. Tavolga, W. N. (ed.). Oxford: Pergamon Press, 1967, p. 244.

Budelli, R., Macadar, O.: Statoacoustic properties of utricular afferents. J. Neurophysiol. 42, 1479-1493 (1979).

Buerkle, U.: Relation of pure tone thresholds to background noise levels in the Atlantic cod. J. Fish. Res. Bd. Can. 25, 1155-1160 (1968).

Chapman, C. J., Hawkins, A. D.: A field study of hearing in the cod. J. Comp. Physiol. 85, 147-167 (1973).

Dijkgraaf, S.: Über die Schellwahrnehmung bei Meeresfischen. Z. Vergl. Physiol. 34, 104-122 (1952).

Dijkgraaf, S., Verheijen, F. J.: Neue Versuchen über das Tonunterscheidungsvermögen der Elritze. Z. Vergl. Physiol. 32, 248-256 (1950).

Enger, P. S.: Single unit activity in the peripheral auditory system of a teleost fish. Acta Physiol. Scand. 59 Suppl. 3, 9-48 (1963).

Evans, E. F.: Cochlear nerve and cochlear nuclei. In: Handbook of Sensory Physiology. The Auditory System, Vol. 5/2. Keidel, W. D., Neff, W. D. (eds.). Berlin: Springer-Verlag, 1975.

Fay, R. R.: Auditory frequency discrimination in goldfish. J. Comp. Physiol. Psychol. 73, 175-180 (1970).

Fay, R. R.: Sound reception and processing in the carp, saccular potentials. Comp. Biochem. Physiol. 49A, 29-42 (1974).

Fay, R. R.: Phase-locking in goldfish saccular nerve fibres accounts for frequency discrimination capacities. Nature 275, 320-322 (1978a).

Fay. R. R.: Coding of information in single auditory nerve fibres of the goldfish, J. Acoust. Soc. Am. 63, 136-146 (1978b).

Fay, R. R., Popper, A. N.: Modes of stimulation of the teleost ear. J. Exp. Biol. 62, 379-387 (1975).

Fay, R. R., Popper, A. N.: Structure and function in teleost auditory systems. In: Comparative Studies of Hearing in Vertebrates. Popper, A. N., Fay, R. R. (eds.). New York: Springer-Verlag, 1980, pp. 3-42.

Fletcher, H.: Auditory patterns. Rev. Mod. Phys. 23, 47-65 (1940).

Furukawa, T.: Sites of termination on the saccular macula of auditory nerve fibres in the goldfish as determined by intracellular injection of procion yellow. J. Comp. Neurol. 180, 807-814 (1978).

Furukawa, T., Ishii, Y.: Neurophysiological studies of hearing in goldfish. J. Neurophysiol. 30, 1377-1403 (1967).

Goldberg, J. M., Brown, P. B.: Responses of binaural neurons of dog superior olivary complex to dichotic tonal stimuli: Some physiological mechanisms of sound localisation. J. Neurophysiol. 32, 613-636 (1969).

Griffin, D. R.: Underwater sounds and the orientation of marine animals, a preliminary survey. Tech. Rep. No. 3, Project NR 162-429 O.N.R. and Cornell Univ., 1-26 (1950).

Hawkins, A. D.: The sensitivity of fish to sounds. Oceangr. Mar. Biol. Annu. Rev. 11, 291-340 (1973).

Hawkins, A. D., Chapman, C. J.: Masked auditory thresholds in the cod, Gadus morhua (L.). J. Comp. Physiol. 103, 209-226 (1975).

Highstein, S. M., Politoff, A. L.: Relation of interspike baseline activity to the spontaneous discharge of primary afferents from the labyrinths of the toadfish, Opsanus tau. Brain Res. 150, 182-187 (1978).

Jacobs, D. W., Tavolga, W. N.: Acoustic frequency discrimination in the goldfish. Anim. Behav. 16, 67-71 (1968).

Kiang, N. Y. S., Watanabe, T., Thomas, E. C., Clarke, L.: Discharge patterns of single fibres in the cat's auditory nerve. Res. Monogr. No. 35. Cambridge, Mass.: M.I.T. Press, 1965.

Kleerkoper, H., Chagnon, E. C.: Hearing in fish, with special reference to Semotilus atromaculatus atromaculatus. J. Fish. Res. Bd. Can. 11, 130-152 (1954).

Lowenstein, O., Roberts, T. D. M.: The equilibrium function of the otolith organs of the thornback ray Raja clavata. J. Physiol. 100, 392-415 (1949).

Lowenstein, O., Roberts, T. D. M.: The localisation and analysis of the responses to vibration from the isolated elasmobranch labyrinth. A contribution to the problem of the evolution of hearing in vertebrates. J. Physiol. 144, 471-489 (1951).

Lowenstein, O., Sand, A.: The mechanism of the semi-circular canal. A study of the responses of single fibre preparations to angular accelerations and to rotations at constant speed. Proc. R. Soc. London Ser. B 129, 256-267 (1940).

Mardia, K. V.: Statistics of directional data. London: Academic Press, 1972.

Moulton, J. M.: Acoustic behaviour of fishes. In: Acoustic Behaviour of Animals. Busnel, R. G. (ed.). Amsterdam: Elsevier, 1963, pp. 655-693.

O'Leary, D. P., Dunn, R. F., Honrubia, V.: Analysis of afferent responses from isolated semi-circular canal of the guitarfish using rotational acceleration white noise inputs. I. Correlation of response dynamics with receptor innervation. J. Neurophysiol. 39, 631-644 (1976).

Piddington, R. W.: Auditory discrimination between compressions and rarefractions by goldfish. J. Exp. Biol. 56, 403-419 (1972).

Popper, A. N., Fay, R. R.: Sound detection and processing by teleost fishes, a critical review. J. Acoust. Soc. Am. 53, 1515-1529 (1973).

Sand, O.: Directional sensitivity of microphonic potentials from the perch ear. J. Exp. Biol. 60, 881-899 (1974).

Sand, O., Enger, P. S.: Evidence for an auditory function of the swimbladder in the cod. J. Exp. Biol. 59, 405-414 (1973).

Sawa, M.: Auditory responses from single neurons of the medulla oblongata in the goldfish. Bull. Jpn. Soc. Sci. Fish. 42, 141-152 (1975).

Schuijf, A.: The phase model of directional hearing in fish. In: Sound Reception in Fish. Schuijf, A., Hawkins, A. D. (eds.). Amsterdam: Elsevier, 1976.

Siegel, S.: Nonparametric Statistics for the Behavioural Sciences. New York: McGraw-Hill, 1956.

Tavolga, W. N.: In: Fish Physiology, Vol. 5. Hoar, W. S., Randall, D. J. (eds.). New York: Academic Press, 1971, pp. 135-205.

Tavolga, W. N.: Signal:noise ratio and the critical band in fishes. J. Acoust. Soc. Am. 55, 1323-1333 (1974).

von Békésy, G.: Experiments in Hearing. New York: McGraw-Hill, 1960.

Walsh, B. T., Miller, J. B., Gacek, R. R., Kiang, N. Y. S.: Spontaneous activity in the eighth cranial nerve of the cat. Int. J. Neurosci. 3, 221-236 (1972).

Wever, E. G., Bray, C. W.: Action currents in the auditory nerve in response to acoustical stimulation. Proc. Nat. Acad. Sci. 16, 344-350 (1930).

Wohlfahrt, T. A.: Untersuchungen über das Tonunterscheidungsvermögen der Elritze, *Phoxinus laevis*. Z. Vergl. Physiol. 26, 570-604 (1939).

Zwicker, E., Flottorp, G., Stevens, S. S.: Critical bandwidth in loudness summation. J. Acoust. Soc. Am. 29, 548-557 (1957).

Chapter 12

Frequency Discrimination in Teleosts—Central or Peripheral?*

Per S. Enger**

1 Introduction

Pitch discrimination in fish is well established although the number of behavioral studies concerned with the topic is fairly limited. Table 12-1 lists some species investigated and the results obtained, including the value for *Cottus scorpius* which will be reported here. In such studies it is important to be aware of the complicated and unpredictable acoustics of small tanks (Parvulescu 1967). For example, when changing the frequency of the sound producing equipment, great fluctuations in sound pressure and particle displacement amplitude will occur as well. Moreover, the relation between sound pressure and particle displacement is not predictable, as it would be in a large body of water, like the open sea. It is generally agreed that particle displacement is the relevant stimulus for the auditory receptor cells. For a fish with a swimbladder, sound pressure is a relevant stimulus as well, since the swimbladder is then acting as a pressure to displacement transformer. This secondary displacement will be transmitted through the surrounding tissues to the inner ear, there stimulating the sensory hair cells. In the ostariophysine species, all of which have a bony connection between the swimbladder and the inner ear, the sense of hearing is particularly good. In nonostariophysine species there seems to be a relation between threshold and hearing range on the one hand, and the anatomical configuration of the peripheral auditory system and the swimbladder on the other (Coombs and Popper 1979).

The ability for directional hearing and pitch discrimination might also be influenced by the same anatomical differences, and particularly by the presence or absence of a swimbladder. For example, one would think that the presence of a swimbladder would make directional hearing impossible or at least hamper it badly, because, irrespective of the true sound source direction, the relevant stimulus would originate from the swimbladder as particle displacement resulting from sound pressure changes. Correspondingly, the accuracy of pitch discrimination might perhaps be better in fish with a

*The scanning electron microscope work in this study has been done at the Electronmicroscopical Unit for the Biological Sciences.
**Institute of Zoophysiology, University of Oslo, Blindern, Oslo 3, Norway.

Table 12-1. Hearing Range and Pitch Discrimination in Some Teleosts

Species	Upper frequency limit (Hz)	Frequency discrimination at given pitch		Reference
Ostariophysi				
Phoxinus laevis	5000-7000	19%	at 1000 Hz	Stetter 1929
		6%	at 1000	Wohlfahrt 1939
		3%	at 400-800	Dijkgraaf and Verheijen 1950
Carassius auratus	3480	3.5%	at 500	Stetter 1929, Jacobs and Tavolga 1968
		3.5%	at 400	Fay 1970
		4.4%	at 1000	Fay 1970
		7%	at 50	Fay 1970
Nonostariophysine				
Gobius niger	800	9%	at 150	Dijkgraaf 1952
		12%	at 300	Dijkgraaf 1952
Corvina nigra	1000	12%	at 385	Dijkgraaf 1952
Sargus annularis	1250	<9%	at 150	Dijkgraaf 1952
		15%	at 300-450	Dijkgraaf 1952
		>30%	at 600	Dijkgraaf 1952
Cottus scorpius	300	19%	at 60	Pettersen 1980
		8%	at 95	Pettersen 1980
		16%	at 140	Pettersen 1980

swimbladder where all the sensory cells are stimulated in the same direction, whereby a better specialization for peripheral frequency discrimination would be possible. Neither of the two assumptions is true. Schuijf, Baretta, and Wildschut (1972) and Olsen (1969) have clearly demonstrated directional hearing in fishes with a swimbladder (also see Schuijf, Chapter 14). The unpublished study of Pettersen (1980) to be presented here shows that a fish lacking swimbladder distinguished pitch just as well as other nonostariophysine species.

2 Frequency Discrimination in a Teleost Without a Swimbladder

Provided that particle displacement is the relevant stimulus for the auditory receptors (cf. Fay and Popper 1975, Buwalda, Chapter 7), it would be advantageous to investigate a species that has no swimbladder and thereby cannot be stimulated through two channels, namely, directly as well as indirectly through the swimbladder. Furthermore, it would be important to measure the stimulus strength, i.e., the displacement amplitude and not the sound pressure.

In recent experiments, Pettersen (1980) studied pitch discrimination in such a fish, the bullhead, *Cottus scorpius,* while measuring particle displacement of the stimulating tone. The fish was supplied with electrodes for recording the heartbeats and placed in a small wide-mesh plastic cage suspended in a larger polyethylene tank. Sound pressure could be measured by a calibrated hydrophone, and particle displacement by velocity meters. The latter were placed in a water tight persphex tubing, dimensioned so that the specific gravity of the tubing with the velocity meter was equal to that of water. A similar measuring system has previously been constructed by Banner (1973). The fish was trained by presenting it a stimulus which consisted of a series of tone pips which alternated in frequency, while a series of tone pips that did not change frequency was the neutral signal. At the end of a series of alternating tone pips, the fish was given a mild electric shock. This procedure is similar to the one used by Jacobs and Tavolga (1968). The response to the shock was a clear, but not very pronounced bradycardia. After three days on the average, each with some 20 test series, the larger specimens had developed bradycardia in response to the alternating tone pips. When trained, the frequency difference between the alternating tone pips was reduced and the threshold for detection of frequency differences was established. Every one of the successfully trained fish were also tested with a series of tone pips, of the same frequency, but alternating in intensity with 4-6 dB. All of these tests gave negative responses. The results are given in Table 12-1. The frequency discrimination was best at 95 Hz with an average value of 8% (extreme values among the five fishes tested were 5.9 and 9.6%), and far better than at 60 Hz with 19% and at 140 Hz with 16%. The audiogram that has also been determined showed the lowest threshold in the 100-125 Hz range; in other words, the best pitch discrimination was found for the optimal hearing range. The values determined correspond well with those obtained from nonostariophysine species possessing a swimbladder (Table 12-1). Thus, the presence of a swimbladder does not necessarily seem to be an advantage with respect to improving pitch discrimination. The orientation of the sensory hair cells have been studied by Bergø (1980). The pattern of cell orientation is the same as that found in many nonstariophysine species with a swimbladder, such as cod (Dale 1976), perch (Enger 1976) and many species investigated by Popper (for review see Fay and Popper 1980; Platt and Popper, Chapter 1). In general, the cell orientation pattern would seem to favor directional hearing rather than pitch discrimination, but as will be seen in the following section, other morphological as well as physiological data support the idea that a crude frequency discrimination takes place peripherally.

3 Evidence for Peripheral Pitch Discrimination

The question of central versus peripheral pitch discrimination in fish is still open. The previous section revealed no differences among the nonostariophysine species with respect to discrimination ability. On the other hand, the Ostariophysi discriminate better than other fish and show a dorsoventral orientation of the hair cells throughout the saccular sensory maculae (Hama 1969, Platt 1977, Neset 1979). This macula does not seem suited for directional hearing, at least not by means of directly detecting the displacement direction; it might in fact be better adapted for frequency discrimination if auditory stimuli reaches the ear from the swimbladder through the Weberian ossicles. If frequency discrimination takes place centrally, and is based on the volley principle, it follows that the central auditory system must be better developed in Ostariophysi than in nonostariophysine species. The larger acoustic areas in the medulla oblongata in the Ostariophysi reflects a generally better sense of hearing. It is hard to believe, however, that two principally different means of pitch discrimination should occur in the two groups of fish, and the arguments given thus far favor the idea that frequency discrimination takes place in the central nervous system.

What are now the evidence and data in support of the opposite view? Enger (1963) found in the bullhead three types of auditory units in the acoustic nerve, each covering a different frequency range. Furukawa and Ishii (1967) could classify acoustic nerve units into two groups according to the best sound frequencies. Fay (1978) found clear differences in response pattern and best frequencies among single units in the saccular nerve of goldfish. Sand and Michelsen (1978) studied the vibration pattern of the saccular otoliths in perch by means of laser technique, when vibrating the fish on a vibrating table, and have given some evidence of a frequency dependent vibration pattern.

The study presented here has its basis in the idea that the sensory maculae will be destroyed by excessive sound stimulation and that the area affected will depend on the sound frequency.

Codfish, about 25 cm long, were placed in an aluminum tube, 71 cm long, with 19 cm outer and 12 cm inner diameter. To both ends were fitted loudspeakers which could be operated in phase or 180° out of phase. The in phase mode was used in the present study, thus producing a high sound pressure and a small particle displacement in the middle of the tube. The fish was held in a wide mesh plastic cage so that the swimbladder was roughly in the middle of the aluminum tube. The maximal sound pressure which could be produced by the equipment was about 80 dB re 1 μbar, which corresponds to a value some 100-110 dB above threshold in the most sensitive frequency range of the codfish, 150-250 Hz. The exposure time was varied from 1 to 5 hr after which the fish was decapitated. The skull was rapidly opened and the brain removed. The neurocranium was divided in the midline by a sharp scalpel, and the two halves were trimmed of connective tissue and muscles so as to contain only utriculus, sacculus, and lagena with the supporting bones. The preparations were fixed in a 0.1 M cacodylate-buffer with 2% glutaraldehyde. The saccular sensory maculae were dissected out not less than 24 hr later and mounted with histoacryl blue on slightly concave, thin plastic pieces and prepared for scanning electron microscopy.

The frequencies tested were 50, 100, 200, and various frequencies between 300 and 400 Hz. Altogether 30 fish have been successfully exposed to intense sound, which means that 60 preparations have been investigated. The criterion required for classifying the epithelium as destroyed by sound exposure was a complete or almost complete lack of sensory cilia. An example is shown in Fig. 12-1, where the destructed areas are seen as small, black patches on the left part of the top picture. Three micrographs at higher magnification are also shown, the left one is taken from a destroyed area. The tops of the sensory cells with the ciliary base plate without cilia are seen. Another example is shown in Fig. 12-2 where, at low magnification in A, the destructed areas appear as two fairly long, dark bands from the middle toward the dorsal side of the macula, separated by a narrow ridge of cells with cilia. In C and D part of the same area is shown at higher magnification. Between destructed patches and along the border of destructed areas, the cilia were usually disorganized, either in that the cilia were not held together but pointed in different directions, as shown in Fig. 12-3B, or in that most of the cilia had the same length and gave the impression of having been

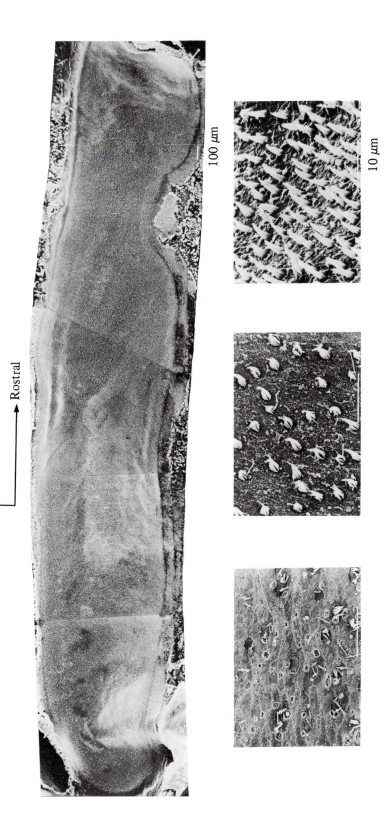

Figure 12-1. Low power micrograph of a complete saccular macula in codfish, exposed to 50 Hz high intensity pure tone (upper graph). White points represent bundles of cilia. Black patches in left half of macula are destructed areas where sensory cilia have disappeared. Destructed and intact areas at higher magnification in lower graphs.

Dorsal

Rostral

100 μm

10 μm

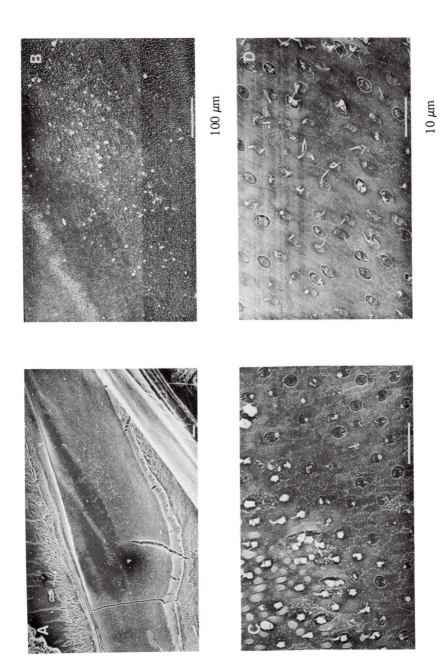

Figure 12-2. (A) Part of saccular macula, exposed to 100 Hz high intensity tone. Dark bands from the middle toward the upper and right side of macula are destructed areas lacking sensory cilia. Rostral end to the left, dorsal side up. (B) Higher magnification, parts of dark bands in upper left corner. (C and D) Details of peripheral and central parts of destructed regions.

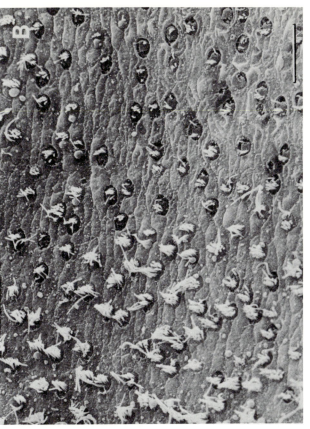

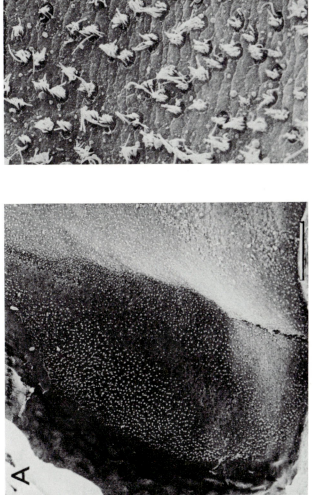

10 μm

100 μm

Figure 12-3. Caudal part of saccular macula after 50 Hz intense tone exposure. Dark patch in the middle of (A) is destructed region, seen in (B) at higher magnification. Note malformed ciliary bundles at the edge of destructed area.

bitten off, as can be seen in Fig. 12-2C. Thus, the areas affected by the intense sound were almost certainly larger than the areas lacking cilia.

The reasons that these affected areas are not included in what is called the destructed regions will be clear from the following. Control preparations, taken from fish that had not been exposed to intense sound, never had areas lacking cilia, even though scattered cells with only the ciliary plate present have been seen. On the other hand, areas with disorganized cilia have been observed. Among the eight control fish studied, one was particularly bad in this respect. The reason for this is not known; neither is the life history of the fish prior to the experiment known. It is possible that rough handling could destroy the sensory cilia. A control experiment to test this, for example, to drop the fish on the floor before killing it, has not been done yet. Micrographs at high magnification of intact and destructed cells are shown in Fig. 12-4.

Figure 12-5 summarizes the results. All the black patches and areas found, as exemplified in Figs. 12-1 and 12-2, are drawn on the same schematic for each of the frequencies tested. The total area that has been destructed in all the specimens together is hatched. For 350 Hz (more correctly 300-400 Hz), the drawing is based on 20 preparations, for 200 Hz on 10 preparations, for 100 Hz on 12 preparations, and for 50 Hz on 10 preparations.

There seems to be a clear tendency for the maculae to be destructed more anteriorly as the sound frequency is increased. This is the strongest argument against a possible criticism that the destruction is due to the handling of the preparation. If the cilia

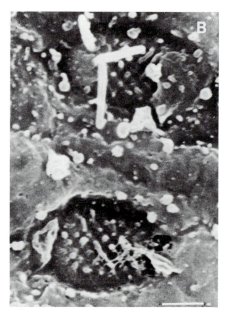

1 μm

Figure 12-4. Intact (A) and destructed (B) hair cells.

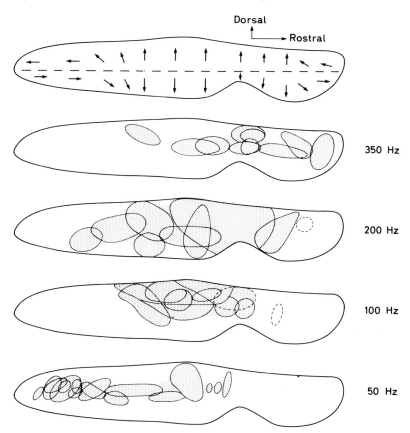

Figure 12-5. Distribution of destructed areas on the saccular macula at sound frequencies indicated. Hatched area represent destructed regions from all fish tested. Top drawing (from Dale 1976) indicates hair cell orientation. Arrows show position of kinocilia with respect to stereocilia.

were rubbed off accidentally during the handling, systematic differences like those shown would not be expected. The reason for exposing twice as many fish to tones of 300-400 Hz than to lower frequencies was that the saccular macula was less apt to be destroyed by the higher frequencies, possibly because the sound pressure applied was higher above threshold at 200 Hz than at 350 Hz.

The central region of the saccular macula is affected by all frequencies. In addition, the higher audible frequencies seem to affect the anterior portion, and the lower frequencies the posterior portion of the macula.

Sand and Michelsen (1978) measured the vertical otolith vibration in perch by laser vibrometer during horizontal vibration along the long axis of the fish. They found that both ends of the otolith vibrated vertically at 40, 90, and 220 Hz, but 180° out of phase. The point of minimum vertical movement varied somewhat with frequency and was measured in the posterior half of the otolith. Comparing their study with the present, one would immediately think that maximal movements would cause most damage and minimum movement would cause little or no damage. In the present investigation on the bullhead, all sound frequencies caused damage in the middle part of the maculae, where the vertical vibrations had the lowest amplitude in the perch. It must be pointed out, however, that in the study of Sand and Michelsen, only vertical vibrations, were measured. The pattern of horizontal vibrations is not known. The two studies do support each other in the conclusion, however, that there is evidence for a coarse frequency discrimination taking place in the peripheral auditory system of fish.

References

Banner, A.: Simple velocity hydrophones for bioacoustic application. J. Acoust. Soc. Am. 53, 1134-1136 (1973).

Bergø, G.: Sensory epithelia in the fish labyrinth. Master Thesis, University of Bergen, 1980.

Coombs, S., Popper, A. N.: Hearing differences among Hawaiian squirrelfish (Family Holocentridae) related to differences in the peripheral auditory system. J. Comp. Physiol. 132, 203-207 (1979).

Dale, T.: The labyrinthine mechanoreceptor organs of the cod Gadus morhua L. Norw. J. Zool. 24, 85-128 (1976).

Dijkgraaf, S.: Über die Schallwahrnehmung bei Meeresfischen. Z. Vergl. Physiol. 34, 104-122 (1952).

Dijkgraaf, S., Verheijen, F.: Neue Versuche über das Tonunterscheidungsvermögen der Elritze. Z. Vergl. Physiol. 32, 248-256 (1950).

Enger, P. S.: Single unit activity in the peripheral auditory system of a teleost fish. Acta Physiol. Scand. 59, Suppl. 210 (1963).

Enger, P. S.: On the orientation of haircells in the labyrinth of perch (Perca fluviatilis). In: Sound Reception in Fish. Schuijf, A., Hawkins, A. D. (eds.). Amsterdam: Elsevier, 1976, pp. 396-411.

Fay, R. R.: Auditory frequency discrimination in the goldfish (Carassius auratus). J. Comp. Physiol. Psychol. 73, 175-180 (1970).

Fay, R. R.: Coding of information in single auditory-nerve fibers of the goldfish. J. Acoust. Soc. Am. 63, 136-146 (1978).

Fay, R. R., Popper, A. N.: Modes of stimulation of the teleost ear. J. Exp. Biol. 62, 379-387 (1975).

Fay, R. R., Popper, A. N.: Structure and function in teleost auditory systems. In: Comparative Studies of Hearing in Vertebrates. Popper, A. N., Fay, R. R. (eds.). New York: Springer-Verlag, 1980, pp. 3-42.

Furukawa, T., Ishii, Y.: Neurophysiological studies on hearing in goldfish. J. Neurophysiol. 30, 1377-1403 (1967).

Hama, K.: A study on the fine structure of the saccular macula of the goldfish. Z. Zellforsch. 94, 155-171 (1969).

Jacobs, D. W., Tavolga, W. N.: Acoustic frequency discrimination in the goldfish. Anim. Behav. 16, 67-71 (1968).

Neset, M.: A morphological and electrophysiological investigation of sacculus and lagena of the catfish (Ictalurus nebulosus). Master Thesis, University of Oslo, 1979.

Olsen, K.: Directional hearing in cod (Gadus morhua L.). 8th Working Group for Fishing Technology. I. F. Meeting, Lowestoft, England, 1969.

Parvulescu, A.: Acoustics of small tanks. In: Marine Bio-Acoustics, Vol. 2. Tavolga, W. N. (ed.). Oxford: Pergamon, 1967, pp. 7-14.

Pettersen, L.: Frequency discrimination in the bullhead, *Cottus scorpius,* a fish without swimbladder. Master Thesis, University of Oslo, 1980.

Platt, C.: Hair cell distribution and orientation in goldfish otolith organs. J. Comp. Neural. 172, 283-298 (1977).

Sand, O., Michelsen, A.: Vibration measurement of the perch otolith. J. Comp. Physiol. 123, 85-89 (1978).

Schuijf, A., Baretta, J. W., Wildschut, J. T.: A fish investigation on the discrimination of sound direction in *Labrus berggylta* (Pisces: Perciformes). Neth. J. Zool. 22, 81-104 (1972).

Stetter, H.: Untersuchungen über den Gehörsinn der Fische, besonders von *Phoxinus laevis* L. und *Amiurus nebulosus* Raf. Z. Vergl. Physiol. 9, 339-447 (1929).

Wohlfahrt, T. A.: Untersuchungen über das Tonunterscheidungsvermögen der Elritze (*Phoxinus laevis* Agass). Z. Vergl. Physiol. 26, 570-604 (1939).

Discussion

JENKINS: In the areas where there was bundle destruction, it looked to me that there were fewer microvilli present. Was that significant?

ENGER: No.

PLATT: How did you remove the otoliths and the otolith membranes from the macula?

ENGER: I did the opposite. I took the macula off the otolith, instead of the other way around. I deflected the macula with fine forceps.

HAWKINS: How did you check the frequency response of the animals after they were subjected to the high intensity stimulation? How much tissue repair did you find after the experiment, and was there any change in the sensitivity of the animals with time?

ENGER: That is an interesting problem, but we have not looked at any of the effects of regeneration.

FINE: Some claim that we can account for frequency discrimination in the goldfish on the basis of the volley principle. It seems to be adequate for the toadfish, but I wonder if the volley principle could explain *all* frequency discrimination in fish? Fish should be able to synchronize their afferent nerve output with the stimulus frequency, and so the volley principle would be operative.

ENGER: The mechanisms for frequency discrimination should be the same for ostariophysines and nonostariophysines. In order to have frequency analysis by the volley principle, one would have to ask a greal deal of the central nervous system, whereas the place principle can actually be shown to exist. However, it seems quite possible that both volley and place principles operate together.

GRAY: This notion of frequency analysis seems to me to be too general a term, and an oversimplification. Do you think there is one single system for coding these auditory signals? You say that sound fundamentally goes up and down in intensity, and on and off, but what about the transients and complex sounds that may be produced? There is a finite onset time, a period of resonance, and a decay time in any transient. For the kinds of frequencies that fish appear to hear best, say about 100 Hz, a period of five cycles is quite a long time. When a fish is listening to a mating call, for example, it may be an adequate amount of time to perform a Fourier analysis. The analysis can be done, given adequate time, by timing the pulses. I am not suggesting that the fish actually measures the time between pulses, but the delay from a compression to a rarefaction may match one part of the fish's spectrum as opposed to another to enable it to make a discrimination.

Is it necessary to assume a single mechanism for frequency analysis? There may be a number of stereotyped behavioral responses under different conditions to different stimuli. Each behavior response may make the recognition of the next signal occur through a different mechanism.

ENGER: Yes, I quite agree with you, but I can only think of two principles, the volley and the place, and both may be operating. From my work, I do not claim that there is no volley principle in operation, but I do show that the place principle can apply.

TAVOLGA: I have always been in favor of a place principle in the fish ear. However, perhaps your data can be interpreted differently. Your standard frequency was 100 Hz, and you equated the intensity of this to your test frequency of, say, 110 Hz. Referring to the steep slope of the audiogram in that region, it might well be that these intensities are in effect quite different to the fish. The steepness of the slope indicates there might be as much as a 3 to 4 dB difference between the reference and text frequencies. The fish could, therefore, be making a discrimination on the basis of intensity rather than frequency.

ENGER: Yes, that could be possible. However, in our studies, the test tones were presented at different intensities in random fashion, so that the fish should not be making an intensity discrimination.

I might add that it was very difficult to condition the cod to respond to intensity differences.

BULLOCK: This is comparable to the problem of color discrimination. One way to eliminate intensity differences is to randomize the intensities so that color becomes the only relevant factor. In the case of sounds, one should randomize both reference and test tones in intensity, so that the fish eventually learns that intensity differences are not significant.

HAWKINS: I should like to return to the question of the involvement of the swimbladder. In work that Chapman and I did, we found that the oscillation of the swimbladder increases with frequency. If there is a close link between the swimbladder and the anterior macula, that would help explain your results.

ENGER: Yes, it may.

BUWALDA: I propose that you repeat these experiments with sculpins, where there is no swimbladder-inner ear link. If your results are the same, then I surrender.

CORWIN: I think your results would be better supported if you used some methods in addition to behavioral audiograms. Perhaps you should use some technique to selectively stain the dead cells, as has been done with some tissue culture methods.

ENGER: I am not sure that the cells are dead, perhaps just disabled. Essentially, the point is that fish exposed to one frequency show an effect on one part of the macula, while another frequency affects another part.

CORWIN: Yes, that, of course, is quite convincing.

POPPER: I am not sure we can equate our results with yours exactly, but we obtained a temporary threshold shift in goldfish after stimulation at 30 to 40 dB (re 1 μbar) for 4 hr. The effects on hearing sensitivity were quite distinct, and in a couple of hours, the sensitivity returned to normal. The experiments [Comp. Biochem. Physiol. 53A, 11-18 (1976)] are not strictly comparable. Our sound stimulus was different, and we used the goldfish. Perhaps we might repeat your kind of a study with the goldfish.

Chapter 13

Mismatch between Sound Production and Hearing in the Oyster Toadfish

Michael L. Fine*

1 Introduction

The spectral properties of the mating calls of both vertebrate and insect species are matched to the most sensitive range of their hearing, a region in which many auditory fibers are tuned (Bullock 1977, Capranica 1978). However, previous single unit studies in fishes have focused on neuron response properties as they relate to auditory processing rather than as a component of an integrated communication system. Because of extensive work (anatomical, physiological, and behavioral) on communication in the oyster toadfish *Opsanus tau* (see Fine, Winn, and Olla 1977 for review), it is an ideal fish species in which to investigate matching of primary auditory neurons to the characteristics of communication signals.

The toadfish produces an agonistic grunt and a courtship boatwhistle by simultaneously contracting paired intrinsic muscles on its swimbladder. The grunt is a short-duration, wide-band sound with a fundamental frequency, equivalent to the muscle contraction rate, of 90 or 100 Hz, while the boatwhistle is a long-duration, tonal call with a fundamental frequency varying about 200 Hz. The fundamental frequency of the boatwhistle undergoes a pronounced seasonal cycle with changes in response to temperature and other factors assumed to be hormonal (Fine 1978). Correlating with the extensive call variation, playbacks of natural boatwhistles as well as synthetic tone bursts with frequencies ranging between 180 and 400 Hz are treated as natural boatwhistles and increase the calling rate of nesting males in the field (Winn 1972).

*Section of Neurobiology and Behavior, Cornell University, Ithaca, New York 14850.
Present address: Department of Biology, Virginia Commonwealth University, Richmond, Virginia 23284.

2 Materials and Methods

Toadfish were anesthetized (ketamine), immobilized (flaxedil), and clamped into a tank with the top of the head above water. Single fibers were then isolated from the saccular nerves (3 M KCl microelectrodes), in response to 300 msec tone bursts from a speaker in air. These tone bursts were phase-locked, had rise-fall time of 5 msec, and were generated once per second. Stimuli and noise were measured with a Celesco LC 34 hydrophone, a B&K 2508 measuring amplifier, and a General Radio wave analyzer with a 3 Hz filter. Although other frequencies were tested, thresholds were determined for standard test frequencies of 25, 30, 40, 60, 90, 120, 150, 200, 250, 300, and 350 Hz. Since units were tested at discrete frequencies, the most sensitive or best-tested frequency (BTF) is assumed to approximate the best frequency (BF) for that neuron.

3 Results

Of the 106 units isolated from 22 fish, all were sensitive to low frequency sounds (Fig. 13-1B). BTFs exhibited a strong modal peak at 40 Hz, and 83% of all units had BTFs of 25, 30, or 40 Hz. Only 5 neurons were tuned to 60 Hz, and there was a small secondary maximum at 90 Hz. Tuning curves took complicated shapes, often having more than one sensitivity peak. For example, a cell with a BTF at 40 Hz might decrease sensitivity at 60 Hz and regain it at 90 Hz. Twenty-five Hz was the lower limit of testing, but the presence of neurons with BTFs at this frequency and the shapes of many tuning curves suggest that the toadfish is sensitive to lower frequencies. Since no units were tuned above 90 Hz, reception of higher frequencies is mediated by the stimulation of neurons in a region where they are relatively insensitive. The upper tails of tuning curves had variable shapes ranging from continuous loss in sensitivity to plateaus to actual increases in sensitivity. Although these discontinuities occurred at different frequencies, they were most common between 200 and 300 Hz.

I attempted to estimate the toadfish's audiogram by employing the threshold of the third most sensitive neuron at each test frequency. Such an operational definition focuses on the fish's most sensitive neurons, yet ensures that a threshold cannot be based on a statistical outlier. The audiogram (Fig. 13-1A) demonstrates a maximal sensitivity of –23 dB re 1 μbar between 40 and 90 Hz. Thresholds increase slightly below 40 Hz and rapidly above 90 Hz. The discontinuities in the tuning curves are reflected in the audiogram's plateau at 3 dB between 200 and 250 Hz. Above 250 Hz sensitivity declines to 23 dB by 350 Hz.

4 Discussion

In those vertebrates studied, the collected thresholds at BF of the most sensitive neurons of a species will approximate its behavioral audiogram (Newman et al. 1977). This generalization does not hold for the toadfish since it is capable of responding to 400 Hz signals (Winn 1972, Fish and Offutt 1972) yet has no neurons with BTFs above 90 Hz. The region of maximal sensitivity covers less than a quarter of the fish's audible range.

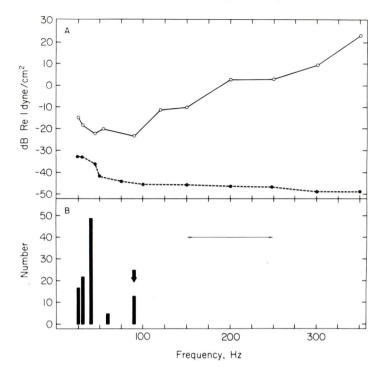

Figure 13-1. Summary of auditory and sound spectral information for the oyster toadfish. (A) Upper curve: Audiogram derived from thresholds of saccular neurons. Lower curve: Background noise measured with 3 Hz bandwidth. (B) Distribution of best tested frequencies for 106 auditory neurons from the saccular branch of the VIIIth nerve. The arrow marks the fundamental frequency of the grunt, and the horizontal bar indicates the range of the fundamental frequency of the boatwhistle during the active part of the mating season.

Such a finding implies that the upper tails of tuning curves assume a greater responsibility for hearing in the toadfish than in other known vertebrates.

In terms of communication, the fish's sensitivity to 40 Hz and lower frequency sound is unclear: The method of recording precludes lateral line involvement. Though Winn (1972) reported that toadfish calling is strongly affected by background noise, we have no hard evidence as to the function of this frequency channel. The fundamental frequency of the grunt, 90 Hz, is matched to the fish's most sensitive auditory region (Fig. 13-1B).

The fundamental frequency of the boatwhistle varies with temperature and assumed hormonal concentration (Fine 1978) over a range of less than 150 Hz to over 250 Hz (Fig. 13-1B). Since neurons with BTFs above 90 Hz were not found, the toadfish appears unique among vertebrates and invertebrates yet examined in not having neurons directly tuned to the fundamental frequency of its mating call. As demonstrated by individual tuning cuves, and the audiogram (Fig. 13-1A), the toadfish possesses a secondary tuning around 200 Hz that could aid in boatwhistle reception. The sound pressure level of the boatwhistle at 1 m is 40 dB re 1 dyne/cm² (114 dB SPL) (Tavolga 1971). With a plateau in auditory threshold at 3 dB between 200 and 250 Hz, it is obvious that the toadfish can hear its call. Detecting the boatwhistle on the tails of neurons tuned to lower frequencies simplifies the design of the auditory system and can be useful if precise frequency discrimination is not required. Because of the large variation in the fundamental frequency of the boatwhistle, precisely tuned neurons, if they existed, would be required to undergo coincidental hormonal and temperature variation in their best frequencies to maintain a match with sound production.

Behavioral and physiological evidence concerning the toadfish mating call are in good agreement:

1. The fundamental frequency of the boatwhistle varies extensively throughout the mating season (Fine 1978).
2. Toadfish will respond to playbacks of 180-400 Hz tone bursts as if they are natural boatwhistles (Winn 1972).
3. Toadfish neurons are most sensitive to frequencies below the fundamental frequency of the boatwhistle.

Thus, it appears unnecessary to postulate a precisely tuned mating call detection system within the frequency domain.

Unlike most communication systems, where frequency spectra of acoustic signals and audiograms are matched, the mismatch between hearing and sound production in the toadfish will minimize the distance over which conspecifics can communicate. This system should hypothetically enable individuals to respond preferentially to relatively loud boatwhistles of nearby callers. By assuming spherical spreading, Myrberg (1978) calculated that a toadfish should be capable of hearing a boatwhistle emitted 75 m away. Yet there is no evidence that a toadfish will respond at such a great distance. Schuijf and Siemelink (1974) have demonstrated that cod can be trained to make a 60° discrimination of (75 Hz) sound under water with its ears at a far-field distance of 5.3 m. This finding has not been generalized to higher frequencies, maximal distances, or most importantly, to unconditioned behavior in teleosts. Popper, Salmon, and Parvulescu (1973) found that squirrelfish would swim toward a speaker emitting conspecific alarm calls when the source was within 2.0 m of the test cage. Fish failed to orient to the source when it was more than 3.0 m distant, which the authors correlated with the acoustic far-field boundary. By playing back boatwhistles, Fish and Offutt (1972) found that male toadfish would not respond to calls with an SPL equivalent to a distance much greater than 2.5 m. They also found that the SPL sufficient to facilitate calling was 15 dB above the fish's threshold to a continuous tone. The presumed evolutionary strategy of the toadfish, namely specialization for short distance communication, is not unreasonable when one considers the likely difficulty of localizing a distant sound underwater and the fact that the frequency-amplitude spectrum of an acoustic signal will be severely distorted with distance in the toadfish's shallow-water habitat.

Although the toadfish may represent an extreme case of mismatch by not having neurons tuned to its mating call, other species have been found where social sounds are not well matched with hearing. Sachs, Sinnott, and Heinz (1978) found that the cowbird is an exception to the usual correspondence in birds since its song makes use of extremely high frequencies (6-11 kHz), while its auditory sensitivity is maximal between 2 and 4 kHz. Likewise Dooling, Zoloth, and Baylis (1978) found that the peak spectral energy in the house finch's song and other social calls is pitched above the most sensitive frequency in the audibility curve. Finally, human beings have peak sensitivity between 2 and 4 kHz, yet most of the energy in our voices is below these frequencies (Geldard 1972). While such mismatches appear to be uncommon, they clearly occur, and more attention should be devoted to uncovering instances of this phenomenon and to understanding the selective pressures favoring its occurrence.

Acknowledgments. This study was supported by an NIH postdoctoral fellowship in the laboratory of Robert R. Capranica, whom I would like to thank for his support in all phases of this project.

References

Bullock, T. H.: Recognition of complex acoustic signals. Life sciences research report 5. Dahlem Konferenzen, Berlin, 1977.

Capranica, R. R.: Symposium on auditory processing and animal sound communication. Fed. Proc. 37, 2315-2359 (1978).

Dooling, R. I., Zoloth, S. R., Baylis, J. R.: Auditory sensitivity, equal loudness, temporal resolving power, and vocalizations in the house finch (*Carpodacus mexicanus*). J. Comp. Physiol. Psychol. 92, 867-876 (1978).

Fine, M. L.: Seasonal and geographical variation of the mating call of the osyter toadfish *Opsanus tau* L. Oecologia 36, 45-47 (1978).

Fine, M. L., Winn, H. E., Olla, B.: Communication in fishes. In: How Animals Communicate. Sebeok, T. A. (ed.). Bloomington: Indiana Univ. Press, 1977, pp. 472-518.

Fish, J. F., Offutt, G. C.: Hearing thresholds from toadfish, *Opsanus tau*, measured in the laboratory and field. J. Acoust. Soc. Am. 51, 1318-1321 (1972).

Geldard, F. A.: The human senses. New York: Wiley, 1972.

Myrberg, A. A.: Ocean noise and the behavior of marine animals: relationships and implications. In: Effects of Noise on Wildlife. Fletcher, J. L., Bushel, R. G. (eds.). New York: Academic Press, 1978, pp. 169-208.

Newman, J. D., et al.: Biological filtering and neural mechanisms—Group report. In: Recognition of Complex Signals. Bullock, T. H. (ed.). Berlin: Dahlen Konferenzen 5, 1977, pp. 279-306.

Popper, A. N., Salmon, A., Parvulescu, A.: Sound localization by the Hawaiian squirrelfishes, *Myripristis berndti* and *M. argyromus*. Anim. Behav. 21, 86-97 (1973).

Sachs, M. B., Sinnott, J. M. Heinz, R. D.: Behavioral and physiological studies of hearing in birds. Fed. Proc. 37, 2329-2335 (1978).

Schuijf, A., Siemelink, M. E.: The ability of cod (*Gadus morhua*) to orient toward a sound source. Experientia 30, 773-774 (1974).

Tavolga, W. N.: Sound production and detection. In: Fish Physiology, Vol. 5. Hoar, W. S., Randall, D. J. (eds.). New York: Academic Press, 1971, pp. 135-205.

Winn, H. E.: Acoustic discrimination by the toadfish with comments on signal systems. In: Behavior of Marine Animals. Current Perspectives in Research. Winn, H. E., Olla, B. L. (eds.). New York: Plenum Press, 1972, pp. 361-385.

Discussion

FAY: Do you think the animals are responding to sound pressure or to particle motion?

FINE: I don't know, but I believe it's displacement.

FAY: If you measure lagenar thresholds in the goldfish, you find that thresholds to pressure and to particle motion are quite different. Perhaps if you made your measurements differently, you might find a match rather than a mismatch.

FINE: Yes, that's a good point. We might be presenting the fish with a sound pressure stimulus, but it really responds to a velocity change.

PARTRIDGE: Did you record from both males and females, and do you have any idea of the sexual condition of the animals when recorded?

FINE: I recorded from both sexes at various times. There was no systematic variation that correlated with sex or sexual condition. I do not believe that these factors are significant here, but I have no data that bear directly on this question.

PLATT: You recorded from a large number of units, how did you select them. Did you pick the biggest, or the ones easiest to get at?

FINE: I recorded from all the units I could get to.

I should like to go back to an earlier point where I compared my audiogram for the toadfish with the one published by Fish and Offutt [J. Acoust. Soc. Am. 51, 1318-1321 (1972)]. The similarities outweigh the differences. Using heart rate conditioning, they found that these fish are most sensitive between 37.5 and 75 Hz. Therefore, I don't think that there is a whole population of high frequency sensitive units that I missed.

BELL: Is there any difference in the distance at which these two signals, the grunt and the boatwhistle, might be important to the fish?

FINE: Grunts are usually produced when the animals are very close together. They are produced if the animals are handled. The question is how far away can they detect a boatwhistle sound? The toadfish is unlike the cod in that he stays in one place, and his acoustic environment is probably quite limited. Myrberg concluded that, with spherical spreading, one toadfish could probably hear another at about 75 m. It is my feeling that the mismatch restricts the fish to respond only to nearby sounds, and these would be close enough to make localization possible. This system allows the toadfish to concentrate on nearby animals. Fish and Offutt found that if the sound intensity was equivalent to that emitted by a fish more than 2.5 m away, the effect on the male's boatwhistle repetition was lost. Popper et al. [Anim. Behav. 21, 86-97 (1973)] found that if they moved their sound source more than 3 m from a squirrelfish, the response to the recorded alarm sound ceased. Although sharks and predatory teleosts may be attracted to sound over large distances and fish have been conditioned to localize far-field stimuli, long-distance auditory communication in fish has not been demonstrated.

MYRBERG: In regard to the calculation you noted, this was an extrapolation, based on the lowest possible sea noise. In a public aquarium, a toadfish boatwhistle can be heard as far as 40 m away from the tank! These sounds are extremely powerful; at least 30 to 40 dB above any other sonic fish species known, at least when the output of individuals has been measured.

There are other means by which a male can increase his reproductive capacities than solely by attracting females. One effect is to increase the female's readiness for reproduction, even though, in the end, some other male might move in and participate in the spawning. Of course he does increase his reproductive chances significantly over those that would exist if he never opened up his mouth at all.

FINE: Fish and Offutt found that the threshold of this fish to a continuous tone was about 15 dB lower than the minimal sound level required to facilitate an increase in calling rate. It appears that to affect calling rate, the stimulus required is strongly suprathreshold, and my main point is that toadfish are more likely, because of the mismatch, to pay attention to animals that are close by.

PART FOUR

Sound Source Localization

Since the pivotal paper by Willem van Bergeijk in 1964 (see reference in Chapter 14), one of the most controversial and significant aspects of fish audition has been the problem of how fish can detect the position of a sound source. Through the use of nearly ideal "free" sound fields and through the development of sophisticated techniques for the synthesis of suitable acoustic environments in the laboratory, recent research has shown that fishes are capable of sound localization in the acoustic far-field, and that the ear is most likely a necessary and sufficient channel for this. In Chapter 14 Schuijf provides a new and exciting analysis of possible localization mechanisms, taking into consideration the variety of peripheral auditory mechanisms found among different species. This is followed in Chapter 15 by an analysis of the directional characteristics of auditory neurons in the cod by Hawkins and Horner. Closely related to these papers are those by Buwalda (Chapter 7), who used psychophysical techniques to analyze directional response characteristics of the cod ear, and Fay (Chapter 5), who has analyzed these properties in the goldfish ear.

Chapter 14

Models of Acoustic Localization

Arie Schuijf*

1 Introduction

It was only several years ago that valid proofs of acoustic localization in teleost fishes were first obtained (see Tavolga 1976, Schuijf and Buwalda 1980 for reviews). Behavioral experiments to demonstrate localization require the production of traveling sound waves that propagate in different directions but do not differ in other qualities. At present, directional stimuli needed to demonstrate acoustic localization can only be obtained in deep water as in a lake or fjord. The inconvenience of doing experiments in the field, to a considerable degree, accounts for the very limited amount of data available today, although such experiments are absolutely necessary.

The only species for which experimental data from field studies are extensively available is the cod, *Gadus morhua*. These experiments (Chapman and Johnstone 1974, Schuijf 1975, Schuijf and Buwalda 1975, Hawkins and Sand 1977, Buwalda, Schuijf, and Hawkins, in press) have provided insight into the relevant and irrelevant stimulus dimensions required for acoustic localization in this species. Therefore, development of models of the acoustic localization system of fishes should be primarily based on data of the cod since we are not sure as to the validity of extrapolating acoustic mechanisms to other teleost species (see Platt and Popper, Chapter 1). To discover new properties of the functional organization of the localization system the best approach may be to continue experimental work on a single species, since a better model is more likely to be obtained for a more complete set of experimental data. The emphasis in the experimental work should then be concentrated on the input-output relations of the whole system and on the study of the organization of the auditory periphery of the species to clarify the interactions between the main subsystems.

On the other hand, the use of a comparative approach can lead to a better understanding of the general principles of auditory space perception in fishes in their natural environment. This approach forces one to use a much simplified description of the

*Laboratory of Comparative Physiology, State University of Utrecht, Jan van Galenstraat 40, 3572 LA Utrecht, The Netherlands.

process as a whole, and the models are less amenable to relating structure and function in any but the most general terms. The comparative approach is also useful to place our knowledge of directional hearing mechanisms in the cod, an economically important species, in its proper perspective among acoustic localization mechanisms in other fishes, for instance, those with other kinds of swimbladder-labyrinth couplings and those without a swimbladder, e.g., the sharks. At present, part of the relevant knowledge is missing, and hypotheses must be created to fill these gaps. This course of action is only permissible if we look on models as a means to make predictions that can be checked against reality in experiments designed to verify these models and obtain the missing knowledge. If the model thus leads to the discovery of an important flow in a major assumption, as was the case with van Bergeijk's (1964) model, it may already have been capable of advancing research.

In order to construct models of the fish's acoustic localization system, which will be considered as a black box, or a collection of black boxes, we will use the principle that the mechanism for acoustic localization must be adapted to take advantage of the modes of wave propagation in the fish's environment. The aim of doing physical analyses of such models based on various simplified environments, rather than the more complex environments, is to search for the potential orientational cues available to the fish.

The next step after developing models of propagation is to select which of these cues are likely to be used by the fish. This problem has several aspects:

1. Can the fish's auditory system detect and process the cues being considered?
2. Is localization actually possible using the(se) cue(s)? Is localization based on these cues efficient?
3. Are there any restrictions on the ability to localize biologically relevant sounds based on the proposed mechanism? For instance, is there directional ambiguity or does the principle fail for transient sounds?

To determine whether the cue is actually detectable, the quantitative aspects of the models cannot be omitted. In discussing the topics above, mathematical precision will be used where it is necessary.

Answers to these questions will ultimately lead to psychophysical models that should conform to the presently known dynamic interrelationships between the acoustic input variables and the location of the sound source as experienced by the fish (which is not an observable quantity, but which may be measured indirectly).

A question that subsequently arises is how the sensory-neural mechanism of the fish filters the directional information from a sound source out of the ambient noises which reach the fish. The models to be discussed are more closely related to the structure of the whole auditory system, such as models of the functional organization of the directional detection process and models of subsystems of the system, than previous models (Sand 1974, Schuijf 1974, 1976a, 1976b). The subsystems of the whole mechanism involved in localization which will be discussed include swimbladder models, otolith mechanics, and the encoding of directional information.

Before examining models of wave propagation in the environment of the fish a summary of the important concepts and results of the model building will be presented.

2 First-Generation Models of Acoustic Localization

One way to describe acoustic localization processes is to consider the whole of the involved system as a *black box*. In a first approximation, the system is only defined by its *output* or, in other words, the response of the animal. It is well known that a suitably chosen behavioral response can serve to measure indirectly the animal's perception (Werner 1968). Thus, as to apparent awareness of the direction from which a sound comes—swimming toward or away from—a sound source can be used to demonstrate that fishes are capable of acoustic localization. It is imperative to further note that physiological studies of perception require that a correlation exists between the studied nervous activity and the localization behavior before they can be used to construct models.

Psychophysical experiments have shown their value in various localization studies on cod (Olsen 1969, Schuijf and Siemelink 1974, Schuijf 1975) and *Leuciscus idus* (ide, Schuijf et al. 1977). For example, by manipulating *particle motion* and *acoustic pressure* at the position of the fish, we could infer that both of these acoustic quantities act as *input quantities* in the cod (Schuijf and Buwalda 1975). Studies using cardiac conditioning show only a directional sensitivity of the fish as a whole (Chapman and Johnstone 1974, Hawkins and Sand 1977).

The physical distance between the ears of a fish apparently plays no role in directional detection, a condition that stands in contrast to man and other mammals (refer to Schuijf and Buwalda 1980). These facts can be summarized in the block diagram of Fig. 14-1, left, where the instantaneous particle motion is a *vector* (directed) *quantity* which is only fully determined by specifying its *three* projections on the x-, y-, and z-axes of a reference frame.

This approach provides: (1) a well-defined description of the stimulus conditions without assuming at this stage how the external stimuli affect the receptors and (2) a realistic linkage of these to the directional responses.

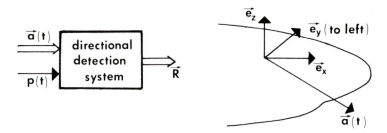

Figure 14-1. Definition of the input and output quantities for the acoustic localization system if the fish and the near surrounding water is considered as one single black box. (Left) Block diagram of the acoustic localization system of fishes with a swimbladder. The directional input is formed by the instantaneous particle motion (acceleration) $\vec{a}(t)$ and the nondirectional input is the acoustic pressure $p(t)$ at time t. $\vec{R}$ denotes the subjective direction experienced by the fish. (Right) The unit vectors $\vec{e}_x$, $\vec{e}_y$, and $\vec{e}_z$ form the basis of a rectangular, body-fixed reference system necessary to describe the input acceleration unambiguously: $\vec{a}(t)$ may thus be decomposed into components along the axes (after Schuijf and Buwalda 1980, Fig. 2-4).

Often the fish's body oscillates by an acoustic field in which particle motion varies only slightly over distances of the order of the dimensions of the body. This is typically the case when the wavelength is large compared with the body (which nearly always applies) and the sound source is more than a few body lengths away. If the mass of the fish's body nearly equals that of the displaced volume of water, then the body will be wholly carried along with the fluid motions which under the above mentioned stimulus conditions, are essentially parallel particle velocities of equal magnitudes at any instant (t). This important theorem (on which nearly all models are based) has been proven in fluid mechanics where "the drag force in a potential flow past a body" is calculated. The derivation takes into account the fact that, in principle, there can exist a velocity of the body relative to what the particle velocity $\vec{v}(t)$ at the position of the body would be if the body were absent, i.e., the unperturbed sound field. Yet, if on average the fish's body possesses the same density as the water, then this relative velocity equals zero. Hence all the mechanical processes inside the body occur in a frame of reference which moves with the external particle velocity $\vec{v}(t)$ relative to a resting frame of reference. This resting (inertial) frame is the obvious choice for describing the external sound fields.

The fish's own frame of reference is in an *accelerated* motion, however. This results in "inertial forces" acting on any mass in the fish's body in the direction opposite to the acceleration. If these forces equal those acting on an equal volume of water, i.e., if the density of the considered body part equals that of water, then there is no net relative motion with respect to the body-fixed frame. This is typically the situation when the cupula of a lateral line neuromast in situ is considered in water through which a sound wave is propagated. Upon local stimulation by a small squirt of water the body as a whole will not be carried along with the flow, and motions of the cupula can occur. Similar arguments hold for lateral line canal organs where the pressure gradients parallel to the canal are probably the adequate stimulus. If, however, these pressure gradients arise from a uniform field of force, as in the case when fish and surrounding water are accelerated equally, then the net displacement is again zero. This may be the explanation why the lateral line is not stimulated in the kinetic antinode (loop) of a standing sound wave (Cahn, Siler, and Fujiya 1973). For models which involve the lateral line, the reader is referred to van Bergeijk (1964), and for a recent review on underwater localization, see Schuijf and Buwalda 1980. From the physical point of view it is likely that the lateral line could be stimulated by the secondary (near) field of a pulsating swimbladder (see Sand, Chapter 23). Perhaps such a coupling may be operative at very short distances to the sound source in specialized forms of directional detection. In this chapter only models of localization involving the labyrinths are considered in view of the total lack of evidence that the lateral line participates in localization of sound sources at distances over a few decimeters.

The otolith organs are stimulated in uniform oscillatory motions as exist in one-dimensional standing waves, whereas the non-mass-loaded hair cells of the lateral line are not. As a first approximation, the inertial force, $\vec{F}$, acting on a body such as an otolith in the accelerated, body-fixed reference frame of the fish is given by

$$\vec{F} = -(m - \rho V_0)\,\vec{a}\,(t) \tag{1}$$

where $\vec{a}(t)$ denotes the particle acceleration[1] in the incident sound wave, or equivalently, the acceleration of the frame moving with the fish. The mass of the otolith is denoted by m, whereas the product of the density of the endolymph ρ and the volume V_0 of the otolith equals the mass of the liquid occupied by the otolith. The driving force $\vec{F}$ equals the sum of three reaction forces: (a) the force to accelerate the otolith and the entrained fluid (remember the relative movement with respect to the macula is the output), (b) the resistance force, and (c) any elastic restoring force (de Vries 1950). From Eq. (1) it follows that *the particle acceleration in the incident wave should be considered the real input and not the displacements.* Only in a special case it does not matter if we substitute the particle displacement for the particle acceleration, that is, if we ignore any torques acting on the otolith (i.e., the cause of otolith rotations) and if the resistance and elastic forces are ignored. In such instances, we can immediately integrate the equation of motion of the otolith organs (see Schuijf 1976b).

Because it is often simpler to think in terms of particle displacement than in acceleration, the models to follow will be worked out in terms of displacements wherever possible. Hair cells of the otolith organs are subjected, under the mentioned conditions, to a forced oscillation about their equilibrium position. This oscillation is, at any instant, the vector combination of (a) the particle displacement in the unperturbed external sound wave at the position of the fish, and (b) the displacement resulting from the swimbladder response to the acoustic pressure. Both kinds of oscillations are strictly additive provided the superposition principle holds which is the case for *small* oscillations (Landau and Lifshitz 1976) as occur here. Accordingly one can consider separately the contributions to the resultant motion for both "conduction modes" (Fay and Popper 1975). To determine the vector sum of both displacements, one must account for the respective directions of the constituent oscillations as well as for their relative timing at the hair cell in question.

The displacement of the hair cells due to the extraneous oscillations (in the inertial mode) are uniform throughout the labyrinths, whereas the pressure-induced displacements from the swimbladder are synchronous, but, in general, neither parallel nor of equal amplitude. If, at a hair cell, we denote the p-induced displacement as a function of time by the vector quantity $\vec{\eta}(t)$, then the transfer of the acoustic pressure $p(t)$ into $\vec{\eta}(t)$ is given approximately by

$$\vec{\eta}(t) = Cp(t)\vec{m} \qquad (2)$$

where the vector[2] $\vec{m}$ of unit magnitude denotes the characteristic local *direction* of $\vec{\eta}(t)$ during a compression. C denotes a proportionality coefficient which is characteristic for the specific position in the labyrinth.

The vector Eq. (2) expresses in precise and general terms the dependence of the p-induced motion as to direction and magnitude on the position in the labyrinths. Of course, each vector $\vec{m}$ has a constant direction in time (for a given point) with respect to the body-fixed frame. In using Eq. (2) we avoid the difficulty that neither $\vec{m}$ nor

---

[1] A common alternative of the symbolic notation of a vector quantity is underlining, e.g., $\underline{a}(t)$.

[2] Avoid confusion between m for otolith mass and $\vec{m}$ for the unrelated vector quantity.

the coefficient C have been measured directly under realistic stimulus conditions. The vector $\vec{m}$ is defined in such a way that, in principle, the positive direction for $\vec{\eta}(t)$ is determined unambiguously, a necessary condition for the sections to follow.

The physiological constants C and $\vec{m}$ can be specified semiquantitatively for a few simple theoretical models of the (functional) swimbladder-labyrinth coupling (Schuijf 1976b). Far below the resonance frequency of the swimbladder (about 1200 Hz in a 16 cm cod at 30 m depth: Hawkins and Sand 1973) the swimbladder behaves as a purely elastic gas volume of equilibrium value V_0. In that case we have, in the present notation, for the spherical swimbladder model

$$\vec{\eta}(t) = \frac{V_0\,p(t)}{p_g c_g^2\,S}\,\vec{m} \tag{3}$$

(ρ_g = density and c_g = velocity of sound in the gas). The surface area $S = 4\pi r^2$ of the concentric sphere passing through the position, where the considered hair cell is, determines in Eq. (3) the decline in amplitude of $\vec{\eta}(t)$ with increasing swimbladder-hair cell distance r. The direction of $\vec{m}$ in Eq. (3) is that of the *inward* normal to the mentioned spherical surface, because during a compression the swimbladder will contract. Equation (3) holds also for elliptical swimbladders if S denotes the surface of the confocal ellipsoid passing through the hair cell (compare Schuijf 1976a).

The gyration model (Schuijf 1976b), on the contrary, predicts that $\vec{m}$ is nearly constant in direction for all points in the labyrinths and, of course, independent of the propagation direction of the incident sound wave.

The window theory (Schuijf 1976b, in particular Figure 5), as another alternative model, predicts that a convergent field of $\vec{m}$ directions toward the bony window will occur. Equation (2) holds in all these cases.

With the aid of Eq. (2) one can determine the relation between the propagation direction of the external wave and the motion of the hair cells and analyze what will be the resultant mechanical input(s) for one or two hair cell unit(s) of arbitrary polarization directions. These topics will be continued in Section 4.2.

An acoustic localization system based on a single directional detector cannot segregate the magnitude and the direction of the particle motion in the sound wave; it only responds to the projection of the displacement vector onto the detector axis. For instance, a strong sound propagating in an insensitive direction of the detector can result in the same component along the detector axis as a weak sound traveling at a small angle from the optimum axis. The vector-weighing model (Schuijf and Buwalda 1975, 1980, Schuijf 1976a) in which a comparison is made between the outputs of two nonparallel vector detectors is a possible solution to the direction/strength ambiguity of the single directional detector. Units from the sacculi (Enger et al. 1973, Sand 1974), the lagena (Fay and Olsho 1979), or from the utriculi (Popper 1978, Platt and Popper, Chapter 1) have been proposed as sites for the hair cell units involved in such a vector weighing. Hawkins and Horner (Chapter 15) suggest that in codfish bidirectional units of the utriculus might be considered as a basis for vector weighing in the horizontal plane. Further experiments will be necessary to test this.

It is now a common argument that vector weighing is impossible if the p-induced displacement is so strong that the direction-conveying particle displacements of the incident wave are not longer detectable. This should not lead one to think that absence of p-induced displacements may be the optimal condition for detecting the propagation direction of the sound. At the hair cell level a moderate "contamination by p-induced displacements" may even be desirable for an unambiguous directional detection (see Section 4).

A vector detecting system cannot detect the propagation direction of a sound wave, since propagation direction and direction of displacement are alternatingly common and opposite. While the orientation of the line along which the particles oscillate about their equilibrium position is determined in space, it is still unknown at which "end" of the line the sound source is located. There is, however, a simple rule which relates the propagation direction with the particle velocity $\vec{v}(t)$ *and* the acoustic pressure $p(t)$ in a traveling sound wave, for the energy is always transported in the propagation direction at a rate[3] $p(t)\vec{v}(t)$ through a unit area. Because energy is essentially a positive quantity it follows that during a compression $\vec{v}(t)$ is in the same direction as the propagation and during a rarefaction in the opposite direction (in which case both terms in the product change sign).

If fishes with a swimbladder use this "phase principle" for acoustic localization, then a change of the algebraic sign of $p(t)$, without altering $\vec{v}(t)$, will result in a 180° change of the directional perception of the fish. This appears to be the case for cod, when using sine waves of 75 Hz (Schuijf and Buwalda 1975), incident approximately along the longitudinal axis. Later experiments show that the cod apparently uses the same principle for discriminating diametrically opposed sound sources in the median and transverse planes (Buwalda et al. 1982). Probably this principle may be generalized for discrimination of any two sound sources at positions which form each other's inversion about the origin of the fish's body-fixed frame (Fig. 14-2).

If the acoustic localization system of fishes with a swimbladder is considered as a black box, then the real acoustic input variables are the particle acceleration, $d\vec{v}/dt$, and the acoustic pressure $p(t)$ (see Schuijf and Buwalda 1980). The motion of the hair cells in their dependence on these basic inputs is, however, a phenomenon that can best be understood in terms of the *external particle displacement, $\vec{\xi}(t)$*, and the *acoustic pressure $p(t)$*. This involves a simplification which holds under special conditions. The model of the immovable otolith is an important case where it is plausible that the "direct" displacement, $\vec{\xi}(t)$, and the p-induced displacement, $\vec{\eta}(t)$ [refer to Eq. (2)] add as the vector sum $\vec{\xi}(t) + \vec{\eta}(t)$ to form the displacement of the hair cell base. (For physicists it is easy to think of a more general situation, as a counterexample, where the simplification is not permissible, see Appendix A). The choice of the external variables $\vec{\xi}(t)$ and $p(t)$ has at least, at the level of the macula, the advantage that quantities of equal physical dimension, i.e., displacements, are compared. Models of acoustic localization in the past were also based on this pair of variables (Schuijf 1976b).

Let us illustrate the principle of phase analysis in terms of the external displacement, $\vec{\xi}(t)$, and acoustic pressure $p(t)$. Consider a free-field situation where a pulsating sphere or monopole sound source (van Bergeijk 1964) is located at some point x on the longitudinal (x-) axis of the fish (hence the x-coordinate for the upper part of

---

[3] This rate is known as the instantaneous intensity (in erg/cm^2 sec) $\vec{I}(t)$, hence $\vec{I}(t) = p(v)\vec{v}(t)$.

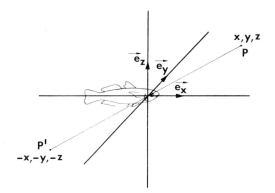

source at	pressure	particle velocity
x, y, z	$P(t) > 0$	
	$P(t) < 0$	
-x, -y, -z	$P(t) > 0$	
	$P(t) < 0$	

Figure 14-2. The symmetry properties of the phase model of acoustic localization. The model implies that any two identical sound sources at points which form each other's inversion about the origin 0 of the body-fixed frame of the fish, like P and P', can only be discriminated as to location by a phase analysis between the particle motion and the acoustic pressure. The origin of the frame forms the center of inversion. See text.

Fig. 14-3 is negative). For harmonic waves of frequency f, the x-component of $\vec{\xi}(t)$ at the position of the fish may be written as

$$\xi_x(t) = A \cos(\omega t + \psi_x) \tag{4}[4]$$

where A = the amplitude of the oscillation; $\omega = 2\pi f$ = the angular frequency in radians/sec; ψ_x = the phase angle of $\xi_x(t)$ relative to $p(t)$.
    Similarly, we have

$$p(t) = B \cos(\omega t) \tag{5}$$

The amplitude constants are positive, as usual, and their values are not independent in this traveling wave situation. Because in the corresponding standing-wave situation

[4] Or, more generally, $\vec{\xi}(t) = A \cos(\omega t + \psi_x)\vec{e}_x$; compare Fig. 14-1, right.

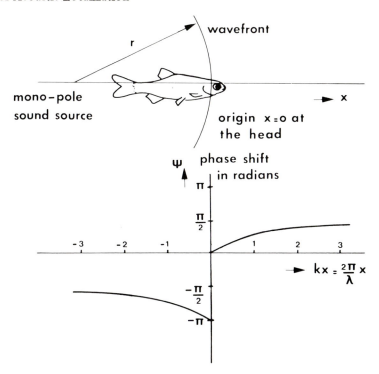

Figure 14-3. Model of the phase information available to a fish in the free field of a monopole sound source. (Upper) Geometry used for the derivation of the graphs in the lower figure. (Lower) Phase difference ψ_x between the particle displacement $\xi_x(t)$ and the acoustic pressure $p(t)$ as function of the position of the sound source (expressed in acoustic distance units $\lambda/2\pi$) relative to the fish. Note that for $kx = 1$, where k denotes the wavenumber $k = 2\pi/\lambda$, the absolute value of x or geometric distance equals $\lambda/2\pi$. Expressed in terms of the dimensionless product kx the graph has general validity for arbitrary frequency. Figure modified after Schuijf and Buwalda (1980).

A and B can be controlled independently (Buwalda, Chapter 7), both situations can be described with appropriate values of A and B which are immaterial to the following argument. From Fig. 14-3 it follows that when plane, traveling sound waves ($x \rightarrow -\infty$) are incident caudally onto the fish then $\xi_x(t)$ *lags* 90° after $p(t)$, since ψ_x tends to $-\pi/2$ for $x \rightarrow -\infty$. For a sound source (acoustically) far in front of the fish $\xi_x(t)$ leads $p(t)$ by 90°. Empirically it can be shown (Buwalda et al. in prep.) that a simple lead/lag phase discrimination for ψ_x accounts for the directional choice behavior (ahead or 180° turn) of a food-conditioned ide (*Leuciscus idus*). Reversals in the directional responses for variable ψ_x occur only at $\psi_x = 0$ and at $\psi_x = -\pi$ or $+\pi$, which correspond, in a mental experiment, to the extreme near-field situations in which a 180° change of $\vec{\xi}$ would occur on passing of the fish through the center of the monopole.

Thus, although the result with the ide was actually obtained in standing wave situations at variable ψ_x values without the presence of a monopole near the fish, the choice behavior for $\psi_x = 0$ would correspond to swimming through a fictitious monopole!

It is likely that similar lead/lag discriminations hold for 180° ambiguity elimination with respect to each of the body axes of the fishes. For instance, that a fish would locate the sound source on the positive $\vec{e}_z$ (dorsal) axis if $0 < \psi_y < \pi$. The discrimination of sources in the zenith and nadir directions by the cod (Buwalda et al. in press) may well be explained in this way.

This so called phase model of directional hearing in fish describes the properties of the angular detections in terms of the *external* sound parameters at the organism level. What is needed is to build models which link this model of the system behavior to the physiological results concerning its subsystems. This is attempted in Section 4.

3 Models of Wave Propagation in Two and Three Dimensions

3.1 Rotating Vectors in Surface Waves

Wave propagation can result in rotating vectors under certain conditions. The rotational sense, clockwise or anticlockwise, in which water particles describe their trajectories during a cycle can provide an orientational cue for the detection of the propagation direction, as we shall see. Later, in Section 4.2, we shall discuss the generalization of this principle for waveforms which differ from simple sine waves.

Surface waves at the water-air interface probably present the simplest example to introduce the essentials of this relationship between propagation direction and rotational sense because they are easily envisaged. Because the principle is largely of a geometrical nature it is immaterial that the periods involved in surface waves are usually much longer than in sound waves.

For deep water, say when the depth exceeds one-half the wavelength of the waves, a water particle below the surface moves very nearly on a circle. In Fig. 14-4 these circles are indicated; however, the positions where the circles are drawn are arbitrary because the particle motions are the same everywhere along the fish's body. Under the crest of the wave a particle moves in the direction in which the wave is traveling whereas a half-period later under a trough the water moves in the backward direction (Elmore and Heald 1969). From Fig. 14-4 it follows that the rotational sense in which the trajectories of the particles are described can be used to infer the propagation directions of the surface waves. It is clear that a fish would experience a reversal of the rotational sense under a roll of 180° about the longitudinal axis. One may feel that if the direction of gravity $\vec{g}$ is taken as an invariant reference direction for the fish (thus an extra input), then the propagation direction may be inferred unambiguously from "measurements" in the fish's own frame of reference: the body-fixed axes $\vec{e}_x$, $\vec{e}_y$ and $\vec{e}_z$ (refer to Section 2). Indeed it can be shown by vector analysis methods that this is possible from the physical point of view.

A fish need not possess a special receptor to sense these circular motions of the particles. Any motion of such a sort can be considered as the result of two harmonical oscillations along different axes in the (vertical) plane through the circular path.

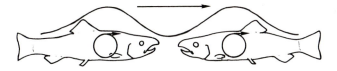

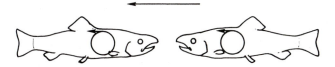

Figure 14-4. Water particles in a surface wave encircle their equilibrium positions in circular paths in first approximation. The sense in which the particles rotate is uniquely determined by the propagation direction of the wave. As such it forms a hypothetical example how rotating displacement vectors can be used as an orientational cue. See text.

If the fish is directed horizontally, as in Fig. 14-4, then a simple description of the particle motion is obtained if we take the body-fixed unit vector $\vec{e}_x$, along the longitudinal axis, and $\vec{e}_z$ along the dorsoventral axis (Fig. 14-1, right). Let us examine the motion of a particle initially located at the origin of the body-fixed frame *if the fish were absent*. The instantaneous position of this particle in its steady-state motion is then fully determined by the endpoint of the *position vector* $\vec{\xi}(t)$ at some instant t.

For the upper left Fig. 14-4 the time dependence of the displacement components is

$$\xi_x(t) = A \sin(\omega t) = A \cos(\omega t - \pi/2) \qquad (6a)$$

and

$$\xi_z(t) = A \cos(\omega t) \qquad (6b)$$

where ξ_x and ξ_z denote the displacement components along $\vec{e}_x$ and $\vec{e}_z$, respectively. The circular frequency ω equals $2\pi f$, where f is the frequency of the surface wave. See Burington (1964) for trigonometric identities.

We see (Fig. 14-4, top left) that the displacement vector $\vec{\xi}(t)$ rotates clockwise: at $t = 0$ the vector points vertically upward [$\xi_x = 0$, $\xi_z = A$ on substituting $t = 0$ in Eqs. (6)] and a quarter period, $T/4$, later the vector points horizontally to the right ($\xi_x = A$, $\xi_z = 0$ for $\omega t = \pi/2$, that is for $t = T/4$). If we compare the phase constants of the motion components along the z- and the x-axis, as specified in Eqs. (6a) and (6b) then it follows that the $\xi_x(t)$ motion leads the $\xi_z(t)$ motion by $-\pi/2$ radians or $-90°$ (i.e., $+90°$ lagging). This value for the relative phase is characteristic for a far field traveling surface wave (here a wave with straight fronts) which is incident onto the tail side of the fish.

If on the contrary the wave travels toward negative x-values in the body-fixed frame, hence a wave generated in front of the fish, *then $\xi_x(t)$ changes sign with respect to $\xi_z(t)$*. This is evident from symmetry considerations: Waves in opposite propagation directions have mirror symmetry in a conceptual vertical plane passing through the origin and perpendicular to the propagation direction of one of the waves. The z-component of the orbital motion is unaltered by this geometrical reflection, whereas the x-component is inverted in the mirror image. Accordingly the phase of the x-component shifts over π radians and for a wave traveling to the left in Fig. 14-4 (fish below left) $\xi_x(t)$ leads the $\xi_z(t)$ motion by $\pi/2$ radians or $+90°$.

The phase *difference* between the x- and z-component as "detected" in the coordinate system of the fish is (also in more general situations than considered here) the property that determines the rotational sense of the rotating vector $\xi(t)$. The converse relationship is well known from the Lissajous figures which result when sine signals are applied simultaneously to the horizontal and vertical channels of an oscilloscope. For the general case where the amplitude for the vertical signal motion is unequal to that for the horizontal signal the resultant pattern is an ellipse. If the horizontal signal has a larger amplitude, an ellipse with a horizontal major axis is formed. This corresponds to the particle motion near the bottom in shallow-water waves where the circular trajectories flatten into elongated ellipses. The pattern degenerates into a straight line when the two signals are either in phase or $180°$ out of phase. These phase differences separate phase regions with opposite rotational senses (refer to Crawford 1968, p. 402).

If, in the mechanical case, the instantaneous rotating displacement vector $\vec{\xi}(t)$ is resolved into two mutually perpendicular components, as before, then a lead/lag comparison between these components would provide a simple means to detect the rotational sense of the path and by that means the propagation direction of waves in one dimension. It is natural to assume that these components are detected by differently oriented directional receptors. This elementary model resembles in performance the effect of a phase shift between particle motion and acoustic pressure on the directional choice behavior of the ide (compare Section 2) if the acoustic pressure would induce a particle displacement oscillation along the z-axis of the fish. Although this is certainly not a final model it points out how the $180°$ periodic effect in the choice behavior of the ide can be accounted for at the receptor level.

For a stimulus variable to be useful as a phase reference it must satisfy some form of invariance under different angles of incidence of the wave (Schuijf 1976a). Because any angle of incidence of a surface wave is equivalent in producing the vertical displacement component at the observation point, this variable meets the requirement. In this (fictitious) change in the stimulus direction it is of course assumed that the (arbitrary) waveform remains unaltered. Because the rotation of the stimulus direction through any angle leaves the vertical component unaltered, i.e., shows cylinder symmetry, this is in particular true for a rotation through $180°$ about the z-axis, that is, if the propagation direction of the wave is inverted. Because of the symmetry property of the vertical component it is logical to study the phase difference, χ_x, of the horizontal displacement: the reverse of the common practice in XY operation of the oscilloscope.

Equations (6a) and (6b) can be rewritten in the more general form

$$\xi_x(t) = A\cos(\omega t + \chi_x) \tag{7a}$$

$$\xi_z(t) = A' \cos(\omega t) \qquad\qquad (7b)$$

where A' is not necessarily equal in amplitude to A. The direction of rotation of the displacement vector is determined by the inequalities

$$-\pi < \chi_x < 0 \qquad \text{for clockwise rotations} \qquad\qquad (8a)$$

$$0 < \chi_x < \pi \qquad \text{for anticlockwise rotations} \qquad\qquad (8b)$$

It is crucial for an understanding of the principle of phase analysis that, *for the fish, the rotation senses of the particle motion,* and the χ-values, *are identical* for the situations that are diagonally opposed in Fig. 14-4. Indeed the horizontal displacement is inverted with respect to the vertical displacement relative to the environment (see above) for waves from opposite directions, but for the compared situations the fish's own coordinate system in space has turned through 180° about the dorsoventral axis. This turn of the body-fixed frame exactly compensates for the 180° phase change in $\xi_x(t)$. To observe in a standard way the rotation sense for the fish we ought to be looking to every fish (e.g., in Fig. 14-4) from its right side. A corollary of all this is that knowledge of the positive direction of a velocity hydrophone in space is necessary if we are to measure phase differences in an acoustic field.

What has been said above on the effect of 180° change in the propagation direction and the effect of a 180° turn of the fish is of a geometrical nature and holds therefore equally for waves of arbitrary wave form, including solitary waves (in the acoustic case various pulse shapes). Only the concept of relative phase must be generalized: a form of coherent detection (see Section 4).

If the surface waves travel in a direction perpendicular to the drawing plane in Fig. 14-4 then a similar phase analysis between the displacement components along the $\vec{e}_y$ and $\vec{e}_z$ axes (Fig. 14-1) of the fish is again a possible solution for the detection.

The essence of the phase comparison between the x-component and the z-component in the surface wave is that a longitudinal wave component is compared with the phase of a transverse wave component. The transverse component provides the phase reference needed to detect the propagation direction without a 180° directional ambiguity. In the parts below we deal with two simple models of environmental acoustics that can be understood as variations on this scheme.

3.2 Sound Propagation in Shallow Water

Sound propagation in shallow water is characterized by repeated reflections at both surface and bottom (Urick 1967, Chapter 6) which, at some distance from the source, can be thought of as a superposition of waves crisscrossing in the layer. The net propagation is horizontally along the layer, but due to the zigzag paths of the constituting waves the sound field has transverse components as well as longitudinal components. We shall study the two-dimensional wave propagation in a water layer to compare the properties of its acoustic field with that of the surface wave in Section 3.1.

The acoustics of a shallow water layer of uniform depth approximates situations encountered in coastal areas, e.g., a flat sandy bay. The potential possibilities for acoustic localization are most easy to assess if the simultaneous time dependencies of the particle displacement $\vec{\xi}(t)$ and of the acoustic pressure $p(t)$ are determined for a periodic point-source (monopole) at distances of at least several times the water depth away from the fish. We shall refer to this situation as a distant source. The question of how the elevation of the sound source with respect to the fish can be detected under those circumstances can then be left aside.

For a distant sound source behind the fish the geometry is therefore almost the same as in Fig. 14-3A. For equal frequencies the curve corresponding to Fig. 14-3B in the case of the shallow water layer tends much more rapidly with distance to its limiting values than in the free field. Because of the existence of transverse (vertical) displacements in the shallow water layer there are potentially more possibilities for timing analysis than in the free field:

1. The horizontal component vs. the acoustic pressure; the only possibility for the situation of Fig. 14-3A in the free field. [This is only relevant for a fish that is sensitive to $p(t)$.] .
2. The vertical component vs. the acoustic pressure: this does not seem a promising model to study.
3. The horizontal component vs. the vertical component: this hypothetical timing analysis will be considered especially in relation to acoustic localization in fishes without swimbladders such as sharks.

The analysis and the results depend on whether (see below) a certain mode of oscillation of the layer participates or not. Aside from the depth of the source, this depends essentially on whether the frequency of the driving source is above or below the natural frequency of the considered oscillation mode for the wave guide.

For the case of a perfectly rigid bottom and a water depth H the frequency

$$f_1 = c/4H \quad (H = \lambda_1/4) \tag{9}$$

is called the *lowest cutoff frequency of the layer* because below this frequency considerable horizontal damping always occurs. If the water depth is greater than one quarter wavelength then the sound field consists at least of the first partial wave, or mode. The higher modes only participate if the driving frequency exceeds an odd integer times the lowest cutoff frequency, specified by Eq. (9).

Above their cutoff frequencies, characteristic wave patterns *travel in a horizontal direction* for this open system which has no outer boundaries so that no horizontal reflections occur. When the modes are driven below their cutoff frequency, however, they do not propagate and vanish exponentially with the horizontal distance, apart from the amplitude attenuation due to cylindrical spreading in the layer.

3.2.2 Driving Frequency Exceeds the Cutoff
Frequency of the First Mode

In nature the frequency band where only the first mode contributes to the sound field at the position of the fish and not the higher modes (for details see Urick 1967) is important to fish hearing. For instance in a layer of depth H = 2 m the cutoff frequency is about 187.5 Hz for c = 1500 m/sec (Eq. 9). The second mode can, however, only participate above three times this frequency, i.e., 562.5 Hz.

The phase difference ψ_x between the horizontal particle displacement $\xi_x(t)$ and the acoustic pressure $p(t)$ tends to the same values $-\pi/2$ and $+\pi/2$ as in the far field of Fig. 14-3B, but these values are approached at shorter range (absolute value of $x > H$ to $2H$ instead of $x > 0.5$ λ, corresponding to $k|x| \simeq 3$). This result increases the importance of the graph in Fig. 14-3B; see, however, also Section 3.2.3.

The third case of a phase analysis (see above) between the horizontal and the vertical displacement produces again elliptic trajectories for the particle motion. The horizontal displacement $\xi_x(t)$ leads, however, the vertical displacement $\xi_z(t)$ in shallow water by 90° if the sound travels rostrad. The value χ_x = 90° corresponds according to Eq. (8b) with counterclockwise rotations. It follows that *for the same propagation direction the displacement vectors rotate in opposite directions in the sound field of the wave guide and in a surface wave.* The explanation for this difference is probably that an upward displacement is produced in the propagation of sound in shallow water by pressure release at the surface, whereas in a surface wave an elevation of the surface results locally in an increase of the hydrostatic pressure and a related downward return force on the water particle. This difference probably inverts the vertical component with respect to the horizontal component, resulting in an inverted rotation sense. In these considerations the fish is assumed to be located at a midwater position, but not at the very surface because there the horizontal component is lacking.

So far the analysis treats only the steady-state situation for continuous driving at a single frequency. The waveform of natural sounds is not necessarily repetitive, and these sounds consist of a spectrum of different frequency components. In the shallow water layer we have the complication that each of the frequency bands (wave groups) in the spectrum travel at its own *group velocity* which we denote here by U. The reason for this is that the modes can only develop when a constructive interference is possible between a pair of plane sound waves propagating in zigzag fashion between surface and bottom. In Appendix B it is derived from Fig. 14-5 that a mode driven by frequency f travels at a specific net speed U in the horizontal direction. The difference in speeds of the spectral bands alters the wave shape of any complex or transient sound during its propagation in the layer.

Before the steady state is established, transient oscillations occur and one can wonder how long it takes for the crisscross waves to reach the receiver. To answer this question let us examine the simple case of a sound source which is off until the instant t = 0 and emits a harmonic oscillation of frequency f thereafter. The crisscross waves will contribute to the sound field at a receiving point at a horizontal distance, r, from the source after a time lapse t which equals

$$t = r/U \tag{10}$$

provided that $r \gg H$.

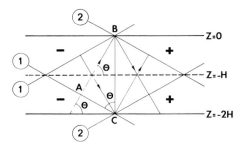

Figure 14-5. The lowest traveling mode in a shallow water layer can in the steady state be conceived as the interference between two systems of plane waves traveling crisscross between surface and bottom. Constructive interference is only possible for an angle θ which depends on the driving frequency (see Appendix B). The mode propagates in the horizontal direction and has a vertical amplitude distribution typical for a standing wave between bottom and surface. See text.

About by this time the transient component of the sound has faded away and the steady state has been reached. The forward front of the outgoing wave, however, has arrived earlier because the high-frequency components arising from the sudden onset of the driving source travel at the (higher) velocity of sound, c, to the receiving point (the water layer is acoustically deep for these frequencies). The forward front arrives therefore at a definite time $t_0 = r/c$ and the transient has nearly ceased after a duration [compare Eq. (10)] :

$$t - t_0 = r/U - r/c \tag{11}$$

With the aid of Eq. (11) it is possible to derive approximately how many cycles are required to reach the steady state (refer to Appendix B). This number of cycles is proportional to r, because of Eq. (10), and depends rather strongly on how near the driving frequency is to the cutoff frequency of the mode. Without any form of sound absorption the steady state is theoretically never reached if the driving frequency equals the cutoff frequency of the layer exactly. For the layer in the example above ($H = 2$ m) with a driving frequency of $f = 190$ Hz and cutoff frequency $f_1 = 187.5$ Hz for the first mode, about 6.6 cycles are required for this mode to build up at $r = 10$ m. At $f = 200$ Hz only 2.5 cycles are required.

What is the relevance of this for acoustic localization? It is clear by now that the shallow water layer has very different acoustic properties as compared to a free field. Not only are the spatial properties of the sound field transformed drastically (cylindrical vs. spherical spreading, see Urick 1967; occurrence of lateral components in addition to longitudinal components), but also the temporal structure of the sounds is changed by the high-pass filtering of the layer. Thus, nonsinusoidal waves are stretched in time during their propagation and also transient beats can occur. This is relevant to the recognition of various fish sounds by fishes where temporal analysis is so important (Myrberg, Spanier, and Ha 1978). To see this we need additional theory.

The cylindrical outgoing pulse wave (in the layer) differs from the spherical freefield wave in that it has a forward front but no backward front: once the wave has

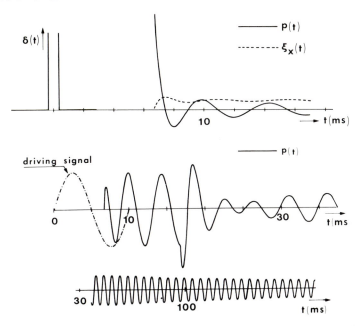

Figure 14-6. If the shallow water layer is driven by a pressure impulse $\delta(t)$ a series of transient waves, each characteristic for the layer is excited simultaneously. The upper trace shows plots of the time dependence of $\xi_x(t)$ and $p(t)$ only due to the first mode of oscillation. The water particles are displaced permanently for such an artificial pulse function. The observation point is 10 m away from the source. Water depth, 2 m. Note the long tail of the $p(t)$ impulse response of the layer. The lower trace shows the theoretical contribution of the first mode to $p(t)$ in response to a pulsating sphere which emits only one cycle ($T = 0.01$ sec). Note that (1) the high-frequency components arrive first, and (2) beats occur which are caused by the fact that the water layer "likes" to oscillate at the cutoff frequency, whereas it is driven at different frequencies. The relevance of the plots is discussed in the main text. For the theoretical background see Appendix B.

reached the observation point, it does not cease, but diminishes comparatively slowly as $t \to \infty$ (Landau and Lifshitz 1976). As any abrupt beginning of the original wave profile is represented mainly in the high-frequency components traveling at the speed $U = c$ (see above), the front propagates at the same speed as these fastest spectral components. Thus, an originally nonperiodic disturbance, which was originally localized in time and space, spreads more and gets an oscillatory tail as it progresses. Similarly in the case of shallow water, *trains of acoustic waves which are modulated both in frequency and amplitude arise in response to a pulse signal.* The effect is the more pronounced the farther away the observation point is (Fig. 14-6).

The phenomenon discussed here has a definite solution in the case of an initial pressure impulse emitted by the point source at time $t = 0$ (modified after Morse and Ingard 1961). In advance, it is remarked that the oscillatory response can, in principle, be constructed from the impulse response of the water layer (between the source and fish location) which acts as a high-pass filter (any initial pulse form can be conceived

Table 14-1. Initial Values of Three Sound Variables[a] on Arrival of the Lowest Partial Wave in Response to a Concentrated Pressure[b] Impulse at Instant $t = 0$ Originating from a Point Source some Distance r Away

Sound variable	Propagation direction	
	Rostrad	Caudad
$\xi_x(t_0)$	Jump to a rostral maximum	Jump to a caudal maximum
$p(t_0)$	Jump to a maximum compression phase	Jump to a maximum compression phase
$\xi_z(t_0)$	Zero start toward dorsal	Zero start toward ventral

[a] Displacements with respect to the body-fixed frame.
[b] Corresponds to a radially outward velocity impulse.

as a sequence of impulses with different "strengths", compare Schuijf 1976a, Fig. 3). The initial values of the impulse response at the receiving point on arrival of the front, as specified in Table 14-1, are pertinent for the temporal interrelationships between the basic sound parameters existing in the early stage of the wave. These initial values are also relatively free of phase distortion which follows thereafter in the wave and are therefore the best representation of the time pattern of the source. The "aging" of a sound wave in response to a pulse (e.g., some fish vocalization) can result in a time-varying spectrum, as could be obtained with a spectrograph, with a lot of natural frequencies characteristic for the acoustics of the layer, rather than for the original sound pulse.

3.2.3 Driving Frequency below the Lowest Cutoff Frequency

If the water depth is less than one-quarter wavelength the steady-state wave amplitude decreases exponentially with the horizontal distance between source and observation point, apart from the amplitude attentuation due to (cylindrical) spreading in the layer. Sound fields without *horizontal damping* are not possible for frequencies below the lowest cutoff frequency. The reason for this is that when the frequency is too low the compressions in the sound field have time to neutralize at the surface. An analogous situation occurs when a traveling wave along the basilar membrane has passed the region of maximum vibration where the local elastic return force is too weak to propagate a wave of the involved frequency. Thus, in the very shallow water layer a forced oscillation below the cutoff frequency forms an evanescent mode rapidly falling off to zero with the radial distance, r. This mode is not traveling but is a special case of a standing wave.

The phase relations between $\xi_x(t)$ and $p(t)$ for sound sources located caudad or rostrad from the fish are such that *a fore-aft discrimination on the basis of a phase analysis appears impossible in a truly steady-state situation* under these stimulus conditions: the phase shifts being $-\pi$ and 0 radians, respectively. The values correspond to the indeterminate case in a free field at zero distance (extreme near field) in Fig. 14-3.

An ostariophysine fish like the ide cannot perform the fore-aft discrimination under these acoustic conditions (see Section 2). It is a matter of speculation whether the experiments of Von Frisch and Dijkgraaf (1935) with *Phoxinus laevis* (minnow) may have failed to demonstrate acoustic localization because the applied periodic sounds with a smooth envelope were likely to result in evanescent modes in shallow water. Besides, the vertical displacements were very dominating over the horizontal displacements which are required to convey the bearing of the source up to a $180°$ uncertainty.

Similarly a phase analysis between $\xi_x(t)$ and $\xi_z(t)$ appears impossible for sine waves below the cutoff frequency of the layer. It must be realized, however, that any kind of modulations of a periodic waveform has spectral components at the harmonics of the carrier frequency. These modulations propagate as transients in the layer. Similar effects occur for noise signals. Banner (1972) showed that juvenile *Negaprion brevirostris* (lemon sharks) are quite capable of locating noise sources in very shallow water. It is likely that there were frequencies above cutoff at the depth at which he worked, although the cutoff frequency would be near the upper limit of the shark's hearing ability.

The acoustic theory of shallow water, originally developed for exploration of the bottom by means of explosive sounds (Pekeris 1948, Brekhovshkikh 1960), has been worked out here to obtain the phase information contained in the sounds. The reader who wishes to acquaint himself with the complex transient phenomena in shallow water is referred to the quoted authors and a standard text on system analysis (e.g., Milsum 1966).

3.3 Wave Propagation under the Surface of the Ocean

Sharks are fishes not having swimbladders but which clearly possess directional hearing (refer to, e.g., Klimley and Myrberg 1979, Nelson and Gruber 1963, Nelson and Johnson 1972, Myrberg 1978, Myrberg et al. 1972, Myrberg, Gordon, and Klimley 1976). Banner (1967) obtained evidence that the acoustic threshold of *Negaprion brevirostris* (lemon shark) is set by acoustic displacements at orders of magnitude from 10^{-10} to 10^{-9} m, depending on the applied frequency. This does not exclude the possibility that sensitivity to pressure can be demonstrated under other stimulus conditions; there is, however, no obvious structure that can serve as a pressure to motion transformer in this fish. Kelly and Nelson (1975) arrived at a similar conclusion for *Heterodontus francisci* (horn shark).

The theory of directional hearing as formulated by Schuijf (1974, 1975, 1976a) cannot explain how a fish can overcome $180°$ directional ambiguity under truly free field conditions without using acoustic pressure as a phase reference. The question arises whether sharks indeed need acoustic pressure as an extra input variable for acoustic localization; the answer is unknown (Nelson 1978).

If they do not require it, then another kind of timing analysis is conceivable between nonparallel components of the particle motion in the pattern of interference between direct and reflected waves (Schuijf 1975). Of course, entirely different principles may possibly apply in the case of sharks, see Corwin 1977 and Chapter 5 for such a view. The principle underlying the above mentioned pattern of interference is illustrated in Fig. 14-7.

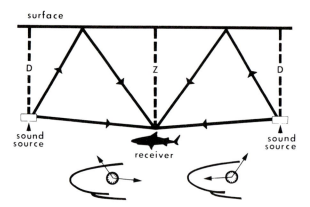

Figure 14-7. Phase analysis theory of sound directionality in sharks (Schuijf 1974). The insets show the contributions of the direct wave and the surface-reflected wave to the particle displacement at the labyrinths of the fish, a quarter period after the arrival of a compression, either from the left or from the right sound source. The indirect displacements in the insets are not directed in the propagation direction at the illustrated instant, as the indirect wave undergoes an inversion in the compression phase on reflection at the soft surface (Schuijf and Buwalda 1980). (Original figure published here courtesy of Dr. A. A. Myrberg, Jr.).

The instantaneous particle displacement at the position of the shark (Fig. 14-7) is, at any time, the sum of two vectors by the parallelogram rule. These displacement vectors are due to (a) the direct sound and (b) the surface reflection.

These constituent displacement vectors oscillate harmonically on the line segments as indicated in the inset figures. The geometrical difference in the stimulation by either of the sound projectors is evident from these insets. The shark, however, only detects the resultant displacement; the theoretical solution is incomplete without an explanation how the phase comparison can be performed on the single resultant vector.

The longer path traveled by the surface-reflected wave causes a time delay of that displacement oscillation with respect to that of the direct wave. It can be shown that the resultant displacement vector describes during one period a complete elliptic trajectory as in Fig. 14-8. Note that the stimulation on the left in Fig. 14-8 corresponds to the left inset figure in Fig. 14-7. The rotational sense for an observer who looks onto the left side of the shark (to the negative y-axis in Fig. 14-1B) is counterclockwise.

It essential to observe that Figure 14-7 has a two-sided or mirror symmetry in the plane through the local vertical at the shark (dotted line) and perpendicular to the drawing plane. *One consequence is that the trajectory of the displacement vector has mirror symmetry in this symmetry plane.* This implies two things: (a) the directions of the major axes of the elliptical trajectories for the stimulus situations in the left and right insets are mirror images of each other, and (b) the rotations, by which the ellipses are described, become inverted.

The crucial point can now be made. Without loss of generality, let us assume that the shark resolves the instantaneous displacement vector into two special, mutually perpendicular components, where $\xi_x(t)$ is along the x-axis and $\xi_z(t)$ along the z-axis of

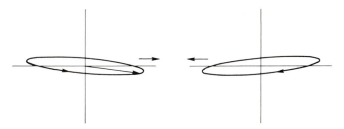

Figure 14-8. Trajectories of a water particle in the sound field of Fig. 14-9 at an observation point 20 m deep and at a radial distance of 100 m from the sound source. Frequency, 75 Hz. (Left) The trajectory if the observation point is located at $X = +100$ m in the source-fixed frame of Fig. 14-9. The vector indicates the instantaneous displacement at an arbitrary instant in the cycle. (Right) The same if the observation point is at $X = -100$ m. The arrows indicate the propagation direction of the sound. The left and right shark in Fig. 14-7 are subjected to similar elliptically polarized motions as in the left and right figures here.

the body-fixed frame. The mirror symmetry in Fig. 14-7 now implies the following relationship:

	left inset	right inset
"horizontal" component	$-\xi_X(t)$	$\xi_X(t)$
"vertical" component	$\xi_Z(t)$	$\xi_Z(t)$

We assume in calculations that the propagation direction of the wave is such that it has a positive projection on the x-axis of the body frame. This explains the plus sign for $\xi_X(t)$ when the sound source is behind the fish. For the same reason light rays are usually incident from the left in figures because the x-axis points by convention to the right.

As a sign change is equivalent to a time shift of half a period for each spectral component, any two direct waves propagating in directions symmetrically about the vertical (line) through the fish *can be discriminated by a temporal comparison between the horizontal and vertical displacements,* the latter being along the zenith direction in the case described here. A special case involves two horizontally traveling direct waves propagating in opposite directions. The same geometrical considerations apply here as in the case of surface waves of an arbitrary waveform (see p. 279).

This model of phase analysis for eliminating directional ambiguity has an important property in common with the models which will be discussed in the next section: The direct wave acts as the *object wave* and the reflected wave (=indirect wave) acts as a *reference wave.* For synchronous or coherent detection of the object wave, a phase coherent reference must be available. The common realization is where the reference wave (field, or signal) comes from the same signal source as the object wave (Fig. 14-9). Full coherency covers the case of signals with different amplitudes and a constant time shift between their waveforms.

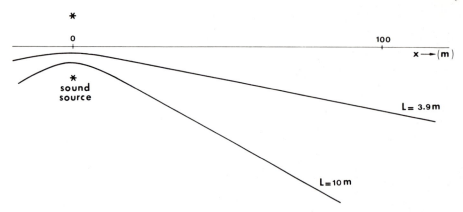

Figure 14-9. The direct and the surface-reflected sound waves show interference at any observation point. The interference effect depends on the difference L in length between the indirect, reflected, and the direct path traveled by the sound and on the frequency of the sound. For sine waves the displacement vector at some point above the hyperbola marked $L = 10$ m rotates clockwise in the figure for any positive value of X, whereas the vector rotates counterclockwise for negative values of X. The depth of the sound source is 10 m. In theory the limiting hyperbola $L = 10$ m is valid for sine waves of 75 Hz. The asterisk over the water denotes the mirror image of the sound source.

We can only understand how a shark can detect the elevation of a sound source in this detection model after studying angular detection by fishes with a swimbladder (Section 4). Anyhow a reflected wave, either from the (soft) surface or the (hard) bottom, is needed as a coherent reference in this model (Schuijf and Buwalda 1980). Since such waves do not exist in a true free field, detection based on this principle will fail there.

Another constraint for detection is that the time delay of the reflected wave must be less than half a period, $T/2$, for single-frequency waves. The reason is that otherwise the rotation sense of the displacement vector (Fig. 14-8) will not match the rotation sense for a receiving position right or left from the sound source in Fig. 14-9. If the difference in path length between the surface-reflected wave and the direct wave is denoted by L then this requires that

$$L < cT/2 \qquad\qquad (12)$$

For a given frequency $1/T$ the region of receiving points where L satisfies the inequality (12) is enclosed between the surface and the hyperbola constituting the locus of the points with constant difference L_{limit} from the foci (i.e., the sound source and its mirror source over the water); e.g., 75 Hz, L_{limit} is given by (12) as 10 m (cf. Fig. 14-9). The elliptic trajectories in Fig. 14-8 correspond to a point at 20 m depth on the hyperbola with $L = 3.9$ m in Fig. 14-9 ($X = \pm 100$ m).

Continuous low-frequency sine waves are not attractive for sharks (e.g., Myrberg 1978). For the localization process itself sine waves appear physically adequate (Nelson 1967). If all frequency components of a (biogene) sound are contained in a band $f < B$ then condition (12) is to be replaced by $L < c/2B$.

4 Models of Acoustic Localization Systems

4.1 Introduction

In this section models of acoustic localization in fishes are introduced, based on the analysis of rotating displacement vectors. We choose here for a black-box approach in constructing the models; for arguments see Section 1. We will compare the predictions of the models with the available empirical data only after their description is completed and the model properties can be derived.

The models will be developed from simple to more complex stimulus situations. Table 14-2 specifies the stimulus situations to be distinguished in regard to the detection problem. The classification in two-dimensional and three-dimensional stimulus directions requires amplification. The propagation direction of the sound stimulus can be completely specified by a vector of unit magnitude which we denote by $\vec{n}$. This unit vector is the direction perpendicular or *normal* to a wave front. Any propagation direction $\vec{n}$ may be represented geometrically by its end-point on a sphere of unit radius around the point of interest: the origin of the body-fixed frame. *Two angles* can fully specify its end-point like the position of a celestial body (altitude and azimuth) to an observer. Mathematically a coding by means of two variables is therefore sufficient to specify $\vec{n}$ with respect to the body-fixed frame in *three*-dimensional space. The classification of direction is therefore in the general case either two- or three-dimensional depending on the attitude of the researcher, e.g., for perception a coding of two variables would, in principle, be sufficient, whereas for the description of the polarization directions of the involved hair cells *three* body-fixed axes are required. Similarly if directional detection in only one plane is considered then the end-points of $\vec{n}$ lie on a circle, but a set of two axes is required to span the plane.

The division at level (1) in Table 14-2 depends on whether the sound source is acoustically far away or rather at a moderate or small acoustic distance (Schuijf and Buwalda 1980).

4.2 Rotating Displacement Vectors as Inputs for the Hair Cells in the Labyrinths of Fishes with Swimbladder

4.2.1 The Basic Equation for the Displacements

A sound field will, in general, cause an instantaneous resultant displacement of the base of some hair cell from its equilibrium position. In the following we make the simplifying (though not necessarily correct, cf. Gallé and Clemens 1976) assumption that this resultant displacement vector $\vec{R}(t)$ determines the neural response of the considered hair cell unit. The assumption also implies simplified otolith mechanics (Section 2), if applicable.

The resultant displacement vector $\vec{R}(t)$ is, at any instant t, equal to the vector sum (p. 273) of $\vec{\xi}(t)$, the particle displacement in the sound wave, and $\vec{\eta}(t)$, the pressure-induced displacement for which Eq. (2) holds. Hence we have

$$\vec{R}(t) = \vec{\xi}(t) + \vec{\eta}(t) \quad \text{or} \quad \vec{R}(t) = \vec{\eta}(t) + \vec{\xi}(t) \tag{13}$$

Table 14-2. Classification of the Models

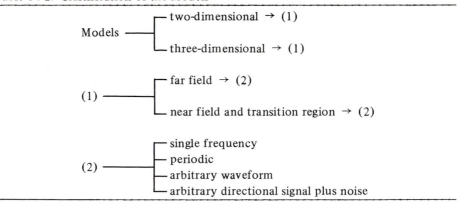

Models	┌ two-dimensional → (1)
	└ three-dimensional → (1)

(1)	┌ far field → (2)
	└ near field and transition region → (2)

(2)	┌ single frequency
	├ periodic
	├ arbitrary waveform
	└ arbitrary directional signal plus noise

according to the commutative law. This formula includes the well-known definition of the addition of vectors by the parallelogram rule. Although vector algebra is of no particular importance for the text below, the reader may refer to textbooks (e.g., Kittel, Knight, and Ruderman 1965, Ch. 2) if the connection between Eq. (13) and its geometric interpretation is not familiar. In vector notation we have the more powerful method to cope with more complex stimulus situations where a geometric illustration is no longer feasible.

4.2.2 The Resultant Displacement for Incident Plane Sound Waves

Let us derive the resultant displacement $\vec{R}(t)$ for a plane, traveling sound wave of frequency f. The propagation direction is $\vec{n}$. Let the acoustic pressure by 1 dyne/cm^2, thus the amplitude factor B in Eq. (5) equals 1. Hence from Eq. (2) we derive that the pressure-induced displacement is given by

$$\vec{\eta}(t) = C\cos(2\pi ft)\vec{m} = C\cos(\omega t)\,\vec{m} \tag{14}$$

This is the precise notation that the sound pressures forces a fluid particle or (hair) cell at some point P to execute a harmonic oscillation along a line through P in the direction $\vec{m}$, a fixed direction for the considered location in the labyrinth.

The value of the amplitude C of the oscillation depends on the particular location and holds only for a sound pressure of 1 dyne/cm^2 (otherwise the amplitude is BC) (see Fig. 14-10).

In the present case the far field equation relates the particle velocity $\vec{v}(t)$ and the acoustic pressure:

$$\vec{v}(t) = [p(t)/\rho c]\,\vec{n} \tag{15}$$

where ρc denotes the wave resistance of water. Equation (15) reads in this case

$$\vec{v}(t) = \frac{\cos(\omega t)}{\rho c}\,\vec{n} \tag{15a}$$

Because the particle velocity is the time derivative of the particle displacement $\vec{\xi}(t)$, it is easy to verify that

$$\vec{\xi}(t) = \frac{\sin(\omega t)}{\rho c \omega}\, \vec{n}$$

satisfies Eq. (15a). We write this result in the more general form

$$\vec{\xi}(t) = D \sin(\omega t)\, \vec{n} \tag{16}$$

where D denotes the amplitude of the displacements at $p = 1$ dyne/cm^2 and at a given value of f. The direction of $\vec{n}$ is constant for the entire fish and therefore all $\vec{\xi}(t)$ oscillate synchronously and parallel or antiparallel to $\vec{n}$ throughout the labyrinth, this in contrast to $\vec{\eta}(t)$. The resultant displacement $\vec{R}(t)$ is found by substituting the expressions for $\vec{\xi}(t)$ and $\vec{\eta}(t)$, Eqs. (16) and (15a) respectively, in Eq. (13). The result is

$$\vec{R}(t) = C \cos(\omega t)\vec{m} + D \sin(\omega t)\vec{n} \tag{17}$$

Two oscillations, one along a line through $\vec{m}$, and one along a line through $\vec{n}$, always will add as a resultant vector $\vec{R}(t)$ which lies in the plane spanned by $\vec{m}$ and $\vec{n}$. *The end-point of $\vec{R}(t)$ describes according to Eq. (8) an ellipse* which is shown in the computerplots of Fig. 14-10 for several directions of $\vec{n}$ and one constant direction of $\vec{m}$. In the plots the ratio D/C equals $1/3$. This is a reasonable estimate of the magnitude of the direct displacement to that of the p-induced component at $f = 100$ Hz for the middle region of the sacculus in cod: an extrapolation from data obtained by Buwalda and Van der Steen (1979) in their compensation experiment.

The direction of the principal axes of the ellipse do, in general, not correspond to the directions $\vec{m}$ and $\vec{n}$, unless $\vec{m}$ and $\vec{n}$ are perpendicular to each other. If $\vec{m}$ and $\vec{n}$ are parallel then the ellipse degenerates into a line segment. The principal axes of the ellipse are not essential for the analysis of the angular detection models below.

The rotation sense of $\vec{R}(t)$ is easily found. Let us follow the elliptical polarized motion of $\vec{R}(t)$ at instants $t = 0$ and $t = T/4$ ($\omega t = \pi/2$). The end-point of $\vec{R}(t)$ is then located at $C\vec{m}$ and $D\vec{n}$, respectively. It follows that we find *the rotation sense of $\vec{R}(t)$ if $\vec{m}$ is turned over the smallest angle so that its direction equals that of $\vec{n}$* (check this in Fig. 14-10).

For the development of a model for the angular detection it is crucial to remark that once during a cycle $\vec{R}(t)$ points in the direction of $\vec{n}$. Additional information is, however, required to detect this instant; this line is picked up again in Section 4.3.

4.2.3 The Resultant Displacement for Incident Spherical Sound Waves

Similar conclusions can be drawn in the case that the source is acoustically not very far away. The forced translatory oscillation of the fish as a whole can now, in analogy with Eq. (4), be written as

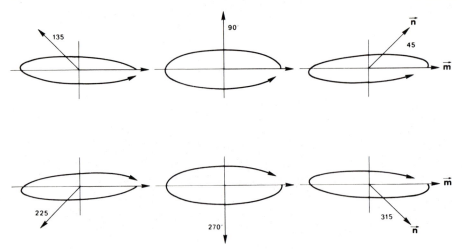

Figure 14-10. The rotation sense, shape, and orientation of the major axis of the elliptical trajectory is illustrated for a series of six angles between the variable propagation direction $\vec{n}$ and the constant direction of $\vec{m}$ to the right. Note that the shape of the ellipses is equal for opposite directions of $\vec{n}$, but that the rotation senses are opposite. A counterclockwise rotation of $\vec{n}$ is mathematically positive and results in a positive rotation sense of the displacement vector up to an enclosed angle of $180°$ when the ellipse changes into a line segment along the direction of $\vec{m}$ or $\vec{n}$ (not drawn). The external sound forms a far-field traveling wave. See text.

$$\vec{\xi}(t) = D \cos(\omega t + \psi)\vec{n} \tag{18}$$

We do not use the subscript x here for the phase shift ψ relative to $\vec{\xi}(t)$ because $\vec{n}$ takes over the role of the $\vec{e}_x$ axis (cf. Fig. 14-1, right). Thus, the phase shift is defined unequivocally without the need to specify coordinate axes. The amplitude D is for the same frequency greater than in Eq. (16) owing to near-field contributions.

If the $\xi(t)$ term in Eq. (17) is replaced by the expression in Eq. (18) then the resultant vector is

$$\vec{R}(t) = C \cos(\omega t)\vec{m} + D \cos(\omega t + \psi)\vec{n} \tag{19}$$

Also for the spherical sound waves the trajectories of $\vec{R}(t)$ are ellipses. Again once during a cycle $\vec{R}(t)$ points in the direction of $\vec{n}$. The instant when this happens does not coincide with the instant that $\vec{\xi}(t)$ attains its extreme magnitude as is the case in the far field. Equation (17) is a special case of Eq. (19) if we substitute the far-field value for ψ of $-\pi/2$ (from Fig. 14-3). The ψ value affects the shape and orientation of the ellipse. Since both of these aspects of the elliptical trajectory are subordinate for the model building below they are not treated in detail. Another reason for omitting these aspects is that ellipses are obtained only for sine waves.

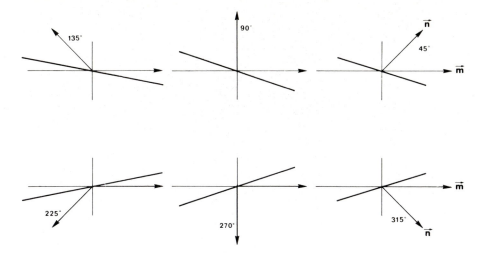

Figure 14-11. The ellipses degenerate into line segments in an extreme near field situation ($\psi = -180°$) for any propagation direction $\vec{n}$. The directions of $\vec{m}$ and $\vec{n}$ in the six figures are the same as in Fig. 14-10. The cue for phase analysis has disappeared for the models in Section 4.3. The conjecture is that this holds generally for any detection system based on $p(t)$ and $\vec{\xi}(t)$. See text.

4.3 Angular Detection by a Comparison of Rotating Vectors

4.3.1 Preliminaries and General Concepts Pertaining to the Models

Buwalda (Chapter 7, Section 4.3) presents carefully contrasted arguments in favor of the idea that the pressure-induced field at the maculae forms a radiating pattern of field or flow lines from one or more "acoustic windows" in the skull, for instance in a nonostariophysine species like the cod. For a schematic illustration of this concept, see Fig. 5 by Schuijf (1976b). The diverging phase of the $\vec{\eta}(t)$ field corresponds probably with the rarefaction phase of the incident sound wave, when the swimbladder expands. Note that in terms of $\vec{m}$ vectors, Eq. (2), this time-independent vector field converges onto the window. Whereas the picture of a field of $\vec{m}$ vectors, where the direction varies from point to point, may be realized anatomically only for windows, we will develop the theory in terms of $\vec{m}$ fields with arbitrary properties regardless of how they arise.

We shall be studying the simultaneous, effective input displacements of two given hair cells as a function of the external sound variables $\vec{\xi}(t)$ and $p(t)$. Only the orientation of these hair cells is taken into account, not their exact morphological location. We shall assume that each hair cell is sufficiently characterized by its position in the labyrinth and by its polarization direction. Thus, we may define the unit vector

$$\vec{h} = \text{the depolarization direction of the considered hair cell} \qquad (20)$$

We shall also need the unit vector $\vec{N}$ which points in the apical direction of the considered hair cell. It follows that $\vec{N}$ denotes the normal to the local tangent plane of the macula. In general the $\vec{m}$ vector does not lie in this tangent plane, unless stated otherwise.

With the introduction of these mathematical assumptions we open the presentation of the models of the angular detection.

4.3.2 A Simplified, Two-Dimensional Model of Directional Encoding Based on a Place Principle

In order to show how the propagation direction $\vec{n}$ of the incident sound wave can be derived from two rotating displacement vectors we shall first study two arbitrary points, located either unilaterally or bilaterally in the labyrinth(s) and satisfying the following simplifying constraints:

1. The $\vec{m}$ vectors at these points are antiparallel
2. The p-induced displacements are of equal magnitude
3. Each of both points corresponds with one $\vec{h}$ vector [refer to Eq. (20)] which lies in a single plane through these points.

All but the first of these restrictions are not necessary in this case, but are imposed to convey the basic ideas more readily. If a single frequency sound propagates parallel to the considered plane in direction $\vec{n}$ then the displacement vectors at both points describe ellipses with opposite rotation senses in parallel planes (see Fig. 14-12A). This difference in rotation sense is caused by the fact that both $\vec{R}$ vectors rotate from their opposite $\vec{m}$ directions over the smallest angle toward the common $\vec{n}$ direction. If and only if the $\vec{m}$ vectors are antiparallel then the $\vec{R}(t)$ will rotate in opposite directions for any incident direction $\vec{n}$.

Twice during a cycle *both displacement vectors* $\vec{R}(t)$ *point simultaneously in the same direction* (see Fig. 14-12B). These parallel or coincident directions correspond to the $+\vec{n}$ and $-\vec{n}$ directions, respectively. The statement is proved by noting that during a zero crossing of $p(t)$ the vector $\vec{\eta}(t)$ vanishes at both points and the common direct displacement $\vec{\xi}(t)$ remains. Obviously this proof *holds for any wave form,* including pulses, as any sound in nature includes at least one condensation and one rarefaction phase (Landau and Lifshitz 1963). Even a pulselike sound has therefore at least one pressure zero crossing as is required for a coincidence. The curve traced by the resultant vector $\vec{R}(t)$ starts and stops as time varies, at the equilibrium position of a hair cell.

Let us examine how the instant of coincidence can be derived from the responses of the primary neurons corresponding with each of the considered points (to be abbreviated as "the units"). The orthogonal projection of the rotating vector $\vec{R}(t)$ on the sensitive axis $\vec{h}$ (Eq. 20) of the unit forms a linear oscillation about the considered point. Obviously the amplitude and phase constants of this mechanical input for the involved hair cell shows a complex dependence on the external qualities $\vec{\xi}(t)$ and $p(t)$ (compare Figs. 14-10 and 14-12). Selection of the instant of a coincidence with a single directional receptor is impossible: information on the instantaneous direction of

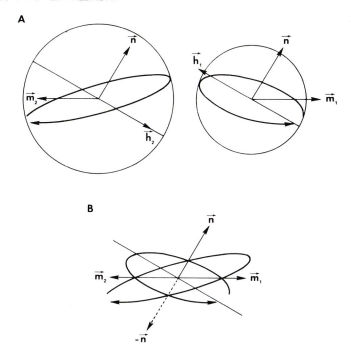

Figure 14-12. Comparison of the trajectories of rotating displacement vectors for two points with opposite $\vec{m}_1$ and $\vec{m}_2$ vectors and the indicated $\vec{n}$ direction ($\psi = -\,135°$ or $-\,3\pi/4$ radians, compare Fig. 14-3). (A) The individual paths for both points. In the illustration the ellipses start at a pressure maximum ($t = 0$) and terminate at the tip of the arrow when $t = (11/12)T$. T denotes the period. The importance of the special hair cell orientation of the pair $\vec{h}_1$ and $\vec{h}_2$ is discussed in Section 4.3.2, point 2. The circumscribed circles play a role as circular time bases in later analysis. Obviously the $\vec{R}(t)$ vectors do not rotate at constant angular velocity. (B) The occurrence of coincidence in direction and magnitude of the individual displacement vectors is clearly seen, if both figures are combined by shifting one of the figures without turning. The two coincidences in one cycle are indicated by $+\,\vec{n}$ and $-\,\vec{n}$. See text.

$\vec{R}(t)$ cannot be separated from its varying magnitude. Even in the special case of $\vec{h}$ and $\vec{n}$ being parallel, the input amplitude to the unit at the instant of coincidence is not necessarily maximal or in any other way characteristic. Hence, the output of such a unit, either in an analog form coding for the instantaneous input amplitude or in a digital phase-locked form, contains no amplitude or timing cue to the instant of coincidence. Not even the fact that the analog outputs of *two* such units, parallel to each other and to $\vec{n}$ and subjected to opposite rotating vectors, are equal to each other in coincidence, provides a sufficient feature: Such input- or output-amplitude equality can also occur at other instants in the cycle, as can easily be seen by considering the (admittedly extreme) case of the rotating vectors describing circles.

Through the output of one unit, however, it is possible to identify the instant when the vector $\vec{R}(t)$, rotating over the smallest angle towards the unit's $\vec{h}$ vector, is perpendicular to its $\vec{h}$. This can be explained as follows for the example of the phase-locking unit: at the very instant that $\vec{R}(t)$ is orthogonal to $\vec{h}$ the unit receives zero input and will not respond. In order to elicit synchronous firing, relatively sharply defined temporal stimulus characteristics are required. Now it is well known that the rate of change in amplitude is highest during a zero crossing of a signal. Thus, shortly after a negative-going zero crossing for the input signal the unit will fire for sufficiently high signal levels.

If it happens that the $\vec{n}$ vector is orthogonal to $\vec{h}$, as is the case in the right side of Fig. 14-12A, then a positive-going zero crossing of the input of unit $\vec{h}_1$ occurs *at the instant of the coincidence*. Note that $\vec{R}(t)$ activates the unit $\vec{h}_1$ shortly after $\vec{R}(t)$ has passed $\vec{n}$. Mutatis mutandis, a similar conclusion holds for unit $\vec{h}_2$ at the left side of Fig. 14-12A. We arrive at the paradoxical conclusion that the instant of coincidence can be identified by a unit $\vec{h}$ which is directed such that it receives no input at the instant of coincidence.

After these preliminaries it is possible to explain how an angular detection could be based on coincidences of rotating vectors. Instead of proposing a special mechanism which can perform such an analysis, it is easier to examine what the conditions are for using a coincidence for angular detection. The box below summarizes the order in which the various points relating to the model are treated. We shall use the same numbering in the main text.

Main characteristics of an angular detection based on coincidences.

1. A coincidence in direction of two rotating vectors is only a distinct event useful for angular detection, if the compared vectors only show a coincidence (or, more general, equivalence) at the instants of the zero crossings of the sound pressure $p(t)$.

2. For the angular detection by means of coincidence the encoding must be organized according to a place principle.

3. The coincidences corresponding to a negative-going zero crossing of $p(t)$ can be identified by the phase-locking properties of the involved hair cell units. This is the principle of phase analysis at the hair cell level for this group of models.

4. Rotating vectors can only exist if:
   (i) $\vec{\xi}(t)$ and $\vec{\eta}(t)$ are simultaneously present and coherent,
   (ii) the signal forms of $\vec{\xi}(t)$ and $\vec{\eta}(t)$, while coherent, do at least partially not share the signal form of $p(t)$.
         [In technical jargon only the far field term of $\vec{\xi}(t)$, which is phase-shifted by $-90°$ (in quadrature) relative to the signal $p(t)$, is effective in producing nondegenerated ellipses. See text.]
   (iii) $\vec{m}$ and $\vec{n}$ are not parallel or antiparallel everywhere.

---

(1) The first condition in the box is only met in the two-dimensional case as in the example of Fig. 14-12, if the displacement vectors $\vec{R}(t)$ rotate in opposite directions. At first sight the elliptical trajectories in Fig. 14-12A seem sufficiently different to allow rotations in the same direction. Yet, in the far field, where $\psi = -\pi/2$ radians (see Fig. 14-3), both paths become congruent. Both $\vec{R}(t)$ vectors are parallel all the time if for any t $\vec{\eta}_1(t) = \vec{\eta}_2(t)$, and the concept coincidence loses its importance. [The C factors (Eq. 2) at both points 1 and 2 may differ, but we will ignore this complication.]

In the plane configuration of Fig. 14-12 angular sectors for $\vec{n}$, where $\vec{R}_1(t)$ and $\vec{R}_2(t)$ rotate in the same directions, are only absent if

$$\vec{m}_2 = -\vec{m}_1 \tag{21}$$

In that case the rotation sense of both $\vec{R}(t)$ vectors will be inverted simultaneously if $\vec{n}$ would be turned through the line defined by the $\vec{m}$ or $-\vec{m}$ vector (compare Fig. 14-10). The condition (Eq. 21) cannot, in general, be satisfied for a bilateral arrangement of acoustic windows, as required in Buwalda's analysis (see Section 4.3.1). This topic is continued in Section 4.3.3.

(2) So far the considerations on the detection of a coincidence started from the knowledge that the hair cell pair $\vec{h}_1$ and $\vec{h}_2$ (Fig. 14-12A) was directed perpendicular to $\vec{n}$. Finding these directions is, of course, an integral part of the angular detection.

To generate a possible solution to this problem we shall compare the inputs for pairs of hair cells, $\vec{h}_1$ and $\vec{h}_2$, divided over two parallel macular planes. We avoid the complication that for any macula the $\vec{m}$ directions, and therefore the resultant displacements $\vec{R}(t)$, may be functions of position in the macula. This implies that the $\vec{m}$ fields are uniform for each macula and antiparallel, because of condition (21), for the pair of maculae we are considering. Such maculae do not exist, not even for a bilateral pair, but this simplified situation is much easier to visualize. (The next more general situation requires mirror symmetry for the $\vec{m}$ fields.)

We shall further assume that the hair cell pairs $\vec{h}_1$ and $\vec{h}_2$ belong to spreading patterns of $\vec{h}$ fields such that their $\vec{h}$ directions vary almost continuously as functions of an arc α measured along the striola (Fig. 14-13). This "fan" type of hair cell orientation is known from otolith organs of several teleost species (see survey in Fay and Popper 1980, Platt and Popper, Chapter 1).

Under the assumption above, every hair cell in one macula is subjected to the same $\vec{R}(t)$. Figure 14-12 applies again for sine waves, where each elliptic path corresponds to any position in one of the two maculae, respectively, left and right, say. We know from paragraphs preceding point (1) that a pair of units, $\vec{h}_1$ and $\vec{h}_2$, which is oriented such that it is orthogonal to $\vec{n}$, is likely to be synchronized just after a coincidence. The model implies that *there is a correspondence between the incident direction $\vec{n}$ and the position of the activated units* over the surface of the macula. An angular detection could thus be based on a place principle. Such a detection is independent of the phase angle ψ (Eq. 19) and of the presence of pressure noise (see Section 4.4). This latter property agrees exactly with Buwalda's finding (Chapter 7) that high pressure noise levels fail to deteriorate the angular resolution of the cod's hearing system. High velocity noise levels, on the other hand, abolish directional discrimination capabilities.

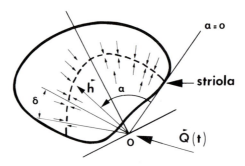

Figure 14-13. This schematic illustration shows how the hair cell polarization directions $\vec{h}$ can be characterized as a function of the arc α along the striola of a generalized macula. See text for the meaning of $\vec{Q}(t)$ and δ.

(3) When the p-induced displacement $\vec{\eta}(t)$ in Eq. (13) passes through a zero in the direction of $\vec{m}$ (=negative-going) then the displacement $\vec{R}(t)$ equals $\vec{\xi}(t)$ in the $\vec{n}$-direction of the traveling wave. The coincidence corresponding to the $+\vec{n}$ direction (Fig. 14-12B) will according to acoustic theory occur when at some instant t

$$p(t) = 0 \quad \text{and} \quad \frac{dp}{dt} < 0 \tag{22}$$

Such instants could be detected by a separate pressure-selective system (see Section 4.4).

In the rotating vector model these coincidences can be detected by the neural comparison of the instantaneous directions of *fictitious rotating vectors* $\vec{Q}(t)$ which lead the actual $\vec{R}(t)$ in their motion by a spatial angle of $90°$. In the example of Fig. 14-12A (right) the fictitious vector will have reached $\vec{h}_1$ at the instant when $\vec{R}_1(t)$ and $\vec{R}_2(t)$ are parallel to $\vec{n}$: the $+\vec{n}$ coincidence. At the same time the fictitious vector $\vec{Q}_1(t)$ reaches $\vec{h}_1$. Within some angle interval δ around this direction for $\vec{Q}_1(t)$ the units with $\vec{h}$ vectors around $\vec{h}_1$ will be activated. The angle interval δ (Fig. 14-13) takes into account variance in the instants of phase-locking caused by several factors. The significance of δ is explained later.

Clearly the hair cells at different angles α are activated successively when the $\vec{Q}(t)$ vector passes. The $\vec{Q}(t)$ activates the hair cells somewhat like the rotating beam of a lighthouse. Which of the two coincidences in one cycle (Fig. 14-12B) corresponds to the actual propagation direction $\vec{n}$ is determined unambiguously by the direction in which $\vec{R}(t)$ rotates from the local direction of the $\vec{m}$ vector toward the first coincidence. At the same instant $\vec{R}_1(t)$ and $\vec{R}_2(t)$ point in the directions $\vec{m}_1$ and $\vec{m}_2$, respectively [this occurs when $\vec{\xi}(t)$ moves through a zero in the same direction as $\vec{n}$]. From these reference directions in the macula the $\vec{R}(t)$ vectors rotate over the smallest angles toward the $+\vec{n}$ coincidence. Recall that there is an unique dependence of the rotation sense of the $\vec{R}(t)$ vectors on $\vec{m}$ and $\vec{n}$.

In the proposed model the *sampling instant* for $\vec{R}(t)$, or more exactly $\vec{Q}(t)$, is obtained by a comparison between the "rotating" activity patterns from two maculae.

(An intramacular comparison based on the same principle is also conceivable for a radially divergent $\vec{m}$ field.) The model forms a realization how $180°$ directional ambiguity can be eliminated without the need for a "neural" phase reference from a pressure-selective system. The segregation of directional and nondirectional acoustic variables $\vec{\xi}(t)$ and $p(t)$ is obtained here by a coherent detection in which the synchronization for sampling at the correct instants ($+\vec{n}$ coincidences, see Eq. 22) is acquired by a differential analysis of mechanical superposition patterns. The role of the p-induced field $\vec{\eta}(t)$ is to form in combination with $\vec{\xi}(t)$ mechanically polarized motions of different rotation senses, which enables the CNS to perform a phase-sensitive angular detection and to differentiate between coincidences at $+\vec{n}$ and $-\vec{n}$.

Finally, it is worth noting that in this model the angular resolution (Buwalda, Chapter 7) is related to the variability δ in angle, for which a coincidence is detected: If $\vec{n}$ diverts over an angle $\Delta\phi$, so does the coincidence direction. By definition this change is discriminated if

$$\Delta\phi \geqslant \delta \qquad (23)$$

where the angle δ (Fig. 14-13) can also be expressed in time units: The rotating vector sweeps the arc δ in the time $\Delta t \simeq \delta/\omega$, hence

$$\delta \simeq \omega\,\Delta t = 360(\Delta t/T) \quad \text{(degrees)} \qquad (24)$$

where T denotes the period. As Buwalda remarks (in Schuijf and Buwalda 1980, Section 3.4.2.3, Timing), Δt may be set by the variability in phase locking of the hair cells. This variability, expressed in the standard deviation of the distribution of phase-locked spike interval periods, appears to be a function of period length, such that $\Delta t/T$ is more or less constant (cf. Fay, Chapter 10). From Eqs. (23) and (24) it would then appear that δ is frequency independent. The same conclusion for $\Delta\phi$ must be considered with some reserve, however, since other factors may affect this parameter.

A constant difference in latency is sufficient for each pair of units involved in the detection of a coincidence. This difference may vary with α (Fig. 14-13).

(4i) It is difficult to tell what it means for two signals to be (partially) coherent. Coherence implies some form of synchronization between the signals. The $\vec{\xi}(t)$ and $\vec{\eta}(t)$ resulting from sound that comes from a single sound source is always coherent, whereas the contributions from independently oscillating sound sources to the total $\vec{\xi}(t)$ and $\vec{\eta}(t)$ are incoherent in general: on this basis sounds can be filtered.

(4ii) One temporal aspect that we still need to consider is why the near-field term of $\vec{\xi}(t)$ is irrelevant for angular detection in this model. The phase shift ψ [Eqs. (4) and (18)] between the component of $\vec{\xi}(t)$ along $\vec{n}$ and $p(t)$ is the essential parameter to discuss this question. Figure 14-11 gives us a hint: when ψ equals $-\pi$ radians in the extreme near field (Fig. 14-3B) then the ellipses degenerate into segments of straight lines (linear oscillations). These oscillations at either of the maculae do not form rotating activity patterns and are inadequate for the detection of coincidences in our model. One can also remark that at the instant of a pressure zero there is no $\vec{\xi}(t)$ contribution left.

It is rather easy to see why the ellipse degenerates into a linear oscillation when $\psi = -\pi$. It is noted that $\vec{R}(t)$ is the resultant of two linear oscillations $\vec{\xi}(t)$ and $\vec{\eta}(t)$ along $\vec{n}$ and $\vec{m}$, respectively [compare Eqs. (13) and (19)]. The resultant vector $\vec{R}(t)$ will oscillate along a line segment if all the parallelograms used to construct $\vec{R}(t)$ from $\vec{\xi}(t)$ and $\vec{\eta}(t)$ have the same diagonal in common. That is the case if the constructing vectors oscillate synchronously in phase or 180° out of phase ($\psi = -\pi$). Apparently this holds generally when the signal forms for $\vec{\xi}(t)$ and for $\vec{\eta}(t)$ are identical or inverted at any instant. Hence to obtain rotating vectors those signal components that are shared are irrelevant.

In the single-frequency case the signal form of $\vec{\eta}(t)$ is determined, according to Eq. (14), by the function cosine. If we split the signal form of $\vec{\xi}(t)$ [see Eq. (18)] in a cosine part and a sine part as follows:

$$\cos(\omega t + \psi) = \cos \psi \cos \omega t - \sin \psi \sin \omega t \qquad (25)$$

then it is clear that the $\cos(\omega t)$ term is shared with the signal form of $\vec{\eta}(t)$ and hence vanishes during a coincidence. The $\sin(\omega t)$ term, however, causes $\vec{R}(t)$ to rotate, provided that $\sin \psi \neq 0$. Thus, in the far field, where $\psi = -\pi/2$, the sine term is the only surviving contribution to $\vec{\xi}(t)$ and thus $\vec{\xi}(t)$ is then completely effective to let $\vec{R}(t)$ rotate.

The phase shift of 90° between the cosine and sine functions is what marks the essential difference in this single-frequency case. These functions are said to be orthogonal to each other (on a time interval of one period). Although passing over from sinusoidal wave forms to arbitrary (orthogonal) wave forms is quite possible and very important, this generalization is, unfortunately, not very tractable without more advanced mathematical techniques (see Ziemer and Tranter 1976).

(4iii) If $\vec{m}$ and $\vec{n}$ are parallel or antiparallel then the ellipse degenerates into a linear oscillation.

4.3.3 Some Remarks on Three-Dimensional Models
Based on Rotating Vectors

The plane model of the last section lacked generality in several ways, in order to simplify the explanation, for instance: (a) The maculae have been chosen to be parallel. (b) Uniform, antiparallel $\vec{m}$ fields have been chosen. Although this simplifies the analysis greatly, Eq. (21) is too stringent for the real anatomical situation.

We will briefly indicate how the foregoing analysis may be adapted to the three-dimensional case. Let the tangent local planes to the maculae at the considered points P_1 (right) and P_2 (left) be characterized by the outward unit normals $\vec{N}_1$ and $\vec{N}_2$ to these planes (see Section 4.3.1) drawn as in Fig. 14-14. The orientation of these "macular planes" is arbitrary, but the configuration has bilateral symmetry, and their line of intersection lies in the median plane: this excludes parallel planes. Also it is assumed that the $\vec{m}$ fields have bilateral symmetry. Ignore for the moment the special directions of $\vec{m}_1$ and $\vec{m}_2$: P_1 and P_2 do not have symmetrical positions.

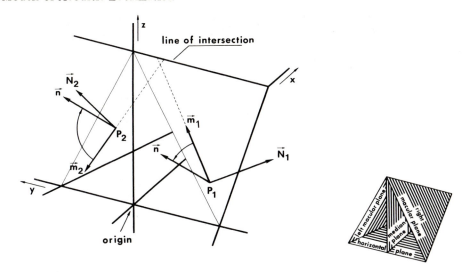

Figure 14-14. Perspective drawing indicating how counterrotating displacement vectors, as required for coincidences, can arise if the macular planes are not parallel. The compared points P_1 and P_2 are not at the same distances from the xy plane, but for clarity, the difference is greatly exaggerated. The $\vec{m}$ vectors for the compared points show a sort of skew symmetry, but the $\vec{m}$ fields themselves have bilateral symmetry. For further details see Section 4.3.3.

First, it is remarked that the trajectory of $\vec{R}_1(t)$ lies again in the plane spanned by $\vec{m}_1$ and $\vec{n}$. This plane will, in general, not coincide with macular plane 1. Only the normal projection of $\vec{R}_1(t)$ on the local tangent plane determines for the response of the hair cell. Thus, we have only to consider the trajectory of the displacement component in the macular plane, which we denote by $\vec{d}_1(t)$. The same holds for the contralateral macular plane and $\vec{R}_2(t)$.

It is very difficult to specify exactly what a coincidence is in a general three-dimensional situation. It is much easier to tell what a sufficient condition is in the present case. We note from inspection of Fig. 14-14 that if the sense of rotation of $\vec{d}_1(t)$ is such that the direction of advance of a right-hand screw corresponds to $\vec{N}_1$, then the same must hold for $\vec{d}_2(t)$ and $\vec{N}_2$ to obtain coincidence. Thus, for the "dorsal" side of the trajectory of $\vec{d}_1(t)$ the motion is directed caudally, while for the left macula the corresponding motion is directed rostrally. It can be shown by vector methods that the condition above can be satisfied for almost all directions of $\vec{n}$ (except for special directions where the ellipses degenerate). A sufficient condition for the $\vec{m}_1$ and $\vec{m}_2$ vectors at two points P_1 and P_2 is here

$$\vec{m}_1 \times \vec{N}_1 = \vec{m}_2 \times \vec{N}_2 \qquad (26)$$

Geometrically the vector or "cross" product $\vec{m}_1 \times \vec{N}_1$ is a vector perpendicular to both $\vec{m}_1$ and $\vec{N}_1$ and such that its direction corresponds to the axial motion of a right-hand screw when it is rotated from $\vec{m}_1$ to $\vec{N}_1$ over the smallest angle. See Kemmer (1976) for a background in vector methods.

If we consider $\vec{m}_1$ (at point P_1), $\vec{N}_1$, and $\vec{N}_2$ in the vector Eq. (26) as known vectors, we can determine the unknown vector $\vec{m}_2$ (at point P_2) from a solution of Eq. (26):

$$\vec{m}_2 = \vec{N}_2 \times (\vec{m}_1 \times \vec{N}_1). \qquad (27)$$

Figure 14-14 illustrates the special case where the $\vec{m}$ vectors are parallel to the macular surface at any point. This condition applies when, for example, the otolith is immovable so that, at its fixed boundary, the displacement of the "fluid" perpendicular to the otolith surface must be zero. Vectors $\vec{m}_1$ and $\vec{m}_2$, satisfying Eqs. (26) and (27), are both perpendicular to the line of intersection of the tangent planes at the compared points P_1 and P_2 (Fig. 14-14). This implies that if we turned the plane with the $\vec{m}_2$ vector about the line of intersection until both macular planes were in the same plane, we would have the plane model of 4.3.2 with antiparallel $\vec{m}$ vectors (Eq. 21). Another conclusion is that, for bilateral $\vec{m}$ fields, P_1 and P_2 cannot form a bilaterally symmetrical pair of points. The geometrical constraints imposed upon the $\vec{m}$ fields can be met by $\vec{m}$ vectors converging onto centers that are located bilaterally. This may be realized in a pair of acoustic windows. Though the $\vec{m}$ fields are probably symmetrical, coincidences can be obtained from a comparison between asymmetrically located hair cells, so that the involved $\vec{m}$ vectors satisfy condition (26). It is difficult to translate this particular model to the real anatomical situation and modifications will be necessary. However, that does not alter the interesting properties of this family of models.

The rotation senses of the displacement vectors at P_1 and P_2 are indicated in Fig. 14-14 by the arrows from $\vec{m}$ to $\vec{n}$. The drawing is simplified in that the line of intersection lies in the median plane, whereas this need not be true. This three-dimensional model can be extended to include unambiguous detection of any possible direction of incidence $\vec{n}$ without the assumption of a configuration with only a few detector axes. The working of the detection process as a whole may, however, be understood in terms of such simple models as vectorial weighing and phase analysis.

4.4 Encoding Schemes Requiring Separate Pressure Processing to Obtain a Neural Phase Reference

These schemes may again be thought of as sampling the resultant displacement vectors $\vec{R}(t)$ at the instants when a negative-going zero crossing of the acoustic pressure occurs [refer to Eq. (22)]. The main differences with the schemes in Sections 4.2 and 4.3 are:

1. The coherent reference signal for sampling the $\vec{R}(t)$ vectors at the correct instant is determined by neural activity from a separate p-selective system.
2. It is not essential that the vectors $\vec{R}(t)$ do rotate. Of course they may do so simply because $\vec{\xi}(t)$ and $\vec{\eta}(t)$ are simultaneously present in parts of the hearing system where the structural adaptations for input selectivity are incomplete or

unnecessary. This may be the case for ostariophysine fish, where the functional anatomy suggests the presence of physically separated pressure and motion sensitive systems and where the $\vec{\eta}(t)$ is so dominating over $\vec{\xi}(t)$ (i.e., the gain of the indirect stimulation mode is very high), that $\vec{\xi}(t)$ specificity (structurally determined motion specificity, see Buwalda, Chapter 7, Section 4.3) is unattainable by mechanical means alone.

3. Sampling during pressure zero crossings is here only a method to get rid of dominating p-induced fields while maintaining the advantage of phase analysis to combat directional ambiguity.

4. The condition (21) that the $\vec{m}$ vectors must be antiparallel in the case of two parallel maculae does not apply for this neural sampling.

The discrimination of negative-going from positive-going zero crossings can most likely be realized by the phase locking of a pressure-sensitive fiber which is determined by the polarization direction to which it belongs (compare Hawkins and Horner, Chapter 15). A problem could be that a *phase error* in the sampling instant of the displacement vector $\vec{R}(t)$ is likely to result in rather large deviations of the detected angular position of the sound source, because the usually strong $\vec{\eta}(t)$ component has not vanished completely. Phase errors in the coherent reference that do not cancel are caused by (1) any time error in the sampling instant due to imperfections in the detection process for the negative-going zero crossing itself, and (2) relative deviations in the phase response of $\vec{\eta}(t)$ at the motion selective system.

A common, very important feature of the time-sampling models in Sections 4.3 and 4.4 is that the angular detection is highly immune for pressure noise. The sampling occurs when the *instantaneous sum* of the pressure contribution $p_s(t)$ due to the signal and the pressure contribution $p_n(t)$ due to the noise equals zero. In that case the $\vec{\eta}(t)$ contribution to $\vec{R}(t)$ disappears and $\vec{\xi}(t)$ remains: the absolute time at which the coincidence occurs does not matter, unless $p_s(t)$ is so weak that sampling in the wrong half of the oscillation period might occur. This is evident from the fact that $p_s(t)$ remains essential for eliminating 180° directional ambiguity.

5 Summary

This chapter is concerned with the model description of acoustic localization in fishes: how the sound reaches the fish in the habitat and subsequently, how the external sound variables relate to the internal vibratory quantities that activate the hair cells. Because it is simpler to think in terms of particle displacements than in accelerations, the models use the particle displacement $\vec{\xi}(t)$ as input or output, although this simplification is not justified in all cases (see p. 271 and Appendix A).

Sections 1 to 3 constitute a view of the role of phase analysis in acoustic localization. In particular the probable relationship between phase analysis and the occurrence of rotating displacement vectors is analyzed. This topic is further explored in Section 4 for the fish's labyrinths.

Continuous sine waves without a clear beginning are good enough for a phase analysis by fish. However, the directional information and phase information contained in transient phenomena at the onset of a sound, although apparently not always required, cannot be discarded. Such transient phenomena are particularly interesting in shallow water. Accordingly, an analysis of the propagation of sound in these conditions is presented, in the minimum detail necessary for an understanding of this complex acoustic habitat. In general it was felt that some discussion was needed concerning the fact that many aquatic environments act as wave guides with very different properties as compared to free fields.

The proposed "time sampling" models of Section 4 explain in a new way why acoustic pressure is required for unambiguous acoustic localization (for fishes with a swimbladder) to act (indirectly) on the receptor sites, and why the acoustic pressure is nevertheless incapable of masking the angular detection operating on the incident direct particle displacements.

The phase shift, normally present between the direct particle displacement and the pressure induced displacement, causes the hair cells to be subjected to rotating displacement vectors. In one of two basic forms of time sampling models the sampling instant is obtained from the detection of a "coincidence": a new concept introduced here. This angular detection based on coincidences forms a spatiotemporal analysis; it can also be looked upon as a place theory for acoustic localization.

Although this topic is not treated explicitly, the models of vectorial weighing and phase analysis are both aspects of the three-dimensional form of angular detection based on coincidences (Section 4.3.3). It seems that the theory links the psychophysical models of acoustic localization with models of the subsystems at the receptor level.

Acknowledgments. The author wishes to express his gratitude for the patience exercised by the editors for him to revise his original time shift model and to thank Drs. R. R. Fay, A. N. Popper, and A. A. Myrberg, Jr., for critically reading the manuscript. The time shift effects remained, but the relevant cue appeared to be different. The help of Dr. Rob Buwalda, my very esteemed colleague, was in many ways decisive for the ultimate form in which the time-sampling models are presented. I wish to express my deep gratitude for his initiatives in this respect and for reading and commenting on the draft. I have special pleasure in expressing my sincerest thanks to Mr. P. Teunis and Mr. C. E. Versteeg for giving support in various ways during the preparation of this chapter. Without their expert assistance the paper would not have appeared at all. The illustrations were produced by them and Mrs. S. van Cornewall. My dearest thanks go to my wife Mrs. W. S. Schuijf-Maarsingh for producing the final typescript and for her devoted assistance. Also Mrs. Uytdehaage-Van Vulpen and Miss Renée Enklaar produced part of the drafts and manuscript.

The journey to Sarasota for presenting the lecture was made possible by the National Science Foundation.

References

Abramowitz, M., Stegun, I. A.: Handbook of mathematical functions. Washington, D.C.: U.S. Govt. Print. Off., 1968.

Banner, A.: Evidence of sensitivity to acoustic displacements in the lemon shark, *Negaprion brevirostris* (Poey). In: Lateral Line Detectors. Cahn, P. H. (ed.). Bloomington: Indiana Univ. Press, 1967, pp. 265-273.

Banner, A.: Use of sound in predation by young lemon sharks, *Negaprion brevirostris*

(Poey). Bull. Mar. Sci. 22, 251-283 (1972).

van Bergeijk, W. A.: Variations on a theme of Békésy: A model of binaural interaction. J. Acoust. Soc. Am. 34, 1431-1437 (1962).

van Bergeijk, W. A.: Directional and non-directional hearing in fish. In: Marine Bio-acoustics, Vol. 1. Tavolga, W. N. (ed.). New York: Pergamon Press, 1964, pp. 281-299.

Brekhovshkikh, L. M.: Waves in Layered Media. New York: Academic Press, 1960.

Burington, R. S.: Handbook of Mathematical Tables and Formulas. New York: McGraw-Hill, 1965.

Buwalda, R. J. A., Schuijf, A., Hawkins, A. D.: On the discrimination in three dimensions of sound waves from opposing directions by the cod. J. Exp. Biol. (in press)

Buwalda, R. J. A., van der Steen, J.: Sensitivity of the cod sacculus to directional and non-directional stimuli. Comp. Biochem. Physiol. 64A, 467-471 (1979).

Chapman, C. J., Johnstone, A. D. F.: Some auditory discrimination experiments on marine fish. J. Exp. Biol. 61, 521-528 (1974).

Cahn, P. H., Siler, W., Fuijya, M.: Sensory detection of environmental changes by fish. In: Responses of fish to environmental changes. Chavin, W. (ed.). Springfield, Illinois: Charles C Thomas, 1973, Ch. 13.

Corwin, J. T.: Morphology of the macula neglecta in sharks of the genus *Carcharhinus*. J. Morphol. 152, 341-362 (1977).

Crawford, F. S.: Waves (Berkeley Physics Course, Vol. 3), 3rd ed. New York: McGraw-Hill, 1968.

Elmore, W. C., Heald, M. A.: Physics of Waves. Tokyo: McGraw-Hill, 1969.

Enger, P. S., Hawkins, A. D., Sand, O., Chapman, C. J.: Directional sensitivity of saccular microphonic potentials in the haddock. J. Exp. Biol. 59, 425-434 (1973).

Fay, R. R., Popper, A. N.: Modes of stimulation of the teleost ear. J. Exp. Biol. 62, 379-387 (1975).

Fay, R. R., Popper, A. N.: Structure and function in teleost auditory systems. In: Comparative Studies of Hearing in Vertebrates. Popper, A. N., Fay, R. R. (eds.). New York: Springer-Verlag, 1980, pp. 3-42.

Fay, R. R., Olsho, L. W.: Discharge patterns of lagenar and saccular neurones of the goldfish eighth nerve: Displacement sensitivity and directional characteristics. Comp. Biochem. Physiol. 62A, 377-387 (1979).

Gallé, H. G., Clemens, A.: The sacculus of *Rana esculenta*: A physiological and physical study of the spatial and temporal properties of an equilibrium system (in Dutch with a summary in English). Thesis, University of Utrecht, Utrecht, 1976.

Hawkins, A. D., Sand, O.: Acoustic properties of the cod swimbladder. J. Exp. Biol. 58, 797-820 (1973).

Hawkins, A. D., Sand, O.: Directional hearing in the median vertical plane. J. Comp. Physiol. 122A, 1-8 (1977).

Kelly, J. C., Nelson, D. R.: Hearing thresholds of the horn shark *Heterodontus francisci*. J. Acoust. Soc. Am. 58, 905-909 (1975).

Kemmer, M. N.: Vector analysis for physicists. Cambridge: Cambridge Univ. Press, 1976.

Kittel, C., Knight, W. D., Ruderman, M. A.: Mechanics. (Berkeley Physics Course, Vol. 1). New York: McGraw-Hill, 1965.

Klimley, A. P., Myrberg, A. A., Jr.: Acoustical stimuli underlying withdrawal from a sound source by adult lemon sharks, *Negaprion brevirostris* (Poey). Bull. Mar. Sci. 29, 447-458 (1979).

Landau, L. D., Lifshitz, E. M.: Fluid mechanics. (Course of Theoretical Physics, Vol. 6), 3rd ed. Oxford: Pergamon Press, 1963.

Landau, L. D., Lifshitz, E. M.: Mechanics. (Course of Theoretical Physics, Vol. 1), 3rd ed. Oxford: Pergamon Press, 1976.

Milsum, J. H.: Biological Control Systems Analysis. New York: McGraw-Hill, 1966.

Morse, Ph. M., Ingard, K. U.: Linear acoustic theory. In: Encyclopedia of Physics, Vol. 11/1. Flügge, S. (ed.). Berlin: Springer-Verlag, 1961.

Myrberg, A. A., Jr.: Underwater sound–Its effect on the behavior of sharks. In: Sensory Biology of Sharks, Skates and Rays. Hodgson, E. S., Mathewson, R. F. (eds.). Arlington: Office of Naval Research, 1978, pp. 391-497.

Myrberg, A. A., Jr., Ha, S. J., Walewski, S., Banburry, J. C.: Effectiveness of acoustic signals in attracting epipelagic sharks to an underwater sound source. Bull. Mar. Sci. 22, 926-949 (1972).

Myrberg, A. A., Jr., Gordon, C. R., Klimley, A. P.: Attraction of free ranging sharks to low-frequency sound, with comments on its biological significance. In: Sound Reception in Fishes. Schuijf, A., Hawkins, A. D. (eds.). New York: Elsevier, 1976, pp. 205-228.

Myrberg, A. A., Jr., Spanier, J. R., Ha, S. J.: Temporal patterning in acoustic communication. In: Contrasts in Behavior. Reese, E. S., Lighter, F. J. (eds.). New York: Wiley, 1978, pp. 137-179.

Nelson, D. R.: Hearing thresholds, frequency discrimination and acoustic orientation in the lemon shark, *Negaprion brevirostris* (Poey). Bull. Mar. Sci. 17, 741-768 (1967).

Nelson, D. R.: Telemetering techniques for the study of free-ranging sharks. In: Sensory Biology of Sharks, Skates, and Rays. Hodgson, E. S., Mathewson, R. F. (eds.). Arlington: Office of Naval Research, 1978, pp. 419-482.

Nelson, D. R., Gruber, S. H.: Sharks: Attraction by low-frequency sounds. Science 142, 975-977 (1963).

Nelson, D. R., Johnson, R. H.: Acoustic attraction of Pacific reef sharks: effects of pulse intermittency and variability. J. Comp. Biochem. Physiol. 42A, 85-89 (1972).

Olsen, K.: Directional hearing in cod (*Gadus morhua*). Coun. Meet. Int. Coun. Explor. Sea (B20) (1969).

Pekeris, C. L.: Theory of propagation of explosive sound in shallow water. Geol. Soc. Am. Memoir 27, 1-117 (1948).

Popper, A. N.: A comparative study of the otolithic organs in fishes. Scanning Electron Microscopy II, 405-416 (1978).

Sand, O.: Directional sensitivity of microphonic potentials from the perch ear. J. Exp. Biol. 60, 881-899 (1974).

Schuijf, A.: Field studies of directional hearing in marine teleosts. Thesis, Univ. of Utrecht, Utrecht, 1974.

Schuijf, A.: Directional hearing of cod (*Gadus morhua*) under approximate free field conditions. J. Comp. Physiol. 98, 307-332 (1975).

Schuijf, A.: The phase model of directional hearing in fish. In: Sound Reception in Fish. Schuijf, A., Hawkins, A. D. (eds.). Amsterdam: Elsevier, 1976a, pp. 63-84.

Schuijf, A.: Timing analysis and directional hearing in fish. In: Sound Reception in Fish. Schuijf, A., Hawkins, A. D. (eds.). Amsterdam: Elsevier, 1976b, pp. 87-112.

Schuijf, A., Siemelink, M. E.: The ability of cod (*Gadus morhua*) to orient towards a sound source. Experientia (Basel) 30, 773-774 (1974).

Schuijf, A., Buwalda, R. J. A.: On the mechanism of directional hearing in cod (*Gadus morhua* L.). J. Comp. Physiol. 98, 333-343 (1975).

Schuijf, A., Buwalda, R. J. A.: Underwater localization–A major problem in fish acoustics. In: Comparative Studies of Hearing in Vertebrates. Popper, A. N., Fay, R. R. (eds.). New York: Springer-Verlag, 1980, pp. 43-77.

Schuijf, A., Visser, C., Willers, A. F. M., Buwalda, R. J. A.: Acoustic localization in an
    ostariophysian fish. Experientia (Basel) 33, 1062-1063 (1977).
Tavolga, W. N.: Acoustic obstacle avoidance in the sea catfish, *Arius felis*. In: Sound
    Reception in Fish. Schuijf, A., Hawkins, A. D. (eds.). Amsterdam: Elsevier, 1976,
    pp. 185-204.
Urick, R.: Principles of underwater sound for engineers. New York: McGraw-Hill,
    1967.
de Vries, H.: The mechanics of the otolith mechanics. Acta Otol. Lar. 38, 262-273
    (1950).
von Frisch, K., Dijkgraaf, S.: Können Fische die Schallrichtung wahrnehmen? Z. Vergl.
    Physiol. 22, 641-655 (1935).
Werner, G.: The study of sensation in physiology: Psychophysical and neurophysio-
    logical correlation. In: Medical Physiology, Vol. 2, 12th ed. Mountcastle, V. B. (ed.).
    St. Louis: Mosby, 1968, pp. 1643-1671.
Ziemer, R. E., Tranter, W. H.: Principles of communications. Boston: Houghton Mif-
    flin, 1976.

Appendix A: Factors Perturbing "Simple" Otolith Mechanics

The motion of an otolith can be treated by what is known as the theory of small
oscillations (Landau and Lifshitz, 1976) of a mechanical system about its position of
equilibrium. The input for the system is the external inertia force due to the acceleration of the body-fixed frame. The output is the displacement of the otolith relative
to the hair cells. The mass of the otolith together with the elastic properties of its suspension determine the natural or resonance frequency of the system. If these elastic
properties are anisotropic then there exist at least two different natural frequencies for
simple periodic oscillations parallel to the macular plane. These free oscillations of the
otolith are heavily damped and are approximately independent of each other if they
occur along two special mutually perpendicular axes. If we take these axes arbitrarily
as the x- and y-axis (not the body-fixed axes here), we can let the z-axis point from the
macula toward the otolith. The orientation of these axes is fixed in space and corresponds to the equilibrium position of the system. This frame serves as reference for
the hair cell and otolith motions. The otolith can execute translatory motions, characterized by the displacement components $\xi_x(t)$ and $\xi_y(t)$ respectively, and oscillatory
rotations with angular displacements $\phi_x(t)$, $\phi_y(t)$ and $\phi_z(t)$ about the x-, y-, and z-axis,
respectively (ξ_z does not produce shearing of the hair cells).

We have "simple" otolith mechanics if the relative displacements between otolith
and the hair cells look like the displacements present at the inputs, both in direction
and in time, except for a constant delay. The output is then a scaled, delayed replica
of the input. Neither of these requirements is fulfilled in real systems.

The temporal distortion consists of two types: neither the amplitude responses nor
the phase response is constant with frequency. For instance, below the natural frequency the phase of the output lags between $0°$ and $90°$. Above the resonance frequency the phase response varies between $90°$ and $180°$ lagging. Thus, far above the
natural frequency the displacement output has an inverted waveform (phase shift of

180°) and looks like the input. If the natural frequencies of the system are in the frequency band of interest and differ considerably for the x and y direction then a differential phase shift affects the resultant direction of the response.

An interesting, important deviation of simple mechanics occurs when a translational acceleration along the z-axis results in a rotation about an axis parallel to the x-axis and, consequently, in an output displacement *perpendicular* to the input. Such a directional transformation seems to exist for the sacculus otolith of cod (K. Horner pers. comm.). In order to show more explicitly the nature of this process we consider the motion of an otolith with a triangular cross section in the yz plane. An acceleration $a_z(t)$ along the z-axis will then, in general, result in a torque $N_x(t)$ about the x-axis. This is the case if the resultant force, to which the fluid pressures on the surface of the otolith reduce, does not pass through the center of mass of the otolith. Hence, the torque would be absent if the otolith body had a plane of symmetry (xz plane) at right angles to the yz plane. The equation for the rotational motion about the x-axis is given by the angular momentum theorum which reads here approximately

$$ I \frac{d^2\phi}{dt^2}x \cong - Ga_z(t) = - G \frac{d^2 \xi^*}{dt^2}z \qquad (A1) $$

where I and G are constants for the otolith-fluid system. The starred variable $\xi_z^*(t)$ denotes the particle displacement in the external sound. The displacement component ξ_y is proportional to ϕ_x and to the shortest distance d between the x-axis and the line parallel to the y-axis through the considered hair cell. It follows that

$$ \xi_y = - (Gd/I)\xi_z^* \qquad (A2) $$

G/I depends on the asymmetry of the cross section and determines together with d the sensitivity to accelerations normal to the macular plane. Equation (A2) gives the transfer function at frequencies several times above the natural frequency for this translational-to-rotational oscillation mode. Large amplitudes are not required for this first-order effect.

Appendix B: The Acoustics of the Shallow Water Layer

First, an expression is derived for the group velocity U of the propagation of the lowest mode in a layer of depth H. Consider a water duct of *twice* the actual depth, and with soft boundaries at[1] $Z = 0$ (actual surface) and $Z = -2H$ (fictitious surface) (see Fig. 14-5). Two rarefaction fronts marked 1 intersect where the actual bottom would be; thus, their vertical motions cancel each other on symmetry grounds whereas the negative sound pressure doubles. These acoustic conditions would also be required if a hard boundary (a perfectly rigid bottom) were there. Each of the compression fronts

---

[1] The capital letters X and Z are used as place coordinates in a resting frame of reference, instead of x and z in a body-fixed frame.

marked 2 is half a wavelength ahead of a parallel wave front 1, such that $|\vec{AB}| = \lambda/2$. At B a rarefaction front and a compression front cancel each other; thus, the acoustic pressure is zero. This is the boundary condition of a sound field at a soft surface: The actual event is that, at the instant the compression wavefront 2 reaches the surface at B, a rarefaction wavefront 1 results from the reflection. In the triangle ABC we have:

$$\sin \theta = \left| \frac{AB}{BC} \right| = \frac{\lambda/2}{2H} = \frac{\lambda}{4H} = \frac{c}{4Hf} \tag{B1}$$

This is the condition for a constructive interference of the crisscrossing waves. It is clear that the sound "ray" AB progresses at the net horizontal propagation speed

$$U = c \cos \theta = c \sqrt{1 - \sin^2 \theta} \tag{B2}$$

along the duct. Substituting Eq. (B1) in Eq. (B2) and using Eq. (9) we find

$$U = c \sqrt{1 - f_1^2/f^2} \tag{B3}$$

where f_1 denotes the lowest cutoff frequency [see Eq. (9)]. Thus, it follows that $U = 0$ for $f = f_1$ when the layer resonates vertically, and $U = c$ for $f \gg f_1$.

Each mode has its own typical vertical distribution. In the first mode there is only one pressure node, located at the surface.

Second, the steady-state forced oscillation of the layer driven by a harmonic sound source is specified. The amplitude of the forced oscillation in the layer is constant. The solution for the lowest mode of oscillation is given by

$$\xi_X(t) = -\gamma \frac{A}{\omega} \sin \left(\frac{\pi Z}{2H} \right) \cos(\omega t - \gamma X) \tag{B4}$$

and

$$p(t) = \rho \omega A \sin \left(\frac{\pi Z}{2H} \right) \sin(\omega t - \gamma X) \tag{B5}$$

where $\omega = 2\pi f$ and $\gamma = (2\pi/c)\sqrt{f^2 - f_1^2} = \omega/c \cos \theta$, and where A is a constant amplitude depending on the vertical location of the sound source and on its strength. It is rather straightforward to derive Eqs. (B4) and (B5) from the formula for the acoustic potential given by Urick (1967, Section 6.3) and by others. These equations, together with one for $\xi_Z(t)$, determine the phase relationships quoted in the main text.

Finally, the transient response of the layer to an impulsive excitation of the lowest oscillation mode is stated. The following equations form the basis for the plots in Fig. 14-6:

$$\xi_X(t) = \omega_1 t_0 A \sin\left(\frac{\pi Z}{2H}\right) \int_{T_0}^{t} \frac{J_1\left(\omega_1 \sqrt{\tau^2 - t_0^2}\right)}{\sqrt{\tau^2 - t_0^2}} \, d\tau \qquad \text{(B6)}$$

and

$$p(t) = \frac{\rho \omega_1 t A}{\sqrt{t^2 - t_0^2}} \sin\left(\frac{\pi Z}{2H}\right) J_1\left(\omega_1 \sqrt{t^2 - t_0^2}\right) \qquad \text{(B7)}$$

provided that the sound wave has arrived at the observation point $t > t_0$ (where $t_0 = X/c$); otherwise, the sound variables are zero. Further specification: $\omega_1 = 2\pi f_1$. The Bessel function of the first order is denoted as usual (Abramowitz and Stegun 1968).

Chapter 15

Directional Characteristics of Primary Auditory Neurons from the Cod Ear

A. D. HAWKINS and KATHLEEN HORNER*

1 Introduction

It is now well established that at least some teleosts have a directional hearing sense (see Schuijf and Buwalda 1980). The cod, *Gadus morhua,* detects a change in the direction of propagation of sound and can discriminate between spatially separated sound sources, both in the horizontal and median vertical planes (Chapman and Johnstone 1974, Schuijf 1975, Hawkins and Sand 1977). Indeed, cod can discriminate between frontally incident and caudally incident sound waves (Schuijf and Buwalda 1975) and even distinguish between diametrically opposed sound sources in both the median vertical and transverse vertical planes (Buwalda, Schuijf, and Hawkins 1981). Thus, the cod discriminates direction in circumstances that are ambiguous or confusing for man, whose directional capabilities in the median vertical plane are poor (Butler 1974).

Fish are unlikely to determine the direction of a sound source using differences in the time of arrival, phase, and intensity at the two ears as do man and the other large mammals (van Bergeijk 1964). The speed of sound in water is so fast that differences in the time of arrival or phase at the two closely placed ears are very small. Moreover, fish are small in comparison to the sound wavelengths of interest to them, and have about the same overall density as the surrounding water, so that sound shadowing is minimal and any differences in stimulus strength at the two ears insignificant. Pumphrey (1950) suggested that the hearing organs of fish, the otolith organs of the labyrinth, essentially detect the movement of the medium that accompanies passage of a sound wave. The fish itself is virtually transparent to sound, and moves with the medium, while the dense mass of the calcareous otolith lags behind, resulting in a shearing force at the site of contact between the otolith and the sensory cells. It has since been confirmed by experiment that the otolith organs detect particle motion, rather than the concomitant sound pressure. Microphonic receptor potentials from the sacculus of the plaice, *Pleuronectes platessa,* changed significantly in amplitude as the

*Marine Laboratory, Victoria Road, Torry, Aberdeen, Scotland.

particle velocity amplitude of an applied sound stimulus was varied, but remained constant as the sound pressure was changed independently of velocity (Hawkins and Mac-Lennan 1976).

Dijkgraaf (1960) pointed out that a particle motion detector is inherently directional in its response. Depending on the suspension of the otolith, and the orientation of the sensory hair cells detecting the shearing force, the otolith organ will respond differently to sound waves incident from different directions. Enger, Hawkins, Sand, and Chapman (1973) demonstrated that the otolith organ was directional by recording microphonic receptor potentials from the sacculus of a vibrated haddock, *Melanogrammus aeglefinus*. Sand (1974), for the perch *Perca fluviatilis,* later reported that the two bilateral sacculi showed different axes of maximum sensitivity in the horizontal plane, set about 40° apart. More recently, Buwalda and van der Steen (1979) have described a dependence of the level of the saccular microphonics upon sound direction in the cod, the axes of sensitivity of the two sacculi differing by 73°.

Schuijf (1976a) has assigned the term "vector weighing" to the determination of direction in a plane by comparison of the outputs of two or more bidirectional detectors aligned on different axes. Such a simple mechanism will show a 180° ambiguity in sound source location, since the direction of sound propagation and the direction of particle motion in a longitudinal traveling wave are alternately common and opposite. However, in theory the 180° ambiguity can be eliminated and the propagation direction determined by comparing the phase of sound pressure with the phase of particle displacement (Schuijf 1976b). Experiments by Schuijf and Buwalda (197 (1975) and Buwalda et al. (1981) have confirmed that this phase relationship is important for directional hearing (see Schuijf, Chapter 14).

Though the otolith organ itself responds to particle motion, the cod detects sound pressure (Chapman and Hawkins 1973), apparently employing the swimbladder as an accessory hearing organ (Sand and Enger 1973). Sound pressures incident upon the swimbladder are transformed into particle motion which then stimulates the otolith organ (Sand and Hawkins 1973). Over a broad range of frequencies the reradiated particle motion is much higher in amplitude than the particle motion of the directly incident sound wave (Sand and Hawkins 1973), so that the sensitivity of the ear is increased.

The direction of the reradiated particle motion is determined by the transmission path between the swimbladder and the ear, rather than the direction of the incident sound wave. The functional coupling between the swimbladder and the ear might therefore interfere with vector weighing of directly received particle motion. For comparison of the phase of the direct and reradiated particle motion to take place within the central nervous system, the vector detectors within the ears must be isolated from the reradiated particle motion. However, peripheral interaction between the direct and reradiated particle motion has not been ruled out, and may even promote the elimination of directional ambiguities (see Schuijf, Chapter 14). There is a clear structural basis for the directionality of the otolith organs. Anatomical studies of the polarized hair cells in the sensory membranes or maculae of the organs have shown well defined patterns of receptor cell orientation with consistent inter- and intramacular differences (Lowenstein, Osborne, and Wersäll 1964, Wersäll, Flock, and Lundquist 1965, Dale 1976). If different parts of the ear are sensitive to motion in different directions then comparison of the nervous response of these groups of receptors might provide the mechanism for vector weighing. Fay and Olsho (1979) have recorded the discharge patterns of lagenar and saccular neurons in the eighth nerve of the goldfish, *Carassius auratus,* in response to vibration in three orthogonal directions. Neurons of both organs "showed similar patterns of directional sensitivity which corresponded reasonably well with hair cell orientation maps, and with the orientations of the maculae themselves. However, the neural data, particularly from the saccule, showed a greater variation than would be predicted from hair cell orientations." The frequency response characteristics of the neurons varied with the direction of vibration.

We have examined the directional characteristics of single units from the ear of the cod. Units from the utriculus and sacculus were studied with the fish stimulated by vibration at different angles in the horizontal plane.

2 Methods

2.1 The Experimental Animals

Juvenile cod weighing between 100 and 500 g, caught on the Scottish west coast, were kept in a recirculating seawater aquarium at 10°C for at least 2 weeks prior to the experiments.

Following initial anaesthesia with a solution of MS 222 (1:15000) and an intraperitoneal injection of Alphaxalone at a dose of 10-20 mg/kg live weight, the fish was placed in a holder with fresh seawater at 10°C irrigating the gills through a mouthpiece, and the dorsal surface of the brain surgically exposed. The fish was rendered decerebrate and a muscle relaxant administered by injection into the lateral superficial flank muscle (*d*-tubocurarine chloride, 3 mg/kg). So prepared, and with minimal bleeding from the cranial blood vessels, the preparation remained viable for up to 12 hr. A full description of the operative technique is given by Horner, Hawkins, and Fraser, Chapter 11).

2.2 The Directional Stimulus

The kinetic component of an underwater sound stimulus was simulated by vibrating the fish in the horizontal plane. The fish was firmly held in its holder placed on top of a turntable running on roller bearings (Fig. 15-1). The angle of azimuth of the fish was altered by rotating the fish on the turntable, and clamping it when the desired angle was reached. The turntable was mounted on a smooth slide which allowed bidirectional linear displacement. The slide or table, incorporating a series of inclined roller bearings, was linked to an electromagnetic vibrator constructed from a J11 transducer and could be driven at frequencies between 15 and 400 Hz. The particle velocity at the head of the fish was monitored by two velocity transducers, one pointing vertically, the other aligned with the long axis of the fish. A further velocity transducer mounted on the vibrating table ensured that the amplitude of vibration of the table itself remained the same with the turntable at different angles.

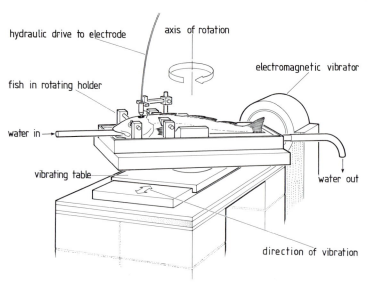

Figure 15-1. The vibrating table. The anesthetized fish was firmly held by eyebars and body clamps in a perspex holder, with the gills irrigated by fresh seawater at 10°C. The holder could be rotated in azimuth and then clamped at any chosen angle. The fish was vibrated by an electromagnetic vibrator.

Several versions of the vibrating table were constructed before a satisfactory performance was obtained. Particular problems included vibration of the electrode (cured with a heavier micromanipulator), vibration of the turntable (cured by introducing a more secure clamp), and excessive vertical motion at the head of the fish (cured by making the turntable and fish holder more rigid).

The driving stimulus was derived from an oscillator via a pulse gate and consisted of a 2 sec long sine wave pulse with a rise and fall time of 40 msec. The pulse was delivered to the vibrator by an attenuator and 100 W power amplifier.

The outputs of all three velocity transducers (Sensor Nederlands, type SM 2) were separately amplified and monitored. The transducers were calibrated by comparison with a calibrated accelerometer (Bruel and Kjaer, type 4319) and also against a displacement measuring system (Kaman, type KD 2300). The final calibration values differed by less than 1 dB. Phase distortion was insignificant above about 30 Hz. Vibrational stimuli were applied at frequencies of 63, 100, 160, and 250 Hz at rms velocity levels ranging from –40 dB/1 cm sec^{-1} to –80 dB/1 cm sec^{-1} (equivalent to free field sound pressures of +64 to +24 db re 1 μbar). For comparison, far field sound pressure thresholds for the cod vary from +10 dB at 470 Hz to –26 dB at 160 Hz (Chapman and Hawkins 1973). However, at its best frequencies the cod detects sound by way of the swimbladder and a more relevant value for the magnitude of vibration might be obtained from a fish without any functional connection between this organ and the ear. The atlantic salmon, *Salmo salar* (Hawkins and Johnstone 1978), gives a particle velocity threshold of –106 dB/1 cm sec^{-1} at its best frequency of 160 Hz, equivalent to a far field sound pressure of –2 dB.

2.3 Physiological Recording

Single unit activity was recorded extracellularly from primary fibers in the various rami of the VIIIth cranial nerve by means of conventional glass microelectrodes filled with 3 M NaCl or 3 M KCl. Electrode impedances ranged from 30-40 MΩ.

The various rami of the auditory nerve are shown in Fig. 15-2. The electrode was slowly advanced into a selected nerve ramus of the left ear by means of a hydraulic microdrive. The fish was vibrated and advancement of the electrode stopped when single unit activity was detected. The angle of vibration was altered in 10°, 15°, or 20° steps and responses recorded for several frequencies. Spike discharges from the units were generally 200 μV to 10 mV in amplitude, and were recorded over a bandwidth of 80 Hz-3 kHz on an FM tape recorder. The number of spikes per unit time was continually registered on a pen recorder throughout the experiment. Other tracks of the tape recorder received a stimulus synchronization pulse, the outputs of several particle velocity transducers, a sound pressure hydrophone signal, and a voice commentary, with appropriate bandwidths in each case. Further details of the recording and analysis techniques are given elsewhere (Horner, Hawkins, and Fraser, Chapter 11).

2.4 Analysis

The directional response of each unit was analyzed by replaying the recorded activity on to a recording oscillograph and counting the number of spikes produced for a given number of stimulus cycles at a number of fixed angles. At the best angle for the unit, that is the angle at which it showed the biggest change from spontaneous activity, a stimulus/response curve was prepared by measuring differences in the response rate for several stimulus amplitudes. From this curve the relative stimulus amplitude (in dB) to produce a given spike discharge rate was determined for each angle and a polar plot prepared, showing the relative sensitivity of the unit at different angles of azimuth. The best angle was then determined by eye, generally by joining the two minima and constructing a line at right angles to this. For some of the units, poststimulus time histograms were prepared by the methods outlined in Horner, Hawkins, and Fraser (Chapter 11). The response of the units at different angles could then be expressed in terms of the degree of synchrony with the stimulus waveform.

Once a number of units had been recorded, and their best angles noted, the circular mean and standard deviation of the angular distribution of the best angles for each particular nerve ramus were calculated, and a Rayleigh test applied to determine whether the best angles were uniformly distributed (Mardia 1972).

3 Results

3.1 General

Many of the units from both the sacculus and utriculus responded to low amplitude vibration. These sensitive units showed spike activity synchronized with the stimulus waveform as described by Horner, Hawkins, and Fraser (Chapter 11). A few units that showed a regular spontaneous discharge were insensitive to vibration regardless of the angle of stimulation. Of 43 vibration sensitive units, 25 were from the sacculus and 18

ANT.

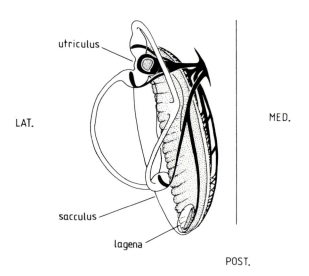

LAT.   MED.

utriculus

sacculus

lagena

POST.

Figure 15-2. Dorsal view of the left ear of the cod, showing the principal rami of the auditory nerve. The rami to the utriculus, and to the anterior vertical and horizontal semicircular canals issue from a common root. The two saccular rami innervate the anterior and posterior parts of the saccular macula. The rami to the lagena and posterior vertical semicircular canal run together for part of their length, and may sometimes run with the posterior saccular ramus. Glass microelectrodes were inserted into the chosen ramus, under close visual control.

from the utriculus. For each of these units the mean number of spikes per unit time, or per stimulus cycle, and sometimes the degree of synchrony with the stimulus waveform, varied consistently with the angle of stimulation in the horizontal plane (that is, with the angle of azimuth). The response of one of these units at several angles is shown in Fig. 15-3. The greatest responses were obtained from the unit with the fish vibrated at azimuth angles of 0 and 180°, with a minimum at 90°. A reversal in the phase of the stimulus cycle at which the spikes occurred was evident in going from 0° to 180°.

As an example of the kind of analysis performed on each unit, Fig. 15-4 is a polar diagram showing the number of action potentials produced per cycle at different angles of azimuth for a single unit stimulated at 160 Hz. This particular example was from the anterior saccular ramus of the auditory nerve. The polar plot is incomplete, since the turntable could not be rotated through a full 360° at that time. However, the plot clearly has a figure-eight shape. Thus, the unit gave a maximal response when the fish was stimulated at about 10° and also 180° removed from this at 190°. The double shape is characteristic of a bidirectional receiver.

For many units, stimulation over 180° of azimuth was sufficient to determine the axis of sensitivity, as shown in Fig. 15-5 (again from the anterior saccular ramus of the left auditory nerve). In Fig. 15-5A the response of the unit is expressed as the number

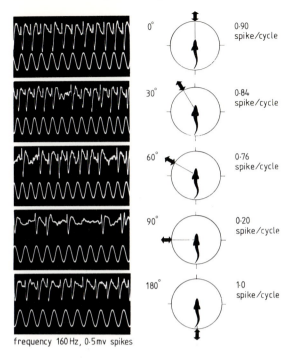

frequency 160 Hz, 0·5mv spikes

Figure 15-3. Response of a unit from the anterior saccular ramus of the left auditory nerve (V-31-8) to stimulation at different angles of azimuth at a frequency of 160 Hz. The mean number of spikes per unit cycle is indicated on the right. Stimulus level, –60 dB/1 cm sec^{-1}. Note that velocity levels in this paper are expressed in dB re 1 cm sec^{-1}.

of spikes per stimulus cycle. Since the number of spikes per cycle does not change linearly with stimulus level the response does not always change dramatically with angle. In Fig. 15-5B the response is given in terms of the relative level of particle velocity to evoke a given number of spikes per cycle. The latter polar plot shows clearly that the magnitude of the unit response depends on the amplitude of the cosine of the angle of stimulation. Indeed, a log-likelihood ratio test (Zar 1974) applied to the data confirmed that there was no significant difference between the observed distribution and that to be expected from a cosine law. (Assuming a best angle of 0°, $G = 11.89$, $df = 11$, critical value at $P = 0.05$ is 19.675.)

The best shape of polar diagram was obtained if the stimulus level was adjusted to evoke full phase-locking (i.e., one spike potential per stimulus cycle) only at the best angle for the unit. At higher stimulus levels full phase locking occurred over a wide range of angles, obscuring the directionality of the unit.

Units with a phase-locked response showed a sharp change in the phase of occurrence of the spikes relative to the stimulus at or about their minima, as illustrated in Fig. 15-6. In Fig. 15-6A the polar diagram shows a unit stimulated at 63 Hz having response maxima at about 0° and 180° with a minimum at just over 90° of azimuth. Poststimulus time histograms for the same unit are presented for different angles of

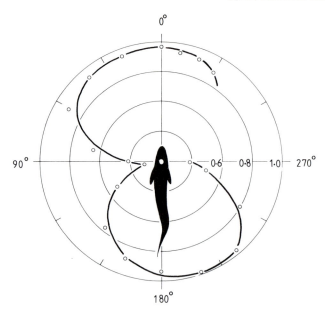

Figure 15-4. Polar diagram for a unit from the anterior saccular ramus of the left auditory nerve (V-17-4) stimulated at 160 Hz. The number of spike potentials per stimulus cycle is plotted radially for different angles of stimulation at a constant amplitude of stimulation. Stimulus level, –62 dB.

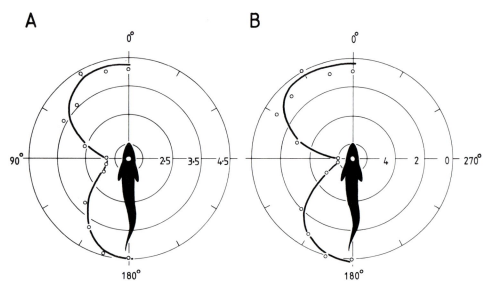

Figure 15-5. Polar diagrams for a unit from the anterior saccular ramus of the left auditory nerve (V-33-1) stimulated at 63 Hz. (A) Response expressed in terms of the number of spikes per stimulus cycle. Stimulus level, –64 dB. (B) Response expressed as the level of particle velocity (dB, relative) to evoke a given number of spikes per cycle.

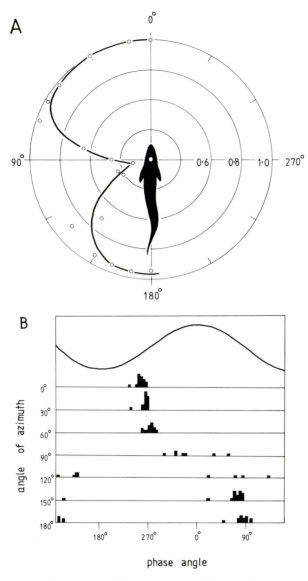

Figure 15-6. Changes in the phase of the spike potentials relative to the stimulus wave-form at different angles of azimuth. (A) Polar diagram of a unit from the anterior sac-cular ramus of the left auditory nerve (V-14-11) stimulated at 63 Hz, with the number of spike potentials per stimulus cycle plotted radially. Note the minimum at about 90°. Stimulus level, −42 dB. (B) Poststimulus time histograms, taken at several angles of azimuth. The histograms were each prepared by counting the number of spike poten-tials occurring in a series of serial time bins during 16 cycles of the stimulus. The parti-cle velocity waveform is provided for comparison. The phase of the stimulus waveform with which the spike potentials coincide reverses about the angle giving the minimal response.

stimulation in Fig. 15-6B, and indicate the precise time of occurrence of the spikes relative to the particle velocity waveform. There is a phase reversal, or 180° change in the phasing of the spike potentials, in going from 60° to 120° azimuth. Such a phase reversal is again characteristic of a bidirectional receiver.

3.2 Saccular Units

Saccular units were most sensitive along the long axis of the fish (0° and 180° in the polar diagrams), or at a small angle offset from this axis, with their minima at right angles. Polar diagrams for saccular units are shown in Figs. 15-4–15-7.

Where a unit was stimulated at several frequencies, approximately the same axis was obtained for the different frequencies. Thus, unit V-17-4, illustrated in Figs. 15-4 and 15-7, showed polar diagrams with a slight offset of 10° to the left of the long axis of the fish at three frequencies, 63, 100, and 160 Hz.

The axes for a total of 25 saccular units from the left ear of the cod are shown in Fig. 15-8C. A distinction is drawn in this figure between units from the anterior and posterior ramus. Each unit typically gives a bilobed response. The mean of the angular distribution for all saccular units was 6.0° ± 13.8 (i.e., 95% of the units had axes falling between 33.1 and 339.0°, assuming a circular normal distribution of axes). The mean for units from the anterior ramus was 8.8° ± 20.7, slightly offset to the left. The mean for units from the posterior ramus was 348.9° ± 38, slightly offset to the

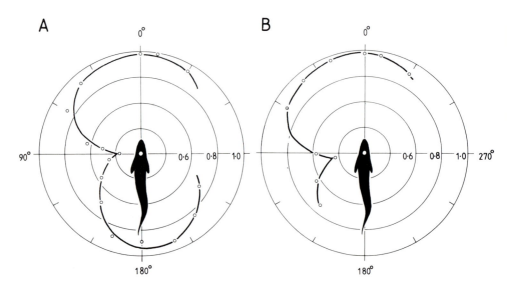

Figure 15-7. Polar diagrams for a unit from the anterior saccular ramus of the left auditory nerve (V-17-4) at two frequencies of stimulation. (A) At 63 Hz. Stimulus level, –52 dB. (B) At 100 Hz. Stimulus level, –57 dB. A polar diagram at a third frequency (160 Hz) for the same unit is shown in Figure 15-4. In all these polar diagrams the number of spike potentials per stimulus cycle is plotted radially for different angles of stimulation.

right. The angular distribution of the saccular units was significantly nonuniform ($P <$ 0.001) by the Rayleigh test (Mardia 1972).

In Chapter 11 Horner, Hawkins, and Fraser pointed out that the units from the sacculus can be classified on the basis of their spontaneous activity, and their response to different frequencies. In this preliminary analysis of the directionality of the units we have made no attempt to compare the directional characteristics of these different classes.

3.3 Utricular Units

Vibration-sensitive units from the utriculus were also strongly directional in their response to stimulation in the horizontal plane, giving a pronounced bilobed polar diagram, as in Fig. 15-9A. In this example, the main axis of the unit is at right angles to the long axis of the fish. At the minima (occurring at about 0° and 180°) there is typically a reversal in the phase of occurrence of the spikes relative to the stimulus waveform, as shown in Fig. 15-9B.

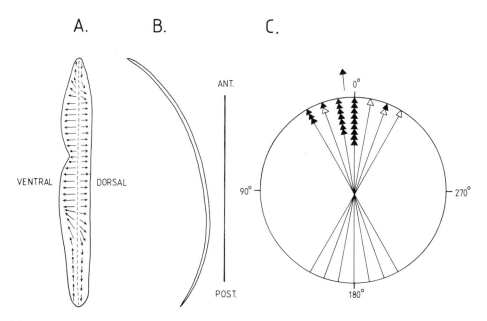

Figure 15-8. (A) Lateral view of the left saccular macula of the cod. Note that the main polarization axes of the sensory hair cells (shown as arrows) are orientated in the vertical plane. Hair cell orientations are taken from Dale (1976). (B) Dorsal view of the corresponding (left) saccular macula of the cod. (C) The angular distribution of 21 units from the saccular rami of the left auditory nerve. Though each unit is bidirectional and can be maximally stimulated from two directions, separated by 180°, its axis is denoted by only one arrow, for clarity. Closed arrows are units from the anterior saccular ramus, and open arrows from the posterior saccular ramus. The mean angle for all saccular units is indicated.

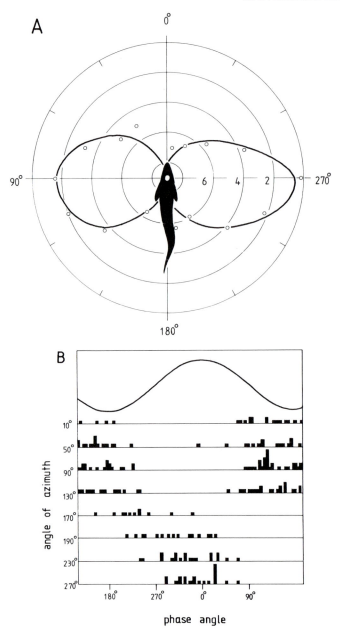

Figure 15-9. The response of a unit from the utricular ramus of the left auditory nerve (V-19-4) to directional stimulation at 63 Hz. Maximum stimulus level, –60 dB. (A) Polar diagram for the unit, with the relative level of particle velocity (dB) to evoke a given spike response plotted radially. Note the minima at about 0° and 180°. (B) Poststimulus time histograms, prepared for 32 stimulus cycles, at different angles of azimuth. The phase of the stimulus waveform with which the spike potentials coincide reverses about the angle of azimuth giving the minimal response.

Not all utricular units showed an axis of sensitivity at right angles to the long axis of the fish. Some were aligned like saccular units, while others were at intermediate angles. This wide angular distribution of utricular units is illustrated in Fig. 15-10, and stands in contrast to the relatively narrow angular distribution of the saccular units. The circular mean of the angular distribution of the utricular units was 18.1° ± 25.8 (i.e., 95% of the units had axes falling between 7.5 and 108.7°, assuming a circular normal distribution of axes). The angular distribution was not significantly nonuniform by the Rayleigh test.

4 Discussion

As expected, most of the primary afferent nerve fibers of both the sacculus and utriculus of cod were strongly directional in their response to horizontal oscillatory motion. It is well established that the otolith organ responds to particle motion, which is inherently directional (Chapman and Sand 1974, Hawkins and MacLennan 1976). Moreover, the sensory hair cell of the vestibular system shows a pronounced morphological asymmetry or functional polarization, which provides it with a well-defined axis of sensitivity (Lowenstein and Wersäll 1959, Flock and Wersäll 1962). In our experiments we were recording the activity of primary afferent nerve fibers synapsing directly with the hair cells, and it is not surprising that these units showed a directional response.

A.   B.

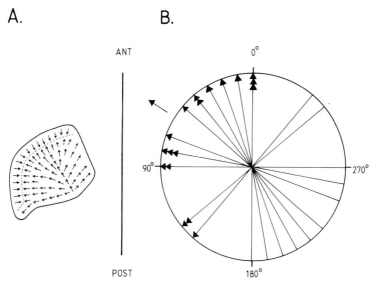

Figure 15-10. (A) Dorsal view of the left utricular macula of the cod. Note that the main polarization axes of the sensory hair cells are orientated at many different angles of azimuth in the horizontal plane. (B) The angular distribution of units from the utricular ramus of the left auditory nerve. Though each unit can be maximally stimulated from two directions, separated by 180°, its axis is denoted by only one arrow for clarity. Note that the widely scattered axes of the units match the great variation in orientation of the utricular hair cells. The mean angle is indicated.

The units were bidirectional, showing two maxima separated by 180°, essentially reflecting the oscillatory nature of the stimulus. However, the spikes were strongly synchronized with only one half or phase of the stimulus cycle and showed a 180° reversal at the minimum angles, indicating that each unit was linked to hair cells having a particular polarity. In the sacculus of the cod, the hair cells are orientated with opposite polarity in the dorsal and ventral halves of the macula (Dale 1976; also see Fig. 15-8). The synchronization of the response of the saccular fibers with only one phase of the stimulus cycle is evidence that each afferent fiber is connected with hair cells from one half of the macula only. Similar observations have been made for afferent fibers from the sacculus of the goldfish (Furukawa and Ishii 1967) with the difference that a further class of fiber was described for the goldfish, responding to both phases of the stimulus. The later fibers might therefore receive inputs from hair cells of opposing polarity. Furukawa (1978) has since confirmed by dye injection into the afferent fibers of goldfish that some of the fibers, responding to one phase, innervate the dorsal half of the macula, others responding to the other phase innervate the ventral half, while fibers responding to both phases innervate both halves.

The significance of the division of the saccular macula into areas of opposing polarity, and the preservation of this polarity in the response of the individual afferent fibers, may lie in the importance of information about the phase of the incident sound waves. Both Piddington (1972) and Schuijf (1976a, 1976b) have stressed the value of phase information to the fish. It should not be assumed, however, that there is a fixed correspondence in phase between the incident sound wave and the afferent nerve spike for all units. The mechanical characteristics of the ear and any latency in excitation of the afferent fiber will introduce phase shifts which may vary with frequency as well as direction of stimulation (see Fay, Chapter 10, for phase data from the goldfish saccular nerve).

Enger et al. (1973) and more recently Buwalda and van der Steen (1979) have investigated the response of the sacculus of gadoid fish to stimulation at different angles of azimuth, by recording microphonic potentials. The results of Enger et al. for the sacculus essentially agree with our own. Buwalda and van der Steen (1979), on the other hand, reported a much more pronounced divergence in azimuth between the principal axes of the paired sacculi. The left sacculus had its angle of optimal sensitivity at +45° (+180°) and the right at -24° (+180°), for comparable sites along the two maculae. None of our saccular units showed such a pronounced divergence.

The mean direction for the comparatively narrow range of best angles observed from saccular fibers did not coincide with the hair cell orientation observed in scanning electron micrographs of the saccular macula. As shown in Fig. 15-8 the majority of hair cells from the middle part of the macula are oriented in a vertical plane, though some cells at the extreme anterior and posterior ends are oriented along the line of the macula. The movements of the saccular otolith may be rather complicated. However, any conclusions about the way the otoliths move must await the results of stimulation in the vertical plane.

The directional responses obtained from vibration sensitive units in the utricular ramus of the auditory nerve corresponded well with the observed orientations of hair cells in the utricular macula (Fig. 15-10). Perhaps the most notable point, however, was that a high proportion of utricular units did respond in a synchronized way to vibration, showing that they would normally be stimulated by sounds. The particle velocity level of the applied stimulus was never more than 66 dB above the particle velocity threshold of the fish, so that the organ could not be said to be overstimulated. In some species, it seems that only the pars inferior of the ear (the sacculus and lagena) is concerned with hearing, the pars superior (the utriculus and semicircular canals) having a static and dynamic equilibrium function (von Frisch 1936, Lowenstein and Roberts 1951). Contrary evidence has, however, suggested that the utriculus is a hearing organ in other species (Dijkgraaf 1949, Enger 1963, Denton and Gray 1979). We suggest that in the cod there is no clear functional division between the utriculus of the pars superior, and the sacculus and lagena of the pars inferior. There may be a wide overlap in the functioning of all three otolith organs, perhaps the main differences among them being directional.

Schuijf (1976a) has developed the theory that in fish the direction of a sound source is determined for any plane by comparison of the outputs of two or more differently aligned bidirectional receivers, or vector detectors. Buwalda and van der Steen (1979) have proposed that the directivity of the sacculus of the cod is all but masked by particle motion reradiated from the swimbladder, and that this organ is unlikely to serve as a functional vector detector.

Our results show that the sacculus and utriculus both contain bidirectional units which might provide a basis for vector weighing. Indeed all three otolith organs may contain such units. The utriculus and lagena, as well as the sacculus, may play a vital role in determining the direction of sounds.

Acknowledgments. We thank Peter Fraser of Aberdeen University for his advice, Charles Robb and Gordon Smith of the Marine Laboratory for valuable technical assistance, and Janice Adair for help with preparation of the manuscript. Our colleagues Colin Chapman and Basil Parrish have commented upon the paper. Kathleen Horner's research was financed by the Northern Ireland Education Department.

References

van Bergeijk, W. A.: Directional and non directional hearing in vertebrates. In: Marine Bio-Acoustics. Tavolga, W. N. (ed.). Oxford: Pergamon, 1964, pp. 281-350.

Butler, R. A.: Does tonotopicity subserve the perceived elevation of a sound? Fed. Proc. 33, 1920-1923 (1974).

Buwalda, R. J. A., Schuijf, A., Hawkins, A. D.: On the discrimination in three dimensions of sound waves from opposing directions by the cod (1981). (in press)

Buwalda, R. J. A., Steen van der, J.: The sensitivity of the cod sacculus to directional and non-directional sound stimuli. Comp. Biochem. Physiol. 64A, 467-471 (1979).

Chapman, C. J., Hawkins, A. D.: A field study of hearing in the cod. J. Comp. Physiol. 85, 147-167 (1973).

Chapman, C. J., Johnstone, A. D. F.: Some auditory discrimination experiments on marine fish. J. Exp. Biol. 61, 521-528 (1974).

Chapman, C. J., Sand, O.: Field studies of hearing in two species of flatfish. Comp. Biochem. Physiol. 47A, 371-385 (1974).

Dale, T.: The labyrinthine mechanoreceptor organs of the cod. Norw. J. Zool. 24, 85-128 (1976).

Denton, E. J., Gray, J. A. B.: The analysis of sound by the sprat ear. Nature Lond. 282, 406-407 (1979).

Dijkgraaf, S.: Untersuchungen-uber die Functionen des Ohrlabyrinths bei Meeresfischen. Physiol. Comp. 2, 81-106 (1949).

Dijkgraaf, S.: Hearing in bony fishes. Proc. R. Soy. London Ser. B. 152, 51-54 (1960).

Enger, P. S.: Single unit activity in the peripheral auditory system of a teleost fish. Acta Physiol. Scand. 59, Suppl. 3, 9-48 (1963).

Enger, P. S., Hawkins, A. D., Sand, O., Chapman, C. J.: Directional sensitivity of saccular microphonic potentials in the haddock. J. Exp. Biol. 59, 425-433 (1973).

Fay, R. R., Olsho, L. W.: Discharge patterns of lagenar and saccular neurones of the goldfish eighth nerve: Displacement sensitivity and directional characteristics. Comp. Biochem. Physiol. 62A, 377-386 (1979).

Flock, A., Wersäll, J.: A study of the orientation of the sensory hairs of the receptor cells in the lateral line organ of fish with special reference to the function of the receptors. J. Cell Biol. 15, 19-40 (1962).

Furukawa, T.: Sites of termination on the saccular macula of auditory nerve fibres in the goldfish as determined by intracellular injection of procion yellow. J. Comp. Neurol. 180, 807-814 (1978).

Furukawa, T., Ishii, Y.: Neurophysiological studies on hearing in goldfish. J. Neurophysiol. 30, 1377-1403 (1967).

Hawkins, A. D., Johnstone, A. D. F.: The hearing of the Atlantic salmon. J. Fish. Biol. 13, 655-673 (1978).

Hawkins, A. D., MacLennan, D.: An acoustic tank for hearing studies on fish. In: Sound Reception in Fish. Schuijf, A., Hawkins, A. D. (eds.). Amsterdam: Elsevier, 1976, pp. 149-169.

Hawkins, A. D., Sand, O.: Directional hearing in the median vertical plane by the cod. J. Comp. Physiol. 122, 1-8 (1977).

Lowenstein, O., Osborne, M. P., Wersäll, J.: Structure and innervation of the sensory epithelia of the labyrinth in the thornback ray. Proc. R. Soc. London Ser. B. 160, 1-12 (1964).

Lowenstein, O., Roberts, T. D. M.: The localisation and analysis of the responses to vibration from the isolated elasmobranch labyrinth. A contribution to the problem of the evolution of hearing in vertebrates. J. Physiol. 114, 471-489 (1951).

Lowenstein, O., Wersäll, J.: Functional interpretation of the electron microscope structure of the sensory hairs in the crista of the elasmobranch *Raja clavata* in terms of directional sensitivity. Nature 184, 1807-1810 (1959).

Mardia, K. V.: Statistics of directional data. London: Academic Press, 1972.

Piddington, R. W.: Auditory discrimination between compressions and rarefactions by goldfish. J. Exp. Biol. 56, 403-419 (1972).

Pumphrey, R. J.: Hearing. In: Physiological Mechanisms in Animal Behaviour. Symp. Soc. Exp. Biol. 4, 1-18 (1950).

Sand, O.: Directional sensitivity of microphonic potentials from the perch ear. J. Exp. Biol. 60, 881-899 (1974).

Sand, O., Enger, P. S.: Evidence for an auditory function of the swimbladder in the cod. J. Exp. Biol. 59, 405-414 (1973).

Sand, O., Hawkins, A. D.: Acoustic properties of the cod swimbladder. J. Exp. Biol. 58, 797-820 (1973).

Schuijf, A.: Directional hearing in the cod under approximate free field conditions. J. Comp. Physiol. 98, 307-332 (1975).

Schuijf, A.: The phase model of directional hearing in fish. In: Sound Reception in Fish. Schuijf, A., Hawkins, A. D. (eds.). Amsterdam: Elsevier, 1976a, pp. 63-86.

Schuijf, A.: Timing analysis and directional hearing in fish. In: Sound Reception in Fish. Schuijf, A., Hawkins, A. D. (eds.). Amsterdam: Elsevier, 1976b, pp. 87-112.

Schuijf, A., Buwalda, R. J. A.: On the mechanism of directional hearing in cod. J. Comp. Physiol. 98, 333-344 (1975).

Schuijf, A., Buwalda, R. J. A.: Underwater localization—A major problem in fish acoustics. In: Comparative Studies of Hearing in Vertebrates. Popper, A. N., Fay, R. R. (eds.). New York: Springer-Verlag, 1980, pp. 43-77.

von Frisch, K.: Über den Gehörsinn der Fische. Biol. Rev. 11, 210-246 (1936).

Wersäll, J., Flock, A., Lundquist, P. G.: Structural basis for directional sensitivity in cochlear and vestibular sensory receptors. Cold Spring Harbor Symp. Quant. Biol. 30, 115-145 (1965).

Zar, J. H.: Biostatistical analysis. Englewood Cliffs, N.J.: Prentice-Hall, 1974.

Discussion

FAY: Do you have any idea of the sensitivity of these fibers, and the relative sensitivity of utricular and saccular fibers?

HAWKINS: The two seem much the same. The stimulus levels we apply are well above threshold by about 20 dB, and up to 60 dB in some cases. This was not unreasonable, since we were only looking for directional responses, and these stimuli were well within the dynamic range of the system.

FAY: Do the thresholds that you measured make physiological sense in terms of what you know of the ability of the animal to orient to a sound source at such sound pressure levels?

HAWKINS: The present report is confined to effects of velocity stimuli, since the results from pressure measurements are incomplete. It does appear that there is greater sensitivity from units stimulated by pressure that from those stimulated by velocity.

FAY: You mentioned the phase angle of the phase locked response. Why is there such a wide distribution of phase angles?

HAWKINS: There is between units, but if you select a particular unit, there is a linear relationship for phase angle against frequency, as has been found in mammals.
    The important point is that there seems to be a very long delay between the stimulus waveform and the actual afferent response. The delay is about 5 or 6 msec. I think that this synaptic delay is what gives rise to the linear relationship between phase and frequency.

PLATT: When I worked with a flatfish, I got a number of stable vibratory units in the utricle. Roughly what proportion of stable units did you find?

HAWKINS: Both in the utriculus and sacculus less than 10% were regularly discharging units. The rest showed irregular spontaneous activity and responded well to vibration.

PLATT: Were these exclusively vibration sensitive?

HAWKINS: They may well serve other functions and respond to linear accelerations. They have a good low frequency response.

PLATT: Could you tell anything about their adaptation?

HAWKINS: There was some adaptation.
    We should remember, however, that there may well be great functional plasticity in these otolith organs in different species. We are just talking about the cod, and it would be unwise to generalize.

FINE: Does latency of firing change with stimulus direction?

HAWKINS: One might expect it to change with stimulus amplitude, and we do in fact observe this. We have not looked for any latency change with direction of stimulation. You do get a very abrupt 180° phase change in the response as you go through the minimum, but this simply reflects the oscillatory nature of the sound.

PART FIVE

Central Nervous System

Until the past few years knowledge of the central projections of the teleost ear has been extraordinarily limited, and the available data were often contradictory. With the advent of new and powerful neuroanatomical tools it has been possible to experimentally analyze the organization and projections of the teleost auditory CNS. These studies are of special importance to physiologists since now, for the first time, we have information on the projections to the various brain regions from which recordings are being made. In Chapter 16, Northcutt provides a historical survey of studies on the auditory CNS in fishes and then describes the various types of brainstem auditory nuclei found in several fish groups. Roberts (Chapter 17) presents similar information for elasmobranchs. More specific data on projections in primitive and modern fishes are described by McCormick (Chapter 18), while Bell (Chapter 19) discusses specific endorgan projections in mormyrids.

Chapter 16

Audition and the Central Nervous System of Fishes

R. GLENN NORTHCUTT*

1 Introduction

It is impossible to discuss the organization and evolution of otic pathways in fishes without discussing their relationship to the lateral line system, as both systems have been claimed to possess a single phyletic origin (Ayers 1892, Wilson and Mattocks 1897, van Bergeijk 1966, 1967) and terminate in the same primary medullar nuclei (Mayser 1882, Herrick 1897, Pearson 1936, Larsell 1967). In examining the development of the lateral line and otic systems in *Salmo,* Wilson and Mattocks (1897) claimed that both systems arose from a single sensory placode and suggested that the inner ear arose phylogenetically by a portion of the peripheral lateral line sinking beneath the skin surface. Mayser (1882) examined the central course of the eighth cranial nerve in cyprinids and claimed that both it and the lateral line nerves terminate in the "acoustic tubercle" of the medulla. Mayser suggested that the lateral line system of fishes should be viewed as an accessory auditory system, based on convergence of the otic and lateral line efferents. The "acoustic tubercle" of fishes thus became known as the acousticolateral area, and the inner ear's proposed origin from a portion of the lateral line system is known as the *acousticolateralis hypothesis.*

Historically, attitudes have shifted back and forth regarding audition in fishes and its anatomical basis. The concensus in the late nineteenth century held that in fishes the primary function of the inner ear, as well as the lateral line system, is audition. As gravistatic functions were linked to the inner ear, it was assumed that only the saccule or its derivative, the lagena, might possess auditory function (Ariëns Kappers, Huber, and Crosby 1960) and that most of the end-organs of the inner ear possess only gravistatic functions. Opinion shifted further, and many researchers (Ariëns Kappers et al. 1960, Larsell 1967, Wever 1976) subsequently accepted that the ear in fishes (except for some teleosts) possesses solely gravistatic functions, despite behavioral experiments by Parker (1903) and von Frisch (1923) which clearly indicated that teleosts hear.

*Division of Biological Sciences, University of Michigan, Ann Arbor, Michigan 48109.

During the first half of this century, it was commonly assumed that audition was a hallmark of land vertebrates, first arising with amphibians (Ariëns Kappers et al. 1960, Larsell 1967). This hypothesis, which assumes that the primitive function of the inner ear was solely gravistatic and that acoustic ability was subsequently acquired, is referred to as the *labyrinth hypothesis* (Wever 1976). Anatomical descriptions of the medullar nuclei receiving primary lateral line and otic efferents clearly reflect this bias. Pearson (1936) provided the first clear description of the acousticolateral area of fishes and concluded that there are two distinct patterns of organization. The first pattern (Fig. 16-1) characterizes lampreys, cartilaginous fishes, and primitive bony fishes and con-

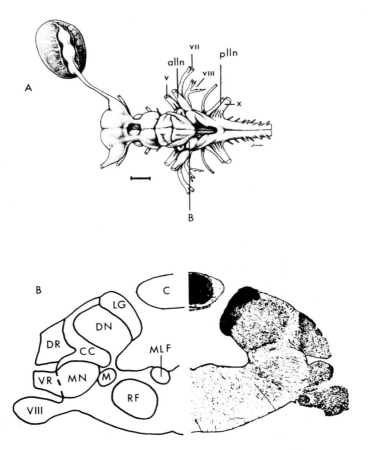

Figure 16-1. (A) Dorsal view of the brain of the thornback guitarfish, *Platyrhinoidis triseriata*. Bar scale equals 5 mm. (B) Transverse section through the medulla illustrating the positions of the nuclei of the octavolateralis area. Abbreviations: alln, anterior lateral line nerve; B, position of transverse section in medulla; C, corpus of cerebellum; CC, cerebellar crest; DN, dorsal octavolateralis nucleus; DR, dorsal root of anterior lateral line nerve; LG, lateral granule cell mass; M, magnocellular nucleus of the ventral octavolateralis column; MLF, medial longitudinal fasciculus; MN, medial octavolateralis nucleus; plln, posterior lateral line nerve; RF, reticular formation; VR, ventral root of anterior lateral line nerve; V, trigeminal nerve; VII, facial nerve; VIII, octaval nerve; X, vagal nerve. (From Bodznick and Northcutt 1980.)

sists of an acousticolateral area divided dorsoventrally into three regions: a dorsal nucleus, a medial or intermediate nucleus, and a ventral nuclear complex. Pearson argued that holostean and teleostean fishes differ from other fishes by not having a dorsal nucleus; however, Larsell (1967) reached the opposite conclusion and argued that all anamniotes possess all three acousticolateral nuclei.

Both Larsell and Pearson reported that the entering octaval fibers terminate in the medial and ventral nuclei of the acousticolateralis area and the entering lateral line nerves in the dorsal and medial nuclei. Thus, there was assumed to be extensive overlap between the entering lateral line and otic primary efferents. Larsell (1967) futher concluded that the origin of audition in amphibians was characterized by loss of the lateral line system but not its primary medullar nucleus, the dorsal acousticolateralis nucleus. He assumed that this nucleus was "captured" by the newly evolving auditory system, and that the "old" dorsal nucleus of fishes became the acoustic cochlear nucleus of land vertebrates.

These observations and the hypotheses they generated were based solely on descriptive anatomical methods. More recent experimental studies (Maler, Karten, and Bennett 1973a, 1973b, Campbell and Boord 1974, Maler, Finger, and Karten 1974, Boord and Campbell 1977, Knudsen 1977, McCormick 1978, Northcutt, 1979a, 1979b, 1980, Boord and Roberts 1980) do not support most of these observations and call into question the earlier hypotheses regarding the evolution of audition in vertebrates.

Many of the recent experimental studies describe the primary efferents of the lateral line and otic nerves and their primary medullar targets. These data and their bearing on earlier hypotheses will be discussed. However, the data also suggest a number of questions regarding the organization of auditory pathways and their functions in fishes. These questions cannot be answered conclusively at this time, but they will be posed, and possible resolutions will be discussed.

2 Peripheral Nerves and Medullar Organization

The lateral line system and inner ear of anamniotes arise from a dorsolateral series of ectodermal placodes (Landacre 1910, 1912, 1927, Stone 1922). The placodes themselves appear to arise by complex interactions of the neural plate, neural crest, and head ectoderm (Hörstadius 1950). In most anamniotes, the earliest evidence of these placodes indicates that they are separate entities; in teleosts, however, the developmental pattern appears to be modified, and the placodes and neural crest are fused as a single field (Landacre 1910). This specialization would appear to account for the early reports that both the lateral line system and inner ear arise from a single embryonic field (Wilson and Mattocks 1897). In most anamniotes, placodes appear rostral and caudal to the otic placode (Fig. 16-2), which sinks beneath the ectodermal surface forming the membranous labyrinth and the ganglion cells of the octaval nerve. The posterior placode migrates onto the trunk and appears to give rise to the mechanoreceptors (neuromasts) of the trunk lateral line, as well as the posterior lateral line nerve which innervates these receptors. The placodes anterior to the otic placode give rise to the lateral line receptors of the head, as well as the anterior lateral line nerve and part of the ganglion of the trigeminal nerve (Landacre 1910, 1927, Stone 1922).

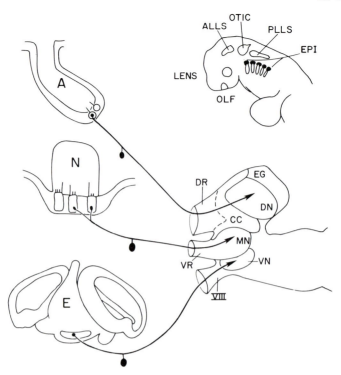

Figure 16-2. Schematic representation of the different types of octavolateralis receptors and their primary sites of termination in the octavolateralis area of the medulla, as
well as the distribution of ectodermal placodes that give rise to these receptors and
those of many of the other special senses. Abbreviations: A, ampullary electroreceptor
of the anterior lateral line nerve; alls, anterior lateral line system placode; CC, cerebellar crest; DN, dorsal octavolateralis nucleus; DR, dorsal root of the anterior lateral line
nerve; E, inner ear; EG, eminentia granularis; epi, epibranchial (taste) placodes; lens,
lens placode of the eye; MN, medial octavolateralis nucleus; N, neuromast mechanoreceptor of the anterior lateral line nerve; OLF, olfactory placode; OTIC, otic placode;
PLLS, posterior lateral line system placode; VN, ventral octavolateralis nuclei; VR,
ventral root of the anterior lateral line nerve; VIII, octaval or otic cranial nerve.

Not only the lateral line system and inner ear arise from placodes; all special sense
organs in vertebrates arise entirely or in part from ectodermal placodes (Fig. 16-2). The
ganglia that innervate taste buds arise from epibranchial placodes located above the gill
slits in anamniotes; the olfactory organs also arise from placodes, as do the lens and
cornea of the eye. If ontogenetic evidence is assumed to indicate phylogenetic sequences, we could just as reasonably argue that taste buds, olfactory organs, or even
part of the trigeminal nerve, arose phylogenetically from the lateral line, as argue that
the inner ear arose from part of the lateral line. The ontogenetic data only suggest
that all special sense organs are related, at least in part, to ectodermal placodes; they
give us no insight into the possibility of phylogenetic ancestry. Thus, the embryologi-
cal data do not support the acousticolateralis hypothesis, i.e., that the inner ear arose

from part of the lateral line system. In fact, the earliest fossil vertebrates (ostracoderm agnathans) possessed both inner ears and lateral line organs (Moy-Thomas and Miles 1971). Thus, there is no fossil evidence to indicate which, if either, of these systems is phylogenetically older.

The lateral line and octaval nerves enter a portion of the alar medulla often termed the acousticolateralis area (Figs. 16-1 and 16-2). This area is more properly termed the octavolateralis area, as the octaval nerve of many fishes contains both acoustic and vestibular fibers. The octavolateralis area of lampreys and cartilaginous, chondrostean, dipnoan, and crossopterygian fishes consists of dorsal, medial, and ventral nuclei (Pearson 1936, Rubinson 1974, Smeets and Nieuwenhuys 1976, Boord and Campbell 1977, McCormick 1978, 1981, Northcutt 1978, 1979a, 1980). The anterior lateral line nerve possesses dorsal and ventral roots (Fig. 16-2) which terminate in the dorsal and medial octavolateralis nuclei, respectively, whereas the posterior lateral line nerve terminates in the medial octavolateralis nucleus (McCready and Boord 1976, Boord and Campbell 1977). This differential termination of the anterior lateral line nerve roots led Boord and co-workers to suggest that the dorsal root of the anterior lateral line nerve consists only of fibers innervating electroreceptive ampullary organs, whereas the ventral root of the anterior lateral line nerve and the posterior lateral line nerve consist of fibers arising from the mechanoreceptive neuromasts of the head and trunk. This anatomical hypothesis was recently confirmed physiologically in the elasmobranch *Platyrhinoidis* by recordings from single fibers in the roots of the anterior lateral line nerve (Bodznick and Northcutt 1980). Thus, the dorsal octavolateral nucleus is the primary target of the entering ampullary fibers of the anterior lateral line nerve. Lampreys, cartilaginous fishes, and primitive bony fishes all possess a dorsal octavolateralis nucleus, suggesting that they are also electroreceptive (McCormick 1978). This was recently confirmed by a series of physiological experiments (Northcutt, Bodznick, and Bullock 1980).

Although earlier studies claimed extensive overlap between the efferents of the lateral line and octaval nerves (Mayser 1882, Herrick 1897, Larsell 1967, Rubinson 1974), more recent studies report little overlap in the primary projections of these nerves (Maler et al. 1973a, 1973b, Boord and Campbell 1977, Korn, Sotelo, and Bennett 1977, McCormick 1978, Northcutt 1979b, 1980, Boord and Roberts 1980). These experimental studies reveal the primary target of the entering octaval fibers to be the ventralmost column of the octavolateralis area (Fig. 16-2); however, this column is generally *not* composed of a single nucleus, as claimed earlier, but consists of a series of nuclei in the rostrocaudal plane (McCormick 1978, Northcutt 1979b, 1980, Boord and Roberts 1980). The exact number and position of these octaval nuclei vary among fishes and will be discussed in detail in the next section.

Before turning to the variation exhibited by the primary octaval nuclei, some remaining details of the octavolateralis area in fishes should be discussed. The lateral line and otic nerves also terminate in a third cell group, termed the eminentia granularis (Maler et al. 1974, Boord and Campbell 1977, McCormick 1978), as well as in the dorsal and medial octavolateralis nuclei. Axons of the eminentia granularis cells project caudally to form the cerebellar crest, a molecular layer capping the lateral surface of the dorsal and medial octavolateralis nuclei (Figs. 16-1 and 16-2). Many cells of the dorsal and medial octavolateralis nuclei send dendrites into their respective cerebellar crests; thus, these cells receive primary lateral line efferents as well as secondary input via the eminentia granularis (Maler et al. 1974, Boord 1977).

Holostean and teleostean fishes do not possess a dorsal octavolateralis nucleus (Pearson 1936, McCormick 1978, Northcutt 1980) and, not surprisingly, holosteans and most teleosts are not electroreceptive (Northcutt et al. 1980, and unpublished observations). However, members of at least three teleostean orders *are* electroreceptive; these fishes appear to have independently evolved new electroreceptors (Bullock 1974). It is probable that these electroreceptive teleosts have specialized a portion of the medial octavolateralis nucleus as an electroreceptive nucleus. Thus, the anterior and posterior lateral line lobes of electroreceptive teleosts are homologous to parts of the medial octavolateralis nucleus in other fishes, not to the dorsal and medial octavolateralis nuclei, respectively, in primitive fishes.

3 Variation in the Octaval Medullar Column

Two distinctly different patterns of nuclear organization characterize the octaval column in fishes (Fig. 16-3A,B). The basic organization of these two patterns will be described and information regarding primary inputs summarized. A comparison of these patterns poses several problems regarding homologies among fishes and between fishes and amphibians.

3.1 Agnathans

In lampreys the octaval column consists of a single nucleus (ventral octavolateralis nucleus) of small to medium-sized cells that run the length of the medulla, beginning at the lateral edge of the cerebellum and ending at obex levels (Figs. 16-3A and 16-4). Three distinct aggregates of much larger neurons are embedded within the ventral octavolateralis nucleus and are termed the anterior, intermediate, and posterior octavomotor nuclei (Figs. 16-3A and 16-4).

Experimental study of the primary projections of the octavus nerve in adult silver lampreys (*Ichthyomyzon unicuspis*) reveals that the entering octaval fibers form ascending and descending limbs (Fig. 16-4) which terminate throughout the length of the ipsilateral ventral octavolateralis nucleus as well as in the ipsilateral cerebellum rostrally (Northcutt 1979a). Degenerating octaval fibers were not traced to the dorsal or medial octavolateralis nuclei, nor to the contralateral cerebellum as reported in earlier descriptive studies (Larsell 1967).

Earlier studies suggested extensive overlap between the lateralis and octaval systems; experimental data (Northcutt 1979a) suggest such overlap in the cerebellum, but it is apparently minimal at the level of the primary octavolateralis nuclei.

There are no physiological data regarding the functions of the octaval system in lampreys. The efferent projections of the octavomotor nuclei (discussed in the next section) suggest that the large cells of these nuclei form connections with the oculomotor nuclei and motor neurons of the spinal cord, and are thus similar to certain vestibular pathways in jawed vertebrates. However, nothing is known about the function of the smaller neurons that constitute the bulk of the ventral octavolateralis nucleus.

The inner ear in lampreys possesses homologues of all maculae, including the macula neglecta, that characterize the inner ear in jawed vertebrates (Lowenstein,

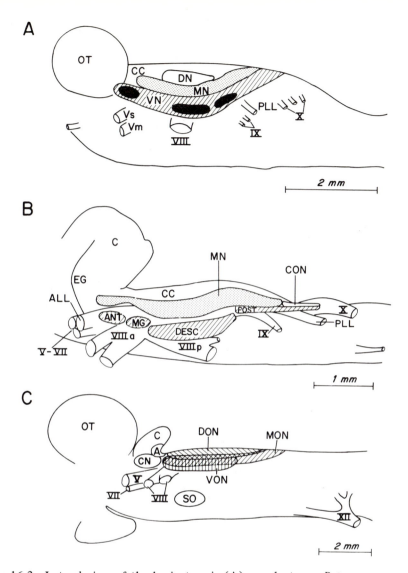

Figure 16-3. Lateral view of the brainstem in (A) a cyclostome, *Petromyzon marinus*, (B) an holostean, *Amia calva* (after McCormick 1978), (C) an anuran, *Rana catesbeiana*, illustrating the position of the octavolateralis and related nuclei. Stippled nuclei receive neuromast input, hatched nuclei receive octaval input, and solid black nuclei in (A) represent positions of anterior, intermediate, and posterior octavomotor nuclei. Abbreviations: A, auricle of the cerebellum; ANT, anterior octaval nucleus; ALL, anterior lateral line nerve; C, body of the cerebellum; CC, cerebellar crest; CN, cerebellar nucleus; CON, caudal octavolateralis nucleus; DESC, descending octaval nucleus; DN, dorsal octavolateralis nucleus; DON, dorsal octaval nucleus; EG, eminentia granularis; MG, magnocellular octaval nucleus; MN, medial ocatavolateralis nucleus; MON, medial or caudal octaval nucleus; OT, optic tectum; PLL, posterior lateral line nerve; POST, posterior octaval nucleus; Vm, motor root of the trigeminal nerve; VN, ventral octaval nucleus; VON, ventral octaval nucleus; Vs, sensory root of the trigeminal nerve; V-VII, trigeminal and facial nerves; VIII, octaval or otic nerve; VIIIa, anterior octaval ramus; VIIIp, posterior octaval ramus; IX, glossopharyngeal nerve; X, vagal nerve; XII, hypoglossal nerve.

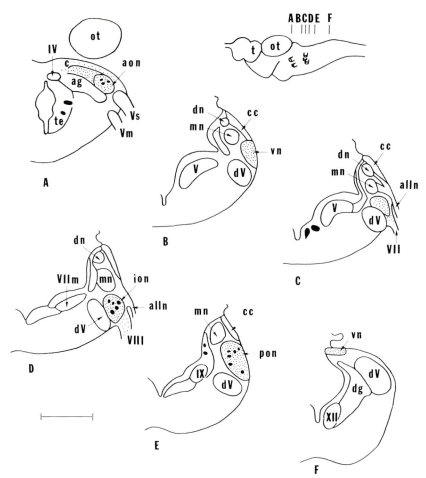

Figure 16-4. Charting of the degenerating octaval afferents in the silver lamprey following removal of the membranous labyrinth. (A-F) Transverse sections through the hindbrain at levels indicated in the lateral view of the brain. Fine stippling indicates degenerating fibers and terminals, large solid black circles indicate individual large neurons. Bar scale equals 500 μm. Abbreviations: ag, alar gray; alln, anterior lateral line nerve; aon, anterior octavomotor nucleus; c, cerebellum; cc, cerebellar crest; dg, dorsal gray; dn, dorsal octavolateralis nucleus; dV, descending trigeminal tract and nucleus; ion, intermediate octavomotor nucleus; mn, medial octavolateralis nucleus; ot, optic tectum; pon, posterior octavomotor nucleus; t, telencephalon, te, tegmentum; vn, ventral octavolateralis nucleus; IV, trochlear motor nucleus; V, trigeminal motor nucleus; Vm, trigeminal motor root; Vs, trigeminal sensory root; VIIm, facial motor nucleus; VII, facial root; VIII, octavus root; IX, glossopharyngeal motor nucleus; XII, hypoglossal motor nucleus. (From Northcutt 1980.)

Osborne, and Thornhill 1968, Thornhill 1972); thus, it is possible that the inner ear in lampreys also possesses both acoustic and gravistatic functions. If lampreys *do* possess acoustic abilities, it is possible that the inner ear was never solely gravistatic in function and that vertebrates possessed acoustic abilities at their origin. However, it is also possible that lampreys possess acoustic properties that have been independently evolved. If initial experiments determine that lampreys hear, these data must be amended with considerable details regarding the peripheral organs involved and their central pathways if we are to determine the ancestry of audition in vertebrates.

Although knowledge of the organization and inputs of the octavolateralis area in lampreys has increased considerably, almost nothing is known about this region in hagfishes. The octavolateralis area in hagfishes is said to be less differentiated than in lampreys (Jansen 1930, Larsell 1967). Only a single nucleus has been recognized, and distinct octavomotor nuclei are not discernible. Both lateral line and octavus nerves are said to terminate within this single nucleus (Larsell 1967); however, it is also claimed that a lateral line system does not even exist in hagfishes (Hardisty 1979). There is similar confusion regarding the existence of a cerebellum (Larsell 1967). Although hagfishes are generally considered the most degenerate living agnathans, they possess brains some four times larger than lampreys of comparable size (Northcutt 1981) as well as well-differentiated brain centers. Examination of the octavolateralis system in hagfishes will likely prove extremely interesting, and such study is clearly needed to achieve a valid perspective on octavolateralis organization in agnathans.

3.2 Gnathostome Fishes

Earlier descriptive studies of the ventral octavolateralis column of nonteleost fishes generally recognized a single nucleus, termed the ventral nucleus (Pearson 1936, Larsell 1967). However, more recent studies of both cartilaginous (Boord and Roberts 1980, Northcutt 1980) and bony fishes (McCormick 1978, 1981, Chapter 18, Northcutt 1979b, 1980, Bell 1981) indicate that the ventral octavolateralis column consists of several nuclei arranged rostrocaudally. In the holostean, *Amia*, McCormick (1978, 1981, Chapter 18) divided the ventral octavolateralis column into four nuclei (Fig. 16-3B) and suggested that the same nuclei also exist in other ray-finned fishes. Subsequent experimental studies of cartilaginous fishes (Boord and Roberts 1980, Northcutt 1980) and a teleost (Northcutt 1979b, 1980) have extended and confirmed this prediction.

In the batoid *Platyrhinoidis* (Fig. 16-4) and the teleost *Gillichthys* (Figs. 16-5 and 16-6), octaval fibers first terminate on dendrites of the large neurons of the octaval magnocellular nucleus, located medial to the entering fibers. The primary octaval fibers bifurcate, forming ascending and descending limbs. The ascending limb courses rostrally, terminating in the anterior octaval nucleus (Figs. 16-4A,B and 16-6A,C) which lies ventral and medial to the medial octavolateralis nucleus. Many of the octaval fibers of the ascending limb continue to an area rostral and dorsal to the anterior octaval nucleus where they terminate in cerebellar or closely related structures (Figs. 16-4B and 16-6A,B). In *Platyrhinoidis,* ascending octaval fibers terminate in the granule cells forming the lower leaf of the auricle (Fig. 16-4B), the suspected homolog of the eminentia granularis in bony fishes (Northcutt 1980, Figs. 16-5A and 16-6A-C).

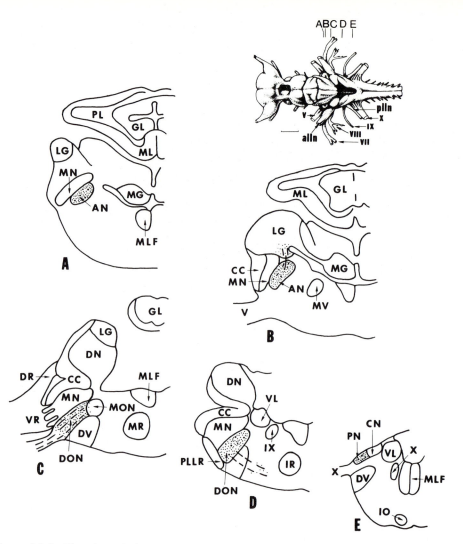

Figure 16-5. Charting of the degenerating octaval afferents in the thornback guitarfish following transection of the octavus nerve proximal to the ganglion. (A-E) Transverse sections through the hindbrain at levels indicated in the dorsal view of the brain. Dashed lines indicate degenerating fibers, and fine stippling indicates degenerating terminals. Bar scale equals 5 mm. Abbreviations: alln, anterior lateral line nerve; AN, anterior octaval nucleus; CC, cerebellar crest; CN, caudal octavolateralis nucleus; DN, dorsal octavolateralis nucleus; DON, descending octavus nucleus; DR, dorsal root of anterior lateral line nerve; DV, descending trigeminal nucleus and tract; GL, granule cell layer; IO, inferior olive; IR, inferior reticular formation; LG, lateral granular layer; MG, medial granular layer of auricle; ML, molecular layer; MLF, medial longitudinal fasciculus; MN, medial octavolateralis nucleus; MON, magnocellular octaval nucleus; MR, medial reticular formation; MV, trigeminal motor nucleus; PL, Purkinje cell layer; plln, posterior lateral line nerve; PLLR, root of the posterior lateral line nerve; PN, posterior octaval nucleus; VL, vagal lobe; VR, ventral root of anterior lateral line nerve; V, trigeminal nerve; VII, facial nerve; VIII, octavus nerve; IX, glossopharyngeal nerve and nucleus; X, vagal nerve and nucleus. (From Northcutt 1980.)

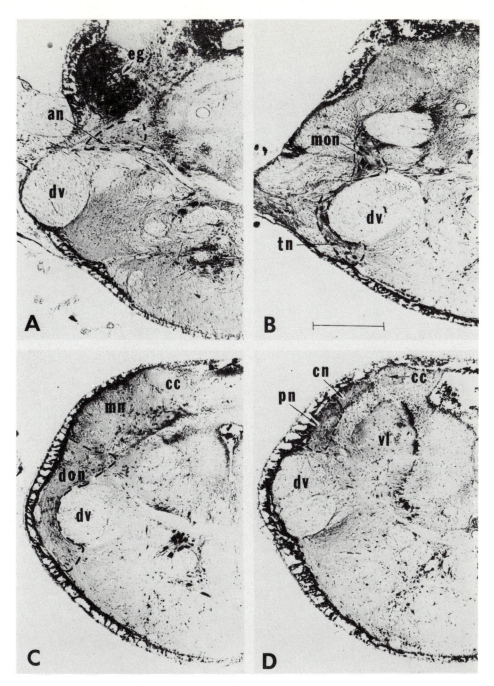

Figure 16-6. Photomicrographs of transverse sections from rostral (A) to caudal (D) of the octavolateralis area of the medulla of the teleost *Gillichthys*. Bar scale equals 400 μm. Abbreviations: an, anterior octaval nucleus; cc, cerebellar crest; cn, caudal octavolateralis nucleus; don, descending octaval nucleus; dv, descending trigeminal tract; eg, eminentia granularis; mn, medial octavolateralis nucleus; mon, magnocellular octaval nucleus; pn, posterior octaval nucleus; tn, tangential octaval nucleus; vl, vagal lobe. (From Northcutt 1980.)

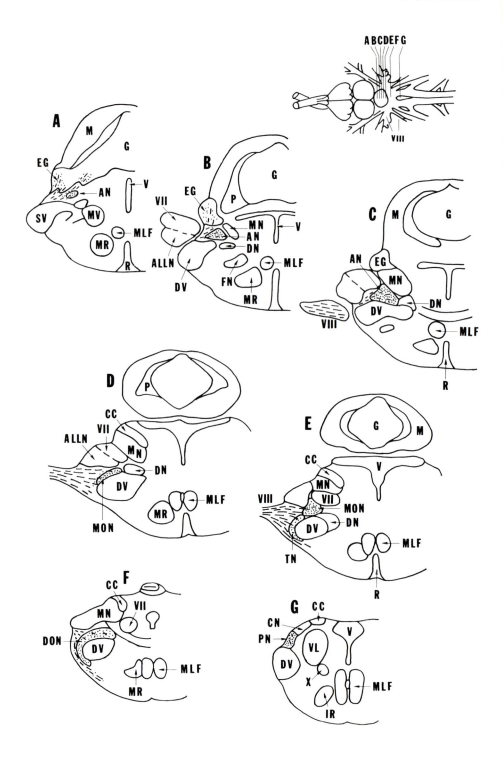

In the teleost *Gillichthys,* the ascending octaval fibers terminate in part of the granule layer of the cerebellar corpus (Fig. 16-6A) and among the granule cells of the more laterally situated eminentia granularis. The medial cerebellar terminal field in *Gillichthys* likely corresponds to the flocculonodular field of other vertebrates.

An earlier study (Northcutt 1980) failed to reveal a similar medial terminal field in *Platyrhinoidis;* however, more sensitive tracing methods now demonstrate that the ascending octaval fibers also turn medially to enter the upper leaf of the auricle where they terminate (Northcutt and Corwin, unpublished observations).

Octaval fibers of the descending limb in both *Platyrhinoidis* (Fig. 16-4) and *Gillichthys* (Figs. 16-6, 16-7) terminate in a descending octaval nucleus and a more caudally situated posterior octaval nucleus.

Teleosts differ from other bony and cartilaginous fishes in possessing a fifth octaval nucleus, the tangential nucleus (Fig. 16-5B). The large cells of this nucleus are similar to those of the magnocellular nucleus (Fig. 16-5B) which surround the lateral edge of the descending trigeminal tract ventrally. This similarity, and its dorsal continuity with the magnocellular nucleus, indicate that the tangential nucleus of teleosts may have developed from the magnocellular nucleus of other fishes.

Primary octaval fibers also terminate directly on parts of the medullar reticular formation in both cartilaginous (Fig. 16-4D) and bony fishes (Fig. 16-7F). Some fibers of the descending limb turn ventrally and medially to terminate among the dendrites of neurons forming the medial and inferior reticular formations. In both *Platyrhinoidis* and *Gillichthys,* the primary octaval fibers are totally ipsilateral. A similar condition has been reported in the holostean *Amia* (McCormick 1980) and in the chain dogfish, *Scyliorhinus* (Boord and Roberts 1980).

Reports also agree that there is little overlap between the lateralis and octaval fibers in their primary sites of termination, except for the magnocellular octaval nucleus. However, a recent and detailed study by Bell (1981; Chapter 19) of the mormyrids *Gnathonemus* and *Brienomyrus* revealed bilateral projections from both lateralis and octaval nerves and far more overlap in their terminal sites than has been observed in other fishes.

Bell traced individual octaval branches from their end-organs to their central terminal sites and recognized utricular and semicircular canal efferents to magnocellular,

◀ Figure 16-7. Charting of the degenerating octavus nerve in the teleost *Gillichthys* following transaction of the nerve medial to the ganglia. (A-G) Transverse sections through the hindbrain at levels indicated in the dorsal view of the brain. Dashed lines indicate degenerating fibers, and stippling indicates degenerating terminals. Abbreviations: ALLN, anterior lateral line nerve; AN, anterior octaval nucleus; CC, cerebellar crest; CN, caudal octavolateralis nucleus; DN, nucleus of the descending trigeminal tract; DON, descending octaval nucleus; DV, descending trigeminal tract; EG, eminentia granularis; FN, facial motor nucleus; G, granular layer of cerebellum; IR, inferior reticular formation; M, molecular layer of cerebellum; MLF, medial longitudinal fasciculus; MN, medial octavolateralis nucleus; MON, magnocellular octaval nucleus; MR, medial reticular formation; MV, trigeminal motor nucleus; P, Purkinje layer of cerebellum; PN, posterior octaval nucleus; R, raphe nucleus; SV, sensory root of trigeminal nerve; TN, tangential octavus nucleus; V, ventricle; VL, vagal lobe; VII, facial sensory root; VIII, octavus nerve; X, vagal lobe. (From Northcutt 1980.)

tangential, and descending nuclei, suggesting that these nuclei receive primarily gravistatic input. Bell did not recognize a distinct anterior octaval nucleus, as reported in many other recent studies, but did recognize a complex rostral nucleus he termed the octaval nucleus. This nucleus receives input from both the lateral line nerves as well as all branches of the octaval nerve. One part of the octaval nucleus receives input solely from the sacculus; thus, input to this area may be solely acoustic, as Bell suggests.

The brainstem in mormyrid fishes is far more complex than in most fishes, and hypertrophy of the octavolateralis area further complicates comparisons. Thus, it is difficult, at best, to recognize exact homologs in the octaval column. It is likely, however, that the mormyrid octaval nucleus corresponds to at least the anterior octaval nucleus and part of the medial octavolateralis nucleus in other fishes. Regardless of the exact homologies, Bell's study demonstrates two very important points: (1) Clearly, there is considerable variation among fishes in the laterality and sites of termination of primary lateralis and octaval fibers; and (2) In at least some fishes, a rostral part of the octaval column likely receives only acoustic input.

Experimental data on the number of octaval nuclei and the primary projections of the octaval nerve in fishes have proliferated in the last few years, but it is still not possible to describe the functional properties of individual octaval nuclei. The differentiation of the octaval medullar column into a series of individual nuclei suggests that these nuclei possess different otic functions and different connections. Extrapolation of the mormyrid data to other fishes would suggest that the anterior octaval nucleus is acoustic and the more caudal nuclei primarily gravistatic. However, it is already established that other taxa (i.e., clupeoids) possess acoustic abilities that do not depend on the saccule but involve other otic end-organs (see Blaxter, Denton, and Gray, Chapter 2). These results argue that different otic end-organs have evolved acoustic properties and that different piscine taxa may possess very different central projections. Thus, acoustic functions may be subserved by different parts of the octaval medullar column in different taxa; fishes may exhibit a variety of octaval patterns; and a given end-organ may project to distinctly different octaval nuclei, resulting in different acoustic nuclei among fishes.

It is also possible that different specialized end organs project to the same acoustic nucleus in various fishes. Platt and Popper (Chapter 1) have suggested that most otic end organs possess multiple functions. This hypothesis is particularly attractive evolutionarily, as it suggests a protoadaptation for acoustic specialization of different otic end-organs. In addition, if most otic end-organs transmit both acoustic and gravistatic information, they may project to both acoustic and gravistatic nuclei. Thus the coding of acoustic and gravistatic information may depend on the type of connections between the ganglion cells and hair cells, and on the nature of the central projections of individual ganglion cells.

If most otic end-organs are shown to possess multiple functions, it is likely that they will project to all primary octaval nuclei but that these nuclei will be conservative, retaining the same function regardless of the specializations of the end-organs.

3.3 Homologs among Primary Octaval Nuclei

Among living anamniotes, there are at least three distinct patterns of primary octaval nuclei in the medulla (Fig. 16-3). Lampreys (Fig. 16-3A) possess an octaval column formed by a single ventral nucleus, within which are embedded three aggregates of larger neurons termed the octavomotor nuclei. All jawed fishes possess an octaval column divided rostrocaudally into anterior, magnocellular, descending, and posterior nuclei (Fig. 16-3B). Finally, amphibians (Figs. 16-3C and 16-8D) possess an octaval column divided into dorsal, medial, and ventral nuclei.

The distribution of octaval nuclei reveals no one-to-one correspondence among anamniotes. Lampreys exhibit the simplest pattern, characterized by a single ventral nucleus, if the octavomotor nuclei are disregarded; however, no other vertebrates possess such obviously large neurons scattered throughout most of the octaval column. Fishes and amphibians possess one or more large-neuron octaval populations (magnocellular and tangential nuclei in teleosts; ventral octaval nucleus in amphibians); these might be homologous to the octavomotor nuclei of lampreys, but it is also possible that octavomotor nuclei in lampreys represent a specific, derived adaptation, not shared by other taxa. No such obvious nuclei have been recognized in hagfishes, but their octaval organization has yet to be experimentally determined.

At present, it can only be suggested that the ventral octaval nucleus of lampreys is the field homolog of the entire octaval column of jawed fishes, and that the gnathostome octaval pattern could have evolved from a pattern similar to that of living lampreys by the subdivision of a single column into a number of distinct octaval nuclei.

There are similar problems in comparing the primary octaval nuclei in jawed fishes with those in amphibians or other tetrapods. Jawed fishes possess four to five distinct octaval nuclei, whereas amphibians possess only three octaval nuclei. Thus, no simple one-to-one comparison is possible.

There are far more data on the anatomy and physiology of octaval systems in amphibians than in other anamniotes (Capranica 1976, Lombard 1980, Northcutt 1980). The dorsal octaval nucleus is known to receive acoustic input, whereas the ventral octaval nucleus receives vestibular input. A third nucleus (medial or caudal octaval nucleus) was recently described (Opdam, Kemali, and Nieuwenhuys 1976), but it is not known whether this nucleus receives auditory and/or vestibular inputs. Given the functional segregation in amphibians, and Bell's recent results in teleosts, the following homologies between amphibians and jawed fishes are possible: (1) The dorsal octaval nucleus of amphibians is homologous to the anterior octaval nucleus of fishes; (2) The ventral octaval nucleus of amphibians is the field homolog of the magnocellular and descending octaval nuclei of fishes; and (3) The medial octaval nucleus of amphibians is homologous to the posterior octaval nucleus of fishes. Additional experimental data regarding the connections and functions of the octaval nuclei in fishes are necessary to test this hypothesis. Equally important, additional data are also needed for amphibians. Most experiments have involved a single genus, representing only one of three orders of living amphibians; it is possible that the octaval pattern exhibited by *Rana* is not typical of other amphibians.

One additional problem should be mentioned. Wever (1976) and Lombard (1980) have proposed that audition may have evolved a number of times independently

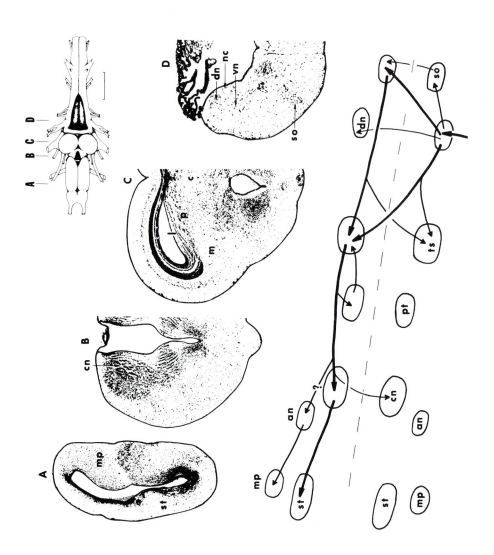

among different vertebrate radiations, based on differences in otic receptors and transduction morphology associated with the inner ear. This scenario further suggests that the different octaval nuclei might not be homologous among vertebrates, as different medullar patterns may have evolved independently. This is clearly a possibility; however, peripheral receptor evolution need not necessitate central reorganization, particularly if Platt and Popper's hypothesis is valid.

4 Higher Order Octaval Projections

Few details are known concerning higher order octaval pathways in fishes. Both descriptive and experimental data on projections of the octavomotor nuclei in lampreys are available (Tretjakoff 1909, Rovainen 1979). Bullock and Corwin (1979) have reported auditory evoked potentials in the telencephalon in several shark species, and Corwin and Northcutt (1980) have identified an auditory center in the midbrain of *Platyrhinoidis*. Knudsen (1977) has reported auditory units in the torus semicircularis of the midbrain in catfishes, and Braford and McCormick (1979) and Finger (1979) have reported toral projections to the diencephalon in *Amia* and *Ictalurus*.

Although these studies demonstrate the existence of higher order octaval centers in all jawed fishes, and suggest that there are a number of similarities among these centers and those in fishes and land vertebrates, there are still insufficient data to establish the ascending systems in any single piscine species. At present, the most complete set of data on higher order octaval pathways in anamniotes is that for anurans (Fig. 16-8). These data will be used to interpret the fragments of information available for fishes and to illustrate several problems regarding the nature of higher order octaval projections in fishes.

4.1 Agnathans

In lampreys, the axons of the octavomotor cells are very large, unmyelinated fibers that are easily observed in nonexperimental material. Axons of the anterior octavomotor cells course rostrally in the midbrain tegmentum, where many decussate across the midline to terminate on cells of the oculomotor nuclei. In contrast, axons of the intermediate and posterior octavomotor cells descend in the medulla, forming ipsilateral and contralateral vestibulospinal pathways. Rovainen (1979) reported that these vestibulospinal fibers form excitatory monosynaptic connections with myotomal motor neurons and lateral interneurons of the spinal cord.

◄ Figure 16-8. Photomicrographs of transverse sections through the brain of the bullfrog, *Rana catesbeiana*, illustrating major auditory centers and their interconnections. The levels of the various sections are indicated in the dorsal view of the brain. Bar scale equals 10 mm. Abbreviations: an, anterior thalamic nucleus; c, commissural toral nucleus; cn, central thalamic nucleus; dn, dorsal octaval nucleus; l, laminar toral nucleus; m, magnocellular toral nucleus; mp, medial pallium; nc, caudal octaval nucleus; p, principal toral nucleus; pt, pretectum; so, superior olive; st, striatum; ts, torus semicircularis; vn, ventral octaval nucleus. (From Northcutt 1980.)

Additional secondary octaval projections in lampreys are presently unknown; however, Heier (1948) claimed that pathways could be traced from the ventral octaval nucleus to a midbrain area, the torus semicircularis, and from the torus to the thalamus. If experimental studies confirm these pathways, they will provide strong indication that the major ascending octaval pathways seen in jawed vertebrates probably arose with the earliest vertebrates.

4.2 Gnathostome Fishes

In land vertebrates (Fig. 16-8), the primary auditory nuclei project bilaterally to a second medullar area (the superior olivary complex), as well as directly to an auditory midbrain area (torus semicircularis or inferior colliculus). If the anterior octaval nucleus of fishes is, indeed, an auditory nucleus, similar projections might exist. In addition, the remaining, more caudally situated nuclei of the primary octaval column should give rise to descending vestibulospinal pathways. This pattern of second-order projections does appear to exist in *Platyrhinoidis* and *Gillichthys* (unpublished observations). Injections of horseradish peroxidase (HRP) into the spinal cord of these genera retrogradely fill the neurons of the magnocellular nucleus and cells in the rostral part of the descending octaval nuclei in *Platyrhinoidis,* the magnocellular, tangential, and descending octaval nuclei in *Gillichthys.* Similarly, in *Platythinoids* and *Gillichthys,* neurons of the anterior octaval nucleus are retrogradely labeled bilaterally only following injections of HRP (unilaterally) into the torus semicircularis. In *Gillichthys,* toral injections also label the superior ólivary nucleus and nucleus of the lateral lemniscus. The former lies just dorsomedial to the lateral lemniscus, and most of the retrogradely labeled cells are ipsilateral to the injected torus. The nucleus of the lateral lemniscus is seen rostrally as numerous filled cells in the midbrain tegmentum, labeled bilaterally. *Amia* reveals similarly labeled cell groups, except for the octaval nucleus (Braford and McCormick 1979). Thus, a superior olivary complex and nucleus of the lateral lemniscus definitely exist in fishes as in land vertebrates, but the projections of the anterior octaval nucleus are in question.

Knudsen (1977) reported that the torus semicircularis in catfish is divided into a lateral nucleus, concerned with lateral line inputs, and a more medial area (nucleus centralis) that receives auditory input. In many land vertebrates (Fig. 16-8), the torus semicircularis projects bilaterally to one or more dorsal thalamic nuclei. The primary projection is to the ipsilateral thalamus, with a sparse projection to the contralateral thalamus via the supraoptic decussation. Similar ascending projections to the diencephalon in *Amia, Carassius,* and *Ictalurus* have also been demonstrated (Braford and McCormick 1979, Finger 1979). An ascending ipsilateral toral pathway projects to the caudal diencephalon, where it terminates adjacent to the central posterior nucleus of the dorsal thalamus, and more ventrally in a division of the glomerular complex of the posterior tuber. Comparable, but sparser, contralateral projections via the supraoptic decussation also occur. Again, these toral connections closely resemble those existing in land vertebrates. In fishes, however, the torus receives both lateral line and auditory inputs, whereas the torus in most land vertebrates receives only auditory input. Thus, we do not know whether lateralis and auditory inputs in fishes converge on a single nucleus at thalamic levels, or maintain modality separation. Given the available anatomical data, we cannot even be sure that both sensory modalities are projecting to the thalamus. This is likely, though, as Bullock and Corwin (1979) have reported auditory evoked potentials in the telencephalon in sharks, and preliminary data indicate that auditory units also occur in the telencephalon of some teleosts (W. Saidel, pers. comm.).

Two distinctly different auditory thalamotelencephalic projection patterns exist in land vertebrates. In amphibians (Fig. 16-8), the main auditory thalamic nucleus (central thalamic nucleus) projects primarily, if not solely, to the ipsilateral striatum. In amniotic vertebrates, the auditory thalamic nucleus (nucleus medialis in reptiles, nucleus ovoidalis in birds, medial geniculate nucleus in mammals) projects sparsely to the ipsilateral striatum (caudate and putamen nuclei) and massively to a portion of the more dorsally situated pallium (dorsal ventricular ridge in birds and reptiles, the auditory isocortex in mammals). Thus, it is possible that auditory projections to the telencephalon in fishes may go primarily to the striatum, if they are like those in amphibians, or to the striatum and pallium, if they are like those in amniotic vertebrates. Given the extreme variation in differentiation and number of telencephalic nuclei in both cartilaginous (Northcutt 1978) and bony fishes (Northcutt and Braford 1980), it is likely that both patterns exist in both groups of jawed fishes. The amphibian pattern may exist in primitive sharks and dipnoan, chondrostean, and polypteriform fishes, and the amniotic pattern in advanced sharks and holostean and teleostean fishes. This prediction is based on two lines of reasoning: (1) the probability that amphibians are not characterized by an aberrant pattern of telencephalic organization, but retain a large number of primitive telencephalic characters common in early jawed fishes; and (2) that the telencephalon in ray-finned fishes exhibits an increasingly complex pallial differentiation from polypteriform to teleostean fishes. A recent analysis of the telencephalic nuclei in ray-finned fishes (Northcutt and Braford 1980) suggests that part of the medial zone of the dorsal area (Dm) in teleosts is homologous to the caudoputamen in land vertebrates, and that the dorsal zone and dorsal part of the lateral zone of the dorsal area (Dd and Dl-d) in teleosts is homologous to the dorsal pallium in land vertebrates. It is these areas in the telencephalon that probably represent the highest auditory centers in teleosts.

Amphibians also possess a third auditory pathway that may characterize other vertebrates as well. Mudry and Capranica (1978) reported auditory activity in the medial pallium of *Rana*; this activity does not result from striatal projections onto the medial pallium, as no such projections exist. The anterior thalamic nucleus, however, projects bilaterally to the medial pallium and may receive auditory input directly from the torus semicircularis or indirectly from the central thalamic nucleus (Fig. 16-8, see Northcutt 1980 for a review of this problem). Thus, the medial pallium of amphibians appears to receive an auditory input whose origin may be thalamic but must still be determined.

This pallial region, generally considered homologous to the hippocampal complex of other vertebrates, should not be considered primarily auditory, however, as it receives a variety of sensory inputs and likely functions as an arousal system (see Northcutt 1981 for a summary of medial pallial connections). Regardless, the existence of these auditory projections in amphibians indicates that a third telencephalic region in fishes may also receive auditory input.

5 Conclusions

Anamniotic vertebrates are characterized by at least three distinct patterns of medullar octaval organization: agnathan, gnathostome fish, and amphibian patterns. Lampreys possess a single ventral octaval nucleus, within which three distinct aggregates of large octavomotor cells are recognized. The octavomotor neurons form vestibulospinal and vestibulo-oculomotor connections. We have no further data regarding connections of their ventral octaval nucleus, nor do we know whether this pattern is also characteristic of hagfishes. The ventral octaval nucleus of lampreys probably represents the primitive condition for vertebrates and is probably a field homolog of all the octaval nuclei in other vertebrates.

All gnathostome fishes examined possess at least four primary octaval nuclei: anterior, magnocellular, descending, and posterior octaval nuclei. Preliminary data suggest that the anterior octaval nucleus is likely an auditory nucleus and that it projects bilaterally to a midbrain auditory center, the torus semicircularis; the magnocellular and, at least, the rostral part of the descending octaval nuclei form descending vestibulospinal pathways. Teleost fishes possess a fifth octaval nucleus, the tangential nucleus, that has likely evolved from the magnocellular nucleus in other fishes; like the magnocellular nucleus, it projects to the spinal cord. The torus semicircularis forms bilateral ascending projections to the diencephalon, and telencephalic auditory areas are established in some sharks and teleosts, but the exact location and connections of these telencephalic centers must still be determined.

All gnathostome fishes possess more octaval nuclei than do amphibians, and no obvious one-to-one correspondence exists. Two possibilities are apparent: (1) Amphibians are characterized by an aberrant octaval pattern that may reflect an evolutionary history of octaval nuclear reduction followed by redevelopment of auditory abilities; or (2) The dorsal octaval nucleus in amphibians is homologous to the anterior octaval nucleus in fishes; the ventral octaval nucleus in amphibians is a field homolog of the magnocellular and descending octaval nuclei in fishes; the medial or caudal octaval nucleus in amphibians is a homolog of the posterior octaval nuclei in fishes.

All available evidence strongly suggests that fishes are characterized by most of the same auditory pathways and centers seen in land vertebrates. Given the incredible range of morphological features in the brains of fishes, it is also likely that they will reveal even more variation in auditory mechanisms than land vertebrates.

Acknowledgments. Original research for this chapter was supported in part by NIH Grants NS11006 and EY02485 and by a Rackham Faculty Research Grant from the University of Michigan. Mary Sue Northcutt assisted in many phases of the research and in the preparation of the manuscript; Drs. Catherine A. McCormick and William M. Saidel critically read the manuscript and offered many helpful suggestions. The author is also grateful to the Guggenheim Foundation for their support during part of the period relating to this work.

References

Ariëns Kappers, C. U., Huber, G. C., Crosby, E. C.: The Comparative Anatomy of the Nervous System of Vertebrates, Including Man. New York: Hafner, 1960.

Ayers, H.: Vertebrate cephalogenesis. II. A contribution to the morphology of the vertebrate ear, with a reconsideration of its function. J. Morphol. 6, 1-360 (1892).

Bell, C. C.: Central distribution of octavolateral afferents and efferents in a teleost (Mormyridae). J. Comp. Neurol. 195, 391-414 (1981).

Bodznick, D., Northcutt, R. G.: Segregation of electro- and mechanoreceptive inputs to the elasmobranch medulla. Brain Res. 195, 313-321 (1980).

Boord, R. L.: Auricular projections in the clearnose skate, *Raja eglanteria*. Am. Zool. 17, 887 (1977).

Boord, R. L., Campbell, C. B. G.: Structural and functional organization of the lateral line system of sharks. Am. Zool. 17, 431-441 (1977).

Boord, R. L., Roberts, B. L.: Medullary and cerebellar projections of the statoacoustic nerve of the dogfish, *Scyliorhinus canicula*. J. Comp. Neurol. 193, 57-68 (1980).

Braford, M. R., Jr., McCormick, C. A.: Some connections of the torus semicircularis in the bowfin, *Amia calva*: A horseradish peroxidase study. Neurosci. Abstr. 5, 139 (1979).

Bullock, T. H.: An essay on the discovery of sensory receptors and the assignment of their functions together with an introduction to electroreceptors. In: Handbook of Sensory Physiology, Vol. III/3. Fessard, A. (ed.). New York: Springer-Verlag, 1974, pp. 1-12.

Bullock, T. H., Corwin, J. T.: Acoustic evoked activity in the brain in sharks. J. Comp. Physiol. 129, 223-234 (1979).

Campbell, C. B. G., Boord, R. L.: Central auditory pathways in nonmammalian vertebrates. In: Handbook of Sensory Physiology, Vol. V/1: Auditory System. Keidel, W. D., Neff, W. D. (eds.). Berlin-Heidelberg-New York: Springer-Verlag, 1974, pp. 337-362.

Capranica, R. R.: Morphology and physiology of the auditory system. In: Frog Neurobiology. Llinas, R., Precht, W. (eds.). Berlin-Heidelberg-New York: Springer-Verlag, 1976, pp. 551-575.

Corwin, J. T., Northcutt, R. G.: Auditory centers in the elasmobranch brain: deoxyglucose localization and evoked potential recording. Neurosci. Abstr. 6, 556 (1980).

Finger, T. E.: A thalamic relay nucleus for the lateral line system in a teleost fish. Neurosci. Abstr. 5, 141 (1979).

Hardisty, M. W.: Biology of the Cyclostomes. London: Chapman and Hall, 1979.

Heier, P.: Fundamental principles in the structure of the brain; a study of the brain of *Petromyzon fluviatilis*. Acta Anat. 5 Suppl. 8, 1-213 (1948).

Herrick, C. J.: The cranial nerve components of teleosts. Anat. Anz. 13, 425-431 (1897).

Hörstadius, S.: The Neural Crest. London: Oxford Univ. Press, 1950.

Jansen, J.: The brain of *Myxine glutinosa*. J. Comp. Neurol. 49, 359-507 (1930).

Knudsen, E. I.: Distinct auditory and lateral line nuclei in the midbrain of catfishes. J. Comp. Neurol. 173, 417-432 (1977).

Korn, H., Sotelo, C., Bennett, M. V. L.: The lateral vestibular nucleus of the toadfish *Opsanus tau*: Ultrastructural and electrophysiological observations with special reference to electrotonic transmission. Neuroscience 2, 851-884 (1977).

Landacre, F. L.: The origin of the cranial ganglia in *Ameiurus*. J. Comp. Neurol. 20, 309-411 (1910).

Landacre, F. L.: The epibranchial placodes of *Lepidosteus osseus* and their relation to the cerebral ganglia. J. Comp. Neurol. 22, 1-70 (1912).

Landacre, F. L.: The differentiation of the preauditory and postauditory primitive lines into preauditory and postauditory placodes, lateralis ganglia and migratory lateral-line placodes in *Amblystoma jeffersonianum*. J. Comp. Neurol. 44, 29-59 (1927).

Larsell, O.: The Comparative Anatomy and Histology of the Cerebellum from Myxinoids through Birds. Minneapolis: University of Minnesota Press, 1967.

Lombard, R. E.: The structure of the amphibian auditory periphery: A unique experiment in terrestrial hearing. In: Comparative Studies of Hearing in Vertebrates. Popper, A. N., Fay, R. R. (eds.). Berlin-Heidelberg-New York: Springer-Verlag, 1980, pp. 121-138.

Lowenstein, O., Osborne, M. P., Thornhill, R. A.: The anatomy and ultrastructure of the labyrinth of the lamprey (*Lampetra fluviatilis* L.). Proc. R. Soc. London Ser. B 170, 113-134 (1968).

Maler, L., Karten, H. J., Bennett, M. V. L.: The central connections of the posterior lateral line nerve of *Gnathonemus petersi*. J. Comp. Neurol. 151, 57-66 (1973a).

Maler, L., Karten, H. J., Bennett, M. V. L.: The central connections of the anterior lateral line nerve of *Gnathonemus petersi*. J. Comp. Neurol. 151, 67-84 (1973b).

Maler, L., Finger, T., Karten, H. J.: Differential projections of ordinary lateral line receptors and electroreceptors in the gymnotid fish, *Apteronotus (Sternarchus) albifrons*. J. Comp. Neurol. 158, 363-382 (1974).

Mayser, P.: Vergleichend anatomische Studien über das Gehirn der Knochenfische mit besonderer Berücksichtigung der Cyprinoiden. Z. Wiss. Zool. 36, 259-364 (1882).

McCormick, C. A.: Central projections of the lateralis and eighth nerves in the bowfin, *Amia calva*. Doctoral Thesis, University of Michigan (1978).

McCormick, C. A.: Central projections of the lateral line and eighth nerves in the bowfin, *Amia calva*, J. Comp. Neurol. 197, 1-16 (1981).

McCready, P. J., Boord, R. L.: The topography of the superficial roots and ganglia of the anterior lateral line nerve of the smooth dogfish, *Mustelus canis*. J. Morphol. 150, 527-538 (1976).

Moy-Thomas, J. A., Miles, R. S.: Paleozoic Fishes, 2nd ed. Philadelphia: Saunders, 1971.

Mudry, K. M., Capranica, R. R.: Electrophysiological evidence for auditory responsive areas in the diencephalon and telencephalon of the bullfrog, *Rana catesbeiana*. Neurosci. Abst. 4, 101 (1978).

Northcutt, R. G.: Brain organization in the cartilaginous fishes. In: Sensory Biology of Sharks, Skates, and Rays. Hodgson, E. S., Mathewson, R. F. (eds.). Arlington: Office of Naval Research, Department of the Navy, 1978, pp. 117-193.

Northcutt, R. G.: Central projections of the eighth cranial nerve in lampreys. Brain Res. 167, 163-167 (1979a).

Northcutt, R. G.: Primary projections of VIII nerve afferents in a teleost, *Gillichthys mirabilis*. Anat. Rec. 193, 638 (1979b).

Northcutt, R. G.: Central auditory pathways in anamniotic vertebrates. In: Comparative Studies of Hearing in Vertebrates. Popper, A. N., Fay, R. R. (eds.). Berlin-Heidelberg-New York: Springer-Verlag, 1980, pp. 79-118.

Northcutt, R. G.: Evolution of the telencephalon in nonmammals. Annu. Rev. Neurosci. 4, 301-350 (1981).

Northcutt, R. G., Bodznick, D. A., Bullock, T. H.: Most non-teleost fishes have electroreception. Proc. Int. Union Physiol. Sci. 14, 614 (1980).

Northcutt, R. G., Braford, M. R., Jr.: New observations on the organization and evolution of the telencephalon of actinopterygian fishes. In: Comparative Neurology of the Telencephalon. Ebbesson, S. O. E. (ed.). New York: Plenum Press, 1980, pp. 41-98.

Opdam, P., Kemali, M., Nieuwenhuys, R.: Topological analysis of the brain stem of the frogs *Rana esculenta* and *Rana catesbeiana*. J. Comp. Neurol. 165, 307-332 (1976).

Parker, G. H.: The sense of hearing in fishes. Am. Naturalist 37, 185-204 (1903).

Pearson, A. A.: The acoustico-lateral centers and the cerebellum, with fiber connections, of fishes. J. Comp. Neurol. 65, 201-294 (1936).

Rovainen, C. M.: Electrophysiology of vestibulospinal and vestibuloreticulospinal systems in lampreys. J. Neurophysiol. 42, 745-766 (1979).

Rubinson, K.: The central distribution of VIII nerve afferents in larval *Petromyzon marinus*. Brain Behav. Evol. 10, 121-129 (1974).

Smeets, W. J. A. J., Nieuwenhuys, R.: Topological analysis of the brain stem of the sharks *Squalus acanthias* and *Scyliorhinus canicula*. J. Comp. Neurol. 165, 333-368 (1976).

Stone, L. S.: Experiments on the development of the cranial ganglia and the lateral line sense organs in *Amblystoma punctatum*. J. Exp. Zool. 35, 421-496 (1922).

Thornhill, R. A.: The development of the labyrinth of the lamprey (*Lampetra fluviatilis* Linn. 1758). Proc. R. Soc. London Ser. B 181, 175-198 (1972).

Tretjakoff, D.: Das Nervensystem von Ammocoetes II. Gehirn. Arch. Midrosk. Anat. Entwicklungsmech. 74, 636-779 (1909).

van Bergeijk, W. A.: Evolution of the sense of hearing in vertebrates. Am. Zool. 6, 371-377 (1966).

van Bergeijk, W. A.: The evolution of vertebrate hearing. In: Contributions to Sensory Physiology, Vol. 2. Neff, W. D. (ed.). Berlin-Heidelberg-New York: Springer-Verlag, 1967, pp. 1-49.

von Frisch, K.: Ein Zwergwels, der kommt, wenn Man ihm pfeift. Biol. Zentralbl. 43, 439-446 (1923).

Wever, E. G.: Origin and evolution of the ear of vertebrates. In: Evolution of Brain and Behavior in Vertebrates. Masterton, R. B., Bitterman, M. E., Campbell, C. B. G., Hotton, N. (eds.). Hillsdale, N.J.: Lawrence Erlbaum, 1976, pp. 89-106.

Wilson, H. V., Mattocks, J. E.: The lateral sensory anlage in the salmon. Anat. Anz. 13, 658-660 (1897).

Discussion

BULLOCK: Would you want to bet that lampreys have acoustic responses?

NORTHCUTT: You really want me to put myself out on a limb. I think there is a good chance that they do. We think they have the homologues of a utricle and a saccule, and something that even looks like a macula neglecta. There are at least two or three major cell types in the right column, and I don't see any reason why they can't have acoustic functions. If they do, then it pushes the evolution of the double function of the inner ear all the way back to the origin of the vertebrates.

BULLOCK: I think the evoked potential techique would be the best way.

NORTHCUTT: I agree that would be a reasonable way. One thing you *don't* want to do is to try to cut lateral line nerves and look for behavioral responses. We have looked at the distribution of lateral line nerves in many species, and it is very difficult to get at the ganglia, and to ablate all of them. I should never want to try this in a behavioral experiment.

BULLOCK: Want to bet on a hagfish?

NORTHCUTT: Hagfish are a real problem. For one, they have a brain size that is four times larger for their body weight than a lamprey. They have a five-layered cortex in the telencephalon, and no trace of a ventricular system. In addition, they appear to have no trace of a lateral line system, and we can't even be sure of the position of the octavolateralis area.

ROBERTS: I was curious about the dorsal nucleus. It is a seductively satisfactory idea that the dorsal nucleus is a pure electroreceptor. Do we have any evidence?

NORTHCUTT: Yes. If you transect the dorsal root of the anterior lateral line nerve, it only goes to the dorsal nucleus and a granular plate of cells that may or may not be the eminentia granularis. While we can say that the primary lateral line input is electroreceptive, these granular masses form the so-called cerebellar crest. It turns out that the electroreceptive, mechanoreceptive, and the eighth nerve go into different segments of the granular masses. Do each of the granular populations project back on to a lobe, or is there an overlap? All we can really say is that the primary inputs into these three nuclei are electroreceptive, mechanoreceptive, and eighth nerve.

GRAY: You were speaking of possible double functions for a receptor, as mentioned by Popper. The receptor response is simple. It responds to a displacement or some time function of displacement. But it may be subjected to many different sources of the same displacement. The entire receptor apparatus determines if the strongest response will be to accelerations, to acoustic vibrations, or other energy forms, just as long as the stimuli are within the frequency response of the system. How does the fish nervous system select or decipher all this information.

NORTHCUTT: There is no simple answer to that. If you kept us all here for a week, we might be able, collectively, to come up with a set of mechanisms.

GRAY: You are bound to have some convergence someplace, aren't you?

NORTHCUTT: Outside of the known example of convergence for the optic tectum, it is an article of faith. I cannot believe that a fish lives its entire life keeping three kinds of information totally segregated.

PARTRIDGE: You probably know of the studies in which cerebellar units have been found to be both mechano- and electroreceptive?

NORTHCUTT: One of the things that has surprised most anatomists who have looked at octavolateralis projections, is that outside of the projection to the eminentia granularis, there are no direct projections of the lateral line system into the cerebellum. The most likely pathway is that the dorsal and medial nuclei project to the tectum. The tectum projects back to another set of cells called the lateral nucleus of the valvula, and these project to the cerebellum. Thus, cerebellar information is of a fourth or fifth order.

FINE: The vestibular branch of the eighth nerve has been traced to the magnocellular vestibular nucleus. Would this imply the likelihood of an acoustic function?

NORTHCUTT: I think that everyone who has looked at both lateral line and eighth nerve projections would claim that that nucleus is probably the only nucleus that gets both lateral line and eighth input, and this nucleus could possess either acoustic or vestibular functions.

FINE: Would you say it is auditory and vestibular, or just vestibular? It arises mostly in the utricle.

NORTHCUTT: How do you define the vestibular branch of the eighth, short of tracing the central projections of each of the otic end organs.

Chapter 17

Central Processing of Acousticolateralis Signals in Elasmobranchs

B. L. ROBERTS*

1 Introduction

At a previous symposium on hearing in fishes organized by Dr. Tavolga, van Bergeijk (1964) put forward some important ideas about the properties of sound in water and imparted to the lateral line system a significant role in the localization of sound sources, thereby appearing to settle the troubling question of the biological significance of these sense organs.

However, with time, doubts have arisen as to the validity of this view with the growing body of evidence that sharks can locate far-field sources (Myrberg, Gordon, and Klimley 1975) and that sound sources cannot be located if the labyrinth is denervated (Schuijf 1975) while surface waves can be detected in *Xenopus* with denervated lateral line organs (Görner 1976). Once again the biological role of the lateral line and its relationship to the sense of hearing is in question (see Sand, Chapter 23).

Our problem of course stems from the fundamental difficulty of interpreting the value of sense organs that we lack ourselves. For olfaction, vision, hearing, and touch we have our own subjective experiences which can be applied to animal studies but these are of little value when we are confronted with novel structures. As Thurber (1949) reminds us, the interpretation of function depends on a personal viewpoint: "If to man the cricket seems to hear with its legs, it is possible that to the cricket man seems to walk on his ears."

One of the features that apparently unites the lateral line sense with audition is the projection of the peripheral nerves to a common center (the statoacoustic center) in the hindbrain. In this paper we shall examine present knowledge of the organization of this center, where central processing of the acousticolateralis input takes place, to see whether these central relationships give clues as to the functioning of the lateral line system.

*The Laboratory of the Marine Biological Association, Citadel Hill, Plymouth, England.

2 The Sense Organs and Their Nerve Supply

Our knowledge of the various sense organs that are grouped as the acousticolateralis system has been reviewed recently for the elasmobranchs (Roberts 1978) and so only a brief summary of the main categories will be given here. The basic component is the sensory hair cell, which is found in all sense organs of this system: in the "free" neuromasts (pit organs) which have a chemo-/mechanoreceptive function; in the organs contained within the lateral line canals that are responsive to water movements around the fish; in the ampullae of Lorenzini (electroreception); and in the various end-organs of the labyrinth that respond to acceleration, sound (vibration), and gravity.

The sensory hair cells of all these end-organs make synaptic connections onto afferent nerve fibers that pass to the hindbrain. The patterns of impulses carried by these afferent fibers encode information about the nature of the peripheral stimulus with respect to amplitude, frequency, and direction. For example, in the case of the lateral line, the hair cells respond to displacements over a range of frequencies up to about 100 Hz (Flock 1965) and to wide stimulus strengths by having hair cells set with differing firing thresholds (Katsuki, Yoshino, and Chen 1950). Consequently the amplitude of any stimulus is reflected in the spectrum of cells activated as well as in their frequency of discharge and latency of response (Sand 1937, Russell and Roberts 1974). The encoding of stimulus direction results from the morphological asymmetry of the sensory hairs (Roberts and Ryan in prep.) and is seen as an increase or decrease of the rate of spontaneous discharge (Sand 1937).

The further processing of all this information depends on the distribution of the nerve fibers that supply the individual hair cells (see Murray and Capranica 1973) and on how these afferent fibers relate to the secondary neurons of the acousticolateralis centers. Table 17-1 gives some idea of the numbers of sense organs and nerve fibers in this system.

The afferent fibers from the labyrinth are confined exclusively within cranial nerve VIII, but those from the pit-organs, lateral-line neuromasts and the receptors of the ampullae of Lorenzini run in association with fibers of other modalities in nerves VII and X. The lateral-line canals of the head are supplied by fibers of the anterior lateral-line division of nerve VII (superficial ophthalmic, buccal and external mandibular branches), and the lateral-line organs of the body, and the pit organs, are supplied by

Table 17-1. Estimates of Sense Organ and Nerve Fiber Numbers

	Number	Fish	Reference
Anterior lateral line nerve fibers content	5,000	*Scyliorhinus*	Paul and Roberts 1977a
Posterior lateral line nerve fibers content	1,000	*Scyliorhinus*	Roberts and Ryan 1971
Nerve VIII fiber content	10,000	*Scyliorhinus*	Montgomery and Roberts 1979
Head lateral line hair cells	8,000	*Carcharhinus*	Tester and Kendall 1969
Body lateral line hair cells	8,000	*Scyliorhinus*	Roberts and Ryan in prep.
Pit organs	77	*Squalus*	Tester and Nelson 1967
Macula neglecta hair cells	47,700	*Carcharhinus*	Corwin 1977

fibers of the posterior lateral-line nerve, which enters the medulla just anterior to, but quite separate from, the vagus (Xth) nerve.

The pore openings to the ampullae of Lorenzini, just like those to the lateral-line canals, are distributed widely over the head, but the sense organs lie deeper in the epithelia of the ampullae, which are grouped within capsules. There are four capsules in *Mustelus,* and these are also innervated by branches of the superficial ophthalmic, external mandibular and buccal branches of nerve VII (Allis 1901).

The end-organs of the labyrinth are innervated by the two branches of cranial nerve VIII. Fibers from the crista of the posterior vertical canal and the maculi of the sacculus, lagena and neglecta enter the posterior ramus, while the anterior ramus is composed of fibers from the cristae of the anterior vertical and horizontal canals and the macula utriculus. These two rami fuse and form a single ganglion that lies adjacent to the brainstem (Fig. 17-1).

Closely associated with this ganglion are the ganglia of the anterior lateral-line nerves, the ganglia of branches of nerve VII and the ganglia of nerve V. Dissections of brains stained with Sudan Black have permitted these neural clusters to be unravelled (Fig. 17-2, McCready and Boord 1976, Boord and Roberts 1980) and have shown that the lateral-line nerves, which convey a mixture of mechanoreceptive and electroreceptive fibers, split into two branches which then reassemble as two divisions (dorsal and ventral) that pass in association into the medulla. It has been suggested that this arrangement groups the afferents from the ampullae of Lorenzini as the dorsal root and separates them from those of mechanoreceptors which enter as the ventral root. However, this attractive interpretation still requires direct anatomical proof.

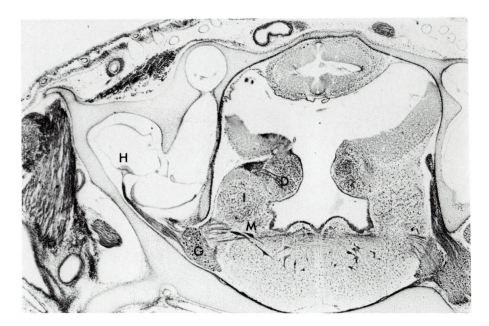

Figure 17-1. Micrograph of transverse section through the head of 59 mm embryo *Scyliorhinus*, passing through labyrinth and showing projection of nerve VIII to hindbrain. H, ampulla of horizontal semicircular canal; G, stato-acoustic ganglion; M, magnocellularis nucleus; I, intermediate nucleus; D, dorsal nucleus.

It is difficult in standard silver-stained material to follow the pathways of these nerves within the brain and although earlier workers postulated various routes with considerable confidence it has been only within the last few years, following the application of experimental neuroanatomical and electrophysiological procedures, that we have been able to state these pathways with some precision (Campbell and Boord 1971, Ilyinsky, Enin, and Volkova 1971, Enin and Ilyinsky 1972, Enin, Ilyinsky, and Volkova 1973, Paul and Roberts 1977a, 1977b, 1977c, Koester and Boord 1978, Montgomery and Roberts 1979, Boord and Roberts 1980). These techniques are now beginning to give us information about secondary and even higher projections (see Northcutt, Chapter 16).

Several main features about the primary projections of the acousticolateralis nerves have emerged from these recent studies. The projections are seen to be entirely ipsilateral and confined almost exclusively to the lateral wall of the medulla in a region that stretches from the obex as far as the cerebellar peduncle. The projection sites of lateral-line and labyrinthine nerves are distinct and the cerebellar corpus receives no primary projection, although significant connections are made with the auricular cerebellum.

3 The Architecture of the Acousticolateralis Centers

To understand the organization of the region of the medulla that is penetrated by these afferent fibers it is necessary first of all to appreciate the general construction of the hindbrain in elasmobranch fishes and, particularly, the morphology of the cerebellum. This part of the brain is complex and, being three-dimensional, is not readily described in words or pictures. But if the reader compares the diagram of the side view of the brain of *Scyliorhinus* (Fig. 17-2) with the diagram of the perspective obtained

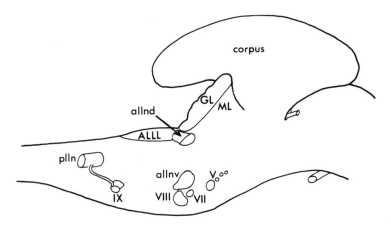

Figure 17-2. Diagram of lateral view of hindbrain of *Scyliorhinus* showing the entry of the acousticolateralis nerves into hindbrain (based on Boord and Roberts 1980). plln, posterior lateral line nerve; allnd, anterior lateral line nerve, dorsal division; allnv, anterior lateral line nerve, ventral division; GL, granular layer; ML, molecular layer; ALLL, anterior lateral line lobe.

when looking down onto the dorsal surface of the medulla (Fig. 17-3), some idea of the main features of this region can be acquired.

It has been customary since Larsell (1967) to consider that the elasmobranch cerebellum is made up of three main divisions, the cerebellar *corpus,* the *auricles* and the *lateral line lobes,* but we shall see that this is an oversimplification. The corpus, which has the same construction as the mammalian cerebellum, receives no primary projection from the acoustico-lateralis fibers which pass only to the auricles and lateral line lobes. The diagram (Fig. 17-3) shows the auricles as earlike lateral protrusions of the brain which join and bridge the midline to form the "lower lip." It is now clear that this medial portion (pars medialis) is distinguishable in organization and function from the earlike extensions (the *upper* and *lower leaves*). Internally the pars medialis is constructed in the characteristic cerebellar fashion with a molecular layer and Purkinje cells. However, the leaves, although covered by a molecular layer, contain only a few Purkinje cells and are mostly filled with granule cells. The axons of the granule cells within the lower leaf pass caudally to constitute the *cerebellar crest* that covers the lateral line lobes (Paul and Roberts 1977a, Boord 1977).

4 The Central Pathways of the Primary Afferent Fibers

In their passage through the brain the afferent fibers make synaptic connections with medullary neurons that compose the nuclear divisions of the acousticolateralis system (Figs. 17-4 and 17-7).

The posterior lateral line nerve fibers run in the lateral wall of the medulla, just beneath the cerebellar crest, from the posterior limit of the Xth nerve to a rostral

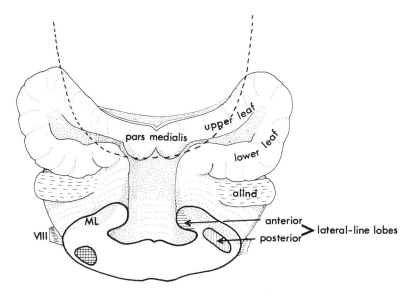

Figure 17-3. Diagram of dorsal view of hindbrain of *Scyliorhinus* to show relationship of auricles and lateral line lobes; the dotted outline indicates the position of the cerebellar corpus.

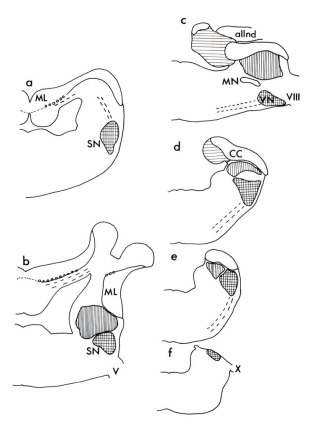

Figure 17-4. Diagrams of representative transverse sections through the hindbrain of *Scyliorhinus* (a-f, rostrocaudal) showing the locations of the acousticolateralis nuclei. The dorsal nucleus is crosshatched; the intermediate nucleus is shown with vertical hatching; the vestibuloacoustic nuclei are crosshatched. Note in (a) and (b) the projection of VIII to pars medialis; in (c), (d), and (e) the projection of VIII to the reticular system. ML, molecular layer; SN, superior nucleus; MN, magnocellularis nucleus; VN, ventral nucleus; CC, cerebellar crest; allnd, dorsal division of the anterior lateral line nerve (based on Boord and Roberts 1980).

position close to the cerebellar peduncle. In the same region of the medullary wall, but lying more medially, are the fibers of the ventral branch of the anterior lateral line nerve; the dorsal branch of this nerve occupies the most medial position of all. The cells and fibers of this medial region constitute the *dorsal nucleus*; the fibers of the posterior nerve and of the ventral anterior nerve project to neurons of the *intermediate nucleus* (Fig. 17-4).

In earlier descriptions (e.g., Kappers, Huber, and Crosby 1936) it was suggested that some of the lateral line fibers projected to the auricles, but field potential experiments were inconclusive about this. However, a recent degeneration study by Koester and Boord (1978) has answered this important quesion by showing that the most lateral granule cells of the lower leaf do receive primary fibers from the posterior lateral line nerve and from the ventral root of the anterior lateral line nerve, whereas the more medial granule cells are supplied by fibers from the dorsal root of the anterior nerve.

The projections of nerve VIII fibers are quite distinct from the lateral line afferent fibers in that they lie most ventrally in the medulla wall in a region which is now subdivided into *superior, ventral* and *magnocellular nuclei* (Fig. 17-4; Montgomery and Roberts 1979). The afferent fibers divide at entry and pass anteriorly as far as the cerebellar auricles and posteriorly to the obex. The relationship with the auricles is particularly interesting because fibers do not enter the leaves but pass only to the granule cells of the pars medialis (Boord and Roberts 1980). It has not yet been possible, either with electrophysiological or anatomical methods, to localize regions within these nuclei which are associated with specific parts of the labyrinth; in particular, we do not know if there is a distinct auditory nucleus in these fishes.

5 Organization of the Secondary Nuclei

The dorsal and intermediate nuclei are covered on their dorsal surface by the cerebellar crest which appears to be homogeneous when examined with the light microscope (Fig. 17-5) but which the electron microscope reveals as being a zone filled with large spiny dendrites (Fig. 17-5), numerous unmyelinated axons, synaptic terminals, and small neurons (Paul, Roberts, and Ryan 1977). The most conspicuous components are the long spiny dendrites which extend from cell bodies (principal neurons) that lie beneath the molecular layer within the confines of the dorsal and intermediate nuclei. The regions between these dendrites are packed with small unmyelinated axons (0.2 μm diameter) that are arranged approximately parallel to each other. We estimate that there are approximately five million of these parallel fibers in this region of *Scyliorhinus*.

Whereas the parallel fibers of the cerebellar corpus run laterally, those of the cerebellar crest are oriented rostrocaudally and may extend for the complete length of the medulla (i.e., around 5 mm). Periodically, en passant, they terminate on spines of the dendrites of the principal cells (Fig. 17-5) and with molecular layer interneurons. These interneurons of the stellate type presumably send axons to the principal cell dendrites.

Additional types of neuron have been described in the lateral line lobes of electric fish and of other teleosts (Rethelyi and Szabo 1973, Caird 1978). Although there are small cells in the elasmobranch lateral line lobes that do not have dendrites that reach the molecular layer, we have not recognized any clear additional neuronal categories.

The cellular organization of the nuclei associated with nerve VIII differs from that of the lateral line lobes primarily in their less close association with the molecular layer, for although dendrites of some of these cells extend to the molecular layer, for the most part, they are locally confined or pass into the intermediate nucleus (Montgomery and Roberts 1979). The neurons within these nuclei are multipolar and exhibit a variety of shapes and sizes, although they cannot easily be divided into distinct categories. In contrast to the dorsal and intermediate nuclei no large principal neuron is easily recognized; the largest neurons within the nerve VIII are those of magnocellularis nucleus (Fig. 17-1), the axons from which pass into Stieda's fasciculus and then descend the spinal cord.

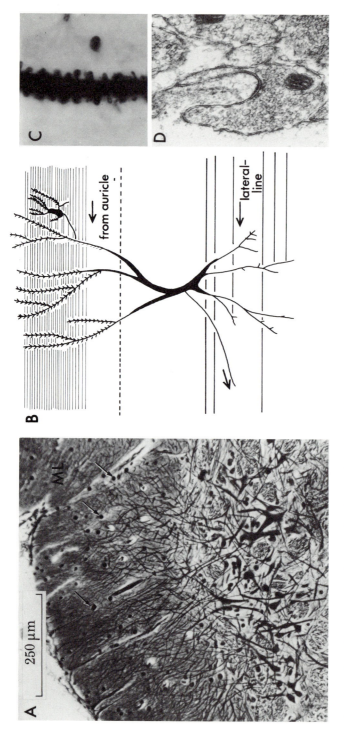

Figure 17-5. The organization of the lateral line lobes. (A) Transverse section through the anterior lateral line lobe showing the molecular layer (ML) covering the dorsal nucleus, the principal neurons of which intermingle. (B) Diagram of arrangement of lateral line lobes, with the spiny dendrites of the principal neurons extending into molecular layer. (C) Light micrograph of Golgi impregnations of molecular layer dendrite of the principal cell, with spines (×1700). (D) Electron micrograph of longitudinal section of principal cell dendrite showing spine bearing a synapse made by parallel fiber axon (×31,600). (A and B based on Paul and Roberts 1977a, C and D from Paul, Roberts, and Ryan 1977.)

6 The Cerebellar-Like Organization of the Acousticolateralis Centers

The similarities in organization between the lateral line (mechano- and electroreceptive) zones and the true cerebellum led Johnston (1902), among others, to suggest that the cerebellum as a whole was phylogenetically elaborated from the medullary regions associated with the lateral line system. However, anatomical studies with the light and electron microscope have shown that the only feature common to the corpus, auricles, and lateral line lobes is the molecular layer, with its numerous parallel fibers, stellate cells, and spiny dendrites of principal neurons. These principal neurons in the lateral line lobes of *Mustelus* were regarded by Houser (1901) as "primitive" Purkinje cells, but there are some important morphological differences, particularly in the submolecular zone (Fig. 17-5). Thus, the cell bodies are multipolar and possess dendrites that lie ventral to the molecular layer. They intermingle rather than form a distinct layer and the spiny dendrites of these neurons overlap with each other rather than lie flattened in one plane, as is so characteristic of cerebellar Purkinje cells. Nevertheless, as in the cerebellum, two afferent inputs impinge on these neurons, provided by the direct lateral line afferent supply and by the parallel fiber pathway from the auricular granule cells.

Although usually associated with motor control, the mammalian cerebellar cortex is involved in sensory processing as it receives major representations from all the senses. Moreover, neural circuits of the complexity and form of the cerebellum are also found in other regions of the brain such as the lateral line lobes, the marginal fiber pyramidal neurons of the teleost optic tectum (Vanegas, Williams, and Freeman 1979), the peduncle lobe of cephalopods (Messenger 1979), and the mammalian hippocampus. The wider occurrence of this particular configuration of neurons suggests, therefore, that it may be utilized for some type of neuronal processing rather than for the production of a specific motor output.

Despite enormous efforts with mammalian material we still have no good explanation for the meaning of the cerebellar circuitry. Considerable emphasis has been placed on the special features of the climbing fiber but the parallel fiber system of long, slowly conducting axons must be highly significant, as it is the feature common to all the cerebellar-type structures. The vast number of synapses present in the cortex has been interpreted as the basis for some kind of memory storage (Marr 1969, Albus 1971). The lateral line lobes would seem to offer good material for further analysis of this problem and could be particularly revealing about the role of the parallel fiber pathway.

7 Processing in the Acousticolateralis Centers

Electrical activation of lateral line nerves and of nerve VIII produces field potentials in the hindbrain (Paul and Roberts 1977b, 1977c, Montgomery and Roberts 1979) which match closely the extent of the nuclei determined in anatomical studies. These fields are produced, however, by the powerful excitatory actions of the afferent fibers on the cell bodies of the principal neurons and there is no sign of corresponding activity of parallel fibers or of molecular layer synapses in the cerebellar crest of the dorsal and intermediate nuclei. The only active processes seen within the molecular layer at these times have been attributed to the propagation of action potentials along the extensive dendritic arborizations of these cells (Fig. 17-6A).

The latency of the evoked fields and of the units recorded within them in response to lateral line nerve stimulation suggests that the input is monosynaptically activating the secondary neurons (Fig. 17-6B). Powerful inputs produce multiple discharges which are followed by long pauses in any spontaneous activity. The mechanism and meaning of this pattern of response remains unclear but a similar silent period has been seen in the response of lobe neurons of the ray *Platyrhinoidis* to an electrical field around the head (Nicholson, Llinas, and Precht 1969).

Many individual neurons in the dorsal nucleus respond to stimulation of both major divisions of the anterior lateral line nerves (superficial opthalmic and buccal) which suggests that there is convergence of the input onto the secondary cells (Fig. 17-6C). Convergence is also indicated by the finding that the secondary neurons show greater variability in their spontaneous activity than is observed in the primary neurons that are the source of the ongoing activity (*Anguilla,* Alnaes 1973), and in the fact that although many afferent fibers have discharges that are strongly modulated by ventilatory movements, such clear patterning is not seen in the activity of the secondary neurons.

The main value of convergence is that it permits comparison between many separate channels so that the signal to noise ratio can be improved and the overall sensitivity enhanced but there is inevitably an accompanying loss of spatial resolution as the

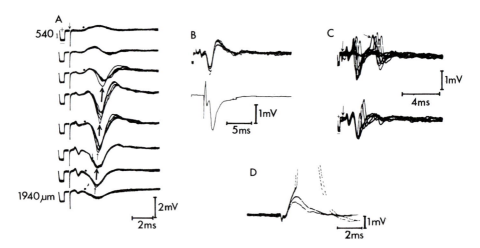

Figure 17-6. (A) Field potential profile recorded at increasing depths of 200 μm stages through the anterior lateral line lobe. Arrow marks a superimposed response that is conducted into molecular layer. (B) Field potential recorded in dorsal nucleus (top) and ventral vestibular nucleus (bottom) in response to stimulation of anterior lateral line nerve and nerve VIII, respectively. (C) Unit in dorsal nucleus evoked by stimulation of superficial opthalmic (top) and buccal (bottom) divisions of anterior lateral line nerve. (D) Intracellular recording from vestibular neuron in response to stimulation of nerve VIII, showing SLDs. (A, B, and C from Paul and Roberts 1977c, B and C from Montgomery 1978.)

receptive fields of the secondary neurons become large (Alnaes 1973). Nevertheless secondary lateral line neurons do retain their directional responses (Caird 1978, Andrianov, Brown, and Ilyinsky 1974) and accurate localization would still be possible if some secondary neurons were to retain small receptive fields or if comparison could be made with information from the cutaneous receptors, which are well tuned to low frequency vibration (Nier 1976).

We have little data on the sensitivity of lateral line organs but it is recognized that the isolated electroreceptors from electrosensory fishes have thresholds to current intensities that are up to two orders of magnitude greater than those determined with behavioral studies for the whole animal. In a recent study of the ampullary end organs of *Scyliorhinus*, Bromm, Hensel, and Tagmat (1976) were able to register threshold currents in the primary afferent discharges similar to those determined for single secondary neurons in the lateral line lobes (Andrianov, Brown, and Ilyinsky 1974) if they computer averaged the signal many times; central convergence within the nervous system would, of course, be a comparable analytical process.

Synchronous excitation of whole nerve bundles by electrical stimulation obviously produces a very abnormal pattern of activation and we might expect to obtain more information about the workings of these centers from the responses to natural stimulation. Unfortunately, there have been very few studies of this type and they have mostly been restricted to the electrosensory fishes (teleosts).

Our information on natural stimulation of the labyrinth is now very complete with respect to the horizontal semicircular canals, with data being available for one species (*Scyliorhinus*) for the behavior of the primary afferent fibers, and vestibular nuclear and cerebellar neurons (Montgomery 1980). There are, however, no published studies on secondary auditory units in elasmobranchs in response to sound. Lowenstein and Roberts (1951) found that primary afferents of the macula utriculi, sacculi, and neglecta would respond to vibrations and follow faithfully frequencies up to 120 Hz, and other authors have commented on the high sensitivity of the macula neglecta (Corwin, Chapter 5). Budelli and Macadar (1979) provide a recent analysis of utricular afferents in *Rhinobatos* and report that some afferents are specialized for acoustic function although they found an overlap of properties with other categories of afferents.

Some of the neurons responding to vibration in *Scyliorhinus* appear to be connected by electrotonic synapses to the afferent fibers for they respond with short latencies to nerve VIII stimulation and intracellular recordings in these neurons (Montgomery 1978) reveal the presence of short latency depolarizations (SLDs), a sign of electrotonic interaction (Fig. 17-6D). Gap junctions, the presumed site of electrotonic interaction, have been observed with the electron microscope in the vestibular nuclei of the toadfish *Opsanus*, where SLDs have also been recorded (Korn, Sotelo, and Bennett 1977). These fast, through-conducting systems probably project to the spinal cord and bring about rapid "escape" movements, which are a fundamental feature of the behavior of aquatic animals. It may well be that the auditory sense as such has developed from this simple stereotyped response to generalized, large amplitude vibration.

There is some literature on auditory unit responses from the brainstem of some teleosts and this suggests that little processing takes place within the hindbrain. It is hard to believe, however, that this would be true for the lateral line system, in view of the complexity of the lateral line lobes.

8 The Efferent Nerve Supply

An important unifying component of the acousticolateralis system, and one which may be important for signal processing, is the efferent supply to the hair cells. For the elasmobranchs an efferent innervation has been reported for all of the various types of end organ, except for the pit organ and the receptors of the ampullae of Lorenzini. An efferent supply is a feature that is found in the labyrinths of other species, including the mammalian cochlea, and in most cases examined so far the efferent supply has been found to have an inhibitory impact on the hair cells. In the case of the dogfish body lateral line, for example, electrical stimulation of the efferent nerve reduces or abolishes spontaneous afferent activity (Russell and Roberts 1972).

The location of the cell bodies that provide the efferent supply has proved elusive although Paul and Roberts (1977d) reported that the efferent neurons for the head mechanoreceptors of the dogfish resided within the dorsal nucleus. However, on several grounds this suggestion appears to be unlikely, particularly now that recent studies of other fishes with HRP have indicated that the efferent cell bodies for both lateral line and labyrinthine receptors lie clustered close to the motor columns of the medulla (see McCormick, Chapter 18).

The activity of the efferent neurons has been closely studied in the case of the lateral line mechanoreceptors of dogfish where it was found that the normally quiescent efferent neurons become active when the fish was stimulated in such a way that body movement resulted (i.e., with cutaneous and vestibular stimulation) and the efferent neurons were rhythmically active when the fish was swimming steadily (Roberts and Russell 1972).

It also became clear from these studies that during steady movement of the fish the efferent system is insufficiently active to bring about much reduction in lateral line sensitivity and would not counteract the rhythmical afferent activity generated during locomotion. However, during more vigorous movements of the kind used for escape or attack the efferent neurons discharge at rates that are capable of modifying lateral line sensitivity. As it is also evident that at these times body movement would strongly stimulate the lateral line organs and possibly deplete the afferent synapses it has been suggested that the inhibitory action of the efferent system might be to prevent this so that the lateral line system retained maximum sensitivity the moment the violent movements had ceased (see Roberts 1978).

Little new data on efferent action have been acquired since this interpretation of efferent function was developed and it is particularly unfortunate that we have little information on the natural activity of efferent neurons in other systems, although work on the vestibular efferent fibers of the frog and goldfish also indicates a relationship between the efferent system and body movement (see Klinke and Galley 1974).

Much experimental work on the labyrinth, both in relation to hearing and balance, is done without recognition of the presence of an efferent system and the possible complications efferent action might impose on the experimental results. For example, in neurophysiological studies made on paralyzed animals there is usually no way of knowing the motor state of the animal and yet it is possible that at least at times the efferent system would be active and so modifying the responses of the hair cells.

9 Summary and Conclusions

This survey of the organization of the secondary nuclei of the acousticolateralis system reveals some significant differences between the various sensory components and some common features. The primary afferents distribute to the hindbrain as summarized in Fig. 17-7 and supply the secondary neurons that form distinct nuclei confined to the ipsilateral hindbrain. The projections of the different systems do not overlap so although the dendrites of some secondary neurons intrude into adjacent projection zones and permit some interaction, processing of the various peripheral inputs occurs separately at this stage; any interchange between these senses most probably takes place therefore at higher levels in the brain. Apart from the separation of lateral line mechanoreceptors from the electroreceptors in the intermediate and dorsal nuclei we cannot yet recognize divisions of the nuclei associated with specific peripheral end-organs nor do we know whether these projections map, in any somatotopic sense, their distribution over the body.

The centers for mechanoreceptors and electroreceptors lying in the lateral line lobes are so alike in their construction, involving an elaborate cerebellar-like molecular layer, that it is probable that they operate similarly and provide a complementary service to the animal. In contrast, the majority of the labyrinthine secondary neurons are not associated with this molecular layer but until we know whether there is a distinct auditory nucleus in elasmobranch fishes and what its relation, if any, is with the cerebellar crest we cannot be sure that the analysis performed by the secondary neurons of the lateral line system is so very different from that carried out by the auditory neurons. However, we can be certain, on the basis of the number of nerve fibers involved (Table 17-1) and after recognizing that many of nerve VIII secondary neurons must relate to vestibular function, that far less of the medulla is devoted to auditory analysis than is involved in handling of information obtained from the lateral line organs.

During the last decade we have begun to accumulate for the elasmobranchs fairly full information on the anatomical organization of this region of the brain. Some anatomical problems remain, of course, particularly in relation to secondary and higher

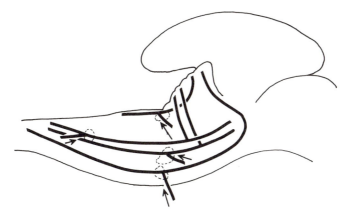

Figure 17-7. Diagram summarizing the projections of the acousticolateralis nerves to the brain. This figure should be compared with Fig. 17-2.

projections, the nature of other inputs to the acoustico-lateral centers and especially to the granule cells of the auricular leaves, the location of the efferent supply and the question of the existence of a distinct auditory nucleus. Nevertheless, the information we now have provides a good basis for neurophysiological exploration which should now be pursued particularly by means of dye-filled microelectrodes so that the electrical properties may be precisely correlated with morphologically specified neurons. With the use of natural stimulation we should be able to collect information about somatotopic organization and explore the possibilities of interaction between the different end-organs, particularly between the electro- and mechanoreceptors, and so obtain some understanding of the meaning of the cerebellar-like construction of this region.

Acknowledgments. I am very grateful to Robert Boord, John Montgomery and Derek Paul for their collaboration in the work described in this chapter and to Roddy Williamson for his comments on the manuscript.

References

Albus, J. S.: A theory of cerebellar function. Month. Biosci. 10, 25-61 (1971).

Allis, E. P.: The lateral sensory canals, the eye-muscles, and the peripheral distribution of certain of the cranial nerves of *Mustelus laevis*. Q. J. Microsc. Sci. 45, 87-236 (1901).

Alnaes, E.: Unit activity of ganglionic and medullary second order neurones in the eel lateral line system. Acta Physiol. Scand. 88, 160-174 (1973).

Andrianov, G. N., Brown, H. R., Ilyinsky, O. B.: Responses of central neurons to electrical and magnetic stimuli of the ampullae of Lorenzini in the Black Sea skate. J. Comp. Physiol. 93, 287-299 (1974).

van Bergeijk, W. A.: Directional and nondirectional hearing in fish. In: Marine Bio-Acoustics. Tavolga, W. N. (ed.). Oxford: Pergamon Press, 1964.

Boord, R. L.: Auricular projections in the clearnose skate, *Raja eglanteria*. Am. Zool. 17, 887 (1977).

Boord, R. L., Roberts, B. L.: Medullary and cerebellar projections of the statoacoustic nerve of the dogfish, *Scyliohinus canicula*. J. Comp. Neurol. 193, 57-68 (1980).

Bromm, B., Hensel, H., Tagmat, A. T.: The electrosensitivity of the isolated ampulla of Lorenzini in the dogfish. J. Comp. Physiol. A 111, 127-136 (1976).

Budelli, R., Macadar, O.: Statoacoustic properties of utricular afferents. J. Neurophysiol. 42, 1479-1493 (1979).

Caird, D. M.: A simple cerebellar system: The lateral line lobe of the goldfish. J. Comp. Physiol. 127, 61-74 (1978).

Campbell, C. B. G., Boord, R. L.: Central pathways of the posterior lateral line nerve in the shark *Mustelus canis*. Am. Zool. 11, 703 (1971).

Corwin, J. T.: Morphology of the macula neglecta in sharks of the genus *Carcharhinus*. J. Morphol. 152, 341-362 (1977).

Enin, L. D., Ilyinsky, O. B.: Representation of the lateral line nerves in the cerebellum of skates. Neurophysiology 4, 192-200 (1972).

Enin, L. D., Ilyinsky, O. B., Volkova, N. K.: Peculiarities in functional organization of the projection zones of the lateral-line organs in the ray midbrain. Neurophysiology 5, 384-391 (1973).

Flock, A.: Electronmicroscopic and electrophysiological studies on the lateral line canal organ. Acta Oto-lar. Suppl. 1-90 (1965).

Görner, P.: Source localization with labyrinth and lateral line in the clawed toad (*Xenopus laevis*). In: Sound Reception in Fish. Schuijf, A., Hawkins, A. D. (eds.). Amsterdam: Elsevier, 1976, pp. 171-184.

Houser, G. L.: The neurons and supporting elements of the brain of a selachian. J. Comp. Neurol. 11, 65-175 (1901).

Ilyinsky, O. B., Enin, L. D., Volkova, N. K.: Electrical activity evoked in the medulla of *R. clavata* by stimulation of the lateral line nerves. Neurophysiology 3, 284-292 (1971).

Johnston, J. B.: The brain of *Petromyzon*. J. Comp. Neurol. 7, 2-82 (1902).

Kappers, C. U. A., Huber, C. C., Crosby, E. C.: The Comparative Anatomy of the Nervous System of Vertebrates Including Man. New York: Macmillan, 1936.

Katsuki, Y., Yoshino, S., Chen, J.: Action currents of the single lateral-line nerve fiber of fish. I. On the spontaneous discharge. Jpn. J. Physiol. 1, 87-99 (1950).

Klinke, R., Galley, N.: Efferent innervation of vestibular and auditory receptors. Physiol. Rev. 54, 316-357 (1974).

Koester, D. M., Boord, R. L.: The central projections of first order anterior lateral line neurons of the clearnose skate, *Raja eglanteria*. Am. Zool. 18, 587 (1978).

Korn, H., Sotelo, C., Bennett, M. V. L.: The lateral vestibular nucleus of the toadfish *Opsanus tau*: Ultrastructural and electrophysiological observations with special reference to electrotonic transmission. Neuroscience 2, 851-884 (1977).

Larsell, O.: The comparative anatomy and histology of the cerebellum from myxinoids through birds. Jansen, J. (ed.). Minneapolis: Univ. Minnesota, 1967.

Lowenstein, O., Roberts, T. D. M.: The localization and analysis of the responses to vibration from the isolated elasmobranch labyrinth. A contribution to the problem of the evolution of hearing in vertebrates. J. Physiol. (London) 114, 471-489 (1951).

Marr, D.: A theory of cerebellar cortex. J. Physiol. (London) 202, 437-470 (1969).

McCready, P. J., Boord, R. L.: The topography of the superficial roots and ganglia of the anterior lateral line nerve of the smooth dogfish, *Mustelus canis*. J. Morphol. 150, 527-538 (1976).

Messenger, J.: The nervous system of *Loligo*. IV. The peduncle and olfactory lobes. Phil. Trans. Roy. Soc. B 285, 275-309 (1979).

Montgomery, J. C.: Dogfish vestibular system: Functional organization of the vestibular nuclei and vestibulo-cerebellum with special reference to horizontal semicircular canal input. Ph.D. Thesis, University of Bristol, 1978.

Montgomery, J. C.: Dogfish horizontal canal system: Responses of primary afferent, vestibular and cerebellar neurons to rotational stimulation. Neuroscience 5, 1761-1769 (1980).

Montgomery, J. C., Roberts, B. L.: Organization of vestibular afferents to the vestibular nuclei of the dogfish. Brain Behav. Evol. 16, 81-98 (1979).

Murray, M. J., Capranica, R. R.: Spike generation in the lateral line afferents of *Xenopus laeris*: Evidence favouring multiple sites of initiation. J. Comp. Physiol. 87, 1-20 (1973).

Myrberg, A. A., Jr., Gordon, C. R., Klimley, A. P.: Attraction of free-ranging sharks by acoustic signals in the near-subsonic range. Tech. Rep. Rosenstiel Sch. Mar. Atmos. Sci., Univ. Miami, No. TR75-4, 40 pp. (1975).

Nier, K.: Cutaneous sensitivity to touch and low frequency vibration in selachians. J. Comp. Physiol. 109, 345-355 (1976).

Nicholson, C., Llinas, R., Precht, W.: Neural elements of the cerebellum in elasmobranch fishes: Structural and functional characteristics. In: Neurobiology of Cerebellar Evolution and Development. Llinas, R. (ed.). Chicago: American Medical Association, 1969, pp. 215-243.

Paul, D. H., Roberts, B. L.: Studies on a primitive cerebellar cortex. I. The anatomy of the lateral-line lobes of the dogfish *Scyliorhinus canicula*. Proc. R. Soc. London Ser. B 195, 453-466 (1977a).

Paul, D. H., Roberts, B. L.: Studies on a primitive cerebellar cortex. II. The projection of the posterior lateral-line nerve to the lateral-line lobes of the dogfish brain. Proc. R. Soc. London Ser. B 195, 467-478 (1977b).

Paul, D. H., Roberts, B. L.: Studies on a primitive cerebellar cortex. III. The projection of the anterior lateral-line nerve to the lateral-line lobes of the dogfish brain. Proc. R. Soc. London Ser. B 195, 479-496 (1977c).

Paul, D. H., Roberts, B. L.: The location and properties of the efferent neurons of the head lateral line organs of dogfish. J. Comp. Physiol. 116, 117-127 (1977d).

Paul, D. H., Roberts, B. L., Ryan, K. P.: Comparisons between the lateral line lobes of the dogfish and the cerebellum: An ultrastructural study. Z. Hirnforsch. 18, 335-343 (1977).

Rethelyi, M., Szabl, T.: Neurohistological analysis of the lateral lobe in a weakly electric fish, *Gymnotus carapo* (Gymnotidae, Pisces). Exp. Brain Res. 18, 323-339 (1973).

Roberts, B. L.: Mechanoreceptors and the behaviour of elasmobranch fishes with special reference to the acoustico-lateralis system. In: Sensory Biology of Sharks, Skates, and Rays. Hodgson, E. S., Mathewson, R. W. (eds.). Arlington: Office of Naval Research, 1978, pp. 331-390.

Roberts, B. L., Russell, I. J.: The activity of lateral-line efferent neurons in stationary and swimming dogfish. J. Exp. Biol. 57, 435-448 (1972).

Roberts, B. L., Ryan, K. P.: The fine structure of the lateral-line sense organs of dogfish. Proc. R. Soc. London Ser. B 179, 157-169 (1971).

Russell, I. J., Roberts, B. L.: Inhibition of spontaneous lateral-line activity by efferent nerve stimulation. J. Exp. Biol. 57, 77-82 (1972).

Russell, I. J., Roberts, B. L.: Active reduction of lateral-line sensitivity in swimming dogfish. J. Comp. Physiol. 94, 7-15 (1974).

Sand, A.: The mechanism of the lateral sense organs of fishes. Proc. R. Soc. London Ser. B 123, 472-495 (1937).

Schuijf, A.: Directional hearing of cod (*Gadus morhua*) under approximate free field conditions. J. Comp. Physiol. 98, 307-332 (1975).

Tester, A. L., Kendall, J. I.: Morphology of the lateralis canal system in the shark genus *Carcharhinus*. Pac. Sci. 23, 1-16 (1969).

Tester, A. L., Nelson, G. J.: Free neuromasts (pit organs) in sharks. In: Sharks, Skates and Rays. Gilbert, P. W., Mathewson, R. F., Rall, D. P. (eds.). Baltimore: Johns Hopkins Press, 1967, pp. 503-531.

Thurber, J.: The Beast in Me and other Animals. New York: Hamish Hamilton, 1949.

Vanegas, H., Williams, B., Freeman, J. A.: Responses to stimulation of marginal fibers in the teleostean optic tectum. Exp. Brain Res. 34, 335-349 (1979).

Discussion

BULLOCK: In talking of cortical organization in the brain, I might mention the cochlear nucleus of mammals. This is especially true in lower primates and many rodents. This cortical structure of molecular and enlarged cells can be associated with acoustic function. In the shark, this acoustical structure may be quite separate from the other parts of the eighth nerve, both anatomically and physiologically.

ROBERTS: Some of the cells in this region of the shark's brain do have molecular layer dendrites. I would not exclude the possibility that there is a discrete nucleus, like the dorsal and intermediate, associated with the molecular layer.

BULLOCK: Primate neurology teaches us that one can expect differences even at the laboratory microscope level, gross anatomical differences among species in the same order. There are important differences in lamination and development of granule cells between anthropoids and prosimians. We should expect such differences between much more diverse orders or classes of vertebrates. In order to get the cerebellum to do more, it may be what it takes is not necessarily another modality, but a different form of information from the same modality. In microelectrode work on the cerebellum in mammals, organization is on a very fine level. Even adjacent tracks may be strikingly different in the peripheral receptive field, sometimes only fifty microns apart. The cerebellum could be doing some kind of analog computation in order to inform the motor system what to do.

NORTHCUTT: Have you looked for any particular topographic organization?

ROBERTS: No, we haven't, but that is obviously an important question that should be investigated. In elasmobranchs, there is considerable convergence of the input. However, there is no loss of directional response at the primary or secondary level. We don't know if there is some sort of map of the body within the system.

NORTHCUTT: From the work of McCormick (Chapter 18), there are evidently projections of the anterior and posterior lateral line, and these may serve as maps simply by going to different portions of the medial nucleus.

BULLOCK: Do the electroreceptors go to one part and the mechanical to another so that each has, in effect, a private line?

ROBERTS: Yes.

PLATT: Do you find any overlap between the intermediate and other nuclei?

ROBERTS: Not in the projection of the primary fibers, but the dendrites of the cells in these nuclei are massive. Dendrites of eighth nerve secondary neurons go into the intermediate nucleus, and intermediate nuclear neuron dendrites go into subacoustic areas, so there is the possibility of interaction.

CORWIN: In some of the work that Northcutt and I did, we used HRP to trace primary projections of eighth nerve fibers. It was hard to determine if there are terminals that synapse in the mechanoreceptor nucleus, or if they proceed beyond that to the medialis of the cerebellum.

Chapter 18

Comparative Neuroanatomy of the Octavolateralis Area of Fishes

Catherine A. McCormick*

The octavolateralis area of fishes has only recently begun to be explored using experimental anatomical techniques. Such studies have concluded that the nuclear organization of the octavolateralis area and the distribution of primary inputs to these nuclei differ substantially from that reported in classic studies of this area. This chapter briefly reviews the organization of the octavolateralis area in light of these recent data, and discusses some of the variations seen within it.

In petromyzontid agnathans, chondricthyans, and the holostean *Amia calva,* lateral line input to the octavolateralis area is largely restricted to the nuclear groups within the dorsal portion of the octavolateralis area, whereas eighth nerve input distributes to a group of more ventrally located nuclei (Table 18-1). It is thus reasonable to hypothesize that the total or nearly total segregation of lateralis and octavus input at the first order level is a generalized, or primitive character (McCormick 1978, 1982). Given this, one can search at the most basic level for two general classes of variation: (1) variation in the nuclear organization of the dorsal lateralis column and the ventral octavus column and (2) variation in the distribution of lateralis and octavus inputs within the octavolateralis area, such that significant overlap of inputs is seen.

Both the lateralis and octavus cell columns show differences in their nuclear subdivisions across taxa. A comparative study of the lateralis cell column reveals that it is composed of at least two nuclei, a nucleus medialis anteriorly, and a nucleus caudalis posteriorly[1] (McCormick 1982; Table 18-1). In addition to nucleus medialis, other nuclei may be present in the anterior portion of the lateralis column. McCormick (1982) recognized three patterns of variation in the organization of the lateralis column (Fig. 18-1), and proposed that these variations reflect the presence or absence of an electroreceptive portion of the lateralis system. Patterns a and c appear to be characteristic of fishes that are known or are believed to be electroreceptive. In pattern a, the rostral lateralis column is composed of nucleus medialis and nucleus dorsalis.

*Department of Anatomy, Georgetown University, Schools of Medicine and Dentistry, Washington, D.C. 20007.

[1]Nucleus caudalis has not been recognized in petromyzontid agnathans (Northcutt 1979a).

Table 18-1. Summary of the Distribution Patterns of the Lateral Line and Eighth Nerves in the Medulla of Some Fishes[a]

	Agnatha Icthyomyzon	Chondrichthyes Mustelus; Platyrhinoides; Scyliorhinus	Osteichthyes Amia	Osteichthyes Gillichthys	Osteichthyes Gnathonemus, Brienomyrus
1° lateralis input	Nuc. dorsalis[b] Nuc. medialis[b]	Nuc. dorsalis[b,d] (anterior lateral line lobe) Nuc. medialis[b,d] (posterior lateral line lobe) Nuc. caudalis[e]	Nuc. medialis[g] Nuc. caudalis[g]		Anterior lateral line lobe[i,j,k] Posterior lateral line lobe[i,j,k] Nuc. octavius[k,l] Nuc. tangentalis[k,l] Nuc. magnocellularis[k,l]
1° octavus input	Nuc. ventralis[c] Anterior octavomotor nuc.[c] Intermediate octavomotor nuc.[c] Posterior octavomotor nuc.[c]	Anterior (superior[f]) octavus nuc.[b] Descending (ventral[f]) octavus nuc.[b] Posterior octavus nuc.[b,f] Nuc. magnocellularis[b,f] (medial[b]) reticular formation[f]	Anterior octavus nuc.[g] Descending octavus nuc.[g] Posterior octavus nuc.[g] Nuc. magnocellularis[g] Medial reticular formation[g]	Anterior octavus nuc.[b,h] Descending octavus nuc.[b,h] Posterior octavus nuc.[b,h] Nuc. magnocellularis[b,h] Nuc. tangentalis[b,h] Medial reticular formation[b,h]	Anterior lateral line lobe[l] Nuc. octavius[j,k,l] Nuc. tangentalis[j,k,l] Nuc. descendens[l] Nuc. magnocellularis[l] Reticular formation[l]

[a] In Amia[g] and Platyrhinoides[b], nucleus magnocellularis may also receive a sparse lateral line input. This table does not reflect hypotheses concerning homologies among the octavolateralis nuclei of different fishes. Similarly named nuclei in different taxa are not necessarily homologous structures (see McCormick 1981a). [b] Northcutt 1980. [c] Northcutt 1979a. [d] Boord and Campbell 1977. [e] Small granule cells in the caudal portion of nucleus caudalis in Mustelus; Boord per. comm. [f] Boord and Roberts 1980; small granule cells in the caudal portion of nucleus ventralis constitute the posterior octavus nucleus in Scyliorhinus. [g] McCormick 1981. [h] Northcutt 1979b. [i] Maler, Karten, and Bennett 1973a. [j] Maler, Karten, and Bennett 1973b. [k] Bell and Russell 1978. [l] Bell 1981.

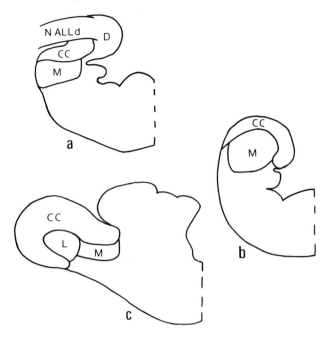

Figure 18-1. Illustration of the three organizational patterns within the lateralis area of fishes. (a) This pattern is characterized by two longitudinal lateralis columns, represented at this level by nucleus medialis and nucleus dorsalis. Nucleus dorsalis comprises the medial lateralis column and is innervated by the dorsal ramus of the anterior lateral line nerve. This pattern is illustrated with a transverse section through the medulla of a chondricthyan, *Squalus acanthias*. This pattern of organization is also present in petromyzontids, polypteriforms, and chondrosteans. (b) This pattern is characterized by a single lateralis column, represented at this level by nucleus medialis. Neither nucleus dorsalis nor the dorsal ramus of the anterior lateral line nerve is present. This pattern is illustrated with a transverse section through the medulla of a holostean, *Amia calva*. This pattern of organization is also present in nonelectroreceptive teleosts. (c) This pattern is characterized by two longitudinal lateralis columns, represented at this level by nucleus medialis and the lateral line lobe. The lateral line lobe comprises the lateral or dorsolateral lateralis column. This pattern is characteristic of electroreceptive teleosts and is illustrated with a transverse section through the medulla of *Ictalurus punctatus*. Abbreviations: CC, cerebellar crest; D, nucleus dorsalis; L, lateral line lobe; M, nucleus medialis; N ALLd, dorsal ramus of the anterior lateral line nerve.

This pattern is found in petromyzontid agnathans, chondricthyans, polypteriforms, and chondrosteans, and thus appears to represent a primitive condition. In chondricthyans, nucleus dorsalis is the termination site of first order fibers from electroreceptors, while primary mechanoreceptor fibers terminate in nucleus medialis and nucleus caudalis (McCready and Boord 1976, Boord and Campbell 1977, Boord pers. comm., Bodznick and Northcutt 1980). Since recent studies have indicated the presence of an electroreceptive ability in at least some members of each of the aforementioned taxa (Pfeiffer 1968, Jørgenson, Flock, and Wersäll 1972, Teeter and Bennett 1976,

Bodznick and Northcutt 1981), it can be hypothesized that the primary electroreceptive and mechanoreceptive inputs in these fish will be found to be segregated, as in chondricthyans, with nucleus dorsalis being a specialized electroreceptive area.

Nucleus dorsalis is not present in the later actinopterygian radiations—the Holostei and the Teleosti—and its absence apparently reflects the loss of the primitive electroreceptive system. Patterns b and c illustrate the organization of the lateralis area in these fish. In pattern b (Fig. 18-1b), nucleus medialis is the only nucleus present in the rostral portion of the lateralis column. This pattern of organization is seen in actinopterygians known or believed to be nonelectroreceptive—the Holostei and most orders of Teleostei.

Pattern c (Fig. 18-1c) is present in three orders of electroreceptive teleosts, the Mormyriformes, the Siluriformes, and the Gymnotidei. In addition to the nucleus medialis, a lateral differentiation of the rostral portion of the lateralis column is present in these fish, and is here termed the lateral line lobe. In a mormyrid and a gymnotid, nucleus medialis and the lateral line lobe have been shown to be first-order mechanoreceptive and electroreceptive areas, respectively (Maler, Finger, and Karten 1974, Bell and Russell 1978). It is likely that this segregation of inputs also characterizes the lateralis area of the Siluriformes. Thus, the lateral line lobe is apparently a specialized electroreceptive area in these three orders of teleosts, each of which is believed to have evolved electroreception independently (Bullock 1974).

Thus, nucleus medialis and, with the exception of agnathans, nucleus caudalis, appear to be present in all fishes, in accordance with the universal presence of the mechanoreceptive portion of the lateralis system. The electroreceptors of primitive fishes are hypothesized to terminate in nucleus dorsalis, located medial or dorsomedial to nucleus medialis. Among the teleosts, those orders that have independently developed electroreception have each also independently hypertrophied a lateral portion of the lateralis column, forming the lateral line lobe. The presence of a specialized portion of the lateralis system, the electroreceptors, is thus reflected centrally by the presence of additional nuclear groups within the lateralis cell column. It can be anticipated that future studies of the lateralis system, especially in teleosts, may reveal more electroreceptive species, and potentially species that possess other types of specializations of the lateralis system, both peripherally and centrally.

Experimental anatomical studies on the projections of the eighth nerve are available for a petromyzontid agnathan, chondricthyans, and several actinopterygians. With the exception of the agnathan (Northcutt 1979a), the organization of the octavus cell column in the majority of fishes studied thus far shows less variation than does the lateralis column. Four first-order octavus nuclei have been experimentally identified in chondricthyans, the holostean *Amia calva,* and a teleost, *Gillichthys mirabilis*: the anterior, magnocellular, descending, and posterior octavus nuclei (Table 18-1). These four nuclei were also identified in a nonexperimental analysis of the octavus cell column in polypteriforms, a chondrostean, and teleosts from each of Greenwood et al.'s (1966) three divisions (Fig. 18-2; McCormick 1982). In addition, both experimental (Northcutt 1979b, 1980, Bell 1981) and nonexperimental (Pearson 1936, Larsell 1967, McCormick 1982) studies in teleosts describe a fifth octavus nucleus not seen in nonteleost fish, nucleus tangentialis (Fig. 18-2, Table 18-1).

Based on these comparative data, it is possible to hypothesize that the octavus area of chondricthyans, nonteleost actinopterygians, and at least some orders of teleosts

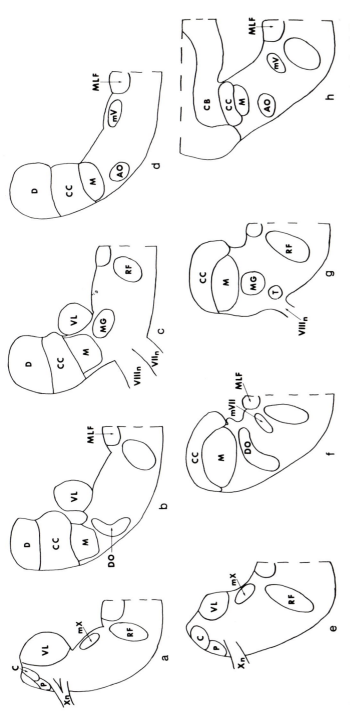

Figure 18-2. Illustration of the organization of the octavus cell column in some gnathostome fishes; transverse sections. (a-d) Illustration of the octavolateralis area of the sturgeon, *Scaphirhynchus platorynchus*, showing the position of the (a) posterior octavus nucleus, (b) descending octavus nucleus, (c) magnocellular nucleus, and (d) anterior octavus nucleus. The octavus column is similarly organized in chondrichthyans, polypteriforms, and holosteans. (e-h) Illustration of the octavolateralis area of the trout, *Salmo gairdneri*, showing the position of the (e) posterior octavus nucleus, (f) descending octavus nucleus, (g) magnocellular nucleus and tangential nucleus, and (h) anterior octavus nucleus. This pattern of organization appears to be typical of many teleosts. Abbreviations: AO, anterior octavus nucleus; C, nucleus caudalis; CB, cerebellum; CC, cerebellar crest; D, nucleus dorsalis; DO, descending octavus nucleus; M, nucleus medialis; MG, nucleus magnocellularis; MLF, medial longitudinal fasciculus; mV, trigeminal motor nucleus; mVII, facial motor nucleus; mX, motor nucleus of the vagus nerve; P, posterior octavus nucleus; RF, reticular formation; T, nucleus tangentialis; VIIn, facial nerve; VIIIn, eighth nerve; VL, vagal lobe; Xn, vagus nerve.

contains at least four first-order nuclei, with teleosts exhibiting an additional nucleus tangentialis. However, the enormous radiation of the teleosts necessarily limits generalizations about these fishes, and it is likely that specializations of the octavus cell column, and of the octavolateralis area in general, will be found. For example, close peripheral contact between the lateralis and otic systems in the Clupeidae has been recently demonstrated (e.g., Blaxter, Denton, and Gray, Chapter 2). Such peripheral specializations might be paralleled by specializations in the central connections of the lateralis and octavus nerves and/or in the general organization of the octavolateralis area. For example, central specializations of the octavolateralis area have been demonstrated in mormyrids (Bell and Russell 1978, Bell 1981, Chapter 19). In this group, lateral line and eighth nerve inputs are not segregated in distinct columns as in other fishes, but rather show some degree of overlap within almost all of the nuclei of the octavolateralis area (Table 18-1). Since earlier radiations of fishes do not show such overlap, mormyrids probably exhibit a derived pattern of organization of the octavolateralis area, possibly relating to specializations at the peripheral level. It is interesting to note that although classic studies (e.g., Pearson 1936, Larsell 1967) considered central overlap of first-order octavus and lateralis fibers to be a generalized feature of the octavolateralis area, on the basis of recent studies (Table 18-1) it seems more plausible to consider this as a derived condition.

References

Bell, C. C.: Central distribution of octavolateral afferents and efferents in a teleost (Mormyridae). J. Comp. Neurol. 195, 391-414 (1981).

Bell, C. C., Russell, C. J.: Termination of electroreceptor and mechanical lateral line afferents in the mormyrid acousticolateral area. J. Comp. Neurol. 182, 367-382 (1978).

Bodznick, D., Northcutt, R. G.: Segregation of electro- and mechanoreceptive inputs to the elasmobranch medulla. Brain Res. 195, 313-321 (1980).

Bodznick, D., Northcutt, R. G.: Electroreception in lampreys: evidence that the earliest vertebrates were electroreceptive. Science 212, 465-467 (1981).

Boord, R. L., Campbell, C. B. G.: Structural and functional organization of the lateral line system of sharks. Am. Zool. 17, 431-441 (1977).

Boord, R. L., Roberts, B. L.: Medullary and cerebellar projections of the statoacoustic nerve of the dogfish, *Scyliorhinus canicula*. J. Comp. Neurol. 193, 57-68 (1980).

Bullock, T. H.: An essay on the discovery of sensory receptors and the assignment of their functions together with an introduction to electroreceptors. In: Handbook of Sensory Physiology, Vol. III/3. Fessard, A. (ed.). New York: Springer-Verlag, 1974, pp. 1-12.

Greenwood, P. H., Rosen, D. E., Weitzman, S. H., Myers, G. S.: Phyletic studies of teleostean fishes, with a provisional classification of living forms. Bull. Am. Mus. Nat. Hist. 131, 339-456 (1966).

Jørgenson, J. M., Flock, Å., Wersäll, J.: The Lorenzian ampullae of *Polyodon spatula*. Z. Zellforsch. 130, 362-377 (1972).

Larsell, O.: *The Comparative Anatomy and Histology of the Cerebellum from Myxinoids through Birds.* Jansen, J. (ed.). Minneapolis: Univ. Minnesota Press, 1967.

Maler, L., Karten, H. J., Bennett, M. V. L.: The central connections of the posterior lateral line nerve of *Gnathonemus petersi*. J. Comp. Neurol. 151, 57-66 (1973a).

Maler, L., Karten, H. J., Bennett, M. V. L.: The central connections of the anterior lateral line nerve of *Gnathonemus petersi*. J. Comp. Neurol. 151, 67-84 (1973b).

Maler, L., Finger, T., Karten, H. J.: Differential projections of ordinary lateral line receptors and electroreceptors in the gymnotid fish, *Apteronotus (Sternarchus) albifrons*. J. Comp. Neurol. 158, 363-382 (1974).

McCormick, C. A.: Central projections of the lateralis and eighth nerves in the bowfin, *Amia calva*. Doctoral Dissertation, Univ. Michigan, 1978.

McCormick, C. A.: Organization and evolution of the octavolateralis area of fishes. In: Fish Neurobiology, Vol. 1. Northcutt, R. G., Davis, R. E. (eds.). Ann Arbor: Univ. Michigan Press, 1981a. (in press)

McCormick, C. A.: Central projections of the lateral line and eighth nerves in the bowfin, *Amia calva*. J. Comp. Neurol. 197, 1-15 (1981).

McCormick, C. A.: Organization and evolution of the octavolateralis area of fishes. In: Fish Neurobiology, Vol. 1. Northcutt, R. G., Davis, R. E. (eds.). Ann Arbor: Univ. Michigan Press, 1982.

Northcutt, R. G.: Central projections of the eighth cranial nerve in lampreys. Br. Res. 167, 163-167 (1979a).

Northcutt, R. G.: Primary projections of VIII nerve afferents in a teleost, *Gillichthys mirabilis*. Anat. Rec. 193, 638 (1979b).

Northcutt, R. G.: Central auditory pathways in anamniotic vertebrates. In: Comparative Studies of Hearing in Vertebrates. Popper, A. N., Fay, R. R. (eds.). New York: Springer-Verlag, 1980, pp. 75-118.

Pearson, A. A.: The acoustico-lateralis centers and the cerebellum, with fiber connections, of fishes. J. Comp. Neurol. 65, 201-294 (1936).

Pfeiffer, W.: Die Fahrenholzchen Organe der Dipnoi und Brachiopterygii. Z. Zellforsch. 90, 127-147 (1968).

Teeter, J. H., Bennett, M. V. L.: Ampullary electroreceptors in sturgeon. Soc. Neurosci. Abstr. 2, 185 (1976).

Discussion

BELL: Have you looked at any clupeid fish?

McCORMICK: No, I have not, but this would be a good group in which to look for variations in primary connections of the octavus and lateralis nerves, and in the general organization of the octavolateralis area. We expect the peripheral specializations (as discussed by Blaxter, Gray, and Denton, Chapter 2) to be reflected in some way in the central nervous system.

Chapter 19

Some Central Connections of Medullary Octavolateral Centers in a Mormyrid Fish

Curtis C. Bell*

1 Introduction

Mormyrid fish have a small gas bladder on each side of the head which abuts the sacculus. As von Frisch (1938) and Stipetic (1939) indicated, the gasbladder probably functions as a pressure to displacement transducer, allowing the fish to sense sound pressure waves. The close apposition to the gasbladder thus implies a hearing role for the sacculus in mormyrids. For most fish, behavioral studies also indicate a role in hearing for the sacculus, and in equilibrium for the utriculus and canals (Popper and Fay 1973).

The central projections of the three otolithic branches of the eighth nerve, the three semicircular canal branches, and the two lateral line nerves have recently been described in mormyrids (Bell 1981). The utricular and canal nerves, but not those from the other end-organs, project densely to the descending, tangential, and magnocellular nuclei. The sacculus nerve, however, projects most densely onto a medial zone of the anterior lateral line lobe (ALLL) (see Maler, Karten, and Bennett 1973, and Bell 1981 for descriptions of medullary nuclei and of ALLL). Afferents from the utriculus, lagena, and the lateral line nerves also end in this lobe but in different zones. Afferents from the semicircular canals to ALLL are sparse.

These patterns of projection and the probable functions of the different eighth nerve end-organs, referred to above, suggest the hypothesis that the equilibrium or vestibular components of the eighth nerve termination region lies within the descending, tangential, or magnocellular nuclei, and that the auditory component includes the ALLL. Utricular, lagenar, and lateral line input to ALLL could be explained if such afferents signal water movement or the displacement component of sound. The possibility of dual functions for eighth nerve end-organs in fish has been suggested previously (see Popper and Fay 1973). The present chapter examines some of the central connections of the mormyrid octavolateral area and provides anatomical support for the above hypothesis about separate vestibular and auditory areas.

*Neurological Sciences Institute, Good Samaritan Hospital and Medical Center, Portland, Oregon 97209.

2 Methods

Retrograde and anterograde transport of horseradish peroxidase (HRP) was used to trace connections. Surgery was done under anesthesia (MS-222, 1:20,000). The HRP (Sigma type IX) was dissolved in a 1% solution of α-lysolecithin (see Frank 1980). Survival times were 3-5 days and the tissue was processed with a modification of the method of Hanker et al. (1977). (See Bell, Finger, and Russell 1981 for a more complete description of methods.)

The cells projecting from the brain to the spinal cord were looked for in two animals by cutting the cord at the fifth spinal segment and placing pieces of gelatin sponge, soaked in the HRP solution, on the cut end of the cord. The central connections of the ALLL were studied by injections into the lobe (three cases), and into the main mesencephalic projection site of ALLL, medialis dorsalis nucleus (three cases) (see Haugede-Carre 1980 for descriptions of mesencephalic nuclei). Two of the ALLL injections included the crista cerebellaris, the layer of large crest cells, the anterior nucleus, and part of the octavius nucleus (Fig. 19-1B,C). In one case the injection was too far posterior to include the crest cell layer. A connection from medialis dorsalis nucleus to the tectum was confirmed by injections into the optic tectum (two cases).

3 Results

3.1 Projections to the Spinal Cord

The great majority of cells projecting to the spinal cord are found in the medullary and mesencephalic reticular formation. Some are also found, however, within restricted parts of the eighth nerve termination region. Cells projecting to the spinal cord from within the eighth nerve termination region include (a) a group of rather large (~ 20 μm) round cells at the lateral margin of the descending nucleus; (b) a group of still larger (~ 25 μm) heteropolar cells located anteriorly and medially in the descending nucleus; (c) the tangential nucleus; and (d) a few cells at the medial border of the magnocellular nucleus. The last group may be an extension of the tangential nucleus rather than part of the magnocellular nucleus proper. The tangential nucleus of mormyrids is probably not homologous to the nucleus of the same name described in other fish by Ramon y Cajal (1908) (see Bell 1981). No labeled cells were seen in the ALLL.

3.2 Afferent and Efferent Connections of ALLL

The eighth and lateral line nerve roots contained labeled fibers in the ALLL injection cases, confirming that these nerves project to ALLL (Bell 1981). Several groups of central cells also project to ALLL as shown by retrograde labeling after ALLL injections. These cell groups include the following: (a) Granule cells of the eminentia granularis. This result supports the suggestion of Maler et al. (1973) that the axons of eminentia granularis cells form the molecular layer of the crista cerebellaris. (b) Cells of the praeeminentialis ventralis nucleus on both sides. This nucleus is closely associated with the praeeminentalis nucleus but is distinct in the small size and close packing of its cells (Figs. 19-1D and 19-3; see Bell et al. 1981). (c) Cells in the descending and magnocellular nuclei on both sides.

The efferent connections of the ALLL can be traced anterogradely after ALLL injections and retrogradely after injections in medialis dorsalis (Fig. 19-1) (see also Haugede-Carre 1980). HRP injections into medialis dorsalis result in retrograde labeling of the large crest cells, cells in the anterior nucleus, and some cells in the octavius nucleus (Figs. 19-1B,C and 19-2B). Cells in the rostromedial octavius nucleus receiving bilateral saccular input (Bell 1981) are among those projecting to medialis dorsalis. The labeling in the ALLL and octavius nucleus is bilateral but predominantly contralateral. The most posterior part of the ALLL, about one-fifth of the anterior-posterior extent, is not labeled even after injections into posterior medialis dorsalis. Similarly, an injection into this part of the ALLL did not label any lemniscal fibers. This part of ALLL lacks crest cells (Bell 1981) and its connections are apparently restricted to the medulla.

Efferent fibers from the ALLL ascend ipsilaterally and contralaterally via the lateral lemniscus, just medial to the much more numerous fibers from the posterior lateral line lobe (PLLL). Fibers originating from ipsilateral ALLL and PLLL course anteriorly in a dorsal component of the lemniscus, whereas those from the contralateral side decussate and ascend in a ventral component (Fig.19-1D). The ALLL fibers terminate most densely in the medialis dorsalis nucleus of the mesencephalon (Fig. 19-2E,F) but apparent collaterals can be seen leaving the lemniscus and entering praeeminentialis ventralis nucleus (Fig. 19-1D) and ventralis posterior nucleus. Such collaterals are seen after both ALLL and medialis dorsalis injections (Fig. 19-1E,F).

On both sides, lateral ALLL projects laterally and medial ALLL medially onto medialis dorsalis nucleus (Fig. 19-1). Many ALLL cells must divide with one branch ascending ipsilaterally and the other contralaterally, because injections into medialis dorsalis result in labeled fibers in the contralateral lemniscus (Fig. 19-1C,D,E). Such fibers terminate in the contralateral medialis ventralis at a point corresponding to the injection site (Fig. 19-2A).

3.3 Afferent and Efferent Connections of Medialis Dorsalis

Other cell groups in addition to the ALLL and octavius nucleus project to medialis dorsalis. These include the following: (a) A nucleus of small cells at the level of anterior ALLL just medial to the ipsilateral lemniscus component (SOL in Fig. 19-1C). These cells are labeled predominately ipsilaterally and are embedded among lemniscal fibers. The nucleus appears homologous to the superior olive in frogs (Wilcyzinski 1978a, 1979b) and holosteans (Braford and McCormick 1979). (b) A second paralemniscal nucleus, at the level of anterior praeeminentialis (IRN in Fig. 19-1E). These cells are labeled predominantly contralaterally and are also embedded among lemniscal fibers. The nucleus appears homologous to the isthmoreticular nucleus (Wilcyzinski 1978a, 1978b, Braford and McCormick 1979). It was termed the paralemniscal nucleus by Haugede-Carre (1980). (c) Cells in contralateral medialis dorsalis at locations corresponding to that of the injection site (Fig. 19-2A), indicating a homotopic commissural connection. (d) Cells in a nucleus near the spino medullary junction, referred to

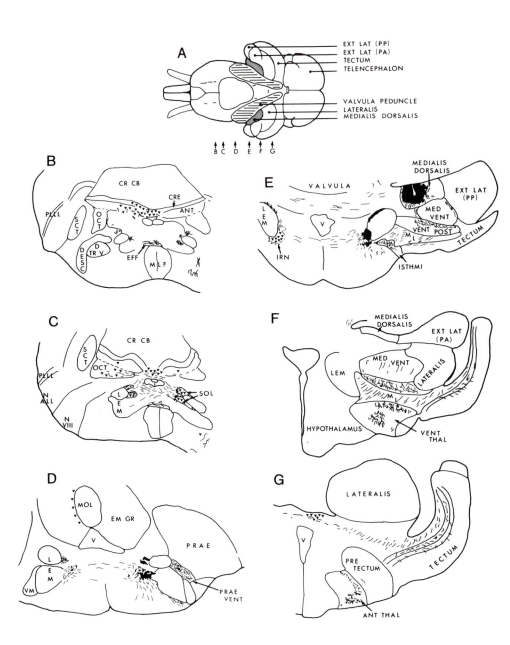

by Maler et al. (1973) as the nucleus of the tract of V, pars funicularis, and by Haugede-Carre (1980) as the subfunicular nucleus. (e) A few cells of the mesencephalic and medullary reticular formation. (f) Cells in the central region of the medial and lateral lobes of the posterior telencephalon, mostly ipsilateral but some contralateral. (g) Small granule-like cells located ipsilaterally near the dorsal midline of the mesencephalon, just posterior to torus longitudinalis (Fig. 19-1G). (h) Cells located contralaterally at the boundary between eminentia granularis and the molecular layer it surrounds (Fig. 19-1D). These efferent cells are neither granule nor Purkinje cells. (i) Basal cells (Nieuwenhuys and Nicholson 1969) in the valvula cerebelli, particularly in its ventral, covered part at the level of PLLL.

Regions to which anterogradely labeled fibers could be traced after medialis dorsalis injections included the following: (a) The optic tectum on both sides (Fig. 19-1F,G). The ipsilateral projection was confirmed by the tectal injections. (b) The mesencephalic and medullary reticular formation. (c) The ipsilateral superior olive. Fibers to this nucleus descend along the lateral ventral margin of the brain stem (Fig. 19-1D). A similar descending pathway is seen in holosteans (Braford and McCormick 1979). (d) Ipsilateral praeeminentialis ventralis nucleus. The axons and the terminals that comprise this projection are finer than those of the lemniscal projection. The fibers descend to praeeminentialis ventralis nucleus via the tectum mark lager (ML in Fig. 19-1E,F). (e) Medialis dorsalis efferents terminate with large boutons on the small round cells of nucleus isthmi, in the same manner as efferents from extrolateralis pars posterior, which also terminate in nucleus isthmi (Haugede-Carre 1979, 1980). Some cells in this nucleus project to the valvula (Finger, Bell, and Russell 1981, Haugede-Carre 1980). (f) The valvula cerebelli, including the region which projects back to medialis ventralis. (g) Two thalamic nuclei (postventral and ventral; Stendell 1914) parts of which at least project to the valvula (Finger et al. 1981), and one thalamic nucleus (anterior) which projects to the telencephalon (Bell unpublished observations).

◄Figure 19-1. Chart of fibers and cells labeled after HRP injection into medial part of medialis dorsalis. (A) Dorsal view of mormyrid brain with valvula cerebelli removed. Coarse horizontal hatching indicates valvula peduncle, fine vertical hatching indicates medialis dorsalis nucleus. (B-G) Caudal to rostral series of transverse sections at levels indicated by arrows and letters in (A). Retrogradely labeled cells are shown with filled triangles. Abbreviations: ALL, anterior lateral line lobe; ANT, anterior nucleus; ANT THAL, anterior thalamic nucleus; CR CB, crista cerebellaris; CRE, crest cell layer, D TR V, descending tract of V; DESC, descending nucleus; EFF, nucleus of lateral line and eighth nerve efferents; EM GR, eminentia granularis; EXT LAT (PA), exterolateral nucleus pars anterior; EXT LAT (PP), exterolateral nucleus pars posterior; IRN, isthmoreticular nucleus; LEMN, lemniscus, ML, tectum mark lager; MED VENT, medialis ventralis; MLF, medial longitudinal fasciculus; MO, molecular layer; N AL, anterior lateral line nerve; N VIII, eighth nerve; OCT, octavius nucleus; PLLL, posterior lateral line lobe; PRAE, praeeminentialis nucleus; SCT, spinocerebellar tract; SOL, superior olive nucleus; PRAE-VENT, praeeminentialis ventralis nucleus; V, ventricle; VM, motor nucleus of V; VENT THAL, ventral thalamic nucleus; VENT POST, ventralis posterior nucleus of mesencephalon.

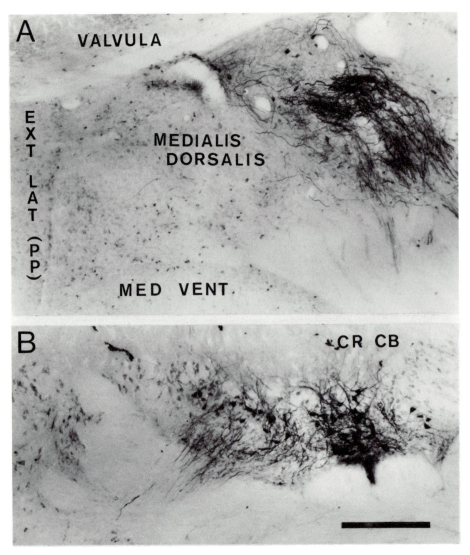

Figure 19-2. Medium power of labeled fibers and cells after HRP injection into medial part of medialis dorsalis. (A) Contralateral medialis dorsalis showing termination of contralateral lemniscal fibers in medial part of medialis dorsalis. Some labeled cells are present in the same region, indicating a commissural connection. (B) Labeled crest cells near the midline of ALLL. Molecular layer dendrites can be seen. Scale bar equals 200 μm. For explanation of abbreviations, see legend to Fig. 19-1.

4 Discussion

The spinal projections of the octavolateral area arise from the descending and tangential nuclei, which receive their input from the utriculus and canals. If, as seems likely,
the spinal projection is concerned with postural control, then the above connection is
consistent with a postural or vestibular role for the utriculus end canals, as well as for
the descending and tangential nuclei. Furthermore, the lack of projections to the spinal
cord or to the oculomotor nuclei from ALLL, the major terminus of saccular afferents,
argues against a vestibular role for the sacculus or for the ALLL.

There are striking parallels between the ALLL and its connections on the one hand,
and the PLLL cortex and its connections on the other. The PLLL cortex receives electroreceptor afferents (Maler et al. 1973, Bell and Russell 1978). The axons of its efferent cells divide and project ipsilaterally and contralaterally via the lemniscus to the
lateralis nucleus of the mesencephalon, in association with ALLL efferents (Bell et al.
1981, Haugede-Carre 1980).

Maler et al. (1973) have suggested that the relation between PLLL and lobus caudalis is similar to the relation between ALLL and eminentia granularis. For example, the
parallel fibers of ALLL come from eminentia granularis whereas those of PLLL come
largely from lobus caudalis. The efferent connection to medialis dorsalis nucleus from
the molecular layer associated with eminentia granularis (see Section 3) is in accord
with this suggestion because there are efferent cells in lobus caudalis that project in a
similar manner to the lateralis nucleus of the mesencephalon (Bell et al. 1981).

The anatomical parallels between the ALLL and PLLL systems are also seen at
the mesenceaphalic level in the connections of the medialis dorsalis and lateralis
nuclei. The commisural, thalamic, telencephalic, tectal, funicular, and valvular pathways of these two nuclei are very similar (Finger et al. 1981, Haugede-Carre 1980).

One further set of similarities is worth a special comment. Nucleus praeeminentialis
is a large nucleus in both mormyrids and gymnotoids. In mormyrids it receives electrosensory input directly from PLLL and indirectly via a large tract (tectum mark
lager) from the lateralis nucleus (Bell et al. 1981, Haugede-Carre 1980). It projects
back to the PLLL, both directly (Bell et al. 1981) and indirectly via lobus caudalis
(Szabo et al. 1979). Similar connections are seen in the electrosensory pathways of
gymnotoids (Maler pers. comm.) and catfish (Finger pers. comm.). The present study
shows that comparable pathways also exist linking ALLL, praeeminentialis ventralis,
and medialis dorsalis (Fig. 19-3). In each case, pathways are provided for returning the
results of higher order processing of sensory information back to earlier stages of
analysis. Haugede-Carre (1979, 1980) describes a connection from the extrolateralis
pars posterior nucleus to a nucleus subpraeeminentialis. Unpublished results of the
present author (Bell) confirm the pathway described by Haugede-Carre (1979, 1980).
These results also indicate that extrolateralis efferents terminate within a sharply
localized region just lateral and dorsal to anterior praeeminentialis ventralis nucleus.
The extrolateralis projection does not overlap with the one from medialis dorsalis, and
only the latter overlaps with the cells which project to ALLL.

The close parallels between the pathways related to PLLL and to ALLL suggest an
evolutionary relationship. Perhaps the central elctrosensory pathways of teleosts
evolved from preexisting pathways related to the analysis of sound or water movement.

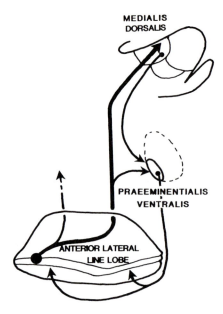

Figure 19-3. Diagram of connections between anterior lateral line lobe, nucleus prae-eminentialis ventralis, and medialis dorsalis nucleus. Dotted line outlines nucleus praeeminentialis.

The projection of ALLL and of the paralemniscal cell groups to the mesencephalon appears homologous to the auditory pathways of mammals and the strong projection of medialis dorsalis to the tectum suggests a role in a spatial localization. The parallels with the electrosensory system, the parallels with the auditory system of higher verte-brates, and the tectal projection all imply that the ALLL has a primary role in sensory analysis rather than in motor control. Physiological experiments must be done, how-ever, to establish what type of sensory analysis is performed.

After this study was completed, the report of Haugede-Carre (1980) appeared, in which the connections of the mormyrid ALLL, octavius nucleus, and medialis dorsalis were also examined. The basic findings in her study and this one are similar. In addition, she found some separation in the parts of medialis dorsalis to which the crest cells, the anterior nucleus, and the octavius nucleus project.

Acknowledgment. This research was supported by a grant from the National Science Foundation (BNS-05096).

References

Bell, C. C.: Central distribution of octavolateral afferents and efferents in a teleost (Mormyridae). J. Comp. Neurol. 195, 391-414 (1981).

Bell, C. C., Finger, T. E., Russell, C. J.: Central connections of the posterior lateral line lobe in mormyrid fish. Exp. Br. Res. (1981) (in press).

Bell, C. C., Russell, C. J.: Termination of electroreceptor and mechanical lateral line afferents in the mormyrid acousticolateral area. J. Comp. Neurol. 182, 367-382 (1978).

Braford, M. R., McCormick, C. A.: Some connections of the torus semicircularis in the bowfin, *Amia calva*. A horseradish peroxidase study. Soc. Neurosci. Abstr. 5, 39 (1979).

Finger, T. E., Bell, C. C., Russell, C. J.: Electrosensory pathways to the valvula cerebelli in mormyrid fish. Exp. Br. Res. (1981) (in press).

Frank, E., Harris, W. A., Kennedy, M. B.: Lysophosphatidyl choline facilitates labeling of CNS projections with horseradish peroxidase. J. Neurosci. Methods 2, 183-189 (1980).

Hanker, J. S., Yates, P. E., Metz, C. B., Rustioni, A.: A new specific, sensitive, and non-carcinogenic reagent for the demonstration of HRP. Histochem. J. 9, 789-792 (1977).

Haugede-Carre, F.: The mesencephalic exterolateral posterior nucleus of the mormyrid fish *Bryenomyrus niger*: Efferent connections studied by the HRP method. Brain Res. 178, 179-184 (1979).

Haugede-Carre, F.: Contribution a l'étude des connexions du torus semicircularis et du cervelet chez certains mormyrides. Thesis. L'université Pierre et Marie Curie, Paris, 1980.

Maler, L., Karten, H. J., Bennett, M. V. L.: The central connections of the posterior lateral line nerve of *Gnathonemus petersi*. J. Comp. Neurol. 151, 57-66 (1973).

Nieuwenhuys, R., Nicholson, C.: A survey of the general morphology, the fiber connections, and the possible functional significance of the gigantocerebellum of Mormyrid fishes. In: Neurobiology of Cerebellar Evolution and Development. Llinas, R. (ed.). Chicago: American Medical Association, 1969, pp. 107-134.

Popper, A. N., Fay, R. R.: Sound detection and processing by teleost fishes: A critical review. J. Acoust. Soc. Am. 53, 1515-1529 (1973).

Ramon y Cajal, S.: Sur un noyau special du nerf vestibulaire des poissons et des oiseaux. Trab. Lab. Invest. Biol. Univ. Madr. 6, 1-20 (1908).

Stendell, W.: Die Faseranatomie des Mormyridengehirns. Abh. Senckenb. Naturforsch. Ges. 36: 3-40 (1914).

Stipetic, E.: Über das Gehörorgan der Mormyriden. Z. Vergl. Physiol. 26, 740-752 (1939).

Szabo, T., Libouban, S., Haugede-Carre, F.: Convergence of common and specific sensory afferents to the cerebellar auricle (auricula cerebelli) in the teleost fish *Gnathonemus* demonstrated by HRP method). Brain Res. 168, 619-622 (1979).

von Frisch, K.: Über die Bedeutung des Sacculus und der Lagena für den Gehörsinn der Fische. Z. Vergl. Physiol. 25, 703-747 (1938).

Wilczynski, W.: Torus semicircularis afferents in the bullfrog, *Rana catesbeiana*. Soc. Neurosci. Abstr. 4, 103 (1978a).

Wilczynski, W.: Connections of the midbrain auditory center in the bullfrog, *Rana catesbeiana*. Thesis, Univ. Michigan, 1978b.

Discussion

PLATT: I think that one of the real evolutionary problems that we face in this field is the question of when the sacculus becomes an equilibrium receptor instead of an auditory receptor. The separation is so clean in fishes, then you have both equilibrium and auditory function in the frog, and a clear change by the time you get to birds and mammals. The otolith receptor in the sacculus is essentially gravistatic. This suggests that somehow the package got reorganized.

BELL: On the other hand, in mammals, as in momyrids, the central projections of canal and utricular afferents show extensive overlap. Similarly, saccular afferents in both of these groups of animals project to central regions which are rather distinct from the projection sites of canal and utricular afferents. Thus, at least some features of the eighth nerve projection pattern in lower forms may be preserved in higher animals.

NORTHCUTT: First, I'd like to congratulate Bell for an elegant study. The differences that he describes are real, and involve more than just nomenclature. As we look at more and more fishes, we should expect to see more and more differences of such magnitude. For instance, in skates, the same pathways that Bell has reported on maintain separation in the midbrain, but converge in the tectum. I think we are going to see many different patterns, and we may have to wait a couple of years before we even recognize what the patterns are. I really would like to see this kind of information on the clupeids.

PART SIX

Sound Production and Acoustical Behavior

As pointed out by Tavolga in Chapter 28, we should not forget that many of the features of the teleost auditory system may relate to acoustic communication. In Chapter 20 Myrberg provides an overview of acoustic behavior and demonstrates the extent of known acoustic communication among fishes. Demski (Chapter 21), in a chapter that could well fit into Part Five of this volume, describes the central control of sound production in fishes, while Mok (Chapter 22) provides data on sound production and use of sounds by a particular species of goby.

Chapter 20

Sound Communication and Interception in Fishes

ARTHUR A. MYRBERG, JR.*

1 Introduction

A most intriguing aspect of social behavior is that eventually the topic of communication must be addressed. This is understandable because specific acts performed by one individual often alter the likelihood that specific acts will be performed by another. An appropriate question then arises: Did such a performance involve communication? Here, the thoughtful worker often encounters a most elusive concept since "communication" means quite different things to different people.

I do not wish to dispute usage of a term whose meaning has obvious value to those who apply it at far different levels of organization (e.g., messenger RNA, the friendly messenger from the telegraph office). Yet many authors in addressing the topic of communication have not defined its usage. Apparently they believed that the process would be explained by "examples in context" or by expecting that ". . . intuitively everyone knows what communication is" (Burghardt 1970).

I sought a clear and useful definition since to define such a process is to establish its boundaries; if such boundaries are diffuse, any discussion will likely be meaningless. Subsequent to an initial review of the literature concerning communication, confusion led me to consider that as a concept, communication was probably meaningless, at least, for those levels of biological organization that are of primary interest to me. However, with the aid of some excellent contributions on the subject (Burghardt 1970, Cherry 1957, Dawkins and Krebs 1978, Klopfer 1977, Marler 1968, 1974, 1977, Sebeok 1977, Tavolga 1974, Wilson 1975) that view changed. Although my present understanding has led to a view somewhat different from that which I initially held, I believe that the concept has heuristic value so long as its use is limited.

I subsequently turned to the bioacoustical literature on fishes. Many studies mention the communication process, but most deal with it in the form of suggestions, speculations, possibilities, or reasonable, but as yet, untested interpretations. Thus, I faced a second dilemma, provide a review similar to that presented by Fine, Winn, and Olla

*School of Marine and Atmospheric Science, University of Miami, Miami, Florida 33149.

(1977) or attempt something else, to describe what could be a reasonable framework upon which analyses of communication might develop in the future. I chose the second course since if achieved, it could provide a means for envisioning new approaches for research aimed at understanding this vexing process, and particularly for research directed at acoustical communication in fishes.

To accomplish these goals, three points must be made: (1) explain my view of communication, (2) provide a framework consistent with that view, and (3) illustrate that framework by examples from the available literature. The remainder of this chapter deals with these efforts. I hasten to add, credit must be given to those whom I have cited for originating many of the ideas presented below. My only contribution has been to synthesize those ideas to attain the aforementioned goals.

2 Communication—Review of Constraints

Most students agree that animal communication is a complex process by which one individual's structure or behavior (i.e., sender, emitter, actor), acting as a stimulus or signal, effects a change in the behavior of another (i.e., receiver, reactor, responder). Yet, the process appears to include more than simply effecting a change since physical force could do that (see Cherry 1957). Communication implies an alternative to physical force, i.e., a relatively lower energy effort (signal) eliciting response than that presumably required to obtain that same response in the absence of the signal. This implication seems readily acceptable by researchers. A change in behavior also infers that some type of information has been transmitted between interactants. It has been argued most effectively, however, that information transfer, without appropriate constraints, is inadequate for establishing a communicatory event since the process includes more than just reducing uncertainty or ambiguity in a social interaction (Burghardt 1970, Dawkins and Krebs 1978, Marler 1974).

2.1 The Primary Constraint—Who Is to Benefit?

Setting aside the idea that information transfer per se is the only central issue and considering instead that communication is a process of adaptive manipulation, the point then arises: For whom is the process adaptive? Is it for both parties in a given interaction or for only one?

Numerous authors consider that communication occurs only when an interaction has provided mutual benefit to the sender and receiver alike. This constraint presupposes that selection has favored those senders that inform receivers to the extent that they are able to predict more surely the sender's behavior. Receivers are accordingly selected to behave as if predicting the future behavior of senders. Communication is thus seen as a means of establishing and maintaining cooperation. This seems simple enough since in the evolutionary sense communication should improve the fitness of the sender and receiver alike by promoting unambiguity, efficiency of action, reduction in physical damage, and synchrony of sexual and other bonding mechanisms. Yet, reasonable evidence exists that social interactions can involve divergent interests on the part of the interactants. A courting male, competing with a second male for the attention of a female, can be rewarded for responding appropriately to a vocal signal that was apparently intended for the female by the second male. In that case, a distinct advantage can well accrue to the signaler when his competitor disrupts the female's approach. Likewise, small fishes certainly do not benefit by closely approaching the "fishing lure" of an angler fish. Yet these instances, which seemingly involve communication, include conflicting, rather than cooperative or mutual interests. Obviously, benefits cannot be ascribed mutually to both parties in such cases. If not, may we consider instead communication as a process involving interactions that possess adaptive advantage for the sender alone? I believe so; but then why should receivers respond if they do not benefit as well? The apparent answer is that *on the average*, benefits do accrue to them. The critical point is that if animals are selected to respond to their environment in ways that are on the average beneficial to themselves, it is likely that other animals can be selected to use such responsiveness for their own benefit as well (Dawkins and Krebs 1978). It is clear that a sender need not benefit everytime it employs specific signals nor that both parties cannot benefit from an interaction. Indeed, *mutual benefits may often characterize specific situations*. Nevertheless, it seems logical, when considering functional significance also as a fundamental issue, that communication is most easily understood and has its greatest heuristic value if the receiver's benefits remain incidental to any definition of the term. It is equally apparent that signal mechanisms may arise either phylogenetically or ontogenetically, the former by appropriate selective processes, the latter by reinforcement acting on the consequences of previous events (Burghardt 1970, see also Losey 1978).

2.2 Who May Interact?

Some workers maintain that communication has value only to interactions between members of the same species. This ignores the extensive literature addressed at interspecific communication. This seems unwarranted since many of the most startling and revealing cases of communication involve members of dissimilar species (Baylis 1979, Losey 1978, Wickler 1968). At the other extreme, the idea that communication occurs within a single organism such as a bat or a porpoise (i.e., echolocation, see Sebeok 1965) leads nowhere since responsiveness to stimuli alone is inadequate either for distinguishing communication or for maintaining its value as a social process.

2.3 Must There Be a Close Relation Between Signal and Response?

That there be a close relation between signal and response cannot be set aside since it usually constitutes the evidence for the occurrence of communication. Although demonstrating such a relation remains today one of the most difficult challenges facing ethologists, accumulated knowledge provides ever-increasing awareness that there seldom exists a strict correlation of "stimulus" with "response." We now realize that a receiver often can be "somehow different" at different times and that a signal's effect may well depend on external factors quite divorced from the signal itself (Hailman

1977, Marler 1974, 1977). The former condition refers to those mechanisms of internal causation represented by motivational factors while the latter refers to the multitude of contextual factors present during any interaction (Smith 1968, 1969). Additionally, signals may have long-lasting, rather than strictly immediate, effects (Schleidt 1973). A signal, for example, may be used to initiate, develop, or even maintain some state of readiness in a receiver. And when the latter eventually responds, it may well be due to the totality of on-going processes and not merely to some immediately preceding signal (Hailman 1977). Thus, the communicatory process cannot be fully understood without considering the physiological state of the organism (Bullock 1961). Finally, signals are often treated atomistically; a single stimulus provided by the sender is considered either to produce or, at least, to control a certain response by the receiver. Yet, impressive evidence exists that responses are often controlled by the summation of quite distinct stimuli or signals in place and time (Curio 1976, Golani 1973, Leong 1969, Rand, Kleerekoper, and Matis 1975, Seitz 1940, Tavolga 1956).

The complications associated with stimulus-response relations can be quite staggering and the demands of the "controlled" experiment seem often to lead us further away from the process we are trying to comprehend. Such experiments probably will remain the basic tools for the behavioral sciences, but their usefulness may become increasingly limited in future attempts at understanding the communication process (see Nelson 1973). In any case, the constraint of "S-R contiguity" remains, but it is one that each worker must deal with carefully in relation to a chosen methodology.

2.4 Must Specialized Signaling and Receptor Mechanisms Be Present?

An additional criterion used to define communication has required that the sender use some specialized structure or behavior to produce a signal which is ". . . an energy output along a single channel and usually within a narrow band of the spectrum of variation of that channel" (Tavolga 1974, p. 64), with reception, in turn, being characterized ". . . by specialized receptor mechanisms and specific, often stereotyped, responses" (Tavolga 1968, p. 274). This criterion, provided by Frings and Frings (1964) and by Tavolga in a series of papers (1968, 1970, 1974), though useful in specific instances, is frequently difficult to assess. Decisions as to the degree of specificity required for acknowledging the process would likely rest with individual investigators using their own understanding of given channels or receptor entities. It is unlikely that a consensus could be obtained among specialists for borderline cases.

2.5 Must Interactants Share a Code?

The final criterion to be considered here is that a shared code must exist between the interactants, as proposed by Klopfer and Hatch (1968), Frings, and Tavolga. This seems reasonable for any definition of communication, but denial of the process does rest on negative evidence.

3 Communication—Its Definition

I propose a definition of communication based on the above considerations. It appears workable within the following analytical framework even though it goes counter to the thoughts of many.

Communication is the process of transferring information between two or more individuals and whose functional aim (i.e., intent) rests solely in obtaining adaptive advantage for the sender (i.e., the initiator of the interaction). Its operation implies the existence of a shared code between the interactants and that the effort required by the sender to accomplish a specific aim is less than that needed to accomplish it without the signal.

The above definition accepts the reality of interspecific communication and in no way denies mutual benefit to the sender and receiver alike. It only considers receiver benefits incidental to the process. It also accepts the reality of specialized mechanisms of emission and reception, but considers them as adjuncts of efficiency and refinement and unnecessary to the basic definition.

The process, by not being restricted to cases of mutual benefit, allows situations involving the application of apparent misinformation or deceptive information by the receiver. Likewise, the challenging new views of mate and competitor assessment, as well as altruistic behavior can stand beside more traditional views, such as arousal and appeasement, as experimentally verifiable functions of communicative interactions. Accordingly, the search for signals mediating such activities is not only plausible but necessary if they are to have meaning in reality. Still another facet of communication recently gaining in interest surrounds the often extreme degree of redundancy provided by a wide assortment of signals. While such redundancy most likely reflects the confounding effects of ambient noise on sensory channels, a not unreasonable idea is that under certain circumstances response may only be elicited by repetitive signaling. Again, I emphasize that intent, as implied here, refers only to the real or perceived advantage to the sender if its signals are received and acted on by another (Burghardt 1970).

4 Interception—A Process Is Proposed

Numerous cases exist where information about an individual (the initiator) reaches another for whom that information was not apparently intended. The receiver, if it reacts, does so for its own benefit and frequently the initiator of the interaction is disadvantaged. Since the beneficiary is clearly not the sender in such cases, communication, as defined above, cannot be considered the operative process. Instead, an equally important, but different, process seems applicable, namely, *interception*. Since its function rests with securing adaptive advantage for the receiver (henceforth called the *interceptor*), the process appears inimical to communication. And yet, both processes may operate simultaneously in a given situation, they simply involve different receivers. This is exemplified by the case mentioned previously involving conspecifics, i.e., one male directing a vocal signal to a female but having that same signal (now referred to as a sign since communication is not involved) intercepted by another male who then actively competes for the female's attention. Although other cases of interception involving conspecifics can be envisioned, far more familiar are those instances involving predators intercepting signs from their prey. The tables can be turned, however; prey can be appropriate interceptors as we shall see later.

Interception does demand that the interceptor be able to interpret signs appropriately so that it may use them to its advantage. It is also apparent that necessary mechanisms involving the process may arise phylogenetically or ontogenetically, as is the case also for communication.

Since this report is primarily a review of communication, I will consider interception further only in a later section where various examples will be described to clarify its role.

5 A Framework for Analyzing Interactions

5.1 The Basic Problems Facing Individuals

Although survival and reproduction are functionally synonymous in genetic terms, each represents a somewhat different set of problems (with intermediates, of course) that must be solved by an individual throughout its lifetime:

A. Survival
    1. obtain security from:
        (a) the physical and chemical environment
        (b) predators of any type
        (c) competitors for resources other than mates
    2. obtain sustenance
B. Reproduction
    3. obtain mates
    4. obtain security from competitors for mates

Survival includes kin as well as self. Although most fishes show no care of eggs and young, some species do and various of these (e.g., cichlids, damselfishes, and toadfishes) have been studied by bioacousticians and others interested in communication. Such care rarely extends for more than a few weeks, however; and as yet there is no clear evidence of adult kin groups existing among fish, as has been demonstrated among various social insects (Wilson 1975) and species of "cooperative breeding" birds (Emlen 1978).

5.2 The Classes of Receivers

If communication is to aid in solving the problems mentioned above, the interests of a sender appear best served if signals are directed to specific types or classes of receivers in each case. These include (1) predators, (2) prey, (3) mates, and (4) two classes of competitors—those who compete directly for mates and those who compete for other resources (some receivers can be included in both classes). For those species whose activities suggest the existence of varying degrees of cooperation, another class must be considered as well—the companions, here defined as individuals involved in a cooperative venture other than the mating act per se. Although companions do exist in "the world of fishes" (e.g., parents guarding young, potential mates defending a common territory, mutual nest cleaning) unequivocal cases of complex cooperation (or coordination), such as structured hunts, communal constructions, and highly organized defense, apparently do not exist. But many fishes do show one phenomenal case of close coordination—schooling.

5.3 The Basic Benefits to Senders

The various benefits listed below could accrue to a sender on communicating with members of the six respective classes by appropriate signals:

Receivers   *Benefits to Senders*

1. predators
2. competitors for mates   { move away rapidly
3. competitors for other resources   { stay away

                                                   arouse
4. mates   attract
5. companions   allow approach by the sender
                                                   coordinate locomotory movements

6. prey   { attract
                                                   { allow approach by the sender

To receive these benefits, it would behoove the sender to provide appropriate information as to identify, readiness (physiological or behavioral) and location. Exceptions may arise when warning of predators or arousing mates (i.e., aiding in the development of reproductive readiness). In such cases, information as to location would probably be detrimental to the sender or not be required. It should also be added that information, as to identity, may or may not be deceptive depending on whether the receiver is a mate, companion, predator, or prey.

6 Application of the Framework to Sound Communication in Fishes

The analytical framework thus becomes evident. Its structure consists of a limited number of basic problems faced by individuals, answers for which may be found through use of the communication process. These answers can provide a number of benefits to an individual (i.e., the sender) initiating a communicatory event with one or more members of six respective classes of receivers. Reciprocity of action may occur but each interaction is best considered as a separate event to simplify analysis. Although exceptions are apparent, if benefits are to be gained, most signals seem to require information as to the sender's identity, location, and readiness to act in a particular manner. No requirement exists as to the nature of the information present in any given signal. It can be true or deceptive, its content being ascribed or interpreted only by actions of the receiver and the benefit(s) obtained by the sender based on those actions.

This framework will now be applied to findings available in the bioacoustical literature on fishes. It will also be used to suggest where future studies may aid us in more fully understanding a most vexing process.

6.1 Signaling Predators

More than 35 families of marine and freshwater fishes (excluding the innumerable families of freshwater catfishes) include species that suddenly produce sounds when disturbed by divers or by being held or prodded mechanically (Table 20-1). Actually,

Table 20-1. Families of Fishes Whose Members Produce "Startle" Type Sounds[a]

Internally produced sounds		Sudden swimming sounds
Ariidae (sea cats)	Myliobatidae (eagle rays)	Albulidae bonefishes)
Batrachoididae (toadfishes)	Ostraciidae (trunkfishes)	Carangidae (jacks)
Balistidae (triggerfishes)	Polynemidae (threadfins)	Clupeidae (herrings)
Carangidae (jacks)	Pomadasyidae (grunts)	Elopidae (ladyfishes)
Characidae (characins)	Pomacentridae (damselfishes)	Ephippidae (spadefishes)
Dactylopteridae (flying gurnards)	Scaridae (parrotfishes)	Gadidae (cods)
Diodontidae (porcupinefishes)	Sciaenidae (drums)	Labridae (wrasses)
Gadidae (cods)	Serranidae (sea basses)	Lutjanidae (snappers)
Holocentridae (squirrelfishes)	Sparidae (porgies)	Polynemidae (thread-fins)
Kyphosidae (sea chubs)	Tetraodontidae (puffers)	Pomadasyidae (grunts)
Labridae (wrasses)	Triglidae (sea robins)	Pomatomidae (bluefishes)
Lutjanidae (snappers)		Scaridae (parrotfishes)
Mullidae (goatfishes)		Sciaenidae (drums)
		Serranidae (sea basses)

[a]Fish and Mowbray 1970, Gray and Winn 1961, Hawkins and Chapman 1966, Hawkins and Rasmussen 1978, Horch and Salmon 1973, Limbaugh 1964, Markl 1971, Moulton 1958, 1963, Protasov and Romanenko 1962, Salmon, Winn, and Sorgente 1968, Schneider and Hasler 1960, Tavolga 1960.

most of the early evidence for sound production in numerous fishes was obtained by subjecting them to a variety of noxious stimuli (Fish 1954, Fish, Kelsey, and Mowbray 1952). More recent studies, using those stimuli, have suggested that sudden, relatively loud sounds are used to startle predators in a manner analogous to the well-documented, visual flash display of various lepidopterans. Such sounds could result in the release of the sender by its captor or, if capture has not occurred, in driving the predator away. That interpretation seems reasonable if one hears a loud sound underwater, for example, the drumbeat of a large grouper confronted in a cave. Yet, no evidence has substantiated that interpretation. Only three cases from the literature note that these sounds even occur in the appropriate context (Table 20-2). Since no mention was made about any apparent effect of the sound on the behavior of the respective predators, it is reasonable to consider that there was none. A fourth case describes observations of sharks attempting to reach groupers hiding in small caves. The sudden, loud drumbeats by the groupers did not appear to lessen the attack by their predators. Yet, we know that sharks will withdraw rapidly from a source if a loud sound (like that of a drumbeat) is suddenly transmitted (Banner 1972, Klimley and Myrberg 1979, Myrberg, Gordon, and Klimley 1978). Such facts are only suggestive, however. Far more appropriate stimuli are provided to a predator by its natural prey then those provided by an unnatural sound source (e.g., a small loudspeaker).

Besides the often mentioned suggestion of startling a predator, various sounds have also been considered as possible alarm signals in the sense of warning nearby individuals that a predator is near. Of course, the question immediately asked is why should an individual "warn" others? This question has been addressed by Harvey and Greenwood (1978) using the literature on birds and mammals. They point out that the best evidence appears to favor a kin-related function, but differences between species as regards the nature and context of apparent alarm calls undermine any unifying function. Although their suggested approaches to tackling the functional and evolutionary questions of such sounds may not be effective in answering the same questions relative to analogous sounds in fishes, these questions should be addressed. The literature already suggests one generality dealing with "startle" or "warning" sounds, and though evidence remains sparse, future studies may establish its validity. Many small fishes produce sounds; yet their repertoires apparently do not include "startle" or "warning" sounds.

Table 20-2. Species Whose Members Produce Apparent "Startle" or "Warning" Sounds When Confronted by a Known Predator. No Effect was Noted by Observers

Family	Species	Confrontation	Sound	Reference
Gadidae	Gadus morhua	"Penned" by a conger eel	Grunts	Brawn 1961
Holocentridae	Holocentrus rufus	Eel appears	Staccatos, grunts	Winn et al. 1964
	Myripristis berndti	Eel approaches	Staccatos, grunts	Salmon 1967
Serranidae	Epinephalus striatus	Shark approaches	Drumbeats	Pers. observ.
	Mycteroperca bonaci	Shark approaches	Drumbeats	Pers. observ.

This could be due to an oversight by workers, but it is more likely that small fishes cannot produce sufficiently loud sounds to deter a nearby predator. This leads to still another point: although reasonable evidence exists that the color patterns of various fishes act as antipredator devices, no such evidence exists for sounds. Is this again due to oversight or possibly to the physical restraints imposed on sounds which preclude use of the modality for purposes of deception(either identity or location)? Although theory can help answer such a question, it cannot replace empiricism. The tactics and strategies used by animals in antipredator defense have provided us with insight into the evolution of adaptive processes. They certainly deserve serious attention by bioacousticians interested in fishes.

6.2 Signaling Prey

If prey are to be receivers in the communication processes, senders (=predators) must provide signals that either attract such animals sufficiently close to allow relative ease in capture or allow the senders themselves sufficient approach so as to achieve the same result. One is reminded of the "lure" of the angler fish and its effect on appropriate prey. Here, certain receivers will perish when communication occurs, but potential receivers will benefit on average from their tendency to respond since the signal, i.e., a small, wiggling "worm," represents a noteworthy food source which should not be ignored. The elegance of the operations involved in such communication processes employing either or both visual and tactile systems has been extremely well elucidated in fishes by Losey (1978) and Wickler (1963, 1968).

I have come upon no example using the acoustical system in an analogous fashion in fishes although one might presume otherwise. Again this suggests that the restrictions on sound production, as regards constraints imposed by bodily structures or the environment, preclude signals that might deceive a receiver about a sender's true identity or location. The costs to the sender may be too high. Losey (1978) has shown, however, that a predator need not necessarily deceive its prey in all cases; rather, action by the receiver depends on the costs it must incur relative to the benefits received. In any case, the elucidation of such constraints to any signaling process would clearly aid in predicting the limits imposed on that process as well as in clarifying the selective forces that control its existence. For example, environmental constraints would appear to be important to any such process (Michelsen 1978). Morton's recent study (1975) demonstrates this fact by showing that habitat acoustics can apparently shape avian signal patterns. My own literature search, dealing with environmental noise and its possible effect on hearing sensitivity in fishes, suggests that our ignorance of underwater habitat acoustics is quite profound (Myrberg 1978a). Although many field studies of underwater sound propagation have been conducted by physical acousticians, few results are even remotely applicable to either the sounds or the habitats that are of interest to most bioacousticians.

Theoretical considerations, while providing some understanding, often cannot handle the nonideal conditions that exist in a complex habitat where, for example, there is little or no compliance with the inverse square law for signal attenuation during spherical spreading. Thus, the effects of habitat acoustics on the signaling systems of fishes and other aquatic creatures would seem to be rewarding subjects for future research. They may actually hold the key not only for making vast strides in understanding communication systems in those animals, but in concert with findings from terrestrial studies, for arriving at principles which will extend across all such systems regardless of origin.

6.3 Signaling Prospective Mates

Sound production accompanies the reproductive activities of a wide variety of fishes (see Fine et al. 1977, Moulton 1963, Tavolga 1971), but relatively little evidence exists that sounds directly influence the behavior of prospective mates (Table 20-3). The most revealing cases have combined behavioral observations of potential mates with sound playback in the absence of a sender. When an appropriate response is shown, such a procedure assures that the role of the given sound is more than simple enhancement of another signal. Absence of response does not eliminate the possibility, however, that sound is a part of a more complex signaling system which is absent at the time.

Despite the wide species divergence of taxa presented in Table 20-3, all species but one, *Glandulocauda inequalis* (Nelson 1965), show territoriality in males (and sometimes females) during the reproductive season. Mating requires a reproductively active female moving to the site of a prospective mate. Courtship often includes specific sounds not produced in other contexts. Based on such noticeable actions, territorial species might well be expected to dominate any such list. Their preponderance suggests that sound production often characterizes species whose sexes are separated by distances over which visual or other signals will be insufficient for communicating information relative to sexual readiness. In these cases, senders could use sounds to advertise their own readiness, arouse reproductive activity in prospective mates, and inform mates as to their locations. Unfortunately, we know precious little about such events. Future studies should examine these possibilities.

If the entries in Table 20-3 reflect a relationship between sounds and the signaling of mates, nonterritorial species whose members form spawing aggregations appear to be less dependent on acoustical signals than their territorial counterparts. Perhaps visual or other signaling systems adequately transfer information across the distances separating prospective mates in the aggregations. Although little evidence is available, intrasexual, aggressive interactions may actually be the source of sounds emanating from such aggregations (e.g., sciaenids, see Dijkgraaf 1947). That does not necessarily apply to nonterritorial species whose reproductive demands require one or both sexes to search for mates over considerable distances.

The single, nonterritorial representative listed in Table 20-3, *Glandulocauda inequalis,* is a small characid fish that apparently does not aggregate during spawning. Instead, solitary males seek out prospective mates in the water column where courtship ensues (Nelson 1964, 1965). Males produce "croaking" sounds at such times near their prospective mates. Since Nelson did not describe female activity during courtship, the effect of such sounds remains unclear. Similar situations are not unknown for territorial species. Low level "grunts" are produced by males just prior to spawning in damselfishes of the genus *Eupomacentrus* (pers. observ.). The effect of that particular sound is also unclear except for apparent synchrony of spawning.

Table 20-3. Signaling Prospective Mates: Species Whose Members Produce Sounds While Courting Mates

Family	Species	Sound (producer)	Playback	Situation	Reference
Anabantidae	*Trichopsis vittatus*	Croak (♂ & ♀) Purr (♀)		Sounds produced during courtship	Marshall 1966 (in Fine et al. 1977)
Batrachoididae	*Opsanus tau*	Boatwhistle (♂)	x	♀♀ attracted to sound source, ♂ increases rate as ♀ approaches	Fish 1972, Gray and Winn 1961, Winn 1964, 1972
Blenniidae	*Chasmodes bosquianus*	Grunt (♂)		♂ increases rate as ♀ approaches	Tavolga 1958b
Centrarchidae	*Lepomis megalotis*	Grunt (♂)	x	♀♀ attracted to sound source	Gerald 1971
	L. macrochirus, L. humilis L. microlophus	Grunt (♂) Pop (♂)		Sounds produced during courtship	Gerald 1971
	L. macrochirus, L. gibbosus	Pharyngeal sound (♂)		Sounds produced during transition from aggression to reproductive behavior	Ballantyne and Colgan 1978a, 1978b, 1978c
Characidae	*Glandulocauda inequalis*	Croak (♂)		Sounds produced during courtship	Nelson 1964, 1965
Cichlidae	*Tilapia mossambica*	Drum (♂) Drum (♂)	x	♀♀ reproductively aroused Sounds produced during courtship	Marshall 1972 Marshall 1971
	Hemichromis bimaculatus	'Sound' (♂)		Sounds produced during courtship	Rowland 1977
Cyprinidae	*Notropis analostanus*	Purr (♂)	x	♀ decreases movement near sound source, ♂ "sounds" as ♀ approaches	Stout 1966, 1975

Family	Species	Sound		Response	References
Gadidae	Gadus morhua, Melanogrammus aeglifinus	Grunt (♂), Knock (♂)		♂ "sounds" as ♀ approaches, ♂ increases rate as spawning climaxes	Brawn 1961, Hawkins and Rasmussen 1978; Hawkins et al. 1967
Gobiidae	Bathygobius soporator	Grunt (♂)	x	♀♀ attracted to sound source, ♂ increases rate as ♀ approaches	Tavolga 1956, 1958a
Pomacentridae	Eupomacentrus partitus, E. planifrons, E. variabilis, E. dorsopunicans, E. leucostictus	Chirp (♂)	x	♂ increases rate as ♀ approaches, ♀ shows no immediate response to sound	Myrberg 1972, pers. observ.

Authors who have used only situation analyses in their studies of intersexual sound production often suggest that sounds enhance associated visual displays. Although that is certainly reasonable, further work with the same species, using the sound playback techniques of Stout (1975) and Schwarz (1974b), may well show additional effects.

The few classical cases of sound communication between prospective mates remain the finest examples of the process (toadfish *Opsanus tau*, Winn 1964, 1972; frillfin goby *Bathygobius soporator,* Tavolga 1956, 1958a; satinfin shiner *Notropis analostanus,* Stout 1966; Delco's results, 1960, involving two additional species of *Notropis* apparently need to be confirmed, see Stout 1975). This small list demands extension not only because of the extreme diversity of fishes, but because we need information on sound as a possible aid for orientation processes, mate selection (i.e., assessment) and the integration of behavioral activities with other processes, such as gonadal activity. The influence of vocalizations on gonadal activity in various avian species has been examined for more than 20 years (Brockway 1965, 1969, Ficken et al. 1960, Lott and Brody 1966); yet, there is only one abstract on the subject in fishes. Sounds of the male, mouthbrooding cichlid *Tilapia mossambica* apparently stimulate gonadal activity on conspecific females (Marshall 1972). One wonders how general such a finding might be: Is the response species-specific and is the entire sound required for response? Future studies could answer these and similar questions.

6.4 Signaling Companions

The coordinated movements of fish in a school provide ample evidence that such animals are taking part in one or more cooperative ventures such as antipredator maneuvers (Kuhlmann and Korst 1967, Myrberg, Brahy, and Emery 1967, Neill and Cullen 1974), obtaining defended resources (Barlow 1974, Robertson et al. 1976), facilitating prey capture (Olla and Samet 1974), or reproductive synchrony (Johannes 1978, Warner and Robertson 1978).

Although vision is important for orienting individuals to their schools (Breder 1959, Franzisket 1960, Hunter 1969, Keenleyside 1955, Shaw 1962, 1970), the small inter-individual distances found in most schools suggest that vibratory signals play an important role, with receivers attending to particle displacements such as those generated by the near-field effect. Evidence for such response comes from the studies by Moulton (1960), Partridge and Pitcher (1980) and Pitcher, Partridge, and Wardle (1976) (also Partridge, Chapter 26). Well oriented movements can be maintained by blinded fish when in a moving school of sighted individuals (Table 20-4).

Schooling has long been a popular subject of research, with studies ranging from movement dynamics and ontogeny to sensory involvement and ecological function. The relationships between mechanoreception and schooling still remain unclear, however, and are certainly worthy of attention.

If communication does occur between companions, one might expect to find it well developed in those cases involving kin. Some years ago, while working with cichlid fishes, I carried out a brief series of tests which were not reported because they were

Table 20-4. Signaling Companions: Species Whose Members Produce Sounds or Probable Near-Field Effects While Interacting in Coordination or Apparent Cooperation

Family	Species	Signal	Response	Reference
Engraulidae	*Anchoviella choerostoma* (blinded individual)	Near field effect (?)	Oriented movements while schooling	Moulton 1960
Gadidae	*Pollachius virens* (blinded individuals)	Near field effect (?)	Oriented movements while schooling	Pitcher et al. 1976; Partridge and Pitcher 1980
Holocentridae	*Holocentrus rufus*	Staccatos	Oriented movement to sound source after sound off	Winn et al. 1964
	Myripristis berndti	Staccatos and grunts	Oriented movement to sound source while sound on	Salmon 1967
	M. berndti *M. argyromus*	Staccatos and grunts	Oriented movement to sound source while sound on	Popper, Salmon, and Parvulescu 1973
	M. violaceus	Growls and grunts	Oriented movement to sound source while sound on	Horch and Salmon 1973

used only to establish a norm for a well-known phenomenon. Many years later I realized the importance that one of the side results might have to those interested in describing the effects of low frequency vibrations. Three pairs of Jack Dempsey cichlids *Cichlasoma octofasciatum* were each tending broods of free-swimming young (7 to 10 days old). The parents were removed and thereafter a small paddle was placed into each aquarium near the school of young resting motionless on the substrate. Minutes after the disturbance, the school began to mill slowly about, a few centimeters above the substrate. The paddle, located about 5 cm above the school, was then moved back and forth at a speed similar to that noted during pelvic fin flickering by hovering parents when disturbances occurred in or near their aquaria. When the paddles was turned "broadside" to the direction of vibration, the school "fell" to the substrate and remained there motionless. If it was turned "side-on" to the direction of vibration, the school did not "fall." A few individuals moved onto the substrate but the majority continued to school above it. Each alternative was repeated a number of times and similar results were obtained with all three schools. Although the design was admittedly crude, the findings suggested that the vibrations produced by the paddle simulated the vibratory signals produced by the parent's (the sender's) pelvic fin flickering, which is often termed a "calling movement." If a parent "calls" when some distance from the young, the latter often swim rapidly below or very near to that parent and only then move rapidly to the substrate. It seems that the rapid approach is controlled by the visual stimuli of the "call" while the sudden movement to the substrate and holding quiet thereafter are controlled by the mechanical portion of the composite signal. Such signals are important during periods of danger (nearby predators, Myrberg 1965) and they deserve more interest than that received to date. In fact, sensitivity to particle motion is probably extremely important for coordinating many behavioral activities of fishes. I am certain that any research pursuits in that wide open field will yield rewarding efforts.

The final examples to be mentioned might best be termed at this time, those involving "possible" companions. The behavior of various soniferous squirrelfishes from both the Atlantic and Pacific oceans has long interested many bioacousticians (Table 20-4). These investigators have established that both territorial (*Holocentrus* spp.) and schooling (*Myripristis* spp.) species move toward sound sources which are transmitting conspecific sounds. Since these particular sounds (staccatos and grunts) are produced during interspecific interactions, and particularly those involving predators, one wonders why an individual fish should take a risk? One suggestion has been made that the response is analogous to "mobbing" in birds and it has the same apparent advantages (Winn 1967). The consistency of the response certainly cannot be disregarded, but testing usually involved a loudspeaker as the sound source. That raises questions: Would a similar magnitude of response be shown to a natural source if a known predator was nearby, would the source or the predator be mobbed, would attracted fish remain near the predator? Answers are not yet available and so "mobbing" by holocentrid fishes remains an enigma. Yet, such clear and consistent response to sound in the face of possible predation poses an intriguing set of circumstances which certainly deserve attention.

6.5 Signaling Competitors

Competitors can best be defined as individuals who control or attempt to control resources required by each. In the present context a competitor is any individual considered to be attempting to control a resource that a sender either controls or is attempting to control at the time of interaction. Competition among fishes centers on three basic resources: mates, space, and food. Mate competition can be examined, at times, apart from the other two resources, but the latter are often intricately intertwined. No effort will be made to separate them here.

6.5.1 Competition for Mates

I found only two reports dealing with sound production during competitive interactions regarding mates (Table 20-5). The first dealt with interactions between territorial males of the bicolor damselfish *Eupomacentrus partitus* (Myrberg 1972). Often when a male was courting an approaching female, a second male would suddenly move to that same female and commence courtship as well. At that time the first male (the sender) attacked the intruder and simultaneously produced a loud "pop" Similar observations have since been recorded from other congeneric species off the coast of Florida (pers. observ.). The second case, though somewhat less clear, seems appropriate considering the context in which the observation was made. Dijkgraaf (1947), while monitoring a small group of nonterritorial, European black croakers, *Corvina nigra*, in a large aquarium, noted that aggressive forays were repeatedly directed at a nearby conspecific (presumed to be a male) by a male which was courting a large female at the time. Such forays were accompanied by series of "knocking" sounds presumably produced by the aggressor. Flight was invariably shown by the fish being attacked. The effect of the sounds used in these two examples, other than possibly enhancing visual display, remains to be explored.

6.5.2 Competitors for Resources Other than Mates

Although situations involving mate competition may seldom be monitored, this is not the case for instances of territorial intrusion or maintenance of interindividual distance. More than 20 studies report sound production at these times; and most provide reason-

Table 20-5. Signaling Competitors for Mates: Species Whose Members Produce Sounds While Interacting Aggressively With Competitors During Active Courtship

Family	Species	Confrontation	Sound	Reference
Pomacentridae	*Eupomacentrus partitus* *E. planifrons* *E. dorsopunicans*	♂'s sound associated with driving other ♂ away from ♀	Pop	Myrberg 1972, Pers. observ.
Sciaenidae	*Corvina nigra*	♂'s sound associated with driving other ♂ away from ♀	Knock	Dijkgraaf 1947

able (though not conclusive) evidence that the attacker was the sound producer (Table 20-6). As noted in previous sections, the most revealing results have been obtained from studies which combined appropriate observation with sound playback. Similar findings have appeared in other studies, however. Almost all reports mention that sound production invariably accompanied motor patterns expressive of highly aggressive motivation in the sender, i.e., ramming, biting, frontal thrusting, and chasing. The sender was often territorial. In those cases where territoriality was not mentioned, the sender was evidently a dominant fish. In such instances, as well as those involving sound playback to a submissive (Stout 1975) or most probably a nonterritorial individual (Schwarz 1974b), behavioral change was apparent in receivers, with avoidance shown either to the sender or to the apparent source of the sound.

A few authors have mentioned variants from this general theme, but only Schneider (1964a) has dealt with one in some detail. Submissive members of two species of clownfish, *Amphiprion xanthurus* (=*A. clarkii*) and *A. polymnus,* produce "shaking" sounds when attacked by dominant individuals. Schneider states that such a sound inhibits aggression directed at the sender. Since there are many species of *Amphiprion,* it would be interesting to determine how widespread this variant is and to examine quantitatively its interesting effects.

The strong correlation between sound production and aggressive motivation in fishes has led to various sounds being used as tools in studies of the causal factors underlying aggression. Perhaps such use will be considered even more in the future. As noted previously, investigators have often relegated sound production to enhancing concomitant visual displays (interestingly never the reverse). Although challenging such an interpretative function is difficult, sound playback used in appropriate circumstances might well demonstrate other, and perhaps even more significant, benefits for the sender.

What might such a benefit be? One that became evident to me sometime ago is that sounds may be used to inform neighbors and potential intruders about the location of a territorial individual. A "keep-out" signal, using either or both courtship and aggressive sounds, if understood by the appropriate receiver, seems a reasonable suggestion. It has long been presumed that birdsong possesses such a function (see Davies 1978) with only recent substantiation of the fact (Krebs, Ashcroft, and Webber 1978). Although the idea of territorial advertisement by sound might be restricted to certain fishes inhabiting specific biotopes, one such biotope is of great interest to me—the coral reef. I plan to carry out a series of field experiments in the near future to test this idea, using one or more species of damselfishes from our nearby reefs. Actually, when considering the number of sedentary species and the number of individuals coexisting on a coral reef, space may well be the most limiting resource for such fishes (Sale 1977, Sale and Dybdahl 1975, 1978, Schoener 1974, Smith 1978, Smith and Tyler 1973). If such a "keep-out" signal was understood by other species as well as conspecifics, a sender could gain an extremely important benefit. This presupposes, of course, that such a sound would be loud enough to be heard by appropriate receivers. Minimal evidence presently suggests that a correlation does exist between the loudness of a signal and the degree of "privacy" apparent in an interaction, i.e., the proximity of a sender to a receiver. For example, premating sounds (grunts) of *Eupomacentrus* spp. are quieter than either the sounds of courtship (chirps) or aggression (pops) (pers. observ.), while sounds of threat in *Amphiprion* spp. are much louder than fighting sounds

Table 20-6. Signaling Competitors for Resources Other than Mates: Species Whose Members Produce Sounds While Showing Moderate to High Levels of Aggression During Interactions. In Most Cases, Authors Mentioned Opponents Move Away or Avoid the Sender; Often the Sound was Considered to Enhance Visual Display

Family	Species	Sound	Reference
Anabantidae	*Trichopsis vittatus*	Croak	Daugherty and Marshall 1971, Marshall 1966 (in Fine et al. 1977)
Ariide	*Arius felis*	Long grunt	Tavolga 1971
Balistidae	*Balistes bursa*	Drum	Salmon et al. 1968
	B. vetula	Drum	Salmon et al. 1968, Steinberg et al. 1965
Batrachoididae	*Opsanus tau*	Grunt	Winn 1972
Cichlidae	*Cichlasoma centrarchus*[a]	Growl	Schwarz 1974a, 1974b
	C. nigrofasciatum	Br-r-r	Myrberg, Kramer,
	Pterophyllum sp.	Tzz-Tzz	and Heinecke 1965
	Hemichromis bimaculatus	Br-r-r; thump	Myrberg et al. 1965, Rowland 1977
	Simochromis diagramma	Br-r-r	Nelissen 1975, 1977
	Tilapia mossambica	Drum	Marshall 1971
Cobitidae	*Botia berdmorei*	Knock; click thump	Klausewitz 1958, Rigley and Marshall 1971
Cyprinidae	*B. hymenophysa*	Knock	Klausewitz 1958
	Notropis analostanus[b]	Knock series	Stout 1966, 1975
Gadidae	*Gadus morhua*	Grunt	Brawn 1961, Hawkins and Rasmussen 1978
	Melanogrammus aeglifinus	Knock; grunt	Hawkins and Chapman 1966, Hawkins and Rasmussen 1978
Gobiidae	*Gobius jozo*	Snore	Kinzer 1961
Holocentridae	*Holocentrus rufus*	Grunt	Winn et al. 1964
	Myripristis berndti	Knock	Salmon 1967
	M. violaceus	Knock; thump	Horch and Salmon
	M. pralinus	Knock; thump	1973
Labridae	*Crenilabrus tinca*	Drum	Protasov and
	C. griseus	Drum	Romanenko
	C. oscellatus	Drum	1962
Mormyridae	*Gnathonemus petersi*	Click	Rigley and Marshall 1973
Pomacentridae	*Amphiprion polymnus*	"Threat," "fight"	Schneider 1973
	A. xanthurus	"Threat," "fight"	

Table 20-6 (continued)

Family	Species	Sound	Reference
	Eupomacentrus partitus[c]	Pop	Myrberg 1972
	E. planifrons	Pop	Thresher 1976
	E. dorsopunicans	Pop	Burke and Bright 1972
	Hypsypops rubicunda	Thump	Limbaugh 1964
Sparidae	*Lagodon rhomboides*	Click	Caldwell and Caldwell 1967
Theraponidae	*Therapon jarbua*[d]	"Threat," drum	Schneider 1964b
	T. theraps[d]	"Threat," drum	

[a] Playback resulted in reduction of aggression by brief resident.
[b] Playback resulted in subordinate avoiding source.
[c] Playback resulted in reduction of courtship by males.
[d] Apparent appeasement sounds produced also by species.

(Schneider 1964a). This point deserves attention in future studies because of its obvious relationship between hearing sensitivity and ambient noise levels (Myrberg 1978a).

Territorial fishes have no exclusive claim to the use of sounds during encounters with competitors, however. Although these species again dominate the entries provided in Table 20-6, their number may reflect the preference for many workers to concentrate their efforts on such fishes. Increased interest in the bioacoustics of nonterritorial species will most probably result in their names appearing in any such future listing.

7 Application of the Framework to Sound Interception in Fishes

7.1 Interceptors

The final class of receivers to be considered are those that I call *interceptors*. As mentioned earlier, I do not consider this class to be an integral part of the communication process since its members are neither required for, nor are they dependent on, that process. Yet under appropriate circumstances they certainly can take advantage of communicative interactions. They benefit by responding to signal(s) and may often force a disadvantage upon the sender by either destroying or disrupting communication.

Interceptors can perhaps be best recognized, however, in contexts quite apart from those involving communication. They include a wide variety of organisms: predators, competitors of any type, and even parasites and prey. In these roles, their response to an appropriate sign will likely be also at the expense of the sender, especially if response is successful.

Since interceptors cover a wide variety of contexts, analysis might best be served by immediately subdividing such a melange. I shall do that, but only to the extent necessary to demonstrate the heuristic value of the process. Since sound reception (used here in its most general sense) is only one of the various channels available to interceptors, I have chosen a simple term that fits the process extremely well when sound is considered—overhearing. If a receiver responds to an acoustical signal in the communicative process, it can be considered to have *heard* the signal, if the same occurs in an interceptive process, the receiver *overheard* the signal.

7.1.1 Overhearing by Predators

Excellent examples of overhearing are provided by the most obvious interceptors—predators. The sounds of struggling fishes or those produced by schools of fishes feeding appear to be particularly attractive signs to sharks, groupers, snappers, and other predacious species (Table 20-7A). To consider such instances as communication would be to destroy any reasonable meaning for the process. However, such instances fit interception well.

7.1.2 Overhearing by Prey

Can one consider situations antithetical to the just-aforementioned example, i.e., prey intercepting signs provided by predators? A fine example involves the blue runner *Caranx crysos* (=*C. fusus*) and its prey, the slippery dick wrasse *Halichoeres bivittatus* (Table 20-7B). Schools of this rapidly moving semipelagic predator are common about coral reefs. They produce rather unique sounds which herald their approach. When such sounds are heard faintly on a hydrophone, slippery dicks dive rapidly into the sand and remain motionless. When the school of blue runners appears on the scene some seconds later, not a single slippery dick can be seen, although numerous individuals had been moving above the substrate just moments before. Apparently the prey had intercepted appropriate acoustical signs from the predator and reacted accordingly.

The above examples (and others to be mentioned below) all show one interesting effect. Playback of natural sounds or synthetic simulations often succeed in eliciting response, at least initially. I maintain that this is a reasonable expectation based on the operation of an interceptive process. Signs that may result in immediate benefit to receivers (interception) are more likely to elicit response than signals that clearly benefit the sender but may or may not benefit the receiver following response (communication). Actually, an interceptive response appears quite analogous to that of an operant in an instrumental learning process.

7.1.3 Overhearing by Competitors

Evidence exists that competitors for mates will respond accordingly when they overhear courtship signals from nearby individuals (Table 20-8). I mentioned previously that among damselfishes of the genus *Eupomacentrus* the courtship sounds (chirps) from one male in the vicinity of a female will be answered almost immediately with competitive courtship by nearby males. Actually this courtship can occur without a female even being nearby. Such a response is quite easy to elicit by sound playback, so long as close attention is paid to the specific features of the sound's temporal structure (Myrberg et al. 1978, Spanier 1979).

Table 20-7. Signs to Interceptors

Family	Species	Sign	Response by interceptors
A. Overhearing by predators[a]			
Alopidae Carcharhinidae Laminidae Orectolobidae Sphyrnidae	Numerous sharks	Sounds of struggling or frenzied fishes or simulations thereof	Rapid attraction to sound source
Carangidae Lutjanidae Scombridae Serranidae Sphyraenidae	Numerous teleosts	Sounds of struggling, frenzied, or feeding fishes, or simulations thereof	Attraction to sound source
B. Overhearing by prey[b]			
Labridae	*Halichoeres bivittatus*	Swimming sounds of rapidly approaching predators (*Caranx crysos*)	Rapid diving from the water column into the sand

[a] Banner 1972, Hashimoto and Maniwa 1971, Maniwa 1976, Myrberg 1978, Myrberg et al. 1969, 1972, 1976, Nelson and Gruber 1963, Nelson and Johnson 1972, Nelson, Johnson, and Waldrop 1969, Richard 1968.
[b] Steinberg et al. 1965, Myrberg pers. observ.

Sound playback, though a powerful tool, cannot answer questions relating to function, however, when used in an experimental situation applicable to different contexts. Rowland (1977) found that males of the cichlid *Hemichromis bimaculatus* held in their own aquaria for long periods show more aggression toward models of conspecifics transmitting aggressive sounds than that shown toward silent models. He interpreted the results as that expected from a male when confronted in its territory with an aggressive conspecific. That interpretation is justified. But did the experiments involve communication or interception? This question is relevant since we are interested in the functional significance of a sound as well as the response of a receiver to that sound. And the answer rests, of course, with the intruder. What type of intruders were the models representing? It is not uncommon for a fish to intrude a territory while chasing another and occasionally sounds will accompany such chasing. If overheard by the territory holder, attack will likely follow and the interaction considered as interception. However, another possibility exists; an intruder could use sound (plus information from other channels) to inform the territory holder of its presence and its highly aggressive nature. Attack by the territory holder would again be likely, but in that instance, the interaction would be considered communication. Neither possibility can be precluded by the present experimental design; hence, no decision can be reached.

Numerous students have repeatedly stressed the importance of context when attempting to understand a social process. Results may be relevant to problems of internal causation and effect if a single experimental design covers a variety of contexts,

Table 20-8. Signs to Interceptors: Overhearing by Competitors

Family	Species	Sign	Response by interceptors	Reference
Batrachoididae	*Opsanus tau*	"Boatwhistling" by a ♂ (court-ship sound)	Nearby ♂♂ moved to source	Winn 1972
		Playback "boatwhistles"	Initiated and facilitated "boatwhis-tling" by ♂♂	Fish 1972 Winn 1967 Winn 1967
Cichlidae	*Hemichromis bimaculatus*	Playback ♂ "thumps" and "purrs" near a territorial ♂ (ag-gressive sound)	♂ increased aggression to-wards source	Rowland 1977
Cyprinidae	*Notropis analostanus*	Playback ♂ "knock series" to an agressive ♂ (aggressive sound)	♂ increased ag-gression to-wards apparent source	Stout 1966, 1975
		Playback ♂ "purrs" to a ♂ in presence of a ♀ (courtship sound)	♂ increased courtship behavior	Stout 1966, 1975
Pomacentridae	*Eupomacentrus partitus* *E. planifrons*	"Chirping" by a ♂ (courtship sound)	Initiated and facilitated competitive courtship in nearby ♂♂	Myrberg 1972, Pers. observ.
	E. partitus, E. planifrons, E. dorsopunicans, E. leucostictus	Playback "chirps"	Initiated and facilitated courtship in nearby ♂♂	Myrberg et al. 1978, Spanier 1979

but they may likely provide little information on functional relevance. A design can-not always be developed that incorporates both types of problems, but early consider-ation of such a possibility may provide fruitful benefits.

7.1.4 Overhearing by Prospective Mates

Prospective mates are other possible interceptors who could use signs from members of the opposite sex for various benefits. One benefit could be to assess individuals and/or their resources for purposes of mate selection. This presumes that mate choice is oper-ating in such instances. This phenomenon, though rarely examined in fishes, is known in some species which produce sounds, e.g., the damselfish *Eupomacentrus partitus*

(Schmale 1979). It also presumes that receivers can perceive individual differences. In the case of sounds, individual recognition has been recently demonstrated by members of *E. partitus* (Myrberg and Riggio in prep.). Another possible benefit might be to locate members of the opposite sex while attending to other matters (e.g., feeding or resting in secure places). In these instances, sound might be an excellent way to accomplish it. Although there is little sense in speculating further, the topics seem worthy of consideration in future work.

8 Closing Statement

Research scientists, wishing to maintain their sanity while studying the complex process of communication must reduce that complexity so that reasonable analysis can proceed. Categorization is one way to accomplish it. Although arbitrary, its value rests in allowing formerly obscured patterns to become evident. If such patterns improve understanding of the process, the categorization has value. If not, the categorization is useless and probably will introduce confusion as well. My attempt to provide a framework on which future studies of communication might develop is based on categorization. I hope that it will be as useful to others as it has been to me.

We must begin with the communication process if we are to understand social behavior since it is one of the bases on which social interactions rest. Another base is that which I have termed interception. Some may consider that "a rose by any other name ...."; but I maintain that this is not the case here. The rules that direct examples to one category or the other appear clearly different and actually somewhat antithetical. Whether the concept of interception will have heuristic value depends on whether it is capable of providing new pathways for understanding. I suggest it will, at least for social interactions among fishes.

Acknowledgments. I thank Donald P. deSylva, James B. Higman, and Robert J. Riggio for their critical reading of the manuscript. Portions of the work discussed here were supported by Grants BNS 77-13843 A02 and BNS 76-80186 from the National Science Foundation.

References

Ballantyne, P. K., Colgan, P. W.: Sound production during agonistic and reproductive behaviour in the pumpkinseed (*Lepomis gibbosus*), the bluegill (*L. macrochirus*), and their hybrid sunfish. 1. Context. Biol. Behav. 3, 113-135 (1978a).

Ballantyne, P. K., Colgan, P. W.: Sound production during agonistic and reproductive behaviour in the pumpkinseed (*Lepomis gibbosus*), the bluegill (*L. macrochirus*), and their hybrid sunfish. II. Recipients. Biol. Behav. 3, 207-220 (1978b).

Ballantyne, P. K., Colgan, P. W.: Sound production during agonistic and reproductive behaviour in the pumpkinseed (*Lepomis gibbosus*), the bluegill (*L. machrochirus*), and their hybrid sunfish. III. Response. Biol. Behav. 3, 221-232 (1978c).

Banner, A.: Use of sound in predation by young lemon sharks, *Negaprion brevirostris*. Bull. Mar. Sci. 22, 251-283 (1972).

Barlow, G. W.: Extraspecific imposition of social grouping among surgeonfishes (Pisces: Acanthuridae). J. Zool. (London) 174, 333-340 (1974).

Baylis, J. F.: Optical signals and interspecific communication. In: Behavioral Significance of Color. Burtt, E. H. (ed.). New York: Garland Press, 1979, pp. 360-377.

Brawn, V. M.: Sound production by the cod (*Gadus callarias* L.). Behaviour 18, 239-255 (1961).

Breder, C. M.: Studies on social groupings in fishes. Bull. Am. Mus. Nat. Hist. 117, 393-482 (1959).

Brockway, B. F.: Stimulation of ovarian development and egg laying by male courtship vocalization in budgerigars (*Melopsittacus undulatus*). Anim. Behav. 13, 575-578 (1965).

Brockway, B. F.: Roles of budgerigar vocalization in the integration of breeding behaviour. In: Bird Vocalizations, Their Relation to Current Problems in Biology and Psychology. Hinde, R. A. (ed.). London: Cambridge University Press, 1969, pp. 131-158.

Bullock, T. H.: The origins of patterned nervous discharge. Behaviour 17, 48-59 (1961).

Burghardt, G. M.: Defining "communication." In: Advances in Chemoreception, Vol. 1. Communication by Chemical Signals. Johnston, J. W., Jr., Moulton, D. G., Turk, A. (eds.). New York: Appleton-Century-Crofts, 1970, pp. 5-18.

Burke, T. F., Bright, T. J.: Sound production and color changes in the dusky damselfish, *Eupomacentrus dorsopunicans* Poey, Family Pomacentridae. Hydro. Lab. J. 1, 21-29 (1972).

Caldwell, D. K., Caldwell, M. C.: Underwater sounds associated with aggressive behaviour in defense of territory by the pinfish, *Lagodon rhomboides*. Bull. S. Calif. Acad. Sci. 66, 69-75 (1967).

Cherry, C.: On Human Communication: A Review, a Survey, and a Criticism. New York: Wiley, 1957.

Curio, E.: The Ethology of Predation. Berlin: Springer-Verlag, 1976.

Daugherty, J., Marshall, J. A.: Sound producing mechanism of the Croaking Gourami, *Trichopsis vittatus* (Pisces, Belontiidae). Am. Zool. 11, 632 (1971).

Davies, N. B.: Ecological questions about territorial behaviour. In: Behavioural Ecology, An Evolutionary Approach. Krebs, J. R., Davies, N. B. (eds.). Sunderland: Sinauer Associates, 1978, pp. 317-350.

Dawkins, R., Krebs, J. R.: Animal signals: Information or manipulation? In: Behavioural Ecology, An Evolutionary Approach. Krebs, J. R., Davies, N. B. (eds.). Sunderland: Sinauer Associates, 1978, pp. 282-309.

Delco, E. A., Jr.: Sound discrimination by males of two cyprinid fishes. Texas J. Sci. 12, 48-54 (1960).

Dijkgraaf, S.: Ein Töne erzeugender Fisch im Neapler Aquarium. Experientia 3, 493-494 (1947).

Emlen, S. T.: The evolution of cooperative breeding in birds. In: Behavioural Ecology, An Evolutionary Approach. Krebs, J. R., Davies, N. B. (eds.). Sunderland: Sinauer Associates, 1978, pp. 245-281.

Ficken, R. W., Van Tienhoven, A. Ficken, M. S., Sibley, F. C.: Effect of visual and vocal stimuli on breeding in the budgerigar (*Melopsittacus undulatus*). Anim. Behav. 8, 104-106 (1960).

Fine, M. L., Winn, H. E., Olla, B. L.: Communication in fishes. In: How Animals Communicate. Sebeok, T. A. (ed.). Bloomington: Indiana Univ. Press, 1977, pp. 472-518.

Fish, J. F.: The effect of sound playback on the toadfish. In: Behavior of Marine Animals: Current Perspectives in Research, Vol. 2, Vertebrates. Winn, H. F., Olla, B. L. (eds.). New York: Plenum Press, 1972, pp. 386-432.

Fish, M. P.: The character and significance of sound production among fishes of the western North Atlantic. Bull. Bingham Oceanogr. Coll. 14, 1-109 (1954).

Fish, M. P., Kelsey, A. S., Mowbray, W. H.: Studies on the production of underwater sound by North Atlantic coastal fishes. J. Mar. Res. 11, 180-193 (1952).

Fish, M. P., Mowbray, W. H.: Sounds of Western North Atlantic Fishes. Baltimore: The Johns Hopkins Press, 1970.

Franzisket, L.: Experimentelle Untersuchung über die optische Wirkung der Streifung beim Preussenfisch (*Dascyllus aruanus*). Behaviour 15, 77-81 (1960).

Frings, H., Frings, M.: Animal Communication. New York: Blaisdell, 1964.

Gerald, J. W.: Sound production during courtship in six species of sunfish (Centrarchidae). Evolution 25, 75-87 (1971).

Golani, I.: Non-metric analysis of behavioral interaction sequences in captive jackals. Behaviour 44, 89-112 (1973).

Gray, G. A., Winn, H. E.: Reproductive ecology and sound production of the toadfish, *Opsanus tau*. Ecology 42, 274-282 (1961).

Hailman, J. P.: Communication by reflected light. In: How Animals Communicate. Sebeok, T. A. (ed.). Bloomington: Indiana Univ. Press, 1977, pp. 184-210.

Harvey, P. H., Greenwood, P. J.: Anti-predator defense strategies: some evolutionary problems. In: Behavioural Ecology, An Evolutionary Approach. Krebs, J. R., Davies, N. B. (eds.). Sunderland: Sinauer Associates, 1978, pp. 129-151.

Hashimoto, T., Maniwa, Y.: Research on the luring of fish schools by utilizing acoustical equipment. In: Modern Fishing Gear of the World. Kristjonsson, H. (ed.). London: Fishing News Books, 1971, pp. 501-503.

Hawkins, A. D., Chapman, C. J.: Underwater sounds of the haddock, *Melanogrammus aeglifinus*. J. Mar. Biol. Assoc. U.K. 46, 241-247 (1966).

Hawkins, A. D., Chapman, C. J., Symonds, D. J.: Spawning of haddock in captivity. Nature 215, 923-925 (1967).

Hawkins, A. D., Rasmussen, K. J.: The calls of gadoid fish. J. Mar. Biol. Assoc. U.K. 58, 891-911 (1978).

Horch, K., Salmon, M.: Adaptations to the acoustic environment by the squirrelfishes, *Myripristis violaceus* and *M. pralinus*. Mar. Behav. Physiol. 2, 121-139 (1973).

Hunter, J. R.: Communication of velocity changes in jack mackerel (*Trachurus symmetricus*) schools. Anim. Behav. 17, 507-514 (1969).

Johannes, R. E.: Reproductive strategies of coastal marine fishes in the tropics. Env. Biol. Fish. 3, 65-84 (1978).

Keenleyside, M. H. A.: Some aspects of the schooling behaviour of fish. Behaviour 8, 133-247 (1955).

Kinzer, J. Über Lautäusserungen der Schwarzgrundel *Gobius jozo*. Terr.-Kunde 7, 7-10 (1961).

Klausewitz, W.: Lauterzeugung als Abwehrwaffe bei der hinterindischen Tigerschmerle (*Botia hymenophysa*). Natur. Volk. 88, 343-349 (1958).

Klimley, A. P., Myrberg, A. A., Jr.: Acoustic stimuli underlying withdrawal from a sound source by adult lemon sharks, *Negaprion brevirostris* (Poey). Bull. Mar. Sci. 29, 447-458 (1979).

Klopfer, P. H., Hatch, J.: Experimental considerations. In: Animal Communication. Sebeok, T. A. (ed.). Bloomington: Indiana Univ. Press, 1968, pp. 31-43.

Klopfer, P. H.: Communication in prosimians. In: How Animals Communicate. Sebeok, T. A. (ed.). Bloomington: Indiana Univ. Press, 1977, pp. 841-850.

Krebs, J. R., Ashcroft, R., Webber, M.: Song repertoires and territory defense in the great tit (*Parus major*). Nature 271, 539-542 (1978).

Kühlmann, D. H. H., Korst, H.: Freiwasserbeobachtungen zum Verhalten von Tobias-fischschwarmen (Ammodytidae) in der westlichen Ostsee. Z. Tierpsychol. 24, 282-297 (1967).

Leong, C.-Y.: The quantitative effect of releasers on the attack readiness of the fish, *Haplochromis burtoni* (Cichlidae: Pisces). Z. Vergl. Physiol. 65, 29-50 (1969).

Limbaugh, C.: Notes on the life history of two californian pomacentrids: Garibaldis, *Hypsypops rubicunda* (Girard), and blacksmiths, *Chromis punctipinnis* (Cooper). Pac. Sci. 18, 41-50 (1964).

Losey, G.: The symbiotic behavior of fishes. In: The Behavior of Fish and Other Aquatic Animals. Mostofsky, D. I. (ed.). New York: Academic Press, 1978, pp. 1-31.

Lott, D. F., Brody, P. N.: Support of ovulation in the ring dove by auditory and visual stimuli. J. Comp. Physiol. Psychol. 62, 311-313 (1966).

Maniwa, Y.: Attraction of bony fish, squid and crab by sound. In: Sound Reception in Fish. Schuijf, A., Hawkins, A. D. (eds.). New York: Elsevier, 1976, pp. 271-282.

Markl, H.: Schallerzeugung bei Piranhas (Serrasalminae, Characidae). Z. Vergl. Physiol. 74, 39-56 (1971).

Marler, P.: Visual systems. In: Animal Communication. Sebeok, T. A. (ed.). Blooming-ton: Indiana Uniersity Press, 1968, pp. 103-126.

Marler, P.: Animal Communication. In: Nonverbal Communication. Krames, L., Pliner, P., Alloway, T. (eds.). New York: Plenum Press, 1974, pp. 26-50.

Marler, P.: The evolution of communication. In: How Animals Communicate. Sebeok, T. A. (ed.). Bloomington: Indiana Univ. Press, 1977, pp. 45-70.

Marshall, J. A.: Sound production by *Tilapia mossambica* (Pisces: Cichlidae). Am. Zool. 11, 632 (1971).

Marshall, J. A.: Influence of male sound production on oviposition in female *Tilapia mossambica* (Pisces, Cichlidae). Am. Zool. 12, 633-664 (1972).

Michelsen, A.: Sound reception in different environments. In: Sensory Ecology, Re-view and Perspectives. Ali, M. A. (ed.). New York: Plenum Press, 1978, pp. 345-373.

Morton, E. S.: Ecological sources of selection on avian sounds. Am. Nat. 109, 17-34 (1975).

Moulton, J. M.: The acoustical behavior of some fishes in the Bimini area. Biol. Bull. 114, 357-374 (1958).

Moulton, J. M.: Swimming sounds and the schooling of fishes. Biol. Bull. 119, 210-223 (1960).

Moulton, J. M.: Acoustic behavior of fishes. In: Acoustic Behavior of Animals. Busnel, R. G. (ed.). New York: Elsevier, 1963, pp. 655-693.

Myrberg, A. A., Jr.: A descriptive analysis of the behavior of the African cichlid fish, *Pelmatochromis guentheri* (Sauvage). Anim. Behav. 13, 312-329 (1965).

Myrberg, A. A., Jr.: Ethology of the bicolor damselfish, *Eupomacentrus partitus* (Pisces: Pomacentridae): A comparative analysis of laboratory and field behavior. Anim. Behav. Monogr. 5, 199-283 (1972).

Myrberg, A. A., Jr.: Ocean noise and the behavior of marine animals: relationships and implications. In: Effects of Noise on Wildlife. Fletcher, J. L., Busnel, R. G. (eds.). New York: Academic Press, 1978a, pp. 169-208.

Myrberg, A. A., Jr.: Underwater sound—Its effect on the behavior of sharks. In: Senso-ry Biology of Sharks, Skates and Rays. Hodgson, E. S., Mathewson, R. F. (eds.). Washington, D.C.: U.S. Govt. Printing Office, 1978b, pp. 391-417.

Myrberg, A. A., Jr., Banner, A., Richard, J. D.: Shark attraction using a video-acoustic system. Mar. Biol. 2, 264-276 (1969).

Myrberg, A. A., Jr., Brahy, B. D., Emery, A. R.: Field observations on reproduction of the damselfish, *Chromis multilineata* (Pomacentridae), with additional notes on general behavior. Copeia 1967, 819-827 (1967).

Myrberg, A. A., Jr., Gordon, C. R., Klimley, A. P.: Attraction of free ranging sharks by low frequency sound, with comments on its biological significance. In: Sound Reception in Fish. Schuijf, A., Hawkins, A. D. (eds.). New York: Elsevier, 1976, pp. 205-228.

Myrberg, A. A., Jr., Gordon, C. R., Klimley, A. P. L.: Rapid withdrawal from a sound source by open-ocean sharks. J. Acoust. Soc. Am. 64, 1289-1297 (1978).

Myrberg, A. A., Jr., Ha, S. J., Walewski, S., Banbury, J. C.: Effectiveness of acoustic signals in attracting epipelagic sharks to an underwater sound source. Bull. Mar. Sci. 22, 926-949 (1972).

Myrberg, A. A., Jr., Kramer, E., Heinecke, P.: Sound production by cichlid fishes. Science 149, 555-558 (1965).

Myrberg, A. A., Jr., Spanier, E., Ha, S. J.: Temporal patterning in acoustical communication. In: Contrasts in Behavior. Reese, E. S., Lighter, F. J. (eds.). New York: Wiley, 1978, pp. 137-179.

Neill, S. R., Cullen, J. M.: Experiments on whether schooling by their prey affects the hunting behaviour of cephalopod and fish predators. J. Zool. (London) 172, 549-569 (1974).

Nelissen, M.: Sound production by *Simochromis diagramma* (Günther) (Pisces: Cichlidae). Acta Zool. Pathol. Antverpiensia 1975, 19-24 (1975).

Nelissen, M.: Sound production by *Haplochromis burtoni* (Günther) and *Tropheus moorii* Boulenger (Pisces: Cichlidae). Ann. Soc. R. Zool. Belg. 106, 155-166 (1977).

Nelson, D. R., Gruber, S. H.: Sharks: Attraction by low frequency sounds. Science 142, 975-977 (1963).

Nelson, D. R., Johnson, R. H.: Acoustic attraction of Pacific reef sharks: Effects of pulse intermittency and variability. J. Comp. Biochem. Physiol. 42A, 85-95 (1972).

Nelson, D. R., Johnson, R. H., Waldrop, L. G.: Responses in Bahamian sharks and groupers to low frequency, pulsed sounds. Bull. S. Calif. Acad. Sci. 68, 131-137 (1969).

Nelson, K.: Behavior and morphology in the glandulocaudine fishes (Ostariophysi, Characidae). Univ. Calif. Publ. Zool. 75, 59-152 (1964).

Nelson, K.: The evolution of a pattern of sound production associated with courtship in the characid fish, *Glandulocauda inequalis*. Evolution 18, 526-540 (1965).

Nelson, K.: Does the holistic study of behavior have a future? In: Perspectives in Ethology. Bateson, P., Klopfer, P. H. (eds.). New York: Plenum Press, 1973, pp. 281-328.

Olla, B. L., Samet, C.: Fish-to-fish attraction and the facilitation of feeding behavior as mediated by visual stimuli in striped mullet, *Mugil cephalus,* J. Fish. Res. Bd. Can. 81, 1621-1630 (1974).

Partridge, B. L., Pitcher, T. J.: The sensory basis of fish schools: relative roles of lateral line and vision. J. Comp. Physiol. 135, 315-325 (1980).

Pitcher, T. J., Partridge, B. L., Wardle, C. S.: A blind fish can school. Science 194, 963-965 (1976).

Popper, A. N., Salmon, M., Parvulescu, A.: Sound localization by the Hawaiian squirrelfishes, *Myripristis berndti* and *M. argyromus*. Anim. Behav. 21, 86-97 (1973).

Protasov, V. R., Romanenko, Y. V.: Significance of certain fish sounds. Zool. J. 41, 1516-1528 (1962).

Rand, G., Kleerekoper, H., Matis, J.: Interaction of odour and flow perception and the effect of parathion on the locomotor orientation of the goldfish *Carassius auratus* L. J. Fish. Biol. 7, 497-504 (1975).

Richard, J. D.: Fish attraction with pulsed, low-frequency sound. J. Fish. Res. Bd. Can. 25, 1441-1452 (1968).

Rigley, L., Marshall, J. A.: Sound production by the loach *Botia berdmorei* (Pisces, Cobitidae). Am. Zool. 11, 632 (1971).

Rigley, L., Marshall, J. A.: Sound production by the elephant-nose fish, *Gnathonemus petersi* (Pisces, Mormyridae). Copeia 1973, 134-135 (1973).

Robertson, D. R., Sweatman, H. P., Fletcher, E. A., Cleland, M. G.: Schooling as a mechanism for circumventing the territoriality of competitors. Ecology 57, 1208-1220 (1976).

Rowland, W. J.: Sound production and associated behavior in the jewelfish, *Hemichromis bimaculatus*. Behaviour 64, 125-136 (1977).

Sale, P. F.: Maintenance of high diversity in coral reef fish communities. Am. Nat. 111, 337-359 (1977).

Sale, P. F., Dybdahl, R.: Determinants of community structure for coral reef fishes in an experimental habitat. Ecology 56, 1343-1355 (1975).

Sale, P. F., Dybdahl, R.: Determinants of community structure for coral reef fishes in isolated coral heads at lagoonal and reef slope sites. Oecologia 34, 57-74 (1978).

Salmon, M.: Acoustical behavior of the menpachi, *Myripristis berndti,* in Hawaii. Pac. Sci. 21, 364-381 (1967).

Salmon, M., Winn, H. E., Sorgente, N.: Sound production and associated behavior in triggerfish. Pac. Sci. 22, 11-20 (1968).

Schleidt, W. M.: Tonic communication: continual effects of discrete signs in animal communication systems. J. Theor. Biol. 42, 359-386 (1973).

Schmale, M.: Sexual selection and reproductive success in males of the bicolor damselfish, *Eupomacentrus partitus* (Pisces: Pomacentridae). Thesis, University of Miami, 1979.

Schneider, H.: Bioakustische Untersuchungen an Anemonenfischen der Gattung *Amphiprion* (Pisces). Z. Morphol. Ökol. Tiere. 63, 453-474 (1964a).

Schneider, H.: Physiologische und Morphologische Untersuchungen zur Bioakustik der Tigerfische (Pisces: Theraponidae). Z. Vergl. Physiol. 47, 493-558 (1964b).

Schneider, H., Hasler, A. D.: Laute und Lauterzeungung beim Süsswassertrommler *Aplodinotus grunniens* Rafinesque (Sciaenidae: Pisces). Z. Vergl. Physiol. 43, 499-517 (1960).

Schoener, T. W.: Resource partitioning in ecological communities. Science 185, 27-39 (1974).

Schwarz, A.: Sound production and associated behavior in a cichlid fish, *Cichlasoma centrarchus*. I. Male-male interactions. Z. Tierpsychol. 35, 147-156 (1974a).

Schwarz, A.: The inhibition of aggressive behavior by sound in the cichlid fish, *Cichlasoma centrarchus*. Z. Tierpsychol. 35, 508-517 (1974b).

Sebeok, T. A.: Animal communication. Science 147, 1006-1014 (1965).

Sebeok, T. A. (ed.): How Animals Communicate. Bloomington: Indiana Univ. Press, 1977.

Seitz, A.: Die Paarbildung bei einigen Cichliden. 1. Die Paarbildung bei *Astatotilapia strigigena*. Z. Tierpsychol. 4, 40-84 (1940).

Shaw, E.: The schooling of fishes. Sci. Am. 206, 128-138 (1962).

Shaw, E.: Schooling in fishes: Critique and review. In: Development and Evolution of Behavior. Aronson, L. R., Tobach, E., Lehrman, D. S., Rosenblatt, J. S. (eds.). San Francisco: Freeman, 1970, pp. 452-480.

Smith, C. L.: Coral reef communities: a compromise view. Env. Biol. Fish. 3, 109-128 (1978).

Smith, C. L., Tyler, J. C.: Population ecology of a Bahamian suprabenthic shore fish assemblage. Am. Mus. Novitates 2525, 1-38 (1973).

Smith, W. J.: Message-meaning analyses. In: Animal Communication. Sebeok, T. A. (ed.). Bloomington: Indiana Univ. Press, 1968, pp. 44-60.

Smith, W. J.: Messages of vertebrate communication. Science 165, 145-150 (1969).

Spanier, E.: Aspects of species recognition by sound in four species of damselfishes, genus *Eupomacentrus* (Pisces: Pomacentridae). Z. Tierpsychol. 51, 301-316 (1979).

Steinberg, J. C., Cummings, W. C., Brahy, B. D., MacBain (Spires), J. Y.: Further bioacoustic studies off the west coast of North Bimini, Bahamas. Bull. Mar. Sci. 15, 942-963 (1965).

Stout, J. F.: Sound communication in fishes with special reference to *Notropis analostanus*. Proc. 3rd Annu. Conf. Biol. Sonar and Diving Mammals 3, 159-177 (1966).

Stout, J. F.: Sound communication during the reproductive behavior of *Notropis analostanus* (Pisces: Cyprinidae). Am. Midl. Naturalist 94, 296-325 (1975).

Tavolga, W. N.: Visual, chemical and sound stimuli as cues in the sex discriminatory behavior of the gobiid fish, *Bathygobius soporator*. Zoologica 41, 49-64 (1956).

Tavolga, W. N.: The significance of underwater sounds produced by males of the gobiid fish, *Bathygobius soporator*. Physiol. Zool. 31, 259-271 (1958a).

Tavolga, W. N.: Underwater sounds produced by males of the blenniid fish *Chasmodes bosquianus*. Ecology 39, 759-760 (1958b).

Tavolga, W. N.: Sound production and underwater communication in fishes. In: Animal Sounds and Communication. Lanyon, W. E., Tavolga, W. N. (eds.). Washington, D.C.: Intelligencer Printing Co., 1960, pp. 93-136.

Tavolga, W. N.: Fishes. In: Animal Communication. Sebeok, T. A. (ed.). Bloomington: Indiana Univ. Press, 1968, pp. 271-288.

Tavolga, W. N.: Levels of interaction in animal communication. In: Development and Evolution of Behavior. Aronson, L. R., Tobach, E., Lehrman, D. S., Rosenblatt, J. S. (eds.). San Francisco: Freeman, 1970, pp. 281-302.

Tavolga, W. N.: Sound production and detection. In: Fish Physiology. Hoar, W. S., Randall, D. J. (eds.). New York: Academic Press, 1971, pp. 135-205.

Tavolga, W. N.: Application of the concept of levels of organization to the study of animal communication. In: Nonverbal Communication. Krames, L., Pliner, P., Alloway, T. (eds.). New York: Plenum Press, 1974, pp. 51-75.

Thresher, R. E.: Field analysis of the territoriality of the threespot damselfish, *Eupomacentrus planifrons* (Pomacentridae). Copeia 1976, 266-276 (1976).

Warner, R. R., Robertson, D. R.: Sexual patterns in the labroid fishes of the Western Caribbean, I: The wrasses (Labridae). Smithsonian Contribs. Zool. (254), 1-27 (1978).

Wickler, W.: Zum problem der Signalbildung, am Beispiel der Verhaltensmimikry zwischen *Aspidontus* and *Labroides* (Pisces: Acanthopterygii). Z. Tierpsychol. 20, 657-679 (1963).

Wickler, W.: Mimicry in Plants and Animals. New York: McGraw-Hill, 1968.

Wilson, E. O.: Sociobiology, The New Synthesis. Cambridge, Mass.: Belknap Press, 1975.

Winn, H. E.: The biological significance of fish sounds. In: Marine Bio-acoustics, Vol. 1. Tavolga, W. N. (ed.). New York: Pergamon Press, 1964, pp. 213-231.

Winn, H. E.: Vocal facilitation and the biological significance of toadfish sounds. In: Marine Bio-acoustics, Vol. 2. Tavolga, W. N. (ed.). New York: Pergamon Press, 1967, pp. 283-303.

Winn, H. E.: Acoustic discrimination by the toadfish with comments on signal systems. In: Behavior of Marine Animals: Current Perspectives in Research, Vol. 2: Vertebrates. Winn, H. E., Olla, B. L. (eds.). New York: Plenum Press, 1972, pp. 361-385.

Winn, H. E., Marshall, J. A., Hazlett, B.: Behavior, diel activities, and stimuli that elicit sound production and reactions to sounds in the longspine squirrelfish. Copeia 1964, 413-425 (1964).

Discussion

TAVOLGA: You talk of the advantages and disadvantages of communication primarily in terms of survival. I suggest that you might emphasize that you must be talking about species, not just individual survival.

MYRBERG: I agree, but survival, in my view, is of the individual who is making the sound. The survival of the species depends on the survival of individuals. We classify individuals into groups, and we call these species. However, we study the processes of reproduction and survival by obtaining data from individuals. A species does not depend upon any single unitary factor, such as the occurrence of a sound, for its survival. Its existence depends on the actions of those individuals of which it is comprised. However, a unitary factor can directly affect individual survival, as, for example, avoidance of predation or attainment of reproductive success. Thus, survival, in the broadest sense, is best understood in terms of the fitness of individuals comprising a species rather than of the species per se.

TAVOLGA: My point is that you cannot apply the probablistic approach to an individual. You can only apply it to a population. The probability of an individual surviving is essentially binary—either he does or he doesn't.

The other point I should like to comment on is your notion of the "interceptor." It is a very nice, useful idea, but its utility depends on your knowledge of the signal system involved. Signals, and especially acoustic signals, are broadcast, i.e., they are not necessarily directed. Say a toadfish is sitting in his beer can and he is essentially broadcasting his signal. This is one of the basic properties of acoustic signals according to Hockett. From one viewpoint the second male is the interceptor and the female is the receptor. From the viewpoint of the second male, however, he is the receptor and the female is the interceptor. The concept and the term are certainly novel and should prove very useful, if one is careful to understand the communication system in question. That is one of the limitations of the concept of the "interceptor," i.e., in order to use it one must have considerable data on the signals and their functions.

MYRBERG: Your last point is well taken; one should certainly exercise care when invoking either interception or communication. So long as limited data allow speculations to remain the primary evidence for such processes, we'll no doubt remain at our present rather uncritical level of understanding. It is time we set aside speculations and concentrate on gathering sufficient information through repeated observations and controlled testing so that appropriate interpretations can be made as to the operative process.

Regarding your point about a toadfish broadcasting its sound, broadcasting is a property based on the physics of sound and the medium through which it is propagated. However, I fail to see how one might gain insight into the selective value of a male's courtship sound by taking the view that another male is the actual intended receiver, while the female stands as interceptor. With sufficient information on courtship behavior in the species, I am confident that differentiating interception from communication will aid our understanding of numerous types of interactions.

Chapter 21

Neural Control of Teleost Sound Production

Leo S. Demski*

1 Introduction

Until recently relatively little was known concerning the neural mechanisms controlling sound production in fishes. This is not surprising considering that prior to World War II there were few experimental studies on even peripheral sonic mechanisms (see papers in Tavolga 1977). Tavolga (1964, 1971) has reviewed the early work of peripheral innervation patterns in the sonic mechanisms of a variety of teleosts and therefore this will not be covered in detail in this review. Some of the information discussed in an earlier review on central mechanisms of sound production in teleosts (Demski, Gerald, and Popper 1973) has been incorporated into this chapter and updated with the results of recent studies including some experiments on electrically evoked croaking in *Micropogon undulatus* (Demski, Williams, and Dulka unpublished observations). Neural mechanisms of sound production are considered under the following functional categories: neuromuscular systems, central motor mechanisms, and higher integrative or control systems. Information in all of these categories is then combined into a model that summarizes many of the most significant results. Although there are a great variety of sonic mechanisms in teleosts, only those systems that use skeletal muscles to vibrate the swimbladder have been considered in this chapter.

2 Neuromuscular Systems

Sonic muscles of teleosts are known to have special characteristics which permit extremely fast contraction rates, in some cases up to 300 to 400/sec (Skoglund 1959, 1961, Packard 1960, Tavolga 1964, 1971, Cohen and Winn 1967, Schneider 1967). An elaborate development of the sarcoplasmic reticulum is one of the most striking features of the sonic muscle in the oyster toadfish, *Opsanus tau,* and it is presumed

*Physiology Group, School of Biological Sciences, University of Kentucky, Lexington, Kentucky 40506; and Gulf Coast Research Laboratory, Ocean Springs, Mississippi 39564.

that this specialization is related to the fast contraction rates of this tissue (Fawcett and Revel 1961). It has also been demonstrated that toadfish and squirrelfish (*Holocentrus rufus*) sonic muscles receive endings from more than one axon (Gainer and Klancher 1965). This polyaxonal innervation has been interpreted as partially accounting for fast synchronization of the muscle.

The sound-producing muscles in many diverse species are innervated by branches of the so-called occipital nerves (Fig. 21-1). The animals studied include the marine catfish, *Arius felis* and *Bagre marinus,* the squirrelfish, *Holocentrus ascenionis,* the slender sea robin, *Prionotus scitulus,* the red hind, *Epinephelus guttatus,* and the toadfish, *O. tau.*

These nerves, which are often double, exit between the vagus and first spinal nerves (Tavolga 1962, 1964, 1971). Tavolga has stated that the sonic muscles in all of these species may be homologous based on this common nerve supply and that the occipital nerves are probably homologous to the hypoglossal nerves of tetrapods. One group of fish, the croakers and drums (Sciaenidae), form a notable exception to this innervation pattern. In these animals the sonic muscles are derived from the lateral body wall and are thought to be innervated by spinal nerves (Schneider and Hasler 1960, Tavolga 1964, 1971). The nerves to the sonic muscles in toadfish have been characterized as containing uniform fibers with a diameter of approximately 10 μm and conduction rates of about 25 to 30 msec at 21°C (Skoglund 1959, 1961). In addition, the fibers are known to be cholinergic and to transport certain glycoproteins at the rapid rate of 96-120 mm/day (Barker et al. 1975).

3 Central Sonic Motor Mechanisms

3.1 Electrophysiology of Toadfish Sonic Motor Systems

Sonic motor neurons have been primarily studied in toadfish (*O. tau* and *O. beta*) where they form a spindle-shaped nucleus extending along the midline from the level of the first spinal nerves to a position below the caudal end of the fourth ventricle or obex (see DMN in Fig. 21-2). At least some of these cells have processes that appear to cross the midline and this may be an important feature necessary for synchronizing the sonic muscles of both sides. Electrical simulation in or near the sonic nucleus in anesthetized toadfish has elicited the same response as stimulation of the occipital nerves. For example, at rates up to 200 Hz one sound pulse results from each stimulation pulse (Demski and Gerald 1972, Fine 1979). The areas from which these "one to one responses" have been evoked are plotted in Fig. 21-2.

The most extensive studies on toadfish sonic motor neurons are those of Pappas and Bennett (1966) and their findings have been incorporated into a hypothetical model of the control mechanisms in toadfish sound production (Demski et al. 1973). Part of this model is reproduced in Fig. 21-3 in order to summarize their results. A detailed description of these electrophysiological findings is beyond the scope of this chapter and the reader is referred to the articles cited above. Pappas and Bennett have demonstrated physiologically and anatomically that the motor cells are coupled electrotonically by way of dendrosomatic and axosomatic tight junctions. The axosomatic junctions are thought to be the endings of fibers originating in a higher region of the

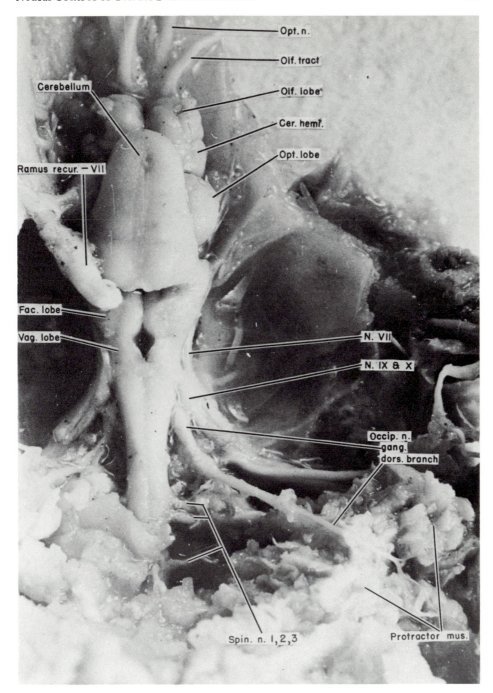

Figure 21-1. Dissection of brain and cranial nerves of formalin-preserved *Galeichthys (Arius) felis.* Right epiotic lamina removed to expose dorsal surface of protractor or sonic muscle and its innervation by dorsal branch of occipital nerve. Large utricular otolith (lapillus) removed (from Tavolga 1962).

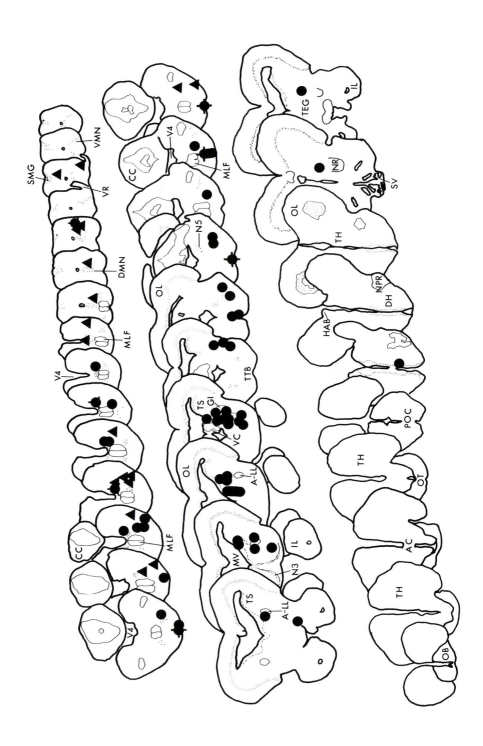

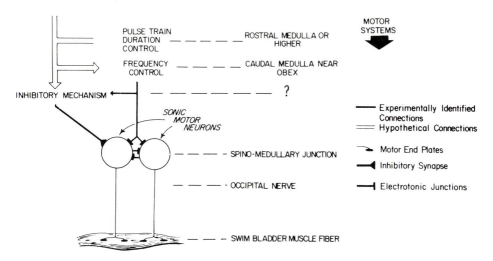

Figure 21-3. A tentative model of sonic motor mechanisms in toadfish (from Demski et al. 1973). See text for details.

brain. Axosomatic junctions having a typical synaptic cleft with numerous clear vesicles have also been found and these are thought to represent chemically mediated inhibitory synapses. Bennett (1971) suggests that electrotonic coupling between motor cells provides a positive feedback system which synchronizes the activation of many cells in a nucleus. This, along with a polyaxonal innervation of the muscle fibers, is undoubtedly important for the maintenance of the fast concentration rates of toadfish sonic muscle. Intracellular recordings from toadfish sonic motor neurons during tetanic stimulation of the spinal cord indicate that there is an initial small depolarization followed by a hyperpolarization which lasts throughout a sustained burst of spikes. This inhibition is thought to be mediated by the chemical-type synapses observed on

◀ Figure 21-2. Distribution of Prussian blue-marked stimulation sites from which sounds have been evoked from the brain in anesthetized *O. beta*. Points from both sides of the brain have been plotted on the same side on tracings of representative frontal sections taken from the olfactory bulbs (lower left) to the spinal cord (upper right). Symbols: triangles, one-to-one responses; circles, grunts; crosses, boatwhistle-like sounds. Overlapping symbols indicate points from which more than one type of sound was evoked. Elongated symbols represent grunt responses evoked from several contiguous sites along the same track. Abbreviations: AC, anterior commissure; A-LL, acousticolateral lemniscus; CC, corpus cerebelli; DH, dorsal hypothalamus; DMN, dorsal (sonic) motor nucleus; GI, ganglion isthmi; HAB, habenula; IL, inferior lobe of the hypothalamus; MLF, medial longitudinal fasciculus; MV, midbrain ventricle; N3, oculomotor nerve, N5, motor nucleus of the trigeminal nerve; MPR, nucleus prerotundus (preglomerulosus); NR, nucleus rotundus (glomerulosus); OB, olfactory bulb; OL, optic lobe; OT, optic tract; POC, postoptic commissures; SMG, supramedullary ganglion cells; SV, saccus vasculosus; TEG, midbrain tegmentum; TH, telencephalic hemisphere; TS, torus semicircularis; TTB, tractus tecto-bulbaris; V4, fourth ventricle, VC, valvula cerebelli; VMN, ventral motor nucleus; VR, ventral root of spinal nerve (from Demski and Gerald 1972).

the motor neurons. The duration of this burst is about 65 msec with a spike frequency of 185/sec. Longer bursts lasting up to 1 sec, can be recorded from the cells following transection of the medulla at a level somewhat behind the cerebellum. Presumably, this procedure removes some type of higher inhibitory influence which may normally control burst duration and therefore the length of sound pulse trains. The duration of short bursts suggests that they may be neurophysiological correlates of the aggressively motivated grunt sound while the duration of the longer burst responses suggests that they may be functionally related to the courtship or boatwhistle call. The discharge frequency of the motor neurons themselves appears to be determined in an area near the obex in the caudal medulla since removal of this area results in a loss of burst activity. The frequency control region appears to be electrotonically coupled to the sonic motor cells, as is indicated by the fact that bursts can be triggered by driving the system backward, in other words, through antidromic activation of the swimbladder nerves. It seems likely that the frequency control area normally activates the motor neurons orthodromically via this same electrotonically coupled system. Additional, physiological evidence suggests that collaterals from this afferent system trigger inhibitory neurons that cause the hyperpolarizations recorded in the motor cells. Some of the basic conclusions reached from these electrophysiological studies are that (1) control of toadfish sonic motor neurons involves both inhibitory as well as excitatory processes; (2) the lower motor system is coupled electrotonically; (3) spike frequency is set in a center near the obex in the lower medulla; and (4) burst duration is controlled by a mechanism located somewhere above the lower medulla.

Electrical stimulation in the vicinity of the proposed frequency control mechanism in lightly anesthetized, partially immobilized toadfish has elicited both grunts and one-to-one responses (Demski and Gerald 1972). These sounds have been evoked both separately and at the same time (Fig. 21-2). It has been suggested that the one-to-one responses result from activation of sonic neurons through the electrotonically coupled afferent fibers while grunts result by triggering the frequency control region and that the latter response may involve the inhibitory system as well (Demski et al. 1973). Hyperpolarizations recorded in motor cells at the beginning of burst responses are consistent with this idea (see above). Simultaneous grunt and one-to-one sounds have also been evoked from higher regions of the medulla as well as from the spinal cord near the caudal end of the sonic motor nucleus (Fig. 21-2). Activation of afferent fibers to the frequency control region and/or the inhibitory area is likely to be involved in triggering the sounds during medullary stimulation. For the spinal stimulation, one-to-one responses may result from direct activation of the motor cells, while the grunts may be triggered by antidromic activation of the frequency control area through the afferent fibers that form electrotonic junctions on the sonic neurons. The inhibitory system may be involved here as well.

Fine (1979) has observed fairly similar effects by stimulating the brain in anesthetized *O. tau.* In addition to the positive areas mentioned above, he found active sites in or near the medial longitudinal fasciculus and suggests that this fiber tract may be an important output pathway for the sonic motor system of toadfish. Recent studies in anesthetized Atlantic croakers, *Micropogon undulatus,* have also indicated that sound production can be evoked by electrical stimulation of areas of the medulla similar to those described above for *O. beta* (see Section 4 for details). This finding suggests the possibility that at least partly similar central mechanisms are involved in controlling sound production in fishes that may have widely different primary motor systems (see discussions in Sections 2 and 4).

Stimulation in the rostral basal medulla near the level of the trigeminal motor nucleus has evoked grunts and boatwhistle-like sounds in anesthetized male toadfish (Fig. 21-2, Demski and Gerald 1972, Fine 1979). In *O. beta,* these evoked boatwhistles frequently occurred as multiple repetitions of the hootlike portion of the call rather than the normal double hoot pattern characteristic of this species (Tavolga 1958, 1960). The location of these positive sites, somewhat situated between the lower motor mechanisms and several higher areas involved in sound production (described in Section 4), suggests that responses evoked from this region could be caused by activation of fibers of passage connecting these latter areas.

3.2 Hormonal Influences on Sonic Motor Systems

There is increasing evidence that sex steroid hormones may influence sound production by directly affecting primary sonic motor neurons as well as nerve cells at higher levels in sound production systems. Arnold, Nottebohm, and Pfaff (1976) have demonstrated that the area of the hypoglossal nucleus that supplies motor neurons to the syrinx in the zebra finch also has neurons that concentrate radiolabeled androgens. It was not possible to identify any particular syringeal motor neuron as also being an androgen-concentrating cell; however, the considerable overlap in the two populations suggests that this is the case. Similarly, Kelley (1980) has also demonstrated that the area of the motor nucleus of cranial nerves IX and X that supplies the larnyx in *Xenopus laevis* also contains androgen-concentrating neurons. It can be suggested that sonic motor neurons in certain fishes may also accumulate sex steroids since in toadfish, for example, normally only males produce courtship calls or boatwhistles yet females can be made to do so using brain stimulation (Demski and Gerald 1974, Fine 1979). Hypothetically, the male steroids could condition the sonic motor neurons by lowering their thresholds for being triggered by normal inputs and brain stimulation may represent a supernormal input that can overcome a relative lack of androgenic conditioning in females. Also possibly related to hormonal effects on toadfish sound production are observations that seasonal changes in the duration of toadfish boatwhistle calls are independent of termperature changes (Fine 1978). Fine suggests that perhaps the call pattern generators are sensitive to sex steroids and that they react to changes in androgen levels by altering call structure. It is reasonable to assume that the effect might be mediated via the proposed pulse-train duration control mechanism (see Section 3.1). It is clear that studies of hormone uptake patterns in the brains of sound-producing fishes are likely to yield considerable information on the neural regulation of sound production.

4 Higher Control Systems

Identification of higher regions of the fish brain which may influence the sonic motor apparatus is based almost exclusively on brain stimulation studies. Consistent grunt and boatwhistle sounds have been evoked by electrical stimulation in the midbrain in the area of the acousticolateral lemniscus, ganglion isthmi, and medial part of the torus semicircularis (region referred to as the sonic midbrain area, SMA) in both anesthetized

(Fig. 21-2, Demski and Gerald 1972, Fine 1979) and unanesthetized free-swimming toadfish (Fig. 21-4, Demski and Gerald 1974). In studies on free-swimming *O. beta,* some of the boatwhistles evoked from the SMA appeared to have the double hoot normally characteristic of this species (Fig. 21-5).

In a recent series of experiments croaking has been evoked by electrical stimulation of the brain in anesthetized Atlantic croakers, *Micropogon undulatus* (Demski, Williams, and Dulka unpublished observations). Seven animals ranging in size from 11½ to 13 cm standard length were used in this study. They were caught in a trawl in the Mississippi Sound in July 1979. The techniques used were very similar to those used with toadfish (see details in Demski and Gerald 1972) and only a few differences need to be mentioned. Due to an equipment failure, single pulses rather than biphasic square-wave pulse pairs were used. Electrodes were approximately the same size as in the toadfish study, but were made from steel wire rather than insect pins and were coated with Insulex instead of varnish. Croaking was recorded using a hydrophone placed in the water surrounding the surgical apparatus holding the fish. The anesthetic used was a solution of 1 to 40,000 MS-222 in seawater. Stimulation sites

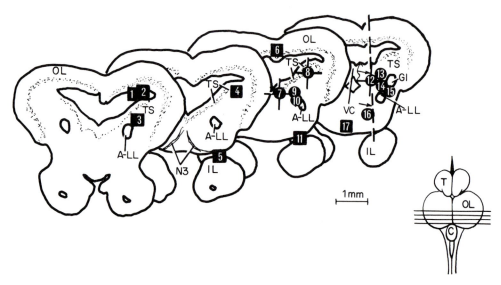

Figure 21-4. Distribution of Prussian blue-marked stimulation sites tested in the midbrain of free-swimming *O. beta.* Sites from both sides of the brain are plotted on one side on a series of representative frontal sections, the level of which is indicated on the dorsal view of a whole brain (lower right). Squares denote sites negative for sound production; circles indicate sites from which grunts were evoked, and crosses denote sites from which boatwhistles were elicited. Overlapping symbols represent points from which more than one type of sound was evoked. The dashed line indicates the track of a movable electrode along which site 12 was located. The arrows indicate the approximate locus of points along this track from which low threshold grunts were elicited. The scale mark was corrected for shrinkage due to tissue processing. Abbreviations: A-LL, acoustico-lateral lemniscus; C, cerebellum; GI, ganglion isthmi; IL, inferior lobe of hypothalamus; N3, oculomotor nerve; OL, optic lobe; T, telencephalic lobe; TS, torus semicircularis; VC, valvula cerebelli (from Demski and Gerald 1974).

were marked using the Prussian blue technique and experimental brains were embedded in paraffin, sectioned in the transverse plane, and counterstained with neutral
red. Identified stimulation points and their respective electrode tracks are reconstructed on a series of representative sections of the midbrain and medulla (Fig. 21-6).
Various frequencies of stimulation were used with currents up to 150 μA; duration
was constant at 1 msec. The details concerning stimulation parameters and the evoked
responses are listed in Table 21-1. Thresholds for evoked croaking seemed to be generally higher than those used to elicit sounds in toadfish. This could be due to many
factors including species differences and such procedural differences as using MS-222

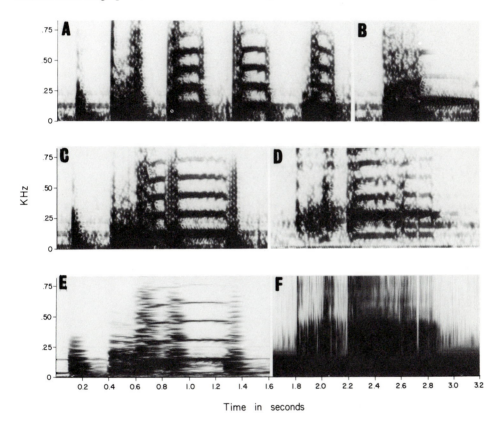

Time in seconds

Figure 21-5. Sonagraphs of boatwhistle sounds evoked by midbrain stimulation in free-
swimming *O. beta*. (A) A series of three boatwhistles preceded by several grunts evoked
by stimulation of site 12 (see Fig. 21-4) at 50 Hz and 10 μA. Analyzing filter was 45
Hz. (B) A single boatwhistle preceded by a long duration grunt component evoked by
stimulation of site 8 at 70 Hz and 80 μA. Filter width was 45 Hz. (C and D) Boatwhistles evoked by stimulation of point 12 at 50 Hz and 10 μA. Note double harmonic
portions preceded by grunts. This complex is characteristic of natural *O. beta* boatwhistles. Filter width was 45 Hz. (E) Same sound as in (C) but analyzed using a very
narrow filter (10 Hz) to emphasize the frequency characteristics of the sound. (F)
Same sound as in (D) but analyzed using a wide band filter (300 Hz) to emphasize the
temporal characteristics of the sound. The continuous horizontal banding in all records
is due to background noise in the laboratory (from Demski and Gerald 1974).

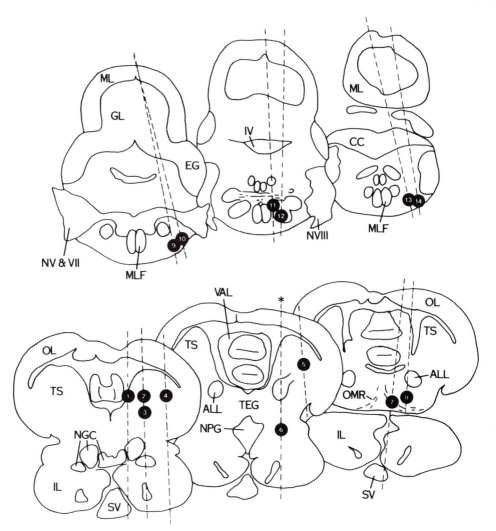

Figure 21-6. Distribution of Prussian blue-marked sites from which croaking was evoked in anesthetized Atlantic croakers, *Micropogon undulatus*. Sites are plotted on representative transverse sections through the midbrain (sites 1-8) and rostral medulla (sites 9-14). Dotted lines indicate reconstructed electrode tracks. Details of stimulation parameters and evoked responses are given in the text and Table 21-1. The asterisk denotes a track in which a low threshold response was evoked above the marked site (#6) (see Table 21-1 and text for explanation). Abbreviations: A-LL, acousticolateral lemniscus; CC, corpus cerebelli; EG, eminentia granularis; GL, granular layer of cerebellum; MLF, medial longitudinal fasciculus; NV and VII, root of trigeminal and facial nerves; NVIII, root of statoacoustic nerve; NGC, nucleus glomerulosus complex; NPG, nucleus preglomerulosus; OL, optic lobe; OMR, oculomotor root fibers; SV, saccus vasculosus; TEG, tegmentum of the midbrain; TS, torus semicircularis; VAL, valvula of cerebellum; IV, fourth ventricle.

Table 21-1. Stimulation Sites and Parameters Used to Evoke Croaking in *Micropogon undulatus*

Point No.	Fish No.	Stimulation parameters		Evoked responses
		Hz	μA	
1	4	75	100	Strong croaking at the termination of the stimulation (after-response).
	4	75	150	Weak croaking during stimulation.
2	4	70	150	Strong croaking as after-response.
3	4	80	150	Strong croaking as after-response (active area, approximately .4 mm).
	4	80	130	Threshold for consistent after-response.
4	4	75	80	Strong croaking after-response (active area, approximately 1 mm with double croaks evoked from the center of the area and single croaks elicited on either side of the center).
5	5	80	150	Croaking as an after-response.
6	3	90	150	Croaking as an after-response (note, croaking was evoked at 50 Hz and 60 μA at an unmarked, more dorsal position on this track).
7	3	90	150	Weak croaking after-response.
		55-90	80	After Prussian blue marking, strong croaking was evoked during stimulation.
8	5	80	150	Croaking as an after-response.
9	7	70	80	Strong croaking during stimulation.
10	1	50	100-150	Croaking, first as after-response, then during stimulation.
11	5	10, 30	150	Croaking during stimulation.
		50, 70	150	Croaking during stimulation.
		70	110	Threshold for consistent croaking during stimulation.
12	3	100	150	Strong croaking during stimulation.
13	1	50	150	Strong croaking during stimulation.
		10	150	Weak croaking during stimulation.
		50	100	Threshold for consistent croaking during stimulation.
14	7	70	150	Strong croaking during stimulation.
		70	70	Threshold for consistent croaking during stimulation.

in place of the urethane used in the toadfish study. As with toadfish, evoked responses occurred both during and at the termination of the stimulation (after-responses). Most responses appeared fairly normal as compared to croaking evoked by gentle squeezing in anesthetized, unoperated fish (Fig. 21-7). In both cases the croaks frequently consisted of 3 to 4 short broad frequency (up to 8 kHz) sound pulses of 20-50 msec duration (Fig. 21-7b,d,e). In a few instances longer sounds up to 20 pulses (Fig. 21-7c) or sounds with only 1 or 2 pulses (Fig. 21-7f,g) were evoked. The short and long croaks were occasionally elicited by squeezing unoperated anesthetized fish (Fig. 21-7a) and have also been recorded from free-swimming unanesthetized animals (Fish and Mowbray 1970). A total of 43 electrode tracks were run in a dorsoventral direction. Fourteen of 28 positive sites were marked and identified histologically.

Sites 1-5 (Fig. 21-6, Table 21-1) were all in the area of the midbrain equivalent to the SMA of toadfish. An unmarked positive site was located above site 6 along the track marked with the asterisk in Fig. 21-6 and this unmarked point was also likely to have been in or near the SMA. The second group of sites (6-8) were in the basal midbrain and their position suggests that a functional system may extend in a ventral direction from the caudal part of the SMA into the reticular formation of the basal tegmentum. Other positive sites (9-14) were found in the ventrolateral medulla in positions analogous to many positive points in toadfish (see Section 3.1). These data in croakers compare quite well with the results of the earlier toadfish studies and suggest that relatively similar sonic mechanisms may exist in the two types of fishes studied, at least at midbrain and medullary levels.

A group of croakers was also subjected to a series of brain transections in an attempt to determine at least grossly, if the midbrain region from which sounds were evoked is critical for normal croaking. The animals were anesthetized in a solution

Figure 21-7. Sound spectrogram of croaks evoked by gentle squeezing of unoperated ▶ (a,b) and by electrical stimulation of the brain in anesthetized (c-g) Atlantic croakers, *Micropogon undulatus*. All sounds are composed of broad-band pulses with frequencies extending up to 8 kHz in some cases. (a) Long croak in an unoperated fish. Note that the sound complex appears to have three components of 1, 16, and 3 pulses, respectively. The interpulse interval is fairly regular, averaging approximately 43 msec in the long portion of the sound. (b) Short croak in an unoperated fish. (c) Long croak evoked by electrical stimulation of site 11 in the medulla at 80 Hz and 150 μA (see Table 21-1 and Fig. 21-6). Note the regularity of the interpulse interval (average of about 47 msec) in the extended part of the sound and the overall similarity of this sound to the more "natural" one shown above in (a). (d) A short croak evoked as an after-response (off response) to electrical stimulation at 70 Hz and 150 μA in an unmarked spot in the caudal midbrain tegmentum. Note similarity to "natural" sound in (b). (e) A short croak evoked as an after-response to stimulation of site 4 in the sonic midbrain area at 75 Hz and 100 μA (see Table 21-1 and Fig. 21-6). (f and g) Double (f) and single (g) pulsed sounds or "brief croaks" resulting from electrical stimulation of an unmarked site in the medulla at 70 Hz and 150 μA. The strong band at approximately 1 kHz most likely represents artifacts caused by the acoustic conditions of the experimental apparatus and/or background noise in the laboratory. Spectrograms were made on a Kay 6061A Sonagraph using a wide-band (300 Hz) filter.

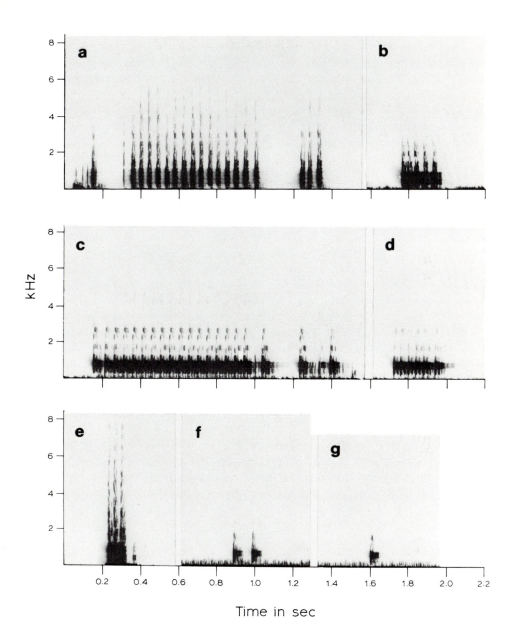

Time in sec

of either 1 to 30,000 or 1 to 40,000 MS-222 and seawater and then tested for croaking in response to gentle squeezing. Those fish that responded underwent a brain transection using a sharpened spatula. Following this, each fish was again tested for croaking. Although the brains of these animals have not been examined histologically, it seemed apparent at the time of surgery that cuts through the rostral part of the midbrain had no striking effect on croaking whereas cuts made in the caudal midbrain near the cerebellum resulted in a loss of or at least a noticeable weakening in sound production. These preliminary observations are thus consistent with the notion of a SMA as defined from the stimulation studies.

The SMA of fishes is adjacent to the midbrain auditory region in the dorsal part of the torus semicircularis as reported for several teleosts and illustrated in Fig. 21-8. This finding of contiguous sound production and acoustic areas in the fish midbrain is similar to the situation observed in a variety of vertebrates including amphibians, reptiles, birds, and mammals (see references in Jürgens and Pratt 1979). Brown (1969) has suggested that the proximity of midbrain sound producing areas to midbrain acoustic regions has important functional implications for the development and maintenance of normal vocal patterns. This may also be the case in sound-producing fishes. Another important point regarding midbrain sonic areas relates to possible hormonal influences. In the zebra finch the midbrain vocal area (nucleus intercollicularis) contains many neurons that concentrate tritium-labeled testosterone (Arnold et al. 1976). Androgen uptake has also been demonstrated in the midbrain sonic area of the chaffinch (Zigmond, Nottebohm, and Pfaff 1973). The suggestion has been made that these androgen sensitive cells in birds are involved in the regulation of singing related to circulating levels of male hormones. I would also like to suggest that a similar situation may be occurring in certain sound-producing fishes such as toadfish in which there are sexual and seasonal differences in call production.

The finding of similar sonic areas in the midbrain of a great variety of vertebrates that employ obviously nonhomologous structures to produce sound raises the question of whether or not these midbrain areas could possibly be homologous. Even in toadfish and croakers the lower sonic motor mechanisms are not likely to be homologous (see Section 2), yet electrical stimulation in a very similar area of the midbrain results in strong sound production in both types. I propose that, at least in fishes, the sonic midbrain areas may be at least partly homologous based on their similar topographic positions and probably rather direct connections with midbrain auditory areas. These sonic regions can be perhaps considered the next link, after midbrain feature analysis, in a system controlling acoustic communication in teleosts and possibly other vertebrates as well. Obviously much more information is needed to adequately test this hypothesis. It should also be mentioned that similar observations have been made concerning the electromotor systems of certain teleosts. For example, electrical stimulation in the torus semicircularis in both knifefish, *Eigenmannia virescens* (Scheich 1977), and stargazers, *Astroscopus y-graecum* (Leonard and Willis 1975), has elicited electric organ discharges. These fishes use vastly different electromotor systems, a situation somewhat analogous to sonic systems of croakers and toadfish. Here again, the common factor may relate to sensory analysis performed by certain toral regions and the functional adaptability of using the information to modulate activity in related motor systems (see also Bullock, Chapter 27).

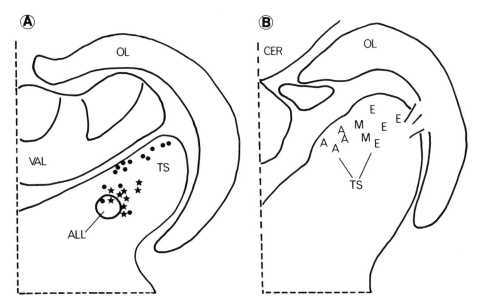

Figure 21-8. Transverse sections through the optic lobe and torus semicircularis of goldfish (A) and catfish, genus *Ictalurus* (B). Dots indicate single units responding to acoustic stimulation alone while stars represent single neurons that responded to both acoustic and visual stimulation. A's represent region where acoustic evoked potentials could be recorded; M's indicate area for potentials evoked by mechanical stimulation to lateral line receptors; E's cover area where responses to electrical stimuli were recorded. Abbreviations: ALL, acousticolateral lemniscus; CER, cerebellum; OL, optic lobe; TS, torus semicircularis; VAL, valvula of cerebellum. A and B were redrawn from Page (1970) and Knudsen (1977), respectively.

With regard to other so called "higher areas" involved in teleost sound production, Fine (1979) located several thalamic sites positive for evoked sound production in toadfish. It can be suggested that these sites may also be related to the integration of sensory information relevant to sound production. The preoptic region (Fig. 21-2) and inferior lobe of the hypothalamus are additional diencephalic areas from which sounds have been evoked in toadfish (Demski and Gerald 1972, 1974, Fine 1979). These areas are involved in reproductive and aggressive behavior in several teleosts (Demski and Knigge 1971, Demski 1973, Demski, Bauer, and Gerald 1975) and the preoptic region contains androgen concentrating neurons in all species studied (Davis, Morrell, and Pfaff 1977, Demski 1978). This latter region appears to be involved in regulating hormonally dependent activities which in some fishes include acoustic signals as components. It is likely that these higher regions mediate their sonic effects via connections to the SMA. This idea is suggested by experiments in squirrel monkeys (Jürgens and Pratt 1979) in which vocalizations evoked by electrical stimulation in widespread areas of the brain were blocked by bilateral lesions in the periaqueductal gray and laterally adjacent tegmentum, an area that appears to be the mammalian equivalent of the toadfish SMA.

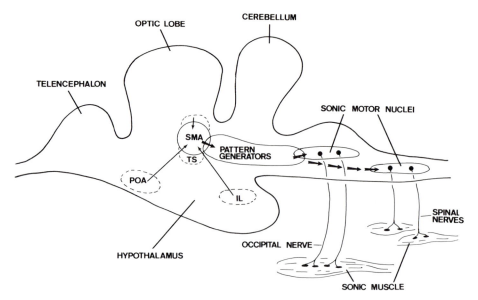

Figure 21-9. Summary model of neural systems involved in the control of sound production in teleosts. The systems are plotted on a hypothetical sagittal section of a fish brain. The occipital nerve system is characteristic of most fishes studied while the spinal sonic motor apparatus is thought to be present in croakers and drums (Sciaenidae). See text for further details. Abbreviations: IL, inferior lobe of hypothalamus; POA, preoptic area; SMA, sonic midbrain area; TS, acoustic area of the torus semicircularis.

5 Summary

Fig. 21-9 summarizes most of the information discussed in this chapter. Sonic motor cells in most of the fishes studied are located in nuclei in the caudal medulla and their axons exit the brain as the occipital nerves which innervate the sonic muscles. The main exception to this pattern is found in croakers and drums (Sciaenidae) which have sonic motor cells presumed to be located in the spinal cord with axons running in spinal nerves. With the great variety of teleosts available it would not be surprising to eventually find more exceptions. As far as is known, and this relates mostly to toadfish, pattern generators are located in the tegmentum, somewhere between the caudal midbrain and level of the obex. The brain stimulation data suggest that reticular systems are involved, and Fine's (1979) results indicate that the medial longitudinal fasciculus may be a possible output pathway for these mechanisms. The sonic midbrain area appears to have a direct influence on lower sonic motor systems. This area probably receives acoustic inputs from nearby auditory regions in the torus semicircularis and perhaps visual and other sensory inputs as well (see Discussion in Demski et al. 1973). The preoptic area and the inferior lobe which are involved in complex sexual and aggressive responses may influence sound production by connections with the sonic midbrain area. In addition, the preoptic area, at the least, and possibly several other regions involved in the central control of sound production in fishes may be influenced by changes in circulating androgens via local hormone-concentrating neurons.

Acknowledgments. The Sonagraphs in this paper were kindly prepared by Dr. Arthur N. Popper. I also wish to thank Mr. Edward J. Williams and Mr. Joseph G. Dulka for their help on the croaker study. The use of facilities and staff personnel at The Gulf Coast Research Laboratory is also gratefully acknowledged. Support provided in part by BioMedical Sciences Research Support Grant RR 07114-08.

References

Arnold, A. P., Nottebohm, F., Pfaff, D. W.: Hormone concentrating cells in vocal control and other areas of the brain of the zebra finch (*Poephila guttata*). J. Comp. Neurol. 165, 487-512 (1976).

Barker, J. L., Hoffman, P. N., Gainer, H., Lasek, R. J.: Rapid transport of proteins in the sonic motor system of the toadfish. Brain Res. 97, 291-301 (1975).

Bennett, M. V. L.: Electric organs. In: Fish Physiology, Vol. 5. Hoar, W. S., Randall, D. J. (eds.). New York: Academic Press, 1971, pp. 347-491.

Brown, J. L.: The control of avian vocalization by the central nervous system. In: Bird Vocalizations. Hinde, R. A. (ed.). Cambridge: Cambridge Univ. Press, 1969, pp. 79-96.

Cohen, M. J., Winn, H. E.: Electrophysiological observations on hearing and sound production in the fish, *Porichthys notatus*. J. Exp. Zool. 165, 355-370 (1967).

Davis, R. E., Morrell, J. I., Pfaff, D. W.: Autoradiographic localization of sex steroid concentrating cells in the brain of the teleost *Macropodus opercularis* (Osteichthyes: Belontiidae). Gen. Comp. Endrocrinol. 33, 496-505 (1977).

Demski, L. S.: Feeding and aggressive behavior evoked by hypothalamic stimulation in a cichlid fish. Comp. Biochem. Physiol. 44A, 685-692 (1973).

Demski, L. S.: Neuroanatomical substrates of reproductive behavior in male sunfish (Genus *Lepomis*). Ann. Biol. Anim. Bioch. Biophys. 18, 831-836 (1978).

Demski, L. S., Bauer, D. H., Gerald, J. W.: Sperm release evoked by electrical stimulation of the fish brain: A functional-anatomical study. J. Exp. Zool. 191, 215-232 (1975).

Demski, L. S., Gerald, J. W.: Sound production evoked by electrical stimulation of the brain in toadfish (*Opsanus beta*). Anim. Behav. 20, 507-513 (1972).

Demski, L. S., Gerald, J. W.: Sound production and other behavioral effects of midbrain stimulation in free-swimming toadfish (*Opsanus beta*). Brain Behav. Evol. 9, 41-59 (1974).

Demski, L. S., Gerald, J. W., Popper, A. N.: Central and peripheral mechanisms of sound production. Am. Zool. 13, 1141-1167 (1973).

Demski, L. S., Knigge, K. M.: The telencephalon and hypothalamus of the bluegill (*Lepomis macrochirus*): Evoked feeding, aggressive and reproductive behavior with representative frontal sections. J. Comp. Neurol. 143, 1-16 (1971).

Fawcett, D. W., Revel, J. P.: The sarcoplasmic reticulum of a fast-acting fish muscle. J. Biophys. Biochem. Cytol. Suppl. 10, 89-109 (1961).

Fine, M. L.: Seasonal and geographical variation of the mating call of the oyster toadfish *Opsanus tau* L. Oecologia 36, 45-57 (1978).

Fine, M. L.: Sounds evoked by brain stimulation in the oyster toadfish *Opsanus tau* L. Exp. Brain Res. 35, 197-212 (1979).

Fish, M. P., Mowbray, W. H.: Sounds of the Western North Atlantic Fishes. Baltimore: Johns Hopkins Univ. Press, 1970.

Gainer, H., Klancher, J. E.: Neuromuscular junctions in a fast-contracting fish muscle. Comp. Biochem. Physiol. 15, 159-165 (1965).

Jürgens, U., Pratt, R.: Role of the periaqueductal grey in vocal expression of emotion. Brain Res. 167, 367-378 (1979).

Kelley, D. B.: Auditory and vocal nuclei in the frog brain concentrate sex hormones. Science 207, 553-555 (1980).

Knudsen, E. I.: Distinct auditory and lateral line nuclei in the midbrain of catfishes. J. Comp. Neurol. 173, 417-432 (1977).

Leonard, R. B., Willis, W. D.: Activation of the electric organ motor nucleus by stimulation in the region of the torus semicircularis in *Astroscopus y-graecum.* Neuroscience Abstr. 755 (1975).

Packard, A.: Electrophysiological observations on a sound-producing fish. Nature 187, 63-64 (1960).

Page, C. H.: Electrophysiological study of auditory responses in the goldfish brain. J. Neurophysiol. 33, 116-128 (1970).

Pappas, G. D., Bennett, M. V. L.: Specialized junctions involved in electrical transmissions between neurons. Ann. N.Y. Acad. Sci. 137, 495-508 (1966).

Scheich, H.: Neural basis of communication in the high frequency electric fish, *Eigenmannia virescens* (jamming avoidance response) III. Central integration in the sensory pathway and control of the pacemaker. J. Comp. Physiol. 113, 229-255 (1977).

Schneider, H.: Morphology and physiology of sound-producing mechanisms in teleost fishes. In: Marine Bio-Acoustics II. Tavolga, W. N. (ed.). Oxford: Pergamon Press, 1967, pp. 135-158.

Schneider, H., Hasler, A. D.: Laute und Lauterzeugung beim Süsswassertrommler *Aplodinotus grunniens* Rafinesque (Sciaenidae, Pisces). Z. Vergl. Physiol. 43, 499-517 (1960).

Skoglund, C. R.: Neuromuscular mechanisms of sound production in *Opsanus tau.* Biol. Bull. 117, 438 (1959).

Skoglund, C. R.: Functional analysis of swim-bladder muscles engaged in sound production of the toadfish. J. Biophys. Biochem. Cytol. Suppl. 10, 187-200 (1961).

Tavolga, W. N.: Underwater sounds produced by two species of toadfish, *Opsanus tau* and *Opsanus beta.* Bull. Mar. Sci. Gulf Carib. 8, 278-284 (1958).

Tavolga, W. N.: Sound production and underwater communication in fishes. In: Animal Sounds and Communication. Lanyon, W. E., Tavolga, W. N. (eds.). Washington, D.C.: Am. Inst. Biol. Sci., 1960, pp. 93-136.

Tavolga, W. N.: Mechanisms of sound production in the ariid catfishes *Galeichthys* and *Bagre.* Bull. Am. Mus. Nat. Hist. 124, 1-30 (1962).

Tavolga, W. N.: Sonic characteristics and mechanisms in marine fishes. In: Marine Bio-Acoustics. Tavolga, W. N. (ed.). Oxford: Pergamon Press, 1964, pp. 195-211.

Tavolga, W. N.: Sound production and detection. In: Fish Physiology, Vol. 5. Hoar, W. S., Randall, D. J. (eds.). New York: Academic Press, 1971, pp. 135-205.

Tavolga, W. N.: Sound Production in Fishes. Benchmark Papers in Animal Behavior, Vol. 9. Stroudsberg, Pa.: Dowden, Hutchinson and Ross, 1977.

Zigmond, R. E., Nottebohm, F., Pfaff, D. W.: Androgen-concentrating cells in the midbrain of a songbird. Science 179, 1005-1007 (1973).

Discussion

WILLOWS: When you stimulated in the motor nuclei, was the response part of a pattern, or was it a rather discrete motor action that *looked* like part of the pattern?

DEMSKI: Most of the time, we got a one-to-one response that looked like we were just activating the cells synchronously with the stimulus. Sometimes, however, we evoked a grunt sound superimposed over the one-to-one response. This grunt looked and sounded like a normal grunt. We may have triggered the frequency control area while simultaneously triggering motor neurons.

WILLOWS: The fact that you occasionally see either all or part of a coordinated response, suggests that there may be at least some part of the motor pattern generator as far down as that. Alternately, the motor units themselves may have a powerful feedback effect on a pattern generator way up forward.

DEMSKI: Bennett [Pappas and Bennett (1966)] stimulated the motor nerve directly. After a short latency period, he obtained neural analogs of the grunt response. Apparently, the motor system can trigger the pattern generator, and thus result in sound production. The frequency control area may be in the rostral part of the motor nucleus. The cells up there are somewhat smaller, and if the fibers are electrotonically coupled to the motor neurons, the stimulus could also antidromically activate the pattern generator system.

CORWIN: Will these animals echo in response to a sound from others?

DEMSKI: Yes, they are noted for doing this, and playback of sounds stimulates males to produce sounds or increase their calling rate.

CORWIN: Is it possible that you might have been stimulating the sound receiving and analyzing system? Some of your stimulus tracks where you evoked boatwhistles were right in the torus semicircularis.

DEMSKI: They were, and I may have been activating the receiving system.

CORWIN: So you might have been going through a loop, from sensory to motor?

DEMSKI: That could indeed be the case. However, it seems to me that, if I were stimulating in an auditory analyzing area, the unpatterned stimulation should disrupt its integrative functions.

Chapter 22

Sound Production in the Naked Goby, *Gobiosoma bosci* (Pisces, Gobiidae)—A Preliminary Study

1 Introduction

To date, *Bathygobius soporator, Gobius jozo,* and *Neogobius melanostomus* are the reported soniferous species within the large ubiquitous perciform family, Gobiidae. The breeding males of *B. soporator* emit low pitched grunts (frequency, 100-150 Hz, non-harmonic; duration, 150-350 msec) as a communicatory signal associated with pre-spawning behavior (Tavolga 1958). Similar hydrodynamic sounds were noted in *G. jozo* and *N. melanostomus* by Kinger and Protasov et al., respectively (in Tavolga 1976). While observing the nest-guarding behavior of a male *Gobiosoma bosci* (naked goby) in a 15-gallon aquarium with six conspecifics (3 females and 3 males; total body length ranges from 28 to 40 mm), a clicking sound produced by this male was recorded. Succeeding additional observations and recordings were thus made. The recording equipment consisted of a hydrophone (Chesapeake Instrument Co., SH-102) and a recorder (Sony TC158SD). The present paper summarizes the results of a preliminary study on the acoustic behavior of this species.

Gobiosoma bosci ranges from Connecticut and Long Island Sound to Campeche, Mexico (Dahlberg and Conyers 1973). It is typically an estuarine species occurring in brackish water areas near sources of fresh water. The microhabitat preference for local populations appears to be large oyster reefs and small isolated oyster clumps that are exposed at low tide which appears to be typical for this species (Hoese 1966, Dahlberg and Conyers 1973). Adults may reach a total length of 53 mm or longer (Dawson 1969). Except for the observation of *G. bosci* nesting behavior made by Dahlberg and Conyers (1973), little behavioral work has been done on this species.

Unproductive sound recordings were made from January to March 1979. The goby sounds were first recorded on June 18 but acoustic activity might have been occurring prior to this date as no effort was made to record goby sounds from April to May. Sound production had continued into October when this manuscript was in preparation. Reproductive activity took place during the observation period June to Octo-

*Harbor Branch Institution, Inc., RR 1, Box 196A, Fort Pierce, Florida 33450.

ber. However, successful spawning was only observed on three occasions, and pre-spawning behavior was missed.

2 Conditions of Sound Production

The clicking sound was observed under the following circumstances and the sounds were produced by the male of this species. No female was noted to produce clicking sounds.

(1) *Sound production while not in visual contact with other fish.* A male staying in-side its shelter or within its territory sometimes produced the clicking sounds without visual contact with other females. The sounds caused by chasing or fighting among other fishes (conspecifics) often lead to sound production by a hidden male that had no visual contact with other fishes. Sound production took place at night when all the lights in the laboratory were turned off.

(2) *Sound production in male-male interactions.* Sound emission was observed when male-male interactions took place at the territorial boundary.

(3) *Sound production in male-female interactions.* A territorial male would approach a gravid female by repeated, short, forward locomotion; its body sank closely to the ground, head remained down, pectorals flipped repeatedly accompanied with no signif-icant changes in position or resulted in only limited forward movement. Repeated sounds occurred throughout these performances. The male was able to move to less than three centimeters from the gravid female without leading to behavioral changes from the latter or an attack by either male or female interactants. It is interesting to note that an approaching male usually led to agonistic response by the female during nonbreeding season (Mok in prep.). When the male was approaching, the female fre-quently responded by remaining in a resting, nonaggressive posture (median fins were not widely spread). The male often chased the female away while producing sounds or the female either swam away or toward the male. The latter response of the female caused the male to return to his shelter. Repetition rate and intensity of the sound increased continuously as the female followed him further into his territory (Fig. 22-1C, arrow indicates these increments). While the female was following the male, her brown ground body color in the areas between the vertical bars converted to light tan; the vertical bars remained distinct (the predominant color pattern for both male and female is characterized by 9-10 dark vertical bars separated by narrow brown inter-spaces). The male would then turn around, face the female, and again make character-istic sounds. In some instances the male hopped while the female was following him. Eventually the male chased the female away. After her departure, sound production usually would last for 1 min or less.

(4) *Sound production during spawning.* The male was relatively inactive in sound production during spawning (actual pairing inside its nests). Only a very limited num-

ber of erratic sounds released by obscure factors other than intrusion were emitted by the spawning male during the temporary spawning female residence (about 2 to 3 hr) inside the male's nest. The spawning male usually, but not always, responded to intrusion by producing sounds. Despite intrusion, the spawning male seldom moved away from its nest in order to drive out the intruders (usually other females). Other females which tried to enter the nest during the spawning were rejected by the spawning male. Repeated lateral tail beatings were performed by the spawning fishes. Sound production was not accompanied by these tail movements. No acoustic activity from the spawning male accompanied the departure of the spawning female from the nest. Whether acoustic activity also takes place in the prespawning stage (when the male allows the ripe female into his nest without chasing her away) remains to be answered.

(5) *Sound production toward intruders.* Intrusion by either males or females into a defended territory usually triggered sound production from the defending territorial male. Intrusion caused the nest-guarding male to patrol its territory more frequently. It moved closely to the intruders and produced sounds before chasing or striking them. Non-nest-guarding males also moved toward intruders and emitted sounds. The repetition rate was not as high as that produced when a male was followed by a gravid female. The responding fish usually stayed quiescent until the territory owner chased it away. These interactions were occasionally observed to end in attack encounters. Sounds were usually produced right after the attack regardless whether the sound producer won or lost the bout.

(6) *Sound production by a bystander.* A male might emit sounds while watching the aggressive interaction between other conspecifics.

3 Description of Sounds

The average duration of sounds is 27 msec (the average was estimated from 11 occurrences of sound). Repetition rate and intensity were variable (Fig. 22-1A-C); repetition rate can reach as high as three sounds per second when the male was followed by a ripe female into his territory. The sounds are nonharmonic. Some sounds could be resolved into two temporal components; energy in the second part concentrates in one or two narrow frequency bands at about 2 or 4 kHz (Figs. 22-4, 22-5, and 22-6; short arrows point to these bands). Frequency composition is highly variable; it may be associated with size and emotional state of the fish among many other possible determinants. The largest male in the aquarium produced a low frequency sound with energy between 1 and 2 kHz just prior to the approach of a ripe female (Fig. 22-2). The frequency increased with repetition rate as the female approached further into his territory (Fig. 22-3). The frequency distribution of these sounds by this male (Fig. 22-3) differs from that of the smaller males; it is characterized by the frequency discontinuance between 2 and 4 kHz and the horizontal bars between 4 and 6 kHz. The sounds of the smaller males have some energy spread to the higher frequency range (Figs. 22-4, 22-5, and 22-6).

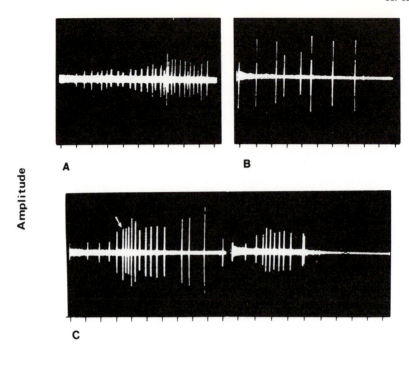

Time (2 seconds/ unit)

Figure 22-1. Oscillograms of sounds produced by *Gobiosoma bosci* males. (A) The sounds of a male responding to an approaching ripe female. (B) The sounds produced in a male-male interaction. (C) The sounds of a male responding to an approaching ripe female.

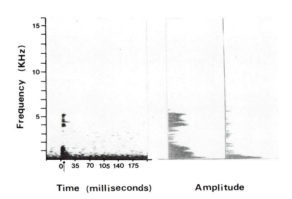

Figure 22-2. Spectrogram of a sound made by the largest male in the aquarium (40 mm TL) prior to the ripe female's approach, narrow band filter (left illustration). A frequency-amplitude section taken at the point indicated by the arrow on the time scale (middle illustration). A frequency-amplitude section of the background noise (right illustration).

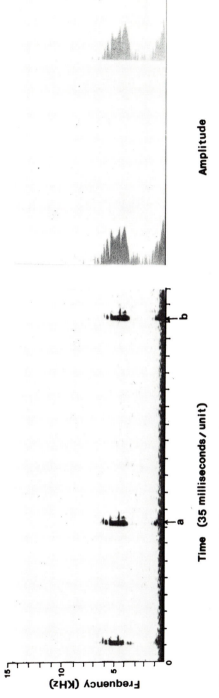

Figure 22-3. Spectrogram of the sounds made by the largest male in the aquarium responding to an approaching ripe female, narrow band filter (left illustration). The frequency-amplitude sections of the second and third sounds (a and b) indicated by the arrows on the time scale (right illustration).

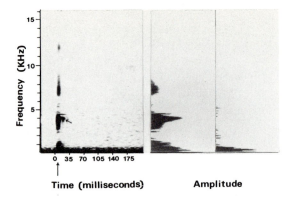

Figure 22-4. Spectrogram of a sound produced in a male-female interaction, narrow band filter (left illustration). A frequency-amplitude section made at the point indicated by the long arrow on the time scale (middle illustration). A frequency-amplitude section of the background (right illustration). Short arrow indicates the secondary portion of the sound.

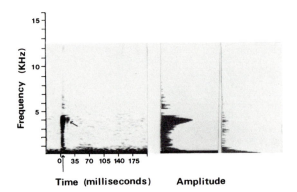

Figure 22-5. Spectrogram of a sound produced in a male-female interaction, narrow band filter (left illustration). A frequency-amplitude section made at the point indicated by the long arrow on the time scale (middle illustration). A frequency-amplitude section of the background noise (right illustration). Short arrow points to the secondary portion of the sound.

Figure 22-6. Spectrograms of some *Gobiosoma bosci* male sounds, narrow band filter ▶ (illustration a); a frequency-amplitude section made at the point indicated by the long arrow on the time scale (illustration b); a frequency-amplitude section of the background noise (illustration c). (A) A sound occurring in a male-male interaction. (B-D) Sounds occurred in male-female interactions. Short arrows point to the secondary portion of the sound.

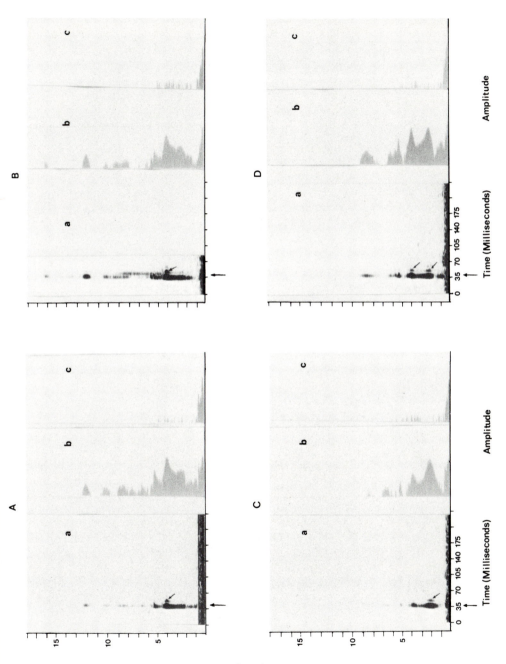

4 Discussion

The *Gobiosoma bosci* male sounds are obviously distinguishable from the *Bathygobius* sounds characterized by their low frequency range, low intensity, and comparatively long duration (Tavolga 1958). On the basis of experimental studies and the characteristic low frequency sounds of the *Bathygobius soporator,* Tavolga (1958) suggested that the sounds might be produced by the squirting of water through the gill openings. The high frequency range of the sounds in *G. bosci* suggests a stridulatory mechanism. Sound emission in this species synchronizes with either sideward shaking of the head or backward pulling of the body associated with slightly upward flipping of the tail.

It is generally agreed that the usable frequency range in most fish does not extend beyond about 2 kHz although ostariophysines have a higher upper range (Tavolga 1971, Myrberg, Gordon, and Klimley 1976). The upper frequency limit of hearing in *Gobius* (Gobiidae) is 800 Hz (Dijkgraaf 1952). The high frequency range of the *G. bosci* male sounds poses a critical question regarding the effectiveness of the sound as a communicatory signal and as to whether *G. bosci* can hear the sounds.

Sound production in *G. bosci* can be released under diverse conditions. The responses of the interactants to these sounds suggest that the sounds are not highly effective in threatening. The body posture and movement which are associated with the approach of a sound-producing male moving toward a female do not indicate a highly aggressive level by themselves (Mok in prep.). Under this particular interacting situation, acoustic signals alone are not associated with aggression. The combination of acoustic signal and fin and body displays carry a specific message to a particular receiver (for example, male versus female). The acoustic signal may be as important as the body display. For instance, it was observed once that as a ripe female followed a sound-producing male, she turned her attention to another sound producer by swimming toward the latter as soon as the first male stopped emitting sounds. This suggests that the sounds alone may serve as an attractive signal to ripe females.

Acknowledgments. The work reported was supported by a postdoctoral fellowship from the Harbor Branch Institution, Inc. Special thanks are due to R. Grant Gilmore, who suggested that *Gobiosoma* would be a fruitful group to study, to Julian Adams, who kindly prepared the sound spectrograms and oscillograms, and to Jush Wallman for the loan of equipment. I also wish to express my gratitude to Joseph Bricker for his technical assistance in electronic problems and his advice on acoustic problems. Robert S. Jones, R. Grant Gilmore, Alan R. Emery, Joseph Bricker, and William N. Tavolga reviewed the manuscript. Michael Clark helped immensely in field collecting. This paper is Contribution No. 180 of the Harbor Branch Foundation.

References

Dahlberg, M. D., Conyers, J. C.: An ecological study of *Gobiosoma bosci* and *G. ginsburgi* (Pisces, Gobiidae) on the Georgia coast. Fish. Bull. 71(1), 279-287 (1973).

Dawson, D. E.: Studies on the gobies of Mississippi Sound and adjacent waters II. An illustrated key to the gobioid fishes. Publ. Gul. Coast Res. Lab. Mus. 1, 1-59 (1969).

Dijkgraaf, S.: Über die Schallwahrnehmung bei Meeresfischen. Z. Vergl. Physiol. 34, 104-122 (1952).

Hoese, H. D.: Habitat segregation in aquaria between two sympatric species of *Gobiosoma*. Publ. Inst. Mar. Sci. Univ. Tex. 11, 7-11 (1966).

Myrberg, A. A., Jr., Gordon, C. R., Klimley, A. P.: Attraction of free ranging sharks to low frequency sound, with comments on its biological significance. In: Sound Reception in Fishes. Schuijf, A., Hawkins, A. D. (eds.). New York: Elsevier, 1976, pp. 205-228.

Tavolga, W. N.: The significance of underwater sounds produced by males of the gobiid fish, *Bathygobius soporator*. Physiol. Zool. 31: 259-271 (1958).

Tavolga, W. N.: Sound production and detection. In: Fish Physiology, Vol. V. Hoar, W. S., Randall, D. J. (eds.). New York: Academic Press, 1971, pp. 135-205.

Discussion

BULLOCK: These sounds are very similar to those of snapping shrimp. Are snapping shrimp found in the same habitat?

MOK: Yes, but we have not done any field recordings. Field recordings of these gobies are very difficult to make, and I have not been successful. It would be important to do that. In the aquarium, these fish are very hardy, and they will breed readily. I have had some for over a year.

BULLOCK: These sounds are supposed to attract females, and I can imagine the females chasing snapping shrimp all over the place.

MOK: And the shrimp might be chasing the gobies.

NORTHCUTT: Are there any other gobies that produce sounds?

MOK: Yes, several, *Bathygobius* here, and *Gobius* in Europe. The Gobiidae is one of the largest perciform families, and many of its species may be sound producers, although we definitely know of four at the present time.

BELL: Do you have any idea of the mechanism of sound production?

MOK: No. We have taken close up, slow motion movies of the animals as they produce sounds, but have no ideas as yet.

POPPER: Perhaps Tavolga can tell us how *Bathygobius* produces its sounds?

TAVOLGA: I don't know.

MOK: The sounds of *Gobiosoma* are quite different from those of *Bathygobius*. In *Bathygobius,* it is a very low frequency, hydrodynamic sound, whereas in *Gobiosoma,* it is much higher.

MYRBERG: Do you know how loud these sounds are?

MOK: No. I have not measured amplitude. They can be picked up by a hydrophone a few inches away, but no measurements were made.

PART SEVEN

Lateral Line System

The lateral line has been the focus of interest of many investigators with pivotal work having been done by Dijkgraaf, Harris, and van Bergeijk. Historically, its role has been described to involve any number of sensory functions including chemoreception and sound detection. In Chapter 23, Sand describes a number of experiments on the lateral line that clarify its controversial role in sound detection within near and far fields. The "convenient" location of lateral line organs on the body surface has resulted in its popularity as a model system for studies of hair cell function that have relevance not only to the lateral line itself, but also to the function of all organs based on the same type of sensory cell. Such studies are described by Strelioff and Sokolich (Chapter 24) and Boston (Chapter 25), all of whom consider the transducer properties of the sensory cell. In Chapter 26 Partridge discusses schooling in fishes and demonstrates that the lateral line is likely to be involved in this behavior.

Chapter 23

The Lateral Line and Sound Reception

OLAV SAND*

1 Introduction

1.1 Early History

A sensory function of the lateral line organ was first postulated by Jacobson (1813) on the basis of its extensive innervation, and he suggested that the organ was sensitive to mechanical stimulation. Knox (1825) corroborated this idea, and proposed that the adequate stimulus was movements of the surrounding water. He stated that ". . . we should consider these organs as organs of touch, so modified, however, as to hold an intermediate place between the sensation of touch and hearing."

At that time the accepted view was that the lateral line formed a system of glands for the production of mucus, and this idea was not abandoned before Leydig (1850, 1851) concluded that the lateral line was a sense organ especially adapted for aquatic life. This work induced numerous studies on the sensory function of the organ, and it was soon universally accepted that the lateral line was a mechanoreceptor. However, the nature of the normal stimulus was heavily disputed, and the situation at the turn of the century was summarized by Parker (1904): ". . . the majority of investigations disagree, some maintaining that the lateral line organs are simply organs of touch (Merkel 1880, de Sède 1884), others that they are organs belonging to an independent class, probably intermediate between touch and hearing (Leydig 1850, 1851, Schulze 1870, Dercum 1880) and, lastly, those that believe them to be accessory auditory organs (Mayser 1881, Bodenstein 1882)." It is quite amazing how well this statement, written 76 years ago, describes the situation today.

Investigations performed during the present century indicate three possible functions of the mechanoreceptive lateral line organs, namely detection of local water movements, surface waves, and low frequency sound waves. The experimental evidence in favor of the first two possibilities is unambiguous, whereas the last is still a

*Department of Physiology, Veterinary College of Norway, P.O. Box 8146 Dep., Oslo 1, Norway.

matter of controversy. Comprehensive surveys of the relevant literature have been given by Dijkgraaf (1963) and Schwartz (1974); therefore, this chapter will review only a few of the more basic experiments.

1.2 Sensitivity to Local Water Movements

Hofer (1908) was the first to show directly that both superficial neuromasts and canal organs were sensitive to weak water currents impinging locally on the fish, and he concluded that the fish could "feel at a distance" with the lateral line. His results have later been confirmed by several authors and extended to numerous species of both fish and amphibians. Dijkgraaf in particular has been a strong advocate for the concept "distance-touch". In a series of papers (see Dijkgraaf 1963) he has shown that local water currents around a swimming fish is the lateral line stimulus responsible for close range obstacle detection. The rheotactic response most fish display in a gross water current is, on the other hand, not dependent on the lateral line, but on visual or tactile stimuli (Lyon 1904, Parker 1904, Dijkgraaf 1934).

1.3 Sensitivity to Surface Waves

In contrast with sound waves, surface waves are transverse waves and the propagation velocity depends on the frequency. The minimum propagation speed is 23 cm/sec at 13.4 Hz, corresponding to a wavelength of 1.7 cm (Schwartz 1967). At higher frequencies the wavelength is even shorter. The particle motion below the surface is furthermore not linear, but elliptical.

In a series of papers Schwartz (1965, 1967, 1970, 1971) and Schwartz and Hasler (1966) have shown beyond doubt that surface-feeding fish are able to detect and locate sources of surface waves, and that the lateral line organs on the head are the sensors involved. Some species respond to surface waves at a depth of several centimeters. The lowest threshold amplitudes were in the order of 2 μm at about 15 Hz, and increased rapidly at frequencies below 8 Hz. Optimal sensitivity was obtained with waves acting parallel to the long axis of the organ.

Xenopus laevis (the clawed toad) also respond to surface waves, and Görner (1976) has recently shown that both the lateral line and the labyrinth are involved in the wave detection by this animal. The threshold for the reaction was estimated to about 0.2 μm.

1.4 Sensitivity to Low Frequency Sound

It is rational to divide previous studies concerning the lateral line and sound reception in three different groups:

(1) Behavior experiments indicating sensitivity to low frequency vibrations or sound (Parker 1904, 1909, Parker and van Heusen 1917, Dye 1921, Kramer 1933, Maliukina 1960, Kuiper 1956, Kleerekoper and Roggenkamp 1959, Backus 1963,

Tavolga and Wodinsky 1963, Wisby, Richard, and Nelson 1964, Wodinsky and Tavolga 1964, Weiss 1967, 1969, Cahn, Siler, and Wodinsky 1969, Cahn, Siler, and Auwarter 1971, Offutt 1974).

(2) Behavior experiments leading to the opposite conclusion (Parker 1902, Hofer 1908, Regnart 1928, von Frisch and Stetter 1932, Reinhardt 1935, Dijkgraaf 1950, 1967, Schuijf and Siemelink 1974).

(3) Electrophysiological experiments, which nearly all show lateral line organs to be sensitive to low frequency vibrations or sound (Hoagland 1933, 1934, Schriever 1935, Sand 1937, Suckling and Suckling 1950, 1964, Katsuki, Yoshino, and Chen 1951, Jielof, Spoor, and de Vries 1952, Kuiper 1956, Harris and van Bergeijk 1962, Suckling 1962, Flock 1965, Horch and Salmon 1973, Tavolga 1977).

A closer look at the data in the first group reveals that in no case has involvement of the lateral line in sound detection been shown beyond doubt. In several of the studies the fish was stimulated by rocking motion of the tank, which will necessarily cause surface waves. Furthermore, in none of the experiments employing extirpation techniques were the lateral line organs completely eliminated without damaging other receptors. Several authors using conditioning techniques have reported dual auditory thresholds in the low frequency range (Tavolga and Wodinsky 1963, Wodinsky and Tavolga 1964, Cahn et al. 1969, 1971, Offutt 1974). However, no extirpation experiments were performed in these cases, and the conclusion that the lateral line is responsible for the lower thresholds is thus speculative.

The studies of the second group are all based on comparison of the sensitivity to sound or vibrations before and after elimination of the lateral line. Dijkgraaf (1967), for example, recorded the interruption of breathing rhythm in flatfish caused by low frequency vibrations. A disk 6 cm in diameter was vibrated at 5 Hz 20 cm from the fish, and the displacement threshold for the response was calculated to be about 0.5 μm. Cutting the lateral line nerves caused no change of this threshold. Theoretical considerations led van Bergeijk (1964) to conclude that the lateral line system was the only receptor which could give directional acoustic information to a fish. However, using a surgical technique developed by Dijkgraaf (1973), Schuijf and Siemelink (1974) showed that complete ablation of the lateral line system in *Gadus morhua* (the Atlantic cod) does not impair directional hearing. On the other hand, unilateral elimination of the pars infereior of the labyrinth abolished the directional response without reducing the sensitivity to sound (Schuijf 1975).

There is thus a complete lack of behavior experiments proving beyond doubt that the lateral line participates in sound reception. This may seem to be in conflict with the overwhelming amount of electrophysiological data showing unambiguously that the lateral line is an extremely sensitive vibration detector. These experiments involved recordings of either action potentials from the lateral line nerve or microphonic potentials from the lateral line canal. The stimulus was either sound, oscillating water currents applied locally, vibrating objects close to the organ, vibrations applied directly to the body of the fish, or direct vibrations of the lateral line cupulae. The conclusion emerging from these studies is that the lateral line organs have optimal sensitivity to vibrational stimuli for the frequency range 50-100 Hz, with threshold values usually in the range of 0.1-0.5 μm.

1.5 General Conclusions from Previous Experiments

The present situation is that a few authors without reserve accept a biologically significant role for the lateral line as a sound detector at low frequencies (Suckling and
Suckling 1964, Weiss 1967, 1969, Offutt 1974, Horch and Salmon 1973), whereas the
majority of the investigators in the field are more cautious. However, it is nearly generally accepted that the lateral line is bound to be stimulated provided the sound intensity is sufficient to cause water displacements above the threshold values determined
from the electrophysiological experiments (Harris and van Bergeijk 1962, van Bergeijk
1964, Tavolga 1971, 1976, Schwartzkopff 1976, Popper and Fay 1973). On the other
hand, the increasing number of fish audiograms obtained under approximate free field
conditions all show thresholds corresponding to water displacements several orders of
magnitude below the reported lateral line thresholds (see for instance Chapman and
Sand 1974).

The lateral line seems thus not to be involved in the detection of low intensity
sounds. However, most biologically significant sounds perceived by fish are well above
threshold intensity for the inner ear, and it is important to clarify if it is possible for
fish to detect intense sound with the lateral line system at some distance from the
source. If this is the case, the fish could extract more information from the sound, and
the directional hearing might be improved.

1.6 Theoretical Considerations

In the electrophysiological experiments cited above, the fish were exposed to stimuli
very different from natural sound stimulation. No measures were taken to avoid surface ripples in the cases where sound energy was transmitted into the water. Furthermore, the fish was usually firmly clamped, which facilitated relative movements between the fish surface and the surrounding water. Such relative movements were of
course predominant when the stimulation was produced by oscillatory objects or
water currents applied locally.

The size of individual lateral line organs, or even most fish, is small compared to the
wavelength of the relevant sound frequencies. The wavelength for 100 Hz is for instance
approximately 15 m. Fish flesh has, furthermore, nearly the same acoustic properties
as seawater, and will vibrate with the same phase and amplitude as the surrounding
water particles when exposed to sound. The specific density of the otoliths is about
2.9 (de Vries 1950), and these structures will therefore lag behind the oscillations of
the tissues, causing shear movements between the otoliths and the sensory maculae.
The otolith organs are thus well adapted to detect far field sound, and the presence of
a swimbladder will only enhance this ability.

The cupulae of the lateral line organs have a specific density of approximately 1.01
(Jielof et al. 1952), or very close to that of the surrounding medium. It is difficult to
imagine how these organs may be stimulated by the gross vibration of fish and neighboring water in a sound field at some distance from the source. Harris and van Bergeijk
(1962) suggested that sound induced swimbladder pulsations might stimulate the
lateral line in the far field. This may be the only possibility for the lateral line to

detect high intensity far field sound. The conditions will be essentially similar even well within the near field of a sound source. However, at extremely short distances the steep displacement gradient and the small curvature of the displacement front will cause relative movements between the fish surface and surrounding water, approaching a stimulus situation employed in many of the cited experiments showing lateral line sensitivity to vibrational stimuli. It may therefore be unsafe to transfer thresholds obtained for such stimuli to fish moving freely at a greater distance from the source.

In an attempt to elucidate these questions, I have recorded from the trunk lateral line nerve in *Rutilus rutilus* (the roach) with implanted electrodes. Three different experimental situations were compared:

1. Near field stimulation caused by a vibrating sphere close to the organ.
2. Stimulation with sound having a low p/v (sound pressure/particle displacement) ratio, causing simultaneous vibration of the fish and surrounding water column.
3. Stimulation with sound having a high p/v ratio.

2 Materials and Methods

2.1 Preparation and Recording

Twelve roaches (*Rutilus rutilus*) ranging in length from 17 to 19 cm were used in the experiments, which were performed under amytal anesthesia (Keys and Wells 1930). This barbiturate given intraperitoneally (50 mg/kg) produces a deep anesthesia for several hours without blocking the respiratory movements.

In the present study it was essential that the fish could move freely with the water mass, and it was therefore necessary to record from the lateral line nerve with implanted electrodes. A superficial incision was made at a right angle to the lateral line nerve at a level close to the pectoral fin. The part of the fish posterior to the operated area was submerged during the implantation procedure. The edges of the cut skin were forced apart, exposing the lateral line nerve only covered by a thin muscle layer. The nerve was dissected free and split into several parallel branches for a length of about 2 mm. The nerve was not cut, and en passant recordings were made using the electrode design shown in Fig. 23-1. This electrode is a modification of a type previously described by Wilkens and Wolfe (1974). The hook electrode was made of 80 μm stainless steel wire centered in a sliding polyethylene tube. One of the split nerve branches was gently drawn into the tapered tip by sliding the tube forward, and the nerve segment inside the tube was then insulated by replacing the saline in the electrode assembly with silicone oil contained in a micrometer syringe. The cut skin was then stitched together over the frontal part of the electrode, whereas the protruding basal part was firmly stitched to the underlying tissue. The wound was finally sealed with tissue glue (Histoacryl blue, B. Braun Melsungen AG).

The tube connecting the electrode to the oil reservoir consisted of heat-shrinkable PVC containing a core of hot-melt plastic, and at this stage heat was applied with a pair of preheated forceps locally to the PVC tube close to the electrode base. The electrode assembly was then freed by a cut through the shrunk part of the tube, leaving

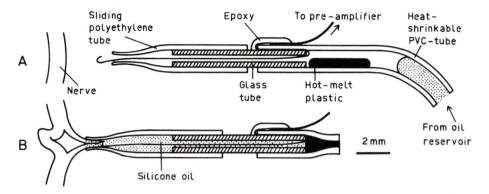

Figure 23-1. Cross section of the electrode assembly. (A) Before recording. (B) During en passant recording in the implanted state. One of the split nerve branches has been drawn into the tip, and the saline in the electrode has been replaced by silicone oil. The tube connecting the electrode to the oil reservoir has been sealed and cut.

the nerve branch trapped inside a small oil compartment completely sealed off from the surroundings. This method made it possible to record action potentials with a reasonable signal to noise ratio for several hours in animals freely positioned in the water.

A block diagram of the experimental setup is shown in Fig. 23-2. The electrode was connected to standard recording equipment. The animals were either stimulated by a glass ball with a radius of 1.7 mm vibrating close to the trunk lateral line, or the fish was positioned centrally in a special acoustic tube for different types of sound stimulation. The ball vibrator and the acoustic tube were driven by gated and attenuated signals from a sine wave generator fed to a power amplifier.

For most frequencies the response was measured as the degree of synchronization between the stimulus and the recorded spikes. The preamplifier was connected to a window discriminator which produced a 2 msec square pulse for each selected spike. These pulses were averaged by a signal analyzer triggered by the sine wave generator. The averaged value for a certain stimulus phase thus gives the probability for occurrence of a spike within ±2 msec of this phase. The height of the original square pulse will represent a probability of 1. I have previously introduced the term "synchronization index" for the difference between the highest and lowest probability for the occurrence of a spike during a stimulus cycle (Sand, Ozawa, and Hagiwara 1975). The synchronization index so defined will have a positive value even without stimulation. The index then depends upon the number of averaged sweeps, approaching 0 for an infinite number. In the present study 64 sweeps were usually averaged. The probability curves were plotted by an x-y writer.

It is important to use the degree of synchronization as the response, since the mean afferent nervous activity during lateral line stimulation at low stimulus intensities is equal to the spontaneous activity (Kroese, van der Zalm, and van den Bercken 1978, Strelioff and Honrubia 1978). However, the synchronization was not pronounced for the highest frequencies tested, and the response at these frequencies was therefore recorded as a change of the impulse rate. This was performed by feeding the square pulses from the discriminator to a voltage-stepper reset at 0.5 sec intervals by a time mark generator. The output from the stepper, giving the impulse rate as a function of time, was recorded by a pen writer.

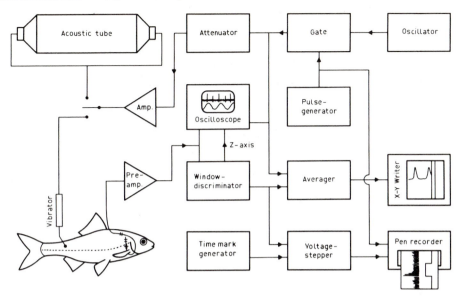

Figure 23-2. Block diagram of the experimental setup. See text for details.

2.2 Vibrating Ball Stimulation

The water displacements caused by a vibrating sphere have been described in detail by Harris and van Bergeijk (1962). I used ball vibrations either parallel to the lateral line canal or at a right angle to the surface of the fish. Figure 23-3 illustrates the water displacements at different points along a straight line in a plane at a distance of $2r$ from the ball center when the ball is vibrated parallel to the line or perpendicular to the plane. The vectors indicate the direction and relative amplitude of the vibrations. The points are marked with the distance to the projection of the ball center onto the plane. It is seen that the vibration pattern is rather complex. I will return to this figure later, after presentation of the relevant results.

The ball vibrator was calibrated using a microscope with stroboscopic illumination, and the values presented are rms. During ball stimulation the fish was submerged in a shallow tank and fixed on its side with the flank from which recordings were performed facing upward. The broad-band vertical velocity noise in this tank was 3.2×10^{-4} cm/sec, and the maximum noise in the horizontal plane was 2.3×10^{-4} cm/sec.

2.3 Acoustic Tube Stimulation

The acoustic tube (Fig. 23-4) was made of aluminum and had a 12 cm bore and a wall thickness of 3.5 cm. An underwater loudspeaker (University Sound, VW-30) was fitted in each end and the distance between the speaker faces was 64 cm. The tank was fitted with a lid for easy loading. The lid was cast in one piece after the top hole was cut, and during operation of the tank the lid was fixed using a steel locking bar with three clamp screws.

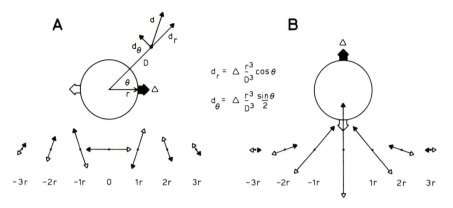

Figure 23-3. Displacement field caused by a vibrating sphere. Equations from Harris and van Bergeijk (1962). The vectors indicate the direction and relative amplitude of the displacements at different points along a straight line in a plane at a distance of $2r$ from the sphere center. (A) Ball vibrations parallel to the line. (B) Ball vibrations perpendicular to the plane. The distances between the points and the projection of the sphere center on to the plane are marked on the figure.

The physics of this type of acoustic tube have been described in detail by Hawkins and MacLennan (1976). Driving the loudspeakers 180° out of phase (displacement mode) causes large and uniform displacements throughout the length of the tank, whereas the sound pressure shows a marked gradient with a pressure null centrally. Driving the speakers in phase (pressure mode) causes large and uniform sound pressure

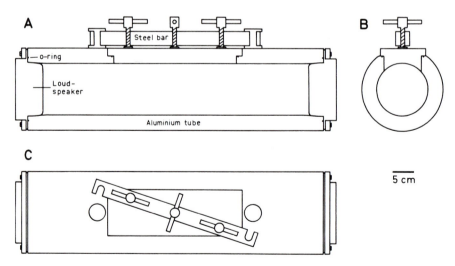

Figure 23-4. Acoustic tube used for sound stimulation with different pressure/displacement ratios. The fish was positioned in the tube center. (A) Length section. (B) Cross section. (C) Top side view. Arrangements for water flow and cable connections are not shown.

combined with very low displacements. In theory, this operation mode should lead to a displacement null in the tank center, though far less pronounced than the pressure null expected when the tank is operated in the displacement mode. Whereas this latter pressure null was detected in the acoustic tube used in the present project, no significant displacement gradient was registered along the tube when the speakers were driven in phase. At 50 Hz the difference in sound pressure/particle velocity ratio in the tank center was approximately 40 dB between the two different operation modes.

The sound pressure in the tank was measured with a calibrated hydrophone, and the presented values are given as dB re 1 μbar rms. The broad-band background noise in the tank was +3.4 dB. The particle velocity was measured with a homemade velocity hydrophone calibrated in a special tank where direct measurement of the water particle motion was obtained using a laser-doppler velocity meter (Durst, Melling, and Whitelaw 1976). The presented values have been recalculated to rms displacements.

3 Results

3.1 Response to Vibrating Ball Stimulation

A typical multiunit recording in the absence of stimulation is shown in Fig. 23-5A. The recording technique was not suited for detecting a possible pattern of the nervous activity in such cases, but the probability for a spike to occur was of course independent of the future stimulus phase (Fig. 23-5B). In the vibration experiments the ball was initially vibrated at 50 Hz parallel to the lateral line and moved slowly straight

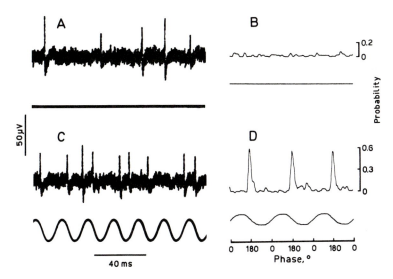

Figure 23-5. (A) Multiunit recording in the absence of stimulation. (B) Probability for the occurrence of spikes relative to the future stimulus phase. (C) Recordings during stimulation with ball vibration parallel to the lateral line canal. (D) Probability for the occurrence of spikes relative to the stimulus phase during ball vibration. The 50 Hz stimulus is monitored on the lower trace of each recording.

above the canal at a vertical distance of $2r$, or 3.4 mm, between the skin surface and the ball center. Usually the spike activity became synchronized to the stimulus for a few separated and rather restricted ball positions. One example of such synchronization is seen in Fig. 23-5C. The corresponding probability plot is given in Fig. 23-5D, which shows probability maxima 360° apart. This was the usual pattern, but occasionally the probability maxima were separated by 180°. In such cases the spike recordings included units innervating the two oppositely orientated populations of hair cells in the stimulated area.

3.2 Directional Sensitivity of the Trunk Lateral Line

It was essential for the continuation of this project to determine both the directional sensitivity and the frequency sensitivity of the trunk lateral line organ. In his review Dijkgraaf (1963) states that the canal organs have optimal sensitivity for local water movements perpendicular to the skin surface, whereas Harris and van Bergeijk (1962) suggest optimal sensitivity to vibrational stimuli parallel to the long axis of the canal. This discrepancy could be due to the fact that the latter authors used a rather large vibrating ball, 1 cm in diameter, positioned at a relatively long distance from the fish. The water motion would therefore be uniform over an area covering several canal pores.

In order to clarify this problem, I moved the vibrating ball along the canal close to a sensitive spot, and compared the response to horizontal vibrations parallel to the canal and vertical vibrations perpendicular to the skin surface. The results from one of these experiments are given in Fig. 23-6. The vertical distance between ball center and canal was $2r$, or 3.4 mm, and the positions of the canal pores are indicated on the figure. It is seen that horizontal vibrations gave maximum synchronization when the ball was straight above one particular pore. At each side of this optimal position there was a point giving null response, and further movement of the ball to either side gave a second, smaller response peak. The phase of these weak responses was identical and 180° shifted relative to the maximum response for the central ball position.

Vertical ball vibration gave a sharp response null for the position showing maximal synchronization to horizontal vibration, and two pronounced response maxima were now detected at either side of this central null position. The response tailed gradually off when the ball was moved further away. However, this time the two symmetrical, optimal positions of the ball produced responses 180° out of phase.

By comparing these results with the estimated water displacements around the vibrating ball for the two different stimulus situations (Fig. 23-3), it is seen that the data fit the idea that the displacement component parallel to the skin surface is the effective stimulus, whereas the canal organ is insensitive to perpendicular vibrations. Horizontal ball vibration gives maximal horizontal displacement just underneath the ball, while at a lateral distance between r and $2r$ the horizontal displacement component will be zero. At a further increased lateral distance we will again get a horizontal component which has the same direction on either side, but opposite to the direction of the large, central horizontal displacement. This pattern thus fits well with both the central response maximum, the peripheral response nulls and the phase reversal seen for horizontal vibration in Fig. 23-6, provided the horizontal displacement component is the effective stimulus.

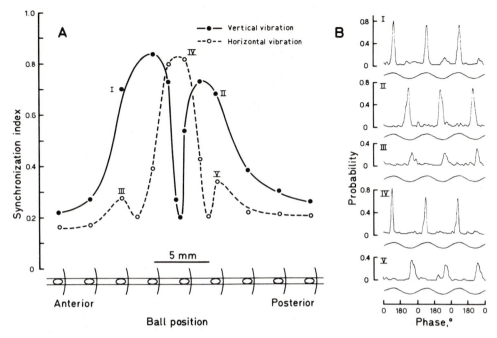

Figure 23-6. (A) Synchronization between spike activity and mechanical stimulation as a function of ball position. The individual canal pores are indicated on the figure, and data for horizontal vibrations parallel to the canal and vertical vibrations perpendicular to the skin are compared. (B) Original probability plots for both types of stimulation recorded at the ball positions indicated by roman numbers in (A). Distance between canal and ball center was $2r$ (3.4 mm). Compare the response nulls and maxima and the response phase shifts with the horizontal displacement components in Fig. 23-3.

For vertical vibrations the horizontal component just underneath the ball will be zero, whereas this component reaches a maximum on either side. However, the direction of the horizontal component on each side is opposite. This pattern of the changes in horizontal displacement component with altered ball position follows closely the response pattern for vertical vibration seen in Fig. 23-6, with a central response null and peripheral response maxima being 180° out of phase. From these results I conclude that the canal lateral line organ is optimally sensitive to water vibrations in a plane parallel to the skin surface.

To further test the directional sensitivity to vibrations in this plane, the ball was positioned just above the pore giving optimal response to horizontal vibrations and the angle between the canal axis and the vibration direction was altered. Figure 23-7 presents results from one of these experiments. Vibrations parallel to the canal corresponds to 0°. It is seen that the response follows a cosine relationship, giving minimum response for vibrations at a right angle to the canal axis.

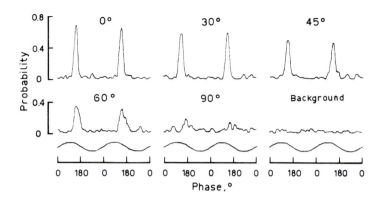

Figure 23-7. Recordings of responses to ball vibration parallel to the skin surface obtained at different angles between the canal axis and the vibration direction. The ball center was positioned 2r above a sensitive pore. Vibrations parallel to the canal corresponds to $0°$.

3.3 Frequency Sensitivity of the Trunk Lateral Line

The frequency response of the organ was tested for the optimal vibration direction parallel to the canal axis. The vibration intensity was taken as the estimated particle displacement at the canal just underneath the ball, using the equations given by Harris and van Bergeijk (1962). While the presence of the fish is bound to influence these values, I have not taken this into account.

Figure 23-8 shows a stimulus-response curve for 50 Hz, and illustrates the method for estimating the approximate threshold values. The hatched area indicates the background level of the synchronization index. The figure also demonstrates the slightly growing phase lag of the response with decreasing stimulus intensity.

Figure 23-9 gives a trunk lateral line vibrogram based on recordings from five specimens. Recordings from several fish showed rather poor mechanosensitivity for unexplained reasons, and these data are not included. It is seen that optimal sensitivity was obtained at 50 Hz, and only one animal responded clearly to 160 Hz. The most obvious response to the highest frequencies was a quickly adapting increase of the impulse rate, whereas the synchronization between stimulus and spike activity was weak and difficult to detect. The thresholds for 140 and 160 Hz are therefore based on the recordings of impulse rate rather than the synchronization index.

3.4 Acoustic Tube Stimulation

The next step was to expose the fish to the different stimulation modes in the acoustic tube, and to compare the response to that obtained using the vibrating ball. The optimal stimulus conditions were selected for these experiments, which means that the fish was positioned parallel to the long axis of the tube and exposed to sound at 50 Hz.

A comparison of this type is presented in Fig. 23-10. The upper record is a control, before the fish was inserted into the tube, and shows the response to a displacement of 1 μm parallel to the canal axis caused by local ball vibration. With the tube in displace-

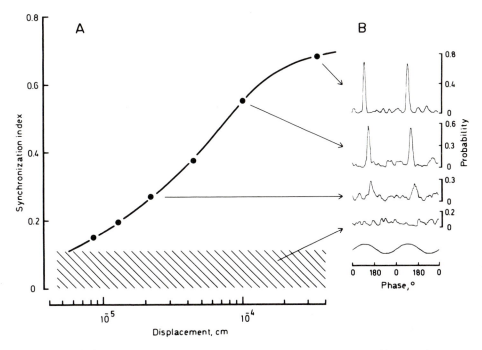

Figure 23-8. (A) Stimulus-response curve for 50 Hz vibration parallel to the canal. Hatched area indicates background level of the synchronization index. The threshold value estimated from this curve was about 5.5×10^{-6} cm. (B) Original probability plots recorded at the indicated stimulus intensities.

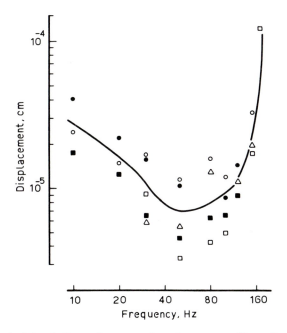

Figure 23-9. Trunk lateral line vibrogram based on recordings from five fish. The curve is drawn through the average threshold at each frequency.

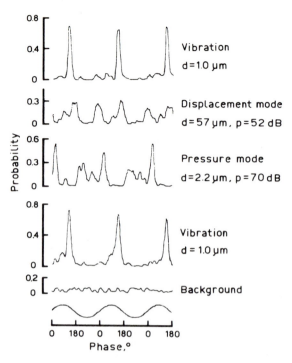

Figure 23-10. Comparison between the response to vibrating ball stimulation (upper and lower recordings) and to sound stimulation at different pressure/displacement ratios (middle recordings). Note the markedly reduced response to sound induced water displacement compared with the response to local ball vibration. During sound stimulation the fish was positioned parallel to the long axis of the acoustic tube. Stimulation frequency was 50 Hz.

ment mode the same fish was stimulated with the immense displacement of 57 μm. The whole water column, including the fish, was vibrated at this amplitude, and the otolith organs must have been excessively stimulated. However, the lateral line response decreased compared to the modest vibration of 1 μm prior to the tank test. It is seen that the tank recordings had the double number of peaks, which is reasonable since the record is based on multi unit activity, originating at several spots along the canal. The bulk stimulation in the tank will affect all these spots, whereas the vibrating ball stimulation is restricted to a few pores.

In this particular fish the area giving maximal response to the ball vibration was located three scales anterior to the anus, or just above the posterior part of the swimbladder. When the acoustic tube was switched to pressure mode, the ratio between sound pressure and particle displacement increased dramatically, and a pressure of 70 dB was obtained at a displacement of only 2.2 μm. It is seen that the lateral line response increased in spite of the reduced displacement. The last recording is a control showing the response to the vibrating ball after termination of the tank experiments.

A similar comparison between ball and tank stimulation was performed on three different fish, and the results were in agreement. In all these cases the recorded spikes were sensitive to stimulation of the lateral line anterior to the anus, where the lateral line runs close to the swimbladder (Fig. 23-11).

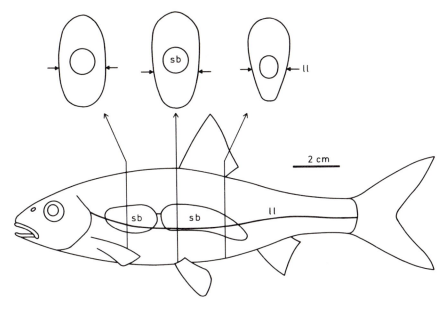

Figure 23-11. Length and cross sections of an 18 cm roach showing the position of the swimbladder (sb) relative to the lateral line (ll). The tracings are based on radiographs and cryosections.

4 Discussion

The present results confirm previous electrophysiological data showing that the trunk lateral line is an extremely sensitive vibration detector. The response to vibrating ball stimulation, causing water displacements relative to the fish surface, displayed optimal sensitivity to frequencies around 50 Hz. The lowest threshold value measured at this frequency was 3.3×10^{-6} cm rms, which agrees well with comparable data reported for other species (Harris and van Bergeijk 1962, Flock 1965, Suckling and Suckling 1964).

The previous reports regarding directional sensitivity of the canal lateral line organs are ambiguous. Katsuki et al. (1951) and Dijkgraaf (1963) claimed that these organs are particularly suited for responding to local water movements at right angles to the body surface of the fish, whereas Harris and van Bergeijk (1962), using vibrational stimuli, suggested that displacement along the canal axis is the effective stimulus. The present data support the latter authors, and it seems safe to conclude that the canal lateral line organ is sensitive to the vibrational displacement component parallel to the canal axis. However, the directional response to local d.c. water flow might be different.

In comparison with the information on lateral line responses to local water displacements, the lack of firm data on the sensitivity to sound stimulation, in a situation where the fish is free to move with the water mass, is striking. The only previous report is by Cahn et al. (1973), who were unable to show that the lateral line was sound sensitive. The vibration sensitivity of the lateral line may show large variation, even between individuals of the same species (Suckling 1967). To judge a possible involvement of the lateral line in sound detection, it is therefore essential to compare the response to sound stimulation with that obtained by local water vibrations for the same specimen. The present data show that the trunk lateral line sensitivity to particle displacement was drastically reduced when the fish was free to move with the water mass, compared to the sensitivity to local water displacements relative to the skin surface. It is even possible that the fish responded to sound pressure, rather than displacement, when the acoustic tank was operated in displacement mode. From this I conclude that the trunk lateral line in the roach is of little use for detection of sound induced water displacements when the fish is at some distance from the source, even at enormous sound intensities.

How close to the sound source must the fish be before the trunk lateral line is stimulated by the incident particle displacement? This distance may be very short indeed, and is likely to depend on the size of both the sound source and the fish. Tavolga (1976) has suggested that *Arius felis* (the sea catfish) uses the lateral line in sound based obstacle detection, and he indicates a distance limit of only a few centimeters. Similar short response distances have been reported by several authors for reactions proposed to depend on lateral line stimulation (Wunder 1927, von Frisch and Dijkgraaf 1935, Reinhardt 1935).

The results given in Fig. 23-10 show that the lateral line response increased when the acoustic tube was switched from displacement to pressure mode. This particular fish was 18 cm long, and Fig. 23-11 shows the position of the swimbladder and the lateral line in a roach of this size. The most sensitive spot along the lateral line canal was situated three scales anterior to the anus, in this particular experiment, and it is seen that the distance between the canal and the bladder surface is about 4 mm in this area. The pressure induced surface displacements of the divided, oddly shaped swimbladder in the cyprinoids are not easily calculated. However, a rough estimate may be obtained applying the equations given by Schuijf (1976) for an ellipsoid gas bladder. The measured pressure threshold was approximately +40 dB, and the particle displacements due to the swimbladder pulsation would then be about $2\text{-}4 \times 10^{-4}$ cm at the lateral line canal just anterior to the anus (Schuijf pers. comm.). This is between one and two orders of magnitude above the displacement threshold determined by ball vibrations for this particular fish. Both Suckling (1962) and Katsuki et al. (1951) have described that the trunk lateral line is very sensitive to vibrations emanating from within the fish. It is therefore likely that the lateral line could be stimulated with high intensity sound via the swimbladder. However, the sensitivity to sound induced swimbladder pulsations seems to be too low to justify the conclusion that the trunk lateral line could be a useful accessory sound detector under natural conditions.

It is important to stress that the above considerations concern the trunk lateral line. The specialized head lateral line system in clupeids, and perhaps other species, may be more sensitive to sound induced swimbladder pulsations (Blaxter, Denton, and Gray, Chapter 2).

The "distance-touch" based on damming phenomena seems mainly to depend on the head lateral line system in most fish (Dijkgraaf 1963). What is then the normal function of the trunk lateral line under natural conditions? Partridge (Chapter 26) has shown that this organ is essential for normal schooling behavior in *Pollachius virens* (saith). The work by Rosen (1959) first demonstrated that the flow about the body of a swimming fish consists of a system of vortices. The long trunk lateral line seems to be ideally suited to detect the vortex patterns generated by the neighbors in a fish school. However, several species with a pronounced schooling behavior lack the trunk lateral line (i.e., the clupeids), whereas this organ is well developed in many solitary species. The current situation does thus not encourage a general statement about the major function of the trunk lateral line under normal conditions.

References

Backus, R. H.: Hearing in elasmobranchs. In: Sharks and Survival. Gilbert, P. W. (ed.). Boston: Heath, 1963, pp. 243-254.

van Bergeijk, W. A.: Directional and nondirectional hearing in fish. In: Marine Bio-Acoustics. Tavolga, W. N. (ed.). Oxford: Pergamon Press, 1964, pp. 281-299.

Bodenstein, E.: Der Seitenkanal von *Cottus gobio*. Z. Wiss. Zool. 37, 121-145 (1882).

Cahn, P. H., Siler, W., Wodinsky, J.: Acoustico-lateralis system of fishes: Tests of pressure and particle velocity sensitivity in grunts, *Haemulon sciurus* and *Haemulon parrai*. J. Acoust. Soc. Am. 46, 1572-1578 (1969).

Cahn, P. H., Siler, W., Auwarter, A.: Acoustico-lateralis system of fishes: Cross-modal coupling of signal and noise in the grunt, *Haemulon parrai*. J. Acoust. Soc. Am. 49, 591-594 (1971).

Cahn, P. H., Siler, W., Fujiya, M.: Sensory detection of environmental changes by fish. In: Responses of Fish to Environmental Changes. Chavin, W. (ed.). Springfield, Illinois: Charles C Thomas, 1973.

Chapman, C. J., Sand, O.: Field studies of hearing in two species of flatfish *Pleuronectes platessa* (L.) and *Limanda limanda* (L.) (Family Pleuronectidae). Comp. Biochem. Physiol. 47, 371-386 (1974).

Dercum, F.: The lateral sensory apparatus of fishes. Proc. Acad. Nat. Sci. Philadelphia, 152-154 (1880).

de Sède, P.: La ligne latérale des poissons osseux. Rev. Scient. (Série 3) 7, 467-470 (1884).

de Vries, H.: The mechanics of the labyrinth otoliths. Acta Oto-Lar. 38, 262-273 (1950).

Dijkgraaf, S.: Untersuchungen über die Funktion der Seitenorgane an Fischen. Z. Vergl. Physiol. 20, 162-214 (1934).

Dijkgraaf, S.: Untersuchungen über die Funktionen des Ohr-labyrinths bei Meeresfischen. Physiol. Comp. Oecol. 2, 81-106 (1950).

Dijkgraaf, S.: The functioning and significance of the lateral-line organs. Biol. Rev. 38, 51-105 (1963).

Dijkgraaf, S.: Biological significance of the lateral line organs. In: Lateral Line Detectors. Cahn, P. H. (ed.). Bloomington: Indiana Univ. Press, 1967.

Dijkgraaf, S.: A method for complete and selective surgical elimination of the lateral-line system in the codfish, *Gadus morhua*. Experientia 29, 737-738 (1973).

Durst, F., Melling, A., Whitelaw, J. H.: Principles and Practice of Laser-Doppler Anemometry. London: Academic Press, 1976.

Dye, W. J. P.: The relation of the lateral line organs of *Necturus* to hearing. J. Comp. Psychol. 1, 469-471 (1921).

Flock, Å: Electron microscopic and electrophysiological studies on the lateral line organ. Acta Oto-lar. Suppl. 199, 1-90 (1965).

Görner, P.: Source localization with labyrinth and lateral line in the clawed toad (*Xenopus laevis*). In: Sound Reception in Fish. Schuijf, A., Hawkins, A. D. (eds.). Amsterdam: Elsevier, 1976.

Harris, G. G., van Bergeijk, W. A.: Evidence that the lateral-line organ responds to near-field displacements of sound sources in water. J. Acoust. Soc. Am. 34, 1831-1841 (1962).

Hawkins, A. D., MacLennan, D. N.: An acoustic tank for hearing studies on fish. In: Sound Reception in Fish. Schuijf, A., Hawkins, A. D. (eds.). Amsterdam: Elsevier, 1976, pp. 149-169.

Hoagland, H.: Electrical responses from the lateral-line nerve of catfish. I. J. Gen. Physiol. 16, 695-714 (1933).

Hoagland, H.: Electrical responses from the lateral-line nerves of catfish. III. J. Gen. Physiol. 17, 77-82 (1934).

Hofer, B.: Studien über die Hautsinnesorgane der Fische. I. Die Funktion der Seitenorgane bei den Fischen. Ber. Kgl. Bayer. Biol. Versuchsstation München 1, 115-164 (1908).

Horch, K., Salmon, M.: Adaptations to the acoustic environment by the squirrelfishes *Myripristis violaceus* and *M. pralinius*. Mar. Behav. Physiol. 2, 121-139 (1973).

Jacobson, L.: Extrait d'un memoire sur une organe de sens dans les raies et les squales. Nouv. Bull. Sci. Soc. Philomotique Paris 3, 332 (1813).

Jielof, R., Spoor, A., de Vries, H.: The microphonic activity of the lateral line. J. Physiol. 116, 137-157 (1952).

Katsuki, Y., Yoshino, S., Chen, F.: Action current of the single lateral-line nerve fiber of fish. II. On the discharge due to stimulation. Jpn. J. Physiol. 1, 179-194 (1951).

Keys, A. B., Wells, N. A.: Amytal anesthesia in fishes. J. Pharmacol. Exp. Ther. 39, 115-128 (1930).

Kleerekoper, H., Roggenkamp, P. A.: An experimental study on the effect of the swimbladder on hearing sensitivity in *Ameiurus nebulosus* (Le Sueur). Can. J. Zool. 37, 1-8 (1959).

Knox, R.: On the theory of the 6th sense in fishes. Edinburgh J. Sci. 2, 12 (1925).

Kramer, G.: Untersuchungen über die Sinnesleistungen und das Orienterungsverhalten von *Xenopus laevis* Dand. Zool. Jb. Physiol. 52, 629-676 (1933).

Kroese, A. B. A., van der Zalm, J. M., van den Bercken, J.: Frequency response of the lateral-line organ of *Xenopus laevis*. Pflügers Arch. 375, 167-175 (1978).

Kuiper, J. W.: The microphonic effect of the lateral line organ. Publ. Biophys. Group, Nat. Lab., Groningen (1956).

Leydig, F.: Über die Schleimkanäle der Knochenfische. Müll. Arch. Anat. Physiol. 170-181 (1850).

Leydig, F.: Über die Nervenknöpfe in den Schleimkanälen von *Lepidoleprus, Umbrina* und *Corvina*. Müll. Arch. Anat. Physiol. 235-240 (1851).

Lyon, E. P.: On rheotropism. I. Rheotropism in fishes. Am. J. Physiol. 12, 149-161 (1904).

Maliukina, G. A.: Hearing in certain Black Sea fishes in connection with ecology and particulars in the structure of their hearing apparatus. (In Russian). Zhurn Obshchei Biol. 21, 198-205 (1960).

Mayser, P.: Vergleichend anatomische Studien über das Gehirn der Knochenfische mit besonderer Berücksichtigung der Cyprinoiden. Z. Wiss. Zool. 36, 259-364 (1881).

Merkel, F.: Über die Endigungen der sensiblen Nerven in der Haut der Wirbeltiere. Rostock, 1880.

Offutt, G. C.: Structures for the detection of acoustic stimuli in the Atlantic codfish, *Gadus morhua*. J. Acoust. Soc. Am. 56, 665-671 (1974).

Parker, G. H.: Hearing and allied senses in fishes. Bull. U.S. Fish Comm. 22, 45-64 (1902).

Parker, G. H.: The function of the lateral line organs in fishes. Bull. U.S. Bur. Fish. 24, 185-207 (1904).

Parker, G. H.: Influence of the eyes, ears, and other allied sense organs on the movements of the dogfish, *Mustelus canis* (Mitchill). Bull. U.S. Bur. Fish. 29, 43-57 (1909).

Parker, G. H., van Heusen, A. P.: The reception of mechanical stimuli by the skin, lateral-line organs and ears in fishes, especially in *Ameiurus*. Am. J. Physiol. 44, 463-489 (1917).

Popper, A. N., Fay, R. R.: Sound detection and processing by teleost fishes: A critical review. J. Acoust. Soc. Am. 53, 1515-1528 (1973).

Regnart, H. C.: Investigations on the lateral sense organs of *Gadus merlangus*. Proc. Univ. Durham Phil. Soc. 8, 55-60 (1928).

Reinhardt, F.: Über Richtungswahrnehmung bei Fischen, besonders bei der Elritze (*Phoxinus laevis* L.) und beim Zwergwels (*Amiurus nebulosus* Raf.). Z. Vergl. Physiol. 22, 570-603 (1935).

Rosen, M. W.: Waterflow about a swimming fish. Tech. Publ. U.S. Naval Test Station, China Lake, Calif., NOTSTP 2298, 1-94 (1959).

Sand, A.: The mechanism of the lateral sense organs of fishes. Proc. R. Soc. London Ser. B 123, 472-495 (1937).

Sand, O., Ozawa, S., Hagiwara, S.: Electrical and mechanical stimulation of hair cells in the mudpuppy. J. Comp. Physiol. 102, 13-26 (1975).

Schriever, H.: Aktionspotensiale des *N. lateralis* bei Reizung der Seitenorgane von Fischen. Pflüg. Arch. Ges. Physiol. 235, 771-784 (1935).

Schuijf, A.: Directional hearing of cod (*Gadus morhua*) under approximate free field conditions. J. Comp. Physiol. 98, 307-332 (1975).

Schuijf, A.: The phase model of directional hearing in fish. In: Sound Reception in Fish. Schuijf, A., Hawkins, A. D. (eds.). Amsterdam: Elsevier, 1976, pp. 63-86.

Schuijf, A., Siemelink, M. E.: The ability of cod (*Gadus morhua*) to orient towards a sound source. Experientia 30, 773-775 (1974).

Schulze, F. E.: Über die Sinnesorgane der Seitenlinie bei Fischen und Amphibien. Arch. Mikr. Anat. 6, 62-88 (1870).

Schwartz, E.: Bau und Funktion der Seitenlinie des Streifenhechtlings *Aplocheilus lineatus*. Z. Vergl. Physiol. 50, 55-87 (1965).

Schwartz, E.: Analysis of surface-wave perception in some teleosts. In: Lateral Line Detectors. Cahn, P. H. (ed.). Bloomington: Indiana Univ. Press, 1967.

Schwartz, E.: Ferntastsinnesorgane von Oberflächenfischen. Z. Morphol. Tiere 67, 40-57 (1970).

Schwartz, E.: Die Ortung von Wasserwellen durch Oberflächenfische. Z. Vergl. Physiol. 74, 64-80 (1971).

Schwartz, E.: Lateral-line mechano-receptors in fishes and amphibians. In: Handbook of Sensory Physiology III/3. Fessard, A. (ed.). Berlin: Springer-Verlag, 1974.

Schwartz, E., Hasler, A. D.: Perception of surface waves by the blackstripe topminnow *Fundulus notatus*. J. Fish. Res. Bd. Can. 23, 1331-1352 (1966).

Schwartzkopff, J.: Comparative-physiological problems of hearing in fish. In: Sound Reception in Fish. Schuijf, A., Hawkins, A. D. (eds.). Amsterdam: Elsevier, 1976.

Strelioff, D., Honrubia, V.: Neural transduction in *Xenopus laevis* lateral line system. J. Neurophysiol. 41, 432-444 (1978).

Suckling, E. E.: Lateral line in fish—Possible mode of action. J. Acoust. Soc. Am. 34, 127 (1962).

Suckling, E. E.: Electrophysiological studies on the trunk lateral line system of various marine and freshwater teleosts. In: Lateral Line Detectors. Cahn, P. H. (ed.). Bloomington: Indiana Univ. Press, 1967.

Suckling, E. E., Suckling, J. A.: The electrical response of the lateral line system of fish to tone and other stimuli. J. Gen. Physiol. 34, 1-8 (1950).

Suckling, E. E., Suckling, J. A.: Lateral line as a vibration receptor. J. Acoust. Soc. Am. 36, 2214-2216 (1964).

Tavolga, W. N.: Sound production and detection. In: Fish Physiology. Hoar, W. S. Randall, D. J. (eds.). New York: Academic Press, 1971, pp. 135-205.

Tavolga, W. N.: Acoustic obstacle detection in the sea catfish (*Arius felis*). In: Sound Reception in Fish. Schuijf, A., Hawkins, A. D. (eds.). Amsterdam: Elsevier, 1976, pp. 185-204.

Tavolga, W. N.: Mechanisms for directional hearing in the sea catfish (*Arius felis*). J. Exp. Biol. 67, 97-115 (1977).

Tavolga, W. N., Wodinsky, J.: Auditory capacities in fishes. Pure tone thresholds in nine species of marine teleosts. Bull. Am. Mus. Nat. Hist. 126, 177-240 (1963).

von Frisch, K., Dijkgraaf, S.: Können Fische die Schallrichtung wahrnehmen? Z. Vergl. Physiol. 22, 641-655 (1935).

von Frisch, K., Stetter, H.: Untersuchungen über den Sitz des Gehörsinnes bei der Elritze. Z. Vergl. Physiol. 17, 686-801 (1932).

Weiss, B. A.: Sonic sensitivity in the goldfish (*Carassius auratus*). In: Lateral Line Detectors. Cahn, P. H. (ed.). Bloomington: Indiana Univ. Press, 1967.

Weiss, B. A.: Lateral-line sensitivity in the goldfish (*Carassius auratus*). J. Audit. Res. 9, 71-75 (1969).

Wilkens, L. A., Wolfe, G. E.: A new electrode design for *en passant* recording, stimulation and intracellular dye infusion. Comp. Biochem. Physiol. 48A, 217-220 (1974).

Wisby, W. J., Richard, J. D., Nelson, D. R., Gruber, S. H.: Sound perception in elasmobranchs. In: Marine Bio-Acoustics. Tavolga, W. N. (ed.). Oxford: Pergamon Press, 1964, pp. 255-268.

Wodinsky, J., Tavolga, W. N.: Sound detection in teleost fish. In: Marine Bio-Acoustics. Tavolga, W. N. (ed.). Oxford: Pergamon Press, 1964, pp. 269-280.

Wunder, W.: Sinnesphysiologische Untersuchungen über die Nahrungsaufnahme bei verschiedenen Knochenfischarten. Z. Vergl. Physiol. 6, 67-98 (1927).

Discussion

CORWIN: Perhaps our general use of the term "lateral line" is a bit misleading. We should make the effort to consider the different types. The system in a clupeid, for example, is quite different from that of *Xenopus*. Lateral line systems really have quite a diverse structure.

SAND: I did the experiments on the trunk lateral line. The head lateral line is embedded in bone and has a rather different anatomy.

CORWIN: We should expect a diversity of function to go along with the diversity of structure.

FINE: Some years ago, Tavolga, Cahn, and others reported that they found thresholds changing at different frequencies, and they assumed that one system was switching off and another taking over, i.e., the lateral line and the auditory system. Perhaps both systems operate simultaneously and the fish is coming up with some sort of consensus, as it were.

TAVOLGA: Back then, we had a very different conception of the function of the lateral line, and how hearing took place in fishes. The possibility that the fish may be switching from one modality to another still exists. Clearly, we are dealing with systems that have a much greater diversity of function than we ever thought they could have. We used to think of the acousticolateralis system as one unit, one modality. Now we see that there are several more or less independent systems.

COOMBS: We are also now thinking more in terms of pressure vs. velocity sensitivity. We used to consider velocity reception as attributable to the lateral line, and pressure as the province of the inner ear. Now we realize that the inner ear is both velocity and pressure sensitive. Lateral line function need not be involved to account for dual thresholds.

SAND: It seems impossible for the trunk lateral line to function as a reasonable sound detector at a distance from the source of more than a few body lengths. Furthermore, it should be noted that the lateral line at such distances is not at all suited for velocity detection, as opposed to the otolith organs. I guess these facts are in conflict with the views of most fish physiologists.

PLATT: The anatomy of the lateral line is comparable to the situation in electroreception. Some fish have electroreceptors with extremely long canals which appear to enhance sensitivity. There are fish that have lateral line systems in very long canals. What effect might that have on pressure vs. displacement detection, as compared to those with exposed lateral organs?

SAND: That is rather a problem. From my experiments, it seemed that the mechanical coupling between neighboring neuromasts was very slight.

GRAY: The orientation of particle displacement has to be in line with the canal for maximum excitation. One visualizes the hair cells as displacement detectors bending in that plane. If the receptors are excited by the expansion and contraction of the swimbladder, then the stimulus would be a vector at right angles to the direction of maximal sensitivity as you described it earlier. Have I missed something?

SAND: I can't really explain why the cells seem to be so sensitive to this stimulation. However, the mechanical effect on the neuromasts could be different for vibrations emanating from inside the fish compared to displacements in the external water.

SCHUIJF: It is an oversimplification to consider the swimbladder as a spherical model. The bladder is more cigar-shaped, and has some flat areas. It would be difficult, but not impossible to calculate the acoustic field it would generate. But in this case, there are extensions and openings at some distances from the center creating complex pressure gradients. Once you work all these out, then you can assume that the body of the fish as a whole is shaken, so that the relative movements of the medium along the body surface can stimulate the canal organs. In *Xenopus,* for instance, the free neuromasts have their maximum sensitivity when the flow is along the flat sides of the cupulae. This seems paradoxical. A plate perpendicular to the flow would have maximum drag. However, with such small structures as free neuromasts, maximum force of laminar flow would be exerted when the flow has maximum contact with the object, and that is when it is flowing *parallel* to the cupula. This gives you a deflection that is proportional to the velocity of the flow. I don't think it is a good idea to use displacement as a measure when you deal with different frequencies. There is a problem here. What is the real input parameter in this case?

DENTON: How do you account for the shape of your frequency response curve?

SAND: The slope of the vibrogram at the low frequency end, between 10 and 40 Hz, is approximately 6 dB/octave. This fits with the theoretical value for a velocity detector. The upper frequency cutoff could be due to mechanical filtering or to inherent properties of the hair cells.

Chapter 24

Stimulation of Lateral-Line Sensory Cells

D. STRELIOFF and W. G. SOKOLICH*

1 Introduction

The lateral line consists of sense organs that are distributed over the skin or found in subepidermal canals of lower vertebrates. These organs are sensitive to water currents and to low frequency vibrations of the water (Jierlof, Spoor, and DeVries 1952, Kuiper 1956, Harris and van Bergeijk 1962, Dijkgraaf 1963, Bauknight, Strelioff and Honrubia 1976, Kroese, Van der Zalm, and Van den Bercken 1978, Strelioff and Honrubia 1978). The sensory hair cells and the basic transduction mechanism in these organs are similar to those in the vertebrate inner ear organs. Due to the relative ease with which lateral line organs and their nerve can be accessed, they have been used for several studies on basic transduction mechanisms of hair cell systems (Harris, Frischkopf, and Flock 1970, Sand 1975, Sand, Ozawa, and Hagiwara 1975, Flock and Russell 1976, Strelioff and Honrubia 1978).

Recordings of the activity of the afferent fibers which innervate hair cell systems demonstrate that they are sensitive to various properties of their environment. The most important of these properties are the ionic concentration gradients across the hair cell membranes (Katsuki 1973, Sand 1975), the motion of the surrounding fluids (Harris and van Bergeijk 1962, Dijkgraaf 1967, Strelioff and Honrubia 1978) and the extracellular potentials (Sand et al. 1975, Strelioff and Honrubia 1978). The relative influence of each of these environmental properties depends on the particular morphological features of the specific hair cell system. To understand how each of the three environmental properties influence the neural activity of the hair cell systems it is instructive to develop a model that incorporates the relevant features of a typical hair cell system. Such a model is described here and is based on the premises that each of the properties affects the membrane potential of the hair cell and that variations of this potential affect the synaptic structures and determine the neural activities of associated fibers.

*Division of Head and Neck Surgery, UCLA School of Medicine, Los Angeles, California 90024.

In order to gain additional insight into the transduction process, experiments were conducted on the lateral line of *Xenopus laevis*, the African clawed toad, to study the characteristics of neural responses to extrinsic electrical stimuli and to combinations of electrical and mechanical stimuli. These experimental results will be presented and the implications of the observed nonlinear processing of these stimuli by the hair cells will be discussed with respect to the model.

2 Model of Hair Cell Function

It is suggested that the features of hair cell systems that are relevant to the determination of their afferent neural activity are those that influence the membrane potential at the afferent synapses near the basal end of the hair cell. These features are ionic concentration gradients across the hair cell membranes, mechanically induced impedance variations at the apical surfaces of the hair cells, extrinsic electrical potentials, and efferent innervation. The means whereby each of these features affects the membrane potentials are best described with the aid of the relatively simple electrical network model illustrated in Fig. 24-1. This model can be used to estimate the relevant electrical currents and membrane potentials and thus to predict the concomitant afferent neural activity. Since only the features of lateral line function which are relevant to the development of the hair cell model will be presented here, the reader is referred to the excellent reviews on lateral line receptors and hair cell transduction by Dijkgraaf (1963), Cahn (1967), Flock (1971), Schwartz (1974) and Russell (1976) for additional background information.

2.1 Effect of Ionic Environment

The main effect of the extracellular ionic environment of hair cells is to determine the electrodiffusion potentials represented by V_e and V_m in the model of Fig. 24-1. Due to their dominant concentrations, the monovalent cations, K^+ and Na^+, play the predominant role in this regard (Bracho and Budelli 1978).

The present body of experimental evidence is consistent with the assumptions that the hair cell membrane potential is dependent on the concentration of K^+ in the external fluids and that the membrane is permeable to a variety of cations. Bracho and Budelli (1978) demonstrated that the dependence of membrane potential on the K^+ and Na^+ concentrations of the extra- and intracellular fluids of the hair cells in the macula sacculi of *Necturus maculosus* is the same as the dependence predicted by the Goldman-Hodgkin-Katz equation (Goldman 1943, Hodgkin and Katz 1949, Moreton 1969). Their data are also consistent with the hypotheses that the membrane ionic selectivity is uniform around the cell and that an active ionic exchange mechanism contributes to the overall maintenance of the membrane potentials. The active energy source is included in V_m of Fig. 24-1. Corey and Hudspeth (1979) demonstrated that the membranes of saccular hair cells of *Rana catesbiana* are permeable to Li^+, Na^+, K^+, Rb^+, Cs^+, Ca^{2+} and at least one small organic cation, tetramethylammonium. Since K^+ is the predominant ion in the extracellular fluids at the apical surface of the hair cell it

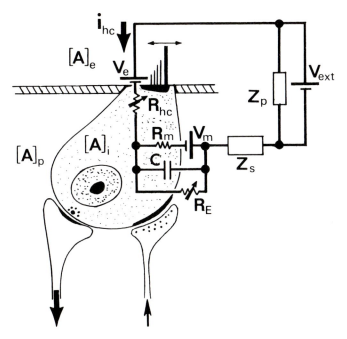

Figure 24-1. Schematic drawing of a hair cell overlayed with an electrical network which simulates the aspects of hair cell function described in the text. The magnitude of the quiescent hair cell current, i_{hc}, is proportional to the sum of the electrodiffusion potential, V_e, at the apical surface; the potential, V_m, produced by active processes and by electrodiffusion at the basolateral surface; and the extrinsic potentials, V_{ext}, produced by sources outside the cell. $[A]_e$, $[A]_p$, and $[A]_i$, the ionic concentrations of the endolymphatic, perilymphatic and intracellular fluids, respectively, play an important role in determination of V_e and V_m. R_m and C simulate the resistive and capacitive properties, respectively, of the basolateral membrane impedance. All impedances external to the hair cell are simulated by Z_s and Z_p. R_{hc}, the membrane impedance of the apical surface, is varied by deflection of the cilia during mechanical stimulation. R_E, the membrane impedance in the vicinity of the efferent synapses, is reduced by efferent activity.

probably carries most of the current through the hair cell (Corey and Hudspeth 1979, McGlone, Russell, and Sand 1979).

The concentration of cations ([A] in Fig. 24-1) in the extra- and intracellular fluids determine the potentials due to electrochemical gradients at the apical and basolateral surfaces of the cell. In vertebrate hair cell systems, the hair cells are located in epithelia which separate fluids of different ionic composition (Smith, Lowry, and Wu 1954, Peterson et al. 1978). The apical ciliated surfaces of the hair cells are in contact with a high K^+, low Na^+ fluid $[A]_e$. This has been shown to be true for the cupulae of the free standing neuromasts in the *Xenopus laevis* lateral line organ (Russell and Sellick 1976) as well as for the auditory and vestibular organs found within the vertebrate inner ear. In contrast, the basolateral surfaces of the hair cells are in contact with a low

K^+, high Na^+ extracellular fluid $[A]_p$. The intracellular composition $[A]_i$ of hair cells is expected to be high in K^+ and low in Na^+ as is found in most cells. Since the cation concentration gradients at the apical and basolateral surfaces of the hair cells are different, it is expected that the membrane potentials V_e and V_m across these surfaces should also be different (Johnstone and Sellick 1972, Honrubia, Strelioff, and Sitko 1976). At the apical surface, the cation concentration gradient is very small and the diffusion potential, V_e, is expected to be very close to zero. At the basolateral surfaces, the K^+ concentration is considerably greater inside than outside the cell resulting in a potential, V_m, of approximately 70 mV. The potential difference, $V_m - V_e$, produces the quiescent component of the current, i_{hc}, which flows through the hair cell as indicated in Fig. 24-1.

The expected effect of an increase in extracellular K^+ at the apical surface of a hair cell is a decrease in V_e, and thus an increase in the potential difference, $V_m - V_e$, the net driving force for i_{hc}. The resulting increase in i_{hc} should increase both the receptor potential due to mechanically evoked changes of R_{hc} (Davis 1965, Strelioff, Haas, and Honrubia 1972, Geisler et al. 1977) and the resting membrane potential at the basal end of the hair cells. Changes in the concentration of the other monovalent cations would also produce changes in hair cell membrane potentials.

It has been demonstrated that both spontaneous and mechanically evoked activity of afferent fibers innervating the surface lateral line organs of both *Xenopus laevis* and *Necturus maculosus* are significantly affected by changes in the ionic concentrations of the fluids bathing the outer surface of the organ (Onoda and Katsuki 1972, Katsuki 1973, Sand 1975). Large (100-200%) increases in the spontaneous rate of the afferent fibers occur when the concentrations of K^+ and Na^+ are increased from 1 to 200 mM. In addition, the gain of the neural response to mechanical stimulation increases by about 20 dB over a 2 to 20 mM concentration range.

The concentration of Ca^{2+} and other divalent anions (Sr^{2+}, Mg^{2+}, and Co^{2+}) also plays an important role in determining the gain of the neural response to mechanical stimulation. Ca^{2+} and Sr^{2+} were found to be effective in increasing the gain of the organs to mechanical stimulation at concentration levels approximately 10 times lower than the levels required for Na^+ and K^+ (Sand 1975). Mg^{2+} and Co^{2+} greatly reduced the gain of the organs for the same concentration ranges as Ca^{2+} and Sr^{2+} and also effectively blocked the effects of Ca^{2+} and Sr^{2+}. These concentrations of divalent anions did not affect the spontaneous rates of the fibers. Since the concentrations of these cations in the normal environment are usually low, it is unlikely that they have much influence on the resting membrane potential or that they contribute significantly to the receptor current (Corey and Hudspeth 1979). It has been suggested that Ca^{2+} has an indirect effect on K^+ permeability and that Ca^{2+} might be necessary for proper operation of the K^+ pumping mechanism (McGlone et al. 1979).

2.2 Effect of Mechanical Stimulation

There is substantial experimental evidence that mechanical stimulation of a hair cell produces a membrane impedance change which is correlated with the displacement of the hair cell cilia (Strelioff et al. 1972, Geisler et al. 1977, Hudspeth and Corey 1977). This impedance change (variation of R_{hc} in Fig. 24-1), which is assumed to occur at the apical surface of the hair cell, modulates the resting current (i_{hc}) through the cell and thus varies the membrane potential in the region of the afferent synapse. It is assumed that the resultant potential variation modulates the spontaneous release of neurochemical transmitter to produce a mechanically evoked neural response.

Numerous studies of mechanically evoked receptor potentials and neural responses from various hair cell systems have enabled the determination of the response characteristics of hair cells to mechanical stimuli. There is no neural response to either constant or varying hydrostatic pressure in the surrounding fields if this pressure is applied from all directions and does not produce deformation of the organ (Bauknight, Strelioff, and Honrubia 1976, Strelioff and Honrubia 1978). In contrast, any shearing force which causes movement of the sensory hairs provides an adequate stimulus for hair cell systems (von Békésy 1960, Flock 1965, Sand, Chapter 23). Due to the morphological orientation of the hair cells within the organ, hair cell systems are directionally sensitive (Lowenstein and Wersäll 1959, von Békésy 1960, Hudspeth and Corey 1977) with the greatest receptor potentials and neural responses being produced when the stimulus is applied in the direction along a line from the center of the stereociliary mass to the kinocilium. Deflection of the stereocilia toward the kinocilium results in a depolarization of the hair cell membrane potential (Hudspeth and Corey 1977) and an increase in the neural activity of the afferent fibers innervating the hair cell (Lowenstein and Wersäll 1959), whereas deflection in the opposite direction produces a hyperpolarization of the membrane potential and a decrease in neural activity. The amplitudes of the receptor potentials measured intracellularly vary from a fraction of a millivolt (Harris et al. 1970, Mulroy et al. 1974, Weiss, Mulroy, and Altman 1974, Sand et al. 1975) to greater than 10 mV p-p (Hudspeth and Corey 1977, Russell and Sellick 1978). Hudspeth and Corey have also found a nonlinear relationship between the deflection of stereocilia and the resulting change in membrane potential. This nonlinear relationship is asymmetrical: the positive response produced by ciliary movements toward the kinocilium exhibits a more gradual approach to a plateau and attains a greater absolute magnitude than does the negative response produced by movements in the opposite direction. Hudspeth and Jacobs (1979) have also demonstrated that deflection of the stereocilia alone is adequate for the production of receptor potentials, whereas deflection of the kinocilium alone is not.

In the *Xenopus laevis* lateral line organ there is a small sustained neural response to sustained cupular displacement produced by a constant flow of water past the organ. The magnitude of this response is proportional to linear water velocity for linear velocities between 1.0 and 16 mm/sec (Strelioff and Honrubia 1978). Görner (1963), Harris and Milne (1966) and Kroese et al. (1978) also provide evidence that the neural response of *Xenopus laevis* lateral line organ is proportional to linear water velocity. Kroese et al. (1978) determined the sinusoidal steady-state response characteristics over a frequency range of 0.1-100 Hz and found that, over the range of 0.1-20 Hz, the gain of the neural response increased at 7.5 ($\pm$0.2) dB/oct. For a velocity detector, a slope of 6 dB/oct is expected. Below 5 Hz, the neural response has a phase lead of approximately 0.7 π radians relative to the water displacement. Above 5 Hz, the response phase gradually changes to a phase lag of 1.6 π radians at 100 Hz. Thus, although the studies of the characteristics of the neural responses to water movements indicate that water velocity and the concomitant cupular displacement are the most effective stimuli for the *Xenopus laevis* lateral line organ, the results of the studies also demonstrate that the behavior of the organ is more complex than that of a simple velocity detector.

Input-output functions, based on the amplitude of modulation of spontaneous neural activity in response to sinusoidal water movements were found to be consistent with a linear relationship between the stimulus and the response over an intensity range from threshold to about +14 dB (Kroese et al. 1978). In contrast, Strelioff and Honrubia (1978) found a power function relationship with a slope of 0.7 between the amplitude of the sinusoidal water movements and the neural response over an intensity range from threshold to +20 dB. However, since only the amplitudes of the fundamental components of the neural response histograms were determined in both studies, it is possible that the waveforms were distorted in accordance with the nonlinear input-output relationship between the ciliary displacement and the receptor potential as determined by Hudspeth and Corey (1977). Further data on the amount of distortion in the waveform of the averaged neural response will be required for a proper comparison of nonlinearities in the receptor potentials with those in the neural responses.

2.3 Effect of Extrinsic Electrical Potentials

External electrical potentials, represented by V_{ext} in Fig. 24-1, affect the magnitude of the resting and mechanically evoked currents through the hair cell and thus influence the neural activity of the afferent fibers. Constant external potentials are produced by electrogenic ion pumps in adjacent supporting cells (Russell and Sellick 1976) or in other structures in the receptor organ such as the stria vascularis in the cochlea. Time-varying external potentials are produced by adjacent hair cells (Strelioff, Sitko, and Honrubia 1976) and by other extrinsic sources within the environment.

Previous studies on the electrical stimulation of hair cell systems in lateral line organs (Katsuki and Yoshino 1952, Murray 1956, Sand et al. 1975, Strelioff and Honrubia 1978) and in the inner ear (Lowenstein 1955, Moxon 1971, Konishi, Teas, and Wernick 1970, Teas, Konishi, and Wernick 1970) have demonstrated that the afferent nerve fibers which innervate hair cells do respond to electrical stimuli. Of these studies, only those of Sand et al. (1975) and of Strelioff and Honrubia (1978) provide strong evidence that the electrical stimuli act directly on the hair cells and not on the afferent fibers. Experiments on electroreceptors (Bennett 1971) provide further indirect evidence that electrical potentials act on the hair cells rather than directly on the nerve fibers (also see Bullock, Chapter 27).

The degree of influence of external electrical potentials on the receptor potential of a hair cell is greatly dependent on the electrical impedances of the particular hair cell system. For example, in the mammalian cochlea where the impedances are low, the stria vascularis has a tremendous influence on neural excitation as well as on generation of receptor potentials. Since the electrical properties of lateral line organs have not received as much attention as those of the cochlea, these properties are unknown and the amount of the electrical interaction between adjacent hair cells cannot be estimated at present.

2.4 Effect of Efferent Activity

Most hair cell systems have efferent fibers which usually terminate on the basal ends of the hair cells (Flock 1971). Efferent activity results in a decrease of membrane impedance and a concomitant hyperpolarization of the hair cell membrane potential (Flock and Russell 1973, 1976), an increase in the magnitude of the mechanically evoked receptor potential (Fex 1959, Konishi and Slepian 1971, Flock, Jørgensen, and Russell 1973) and inhibition of spontaneous and evoked neural activity (Galambos 1956, Russell 1968, Weiderhold and Kiang 1970). Based on their experimental findings on efferent effects in lateral line systems Flock et al. (1973) proposed that efferent inhibition is the result of a decrease in hair cell membrane impedance (R_E in Fig. 24-1) at the efferent synapses which shunts excitatory current away from afferent synapses. The experimental findings on the effects of efferent activity on the function of the mammalian cochlea (Galambos 1956, Fex 1959, Konishi and Slepian 1971, Weiderhold and Kiang 1970) are consistent with predictions of a cochlear model based on this current shunting hypothesis of efferent effects on hair cell function (Geisler 1974).

2.5 Passive Electrical Properties of Hair Cell Membranes

Since the predictions of the model of Fig. 24-1 are dependent upon the electrical properties of the hair cell it is essential to obtain quantitative estimates of the specific membrane resistance (R) and capacitance (C), the membrane time constant (τ), and the membrane potential (V_m). In the canal organs of the *burbot* (Flock et al. 1973) the values of R, C, τ, and V_m were found to be 100-1000 Ω-cm^2, < 0.3 μF/cm^2, < 0.2 msec, and 20-60 mV, respectively. Russell and Sellick (1978) obtained 733 Ω-cm^2, 0.9 μF/cm^2, 0.3-0.8 msec, and 20-45 mV, respectively, for corresponding values of the inner hair cell properties of the guinea pig cochlea. These values of R and C are in reasonable agreement with 1000 Ω-cm^2 and 1 μF/cm^2, the corresponding values for cell membranes (Cole 1968).

2.6 Model Predictions of Responses to Complex Stimuli

Due to the difficulty of mathematically describing an electrical network which has time-varying resistances as well as reactive elements (e.g., Weiss et al. 1974) a simplified model for hair cell function has been derived (Fig. 24-2). Since the model predictions to be tested do not depend on stimulus frequency, the capacitance of the hair cell membrane need not be included. In addition, all series resistances, including that of the hair cell membrane, are lumped into Z_s. It is also assumed that V_e in the model of Fig. 24-1 is small relative to V_m and that R_E is fixed. The simple equation shown below the network allows computation of the current, i_{hc}, through the basal portion of the hair cell for both mechanical (ΔR_{hc}) and electrical (ΔV_{ext}) stimulation.

If V_{hc} and Z_s are constant during electrical stimulation then the equation predicts that changes in i_{hc} would be directly proportional to changes in V_{ext}. On the other

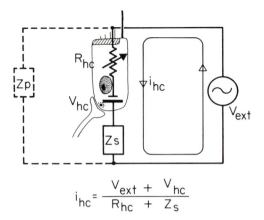

$$i_{hc} = \frac{V_{ext} + V_{hc}}{R_{hc} + Z_s}$$

Figure 24-2. Simplified version of the electrical network model illustrated in Fig. 24-1. See text for details.

hand, changes in i_{hc} resulting from changes in R_{hc} during mechanical stimulation would be inversely proportional to the changes in R_{hc} and, if changes in R_{hc} are large with respect to Z_s, the relation between changes of R_{hc} and i_{hc} should be nonlinear. If the amplitude of the neural response is proportional to the amplitude of the current variation then responses to electrical stimuli should be linear and the responses to mechanical stimuli should be nonlinear.

The main objectives of the experiments reported here were to determine the characteristics of the neural response to mechanical, electrical, and combinations of mechanical and electrical stimuli. In order to study the interaction between two stimuli the technique used was that of varying the operating point of the hair cells with a low frequency modulating sinusoidal stimulus and measuring the effects of this variation on the neural response to a high frequency "test" stimulus. For an electrical modulating stimulus the response to an electrical test stimulus should be independent of the modulator whereas the response to a small amplitude mechanical test stimulus should be directly proportional to the magnitude of the modulator. In contrast, a mechanical modulating stimulus should produce nonlinear responses for both electrical and mechanical test stimuli. These stimulus combinations would thus help to characterize the neural responses to electrical and mechanical stimuli.

3 Methods

3.1 Morphology of the Preparation

Xenopus laevis is an aquatic anuran with about 200 free-standing lateral line organs arranged in rows on the head and body (Fig. 24-3a). Morphological studies (Murray 1955, Flock 1967, Flock and Russell 1976) of this system have shown that each organ or stitch consists of 3 to 12 neuromasts and is covered by a gelatinous structure called a cupula (Fig. 24-3b-d). Each of the neuromasts contains between 30 and 60

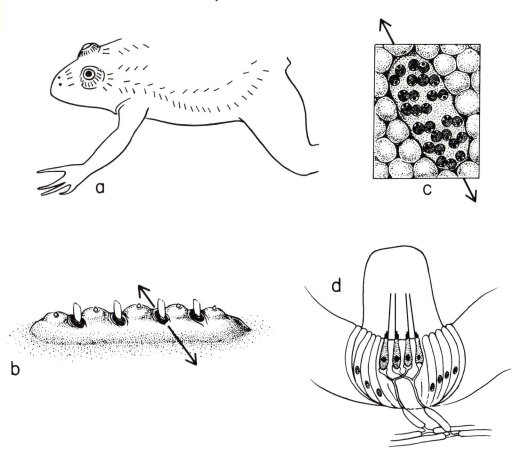

Figure 24-3. Diagrams illustrating the distribution of lateral line organs of the *Xenopus laevis* (a), a single end organ with the hair cells of the four neuromasts covered by cupulae (b), the morphological orientation of the hair cell kinocilia in a single neuromast (c), and the afferent innervation of a stitch by two myelinated fibers (d). The organ is most sensitive to stimulation parallel to the longitudinal axis of the neuromast and perpendicular to the longitudinal axis of the stitch, as indicated by the arrows in (b) and (c). (From Strelioff and Honrubia 1978).

hair cells which are oriented so that the kinocilia of adjacent cells are on opposite sides of the apical surfaces of the cells.

Each stitch is innervated by two large diameter (8-15 μm) myelinated fibers, several smaller (3-5 μm) myelinated fibers and a few fine (0.25-0.8 μm) unmyelinated fibers corresponding to the afferent, efferent, and sympathetic nerves, respectively (Russell 1976). During stimulation with sinusoidal water movements the responses of the two afferent fibers innervating each stitch are approximately 180° out of phase (Görner 1963, Bauknight et al. 1976). Although no direct anatomical evidence is available, this finding indicates that each afferent fiber innervates hair cells of only one orientation.

3.2 Dissection and Mounting of the Preparation

After cooling the toad to near 0°C in order to permit decapitation and pithing of the animal without damaging the lateral line organ, a circular piece of the skin, 3 cm in diameter, was dissected from the side of the animal. A 2 cm length of nerve innervating the medial lateral row of stitches was dissected away from the skin and the nerve branches to all stitches but one were cut to allow recording from only the stitch selected for study. The connective tissue was carefully removed from the 1 cm length near the cut end of the nerve in order to improve the signal-to-noise ratio for recording the neural activity.

The dissected piece of skin was mounted between two fluid-filled chambers designed to minimize the motion of the skin during mechanical stimulation (Fig. 24-4) A thin layer of Ringer solution covered the upward-facing inside surface of the skin. The cut end of the nerve was draped over the two chlorided silver recording electrodes and adjusted to provide a different spike amplitude from each of the two afferent fibers. The upper chamber was then filled with mineral oil and sealed. The water-filled lower chamber was perfused with oxygenated water every 10 min in order to maintain the spontaneous rate and preserve the sensitivity of the preparation.

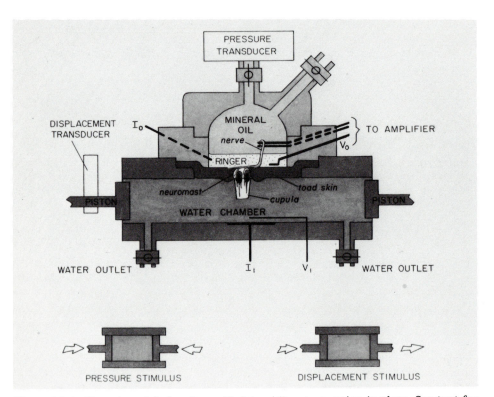

Figure 24-4. Experimental chamber with lateral line preparation in place. See text for detailed description. (From Streiloff and Honrubia 1978).

3.3 Stimulation of the Preparation

Mechanical stimulation of the preparation consisted of water displacements generated by two pistons driven with electromagnetic drivers (Fig. 24-4). When driven in the same direction the two pistons produced water movements with a minimum pressure variation at the center of the chamber where the preparation was located. The preparation was always mounted so that the water movement was in the direction of the greatest sensitivity of the lateral line organ. Sinusoidal water movements with amplitudes up to 200 μm peak-to-peak over a frequency range of 0-120 Hz could be generated at the center of the chamber. The output of a strain gauge displacement transducer, attached to one of the pistons, was recorded on FM tape and served to infer the movements of the water past the organ.

Electrical stimulation consisted of current passed between a chlorided silver disk (I_1) in the water chamber and a chlorided silver-wire ring (I_0) attached to the inside of the upper chamber (Fig. 24-4). The current was monitored by measuring the voltages across resistors in series with the output terminals of the current source. The voltage developed across the skin was monitored with chlorided silver wire electrodes V_0 and V_1. Electrode V_0 was connected to the system ground and served as the reference for recording neural responses as well as for monitoring the voltage applied across the skin. The current and voltage signals were recorded on FM tape for off-line processing. Whenever two stimuli of different frequencies were presented simultaneously both stimulus waveforms were always gated at the positive-going zero-crossings of the electrical signals.

3.4 Neural Recording and Data Collection

The amplitudes of the action potentials recorded from the nerve were as large as 100 μV when the nerve, with connective tissue removed, was draped across the two silver wire electrodes separated by a distance of 4 mm. Careful adjustment of the contact points between the nerve and electrodes enhanced the difference in amplitude between the action potentials recorded from the two afferent fibers and allowed reliable separation of the two spike trains. Differential recording and band-pass filtering of the response waveforms between 300 and 1000 Hz resulted in a 60 dB reduction in the amplitude of the stimulus signal picked up by the nerve electrodes during electrical stimulation at low stimulus frequencies and permitted reliable recording of the neural responses at stimulus frequencies up to 150 Hz. The response waveforms were amplified by 1000 to 5000 before they were recorded on FM tape and were applied to a window discriminator which separated the two spike trains and generated the appropriate pulses for on-line computer processing.

The number of spikes collected for each cycle histogram varied from 500 spikes for suprathreshold responses to 2000 spikes for near-threshold responses. PST histograms were computed from data records containing 2000-8500 spikes in order to improve the signal-to-noise ratios in the resultant histograms for the neural responses to the combined stimulus waveforms. Two or three data records, obtained under the same stimulus conditions, were often combined to compute a single PST histogram.

3.5 Data Analysis

The frequency characteristics of the neural responses to sinusoidal electrical stimuli were obtained from the amplitude and phase of the fundamental component of the discrete Fourier transform of each cycle histogram. The amplitude and phase of each current stimulus signal was determined either from oscilloscope displays or from signal-averaged waveforms of the electrical signals that had been recorded on FM tape. The relative gain of the neural response was computed as the ratio of the response amplitude to the stimulus amplitude and expressed in decibels. The phase was computed as the difference between the response and stimulus phases.

Discrete Fourier transforms of 250 msec PST histogram segments were used to analyze the neural responses to combined 4 Hz and 40 Hz stimuli. Since 40 Hz is a harmonic of 4 Hz, all the energy of each frequency component in the response occurred exactly at a harmonic of the low frequency and there was no spread of energy to other frequencies. This choice of stimulus frequencies enabled the reliable determination of the distortion components in the neural responses.

4 Results

4.1 Frequency Characteristics of Neural Responses to Sinusoidal Electrical Stimuli

The gain and phase of the time-locked neural responses to sinusoidal electrical currents passed through the skin were measured as functions of frequency in four preparations. It was found that the gain of the system increases at approximately 6 dB/oct below 4 Hz, remains relatively flat between 4 Hz and 15 Hz, and decreases at approximately 6 dB/oct between 15 Hz and 60 Hz (Fig. 24-5, top). Above 60 Hz the slope appears to change to 12 dB/oct. A phase lead of approximately $30°$ at 1 Hz increases to nearly $60°$ between 3 and 4 Hz (Fig. 24-5, middle). Above 4 Hz the phase decreases very rapidly with frequency and is $0°$ at 9 Hz and $-350°$ at 150 Hz. Between 50 and 150 Hz the phase characteristic is consistent with a constant time delay of 5.0 msec. When the phase corresponding to a 5.0 msec delay time is subtracted from the measured phase the resulting phase difference asymptotes to a lag of $90°$ at frequencies greater than 60 Hz (Fig. 24-5, bottom).

The resulting magnitude and phase functions are consistent with a minimum phase band-pass network having a pole and a zero at 4.0 and 15 Hz, respectively. The phase and magnitude characteristics of such a network are in good agreement with the experimental data up to 60 Hz. Above 60 Hz, however, the magnitude decreases more rapidly than predicted.

4.2 Electrical Properties of the Skin

In order to ascertain whether the characteristics of the neural responses to electrical stimuli could be influenced by the electrical properties of the skin, the impedance of

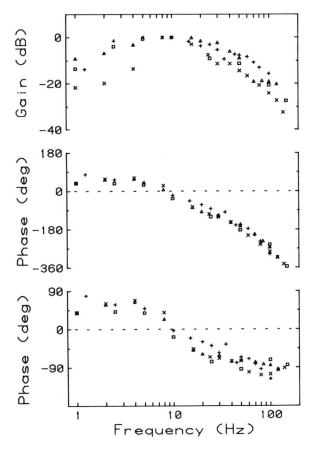

Figure 24-5. Gain (top) and phase (middle) of the fundamental component of neural responses to sinusoidal electrical currents versus stimulus frequency. The corrected phase, determined by subtracting a phase corresponding to a 5.0 msec delay from the raw phase, is shown at the bottom. Each symbol corresponds to data from one preparation.

the skin was measured as a function of frequency in three preparations. The magnitude and phase of the impedance are shown in the upper and lower plots, respectively, of Fig. 24-6. The impedance of a 1.0 cm^2 area of skin decreases from 15 ± 2 kΩ below 2 Hz to 1.8 ± 0.5 kΩ at 100 Hz. On the basis of data not shown, it is estimated that the impedance magnitude would asymptote to 1.2 ± 0.4 kΩ at high frequencies. There is a concomitant frequency-dependent phase lag of the voltage relative to the current which has a maximum value of about 52° at 20 Hz and which decreases to 0° above and below that frequency. The electrical properties of the toad skin can be modeled by the lead network shown in the inset of Fig. 24-5 and having values of 1.2 kΩ, 14.0 kΩ, and 1.8 μF for R_1, R_2, and C, respectively. The electrical network has a pole at 6.3 Hz and a zero at 80 Hz.

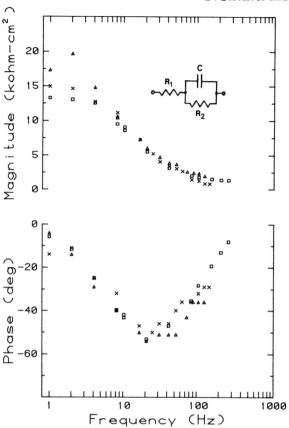

Figure 24-6. Magnitude and phase of the electrical impedance of the *Xenopus laevis* skin versus frequency. Each symbol corresponds to data from one preparation.

4.3 Neural Response Characteristics for Combined Stimuli

The time-locked neural responses to combinations of electrical and mechanical stimuli were studied in 10 preparations. Due to a limited dynamic range of about 18 dB in the determination of the neural response amplitudes from PST histograms, it was found that the best data records were obtained for stimulus combinations where each stimulus alone produced a response of 60 to 90% modulation of the spontaneous activity.

The general characteristics of the neural responses to combined 4 Hz and 40 Hz stimuli are similar for all combinations of the electrical (E) and mechanical (M) stimuli. A representative example (Fig. 24-7) illustrates the electrical current stimuli (left) and the neural response (right) for 4 Hz E (top), 40 Hz E (middle) and 4 Hz E + 40 Hz E (bottom). The enclosed 250 msec interval corresponds to the cycle of the 4 Hz stimulus during which the neural data shown on the right were averaged. In order to enable the easy comparison of the neural responses the vertical scales for all three PST histograms are the same and the number of counts/bin for each histogram reflects the neural activity during 1070 stimulus presentations.

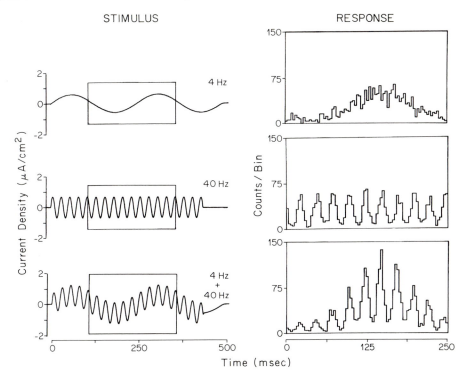

Figure 24-7. Waveforms of electrical current stimuli (left) and PST histograms of neural responses (right) to 4 Hz (top), 40 Hz (middle), and the combination of 4 Hz and 40 Hz (bottom) stimuli. The enclosed 250 msec interval corresponds to the 4 Hz cycle during which the neural data on the right were averaged. The mean firing rate during the averaged interval and the number of spikes per histogram were 7.2 ips, 997 spikes (top); 8.6 ips, 2376 spikes (middle); 10.3 ips, 2710 spikes (bottom). Response magnitudes of the fundamental components for the top and middle histograms are 89.2 and 85.5% modulation, respectively. The magnitudes and phases of the response components in the DFT of the bottom histogram are listed in Table 24.1.

The main features of neural responses to M + M, E + E or E + M combined stimuli are (1) the peak-to-peak amplitude of the 40 Hz component in the response is modulated above and below its normal level by the 4 Hz stimulus, (2) the instantaneous firing rate (averaged over one 40 Hz cycle) is modulated by the 4 Hz stimulus and (3) the overall mean firing rate averaged over a single 4 Hz cycle is increased above the average rate of response to either stimulus presented alone.

In order to quantify the responses to combined stimuli, discrete Fourier transforms (DFT) of the PST histograms were computed. The frequency components of the Fourier analysis which had magnitudes above the noise level were at 4, 40, 8, 80, 44, and 36 Hz corresponding to the frequencies of the fundamentals, the second harmonics, and the sum and difference of the two stimulus frequencies, respectively. The magnitudes and phases of the Fourier components of the neural response to the 4 Hz E + 40 Hz E stimulus combination illustrated in Fig. 24-7 are listed in Table 24-1. The

Table 24-1. Magnitudes and Phase of the Energy Components in the DFT of the PST Histogram of the Neural Response of a Combination of 4 Hz, 1.2 $\mu A/cm^2$ p-p and 40 Hz, 1.4 $\mu A/cm^2$ p-p Electrical Stimuli[a]

Frequency (Hz)	Magnitude (% mod.)	Phase (°)
4	82.9	237
40	75.5	187
8	12.3 (17.2)	40 (24)
80	8.4 (14.3)	−77 (−76)
36	27.2 (31.3)	45 (40)
44	35.5 (31.3)	−17 (−26)

[a]The PST histogram is illustrated in the lower right of Fig. 24-7. The values in parentheses are computed from the DFT magnitudes and phases at 4 Hz and 40 Hz using the formulas listed in Table 24-3. See text for details.

magnitudes are expressed as % modulation (magnitude $\times$ 100/mean firing rate) and the phases are expressed relative to an arbitrary reference. Magnitudes and phases of the Fourier components of the neural responses to 4 Hz M + 40 Hz M and to 4 Hz E + 40 Hz M are listed in Table 24-2.

As the tabulated data show, consistent findings with regard to the Fourier components are that: (1) the square of the amplitude of each fundamental component is approximately equal to four times the amplitude of the second harmonic, (2) the product of the amplitudes of the two fundamental components is approximately twice the sum of the amplitudes of the components at the sum and difference frequencies,

Table 24-2. Magnitudes and Phases of the Energy Components in the DFTs of PST Histograms of Neural Responses to Stimulus Combinations (see Table 24-1)

Frequency (Hz)	Magnitude (% mod.)	Phase (°)
A. 4 Hz, 16 μm p-p and 40 Hz, 59 μm p-p mechanical stimuli		
4	89.5	109
40	94.7	−22
8	23.9 (20.0)	121 (128)
80	22.2 (22.4)	208 (226)
36	43.3 (42.4)	−42 (−41)
44	48.4 (42.4)	−10 (−3)
B. 4 Hz, 0.52 $\mu A/cm^2$ p-p electrical and 40 Hz, 10 μm p-p mechanical stimuli		
4	59.0	76
40	88.3	−87
8	11.6 (8.7)	68 (62)
80	23.6 (19.5)	99 (96)
36	24.5 (26.0)	−63 (−73)
44	31.3 (26.0)	257 (259)

and (3) the phase relations among the components are consistent with the predictions of simple formulas. Through a trial and error process it was determined that the characteristics of the magnitudes and phases of the Fourier components are consistent with a quadratic type of nonlinear relationship between the input stimuli and the output responses. That is, if the stimulus is expressed as

$$S(t) = a_L \sin(w_L t + \theta_L) + a_H \sin(w_H t + \theta_H) \tag{1}$$

then the response could be expressed as

$$R(t) = R_0 + U(t) + cU^2(t) \tag{2}$$

where $U(t)$ is a linear transformation of $S(t)$. For electrical stimuli the relationship between $S(t)$ and $U(t)$ is determined by the frequency response characteristics described in Fig. 24-5. Specifically for

$$U(t) = A_L \sin(w_L t + \phi_L) + A_H \sin(w_H t + \phi_H) \tag{3}$$

the relative magnitudes and phases of the response components are given in Table 24-3.

An important finding was that a value of 0.5 was consistently found to be the best estimate of c in Eq. (2) for all of the available data records. The computed amplitudes and phases of the second harmonic and sum and difference response components for $c = 0.5$ are given in parentheses beside the measured values in Tables 24-1 and 24-2. For each data record the amplitudes and phases of the 4 and 40 Hz components were used in computing the values of the other components. Differences between computed and measured magnitude and phase values were less than 6% modulation and 18°, respectively.

5 Discussion

The data presented here extend the previously available description (Strelioff and Honrubia 1978) of the neural response characteristics of the *Xenopus laevis* lateral line organ to electrical stimulation. These data demonstrate that the organ is sensitive to sinusoidal electrical stimuli over a frequency range encompassing at least two decades with the most sensitive region occurring between 4 and 15 Hz. By comparing the frequency dependence of the neural responses to electrical stimuli with the electrical impedance of the toad skin it can be concluded that the 6 dB/oct gain increase of the neural response below 4 Hz is the result of the properties of the hair cell system rather than the electrical impedance of the skin because the impedance is constant below 4 Hz. At present it is not known whether this frequency dependent response gain increase is due to the electrical properties of the hair cell membrane or to some other aspects of the neural excitation process.

Since the impedance of the skin and the gain of the neural response both decrease at approximately 6 dB/oct above 15 Hz, it is possible that the properties of the skin

Table 24-3. Magnitudes and Phases of the Resulting Energy Components When Two Sinusoidal Waveforms With Frequencies f_L and f_H are Passed Through a Quadratic Type of Nonlinearity

Frequency	Magnitude	Phase
f_L	A_L	θ_L
f_H	A_H	θ_H
$2f_L$	$cA_L^2/2$	$2\theta_L - 90°$
$2f_H$	$cA_H^2/2$	$2\theta_H - 90°$
$f_H - f_L$	$cA_L A_H$	$\theta_H - \theta_L +90°$
$f_H + f_L$	$cA_L A_H$	$\theta_H + \theta_L -90°$

play a role in determining the frequency characteristics of the neural responses to extrinsic electrical potentials in this frequency range. If the impedance of the skin decreases more rapidly with increasing frequency than does the impedance of the hair cell membrane, it would shunt the extrinsic currents away from the hair cells at higher frequencies and thus would reduce the magnitude of the neural response. Direct measurement of the intracellular hair cell potentials produced by extrinsic currents will be necessary to determine whether this is, in fact, the cause of the high frequency roll-off.

The frequency independent time delay of 5.0 msec is probably due to the time constant of the neural excitation process at the afferent synapses and to the propagation time of the nerve impulses along the nerve between the hair cells and the recording electrodes (approximately 2 cm). This delay time is close to the 4 msec delay estimated by Harris and Flock (1967) for the *Xenopus* preparation. A synaptic delay time of about 1 msec (Katz and Miledi 1965), a neural propogation velocity of 5.5 m/sec along the nerve branches within the neuromast (Pabst 1977), and 17.4 m/sec along the stem fiber (Görner 1967) would result in a total delay of approximately 4.0 msec. The delay time reflected in our data is thus in reasonable agreement with the measurements of others.

In the model of hair cell function illustrated in Fig. 24-1 it is assumed that, during electrical stimulation, the variation in neural response (i.e., the variation in mean firing rate) is proportional to the variation in membrane potential produced by the flow of extrinsic current through the hair cell. On the other hand, during mechanical stimulation, the variation in neural response is assumed to be proportional to the variation in membrane potential produced by modulation of the steady-state intrinsic current by the mechanically induced impedance changes. The ratio of the magnitudes and the difference between the phases of the neural responses to mechanical and electrical stimuli determined as functions of frequency should therefore give the gain and phase characteristics of the mechanics of the organ and the mechanoelectrical transduction of the hair cell system.

The only available data on the frequency response characteristics of the *Xenopus laevis* lateral line for mechanical stimuli are given by Kroese et al. (1978, Fig. 2). According to their data, the gain increases at approximately 6 dB/oct from 0.1 to 4 Hz and at 12 dB/oct to about 10 Hz. The gain curve has a maximum at 20 Hz and begins to decrease with a slope of 9-18 dB/oct at frequencies between 25 and 40 Hz. The corresponding phase has a constant lead of about 125° up to 5 Hz and then gradually changes to a phase lag of about 290° at 100 Hz. Thus, when the ratio of the gains and difference between the phase of the neural responses to mechanical and electrical responses are determined, it is found that, at frequencies below 4 Hz, the gain and phase of the organ mechanics and the mechanoelectrical transduction process are approximately constant. At these frequencies the phase of the mechanical response leads the phase of electrical response by approximately 90°. Between 4 Hz and 50 Hz the gain increases by about 25 dB and the phase lead decreases to nearly 0°. Above 50 Hz both gain and phase remain constant. There is no simple minimum-phase system which has such a transfer characteristic. Since the mechanical-to-electrical transformation must account for the characteristics of the waves generated by a vibrating sphere, the viscous coupling of the water to the cupula, and, finally, the mechanics of the cupula and hair cell bundles, it is not surprising that the transformation is not simple. The general increase of gain and the 90° phase lead at low frequencies are, nonetheless, consistent with the conclusion that velocity-dependent viscous forces act between the moving water and the cupula and determine the main characteristics of the response of the lateral line organ to water vibrations (Görner 1963, Harris and Milne 1966, Strelioff and Honrubia 1978).

By definition, the response of a linear system to two stimuli presented simultaneously is the sum of the responses to each stimulus presented alone. Since all combinations of electrical, mechanical, or electrical and mechanical stimuli used in the present experiments resulted in a modulation of the neural response to the high frequency stimulus by the low frequency stimulus rather than a linear sum, the experimental results are not consistent with a linear model. The nonlinear relationship between the electrical stimulus and the neural response is due to a nonlinear current-voltage relationship of the hair cell membrane, to a nonlinear relationship between membrane potential and neural response, or to a combination of both. Measurements made by Corey and Hudspeth (1979) under voltage clamp conditions show that hair cells in the sacculus of the bullfrog have a nonlinear current-voltage relationship for membrane potentials between -45 and -65 mV. Corey and Hudspeth also demonstrated that the magnitude of the receptor current during mechanical stimulation was directly proportional to the membrane potential. Thus there is direct experimental evidence that one source of our observed stimulus-response nonlinearity could reside in the membrane properties of the hair cells. At present there are no direct experimental data demonstrating the exact relationship between changes in the hair cell membrane potential and the resulting neural response.

In addition to the nonlinearity in the response to electrical currents, the neural response should exhibit a nonlinearity due to the mechanoelectrical transduction process as demonstrated by Hudspeth and Corey (1977). However, the available data are not adequate to permit a conclusive determination of whether the neural responses to sinusoidal mechanical stimuli exhibit different nonlinearities than do the responses to electrical stimuli.

The magnitudes and phases of the distortion products present in the neural responses to both electrical and mechanical stimuli are consistent with a quadratic nonlinearity. However, since the coefficient of the quadratic term was consistently found to be close to 0.5, it is possible that the nonlinear relationship may be exponential. Due to the limited dynamic range of the present measurements, all that can be said is that the experimental data are consistent with both quadratic and exponential nonlinearities.

During stimulation with either two mechanical or two electrical stimuli the frequency components add before they pass through the nonlinear portion of the hair cell system and produce the neural response. On the other hand, when mechanical and electrical stimuli are presented simultaneously, the mechanical stimulus goes through the mechanoelectrical transduction process before combining with the electrical stimulus and producing the neural response. Since the responses to combined mechanical and electrical stimuli are similar to the responses in stimuli of the same type, it is likely that both types of stimuli combine before passing through the major nonlinearity of the hair cell system. This conclusion is consistent with the assumptions that both electrical and mechanical stimuli produce neural responses through their effects on the hair cell current and membrane potential and that these effects are combined before the action potentials are generated. The implication of the finding is that the nonlinearity must be in the relationship between the membrane potential and the initiation of action potentials in the nerve.

In conclusion, our experimental findings are consistent with the basic assumption of the model of hair cell function that physiological stimuli and environmental properties affect neural activity by changing the resting current through the hair cells. In addition to providing information on the frequency characteristics of neural responses of the *Xenopus laevis* lateral line system to electrical stimuli the present data are consistent with the existence of an exponential nonlinear relationship between the electrical stimulus and the neural response. Since this nonlinearity is also indicated by the neural responses to mechanical stimuli it is concluded that the nonlinearity exists at a stage of the transduction process which follows the generation of the receptor potential. It is suggested that the nonlinearity is characteristic of the relationship between the variations of the membrane potential at the afferent synapses and the generation of the action potentials in the nerve.

Acknowledgments. This research was supported by grants from NINCDS and the Deafness Research Foundation.

References

Bauknight, R. S., Strelioff, D., Honrubia, V.: Effective stimulus for the *Xenopus laevis* lateral-line hair-cell system. Laryngoscope 86, 1836-1844 (1976).

Bennett, M. V. L.: Electrolocation in fish. Ann. N.Y. Acad. Sci. 188, 242-269 (1971).

Bracho, H., Budelli, R.: The generation of resting membrane potentials in an inner ear hair cell system. J. Physiol. 281, 445-465 (1978).

Cahn, P.: Lateral Line Detectors. Bloomington: Indiana Univ. Press, 1967.

Cole, K. S.: Membrane, Ions and Impulses. Berkeley: University of California Press, 1968.

Corey, D. P., Hudspeth, A. J.: Ionic basis of the receptor potential in a vertebrate hair cell. Nature 281, 675-677 (1979).

Davis, H.: A model for transducer action in the cochlea. Cold Spring Harbor Symp. Quant. Biol. 30, 181-190 (1965).

Dijkgraaf, S.: The functioning and significance of the lateral line organs. Biol. Rev. 38, 51-105 (1963).

Dijkgraaf, S.: Biological significance of the lateral line organs. In: Lateral Line Detectors. Cahn, P. (ed.). Bloomington: Indiana Univ. Press, 1967, pp. 83-96.

Fex, J.: Augmentation of the cochlear microphonics by stimulation of efferent fibers to cochlea. Acta Oto-Laryngol. 50, 540-541 (1959).

Flock, Å.: Electromicroscopical and electrophysiological studies on the lateral line canal organ. Acta Oto-Laryngol. Suppl. 199, 1-90 (1965).

Flock, Å.: Ultrastructure and function in the lateral line organs. In: Lateral Line Detectors. Cahn, P. (ed.). Bloomington: Indiana Univ. Press, 1967, pp. 163-197.

Flock, Å.: Sensory transduction in hair cells. In: Handbook of Sensory Physiology, Vol. 1. Lowenstein, W. (ed.). Berlin-Heidelberg-New York: Springer-Verlag, 1971, pp. 396-441.

Flock, Å., Russell, I. J.: The postsynaptic action of efferent fibers in the lateral line organ of the burbot Lota lota. J. Physiol. 235, 591-605 (1973).

Flock, Å., Jørgensen, J. M., Russell, I. J.: The physiology of individual hair cells and their synapses. In: Basic Mechanisms in Hearing. Møller, A. (ed.). New York: Academic Press, 1973, pp. 273-306.

Flock, Å., Russell, I. J.: Inhibition by efferent nerve fibers: Action on hair cells and afferent synaptic transmission in the lateral line canal organ of the burbot Lota lota. J. Physiol. 257, 45-62 (1976).

Galambos, R.: Suppression of auditory nerve activity by stimulation of efferent fibers to cochlea. J. Neurophysiol. 19, 424-437 (1956).

Geisler, C. D.: Model of crossed olivocochlear bundle effects. J. Acoust. Soc. Am. 56, 1910-1912 (1974).

Geisler, C. D., Mountain, D. C., Hubbard, A. E., Adrian, H. O., Ravindran, A.: Alternating electrical-resistance changes in the guinea-pig cochlea caused by acoustic stimuli. J. Acoust. Soc. Am. 61, 1557-1566 (1977).

Goldman, D. E.: Potential, impedance and rectification in membranes. J. Gen. Physiol. 27, 37-60 (1943).

Görner, P.: Untersuchungen zur Morphologie und Elektrophysiologie des Sietenlinienorgans vom Krallenfrosches (Xenopus laevis Daudin). Z. Vergl. Physiol. 47, 316-338 (1963).

Görner, P.: Independence of afferent activity from efferent activity in the lateral line organ of Xenopus laevis Daudin. In: Lateral Line Detectors. Cahn, P. (ed.). Bloomington: Indiana Univ. Press, 1967, pp. 199-214.

Harris, G. G., van Bergeijk, W. A.: Evidence that the lateral line organs responds to water displacements. J. Acoust. Soc. Am. 34, 1831-1841 (1962).

Harris, G. G., Milne, D. C.: Input-output characteristics of the lateral line sense organ. J. Acoust. Soc. Am. 40, 32-42 (1966).

Harris, G. G., Flock, Å.: Spontaneous and evoked activity from the Xenopus laevis lateral line. In: Lateral Line Detectors. Cahn, P. (ed.). Bloomington: Indiana Univ. Press, 1967, pp. 135-161.

Harris, G. G., Frischkopf, L. S., Flock, Å.: Receptor potentials from hair cells of the lateral line. Science 167, 76-79 (1970).

Hodgkin, A. L., Katz, B.: The effect of sodium ions on the electrical activity of the giant axon of the squid. J. Physiol. 108, 37-77 (1949).

Honrubia, V., Strelioff, D., Sitko, S.: Physiological basis of cochlear transduction and sensitivity. Ann. Otol. Rhinol. Largyngol. 85, 697-710 (1976).

Hudspeth, A. J., Corey, D. P.: Sensitivity, polarity, and conductance change in the response of vertebrate hair cells to controlled mechanical stimuli. Proc. Nat. Acad. Sci. 74, 2407-2411 (1977).

Hudspeth, A. J., Jacobs, R.: Stereocilia mediate transduction in vertebrate hair cells. Proc. Nat. Acad. Sci. 76, 1506-1509 (1979).

Jierlof, R., Spoor, A., DeVries, H.: The microphonic activity of the lateral line. J. Physiol. (London) 116, 137-157 (1952).

Johnstone, B. M., Sellick, P. M.: The peripheral auditory apparatus. Q. Rev. Biophys. 57, 1-57 (1972).

Katsuki, Y.: The ionic receptive mechanism in the acoustico-lateralis system. In: Basic Mechanisms in Hearing. Møller, A. (ed.). New York: Academic Press, 1973, pp. 307-334.

Katsuki, Y., Yoshino, S.: Response of the single lateral line nerve fiber to the linearly rising current stimulating the end-organ. Jpn. J. Physiol. 2, 219-231 (1952).

Katz, B., Miledi, R.: The measurement of synaptic delay, and the time course of acetylcholine release at the neuromuscular junction. Proc. R. Soc. London Ser. B 161, 483-495 (1965).

Konishi, T., Teas, D. C., Wernick, J. S.: Effects of electrical current applied to cochlear partition on discharges in individual auditory-nerve fibers: I. Prolonged direct-current polarization. J. Acoust. Soc. Am. 47, 1519-1526 (1970).

Konishi, T., Slepian, J. Z.: Effects of the electrical stimulation of the crossed olivo-cochlear bundle on cochlear potentials recorded with intracochlear electrodes in guinea pigs. J. Acoust. Soc. Am. 49, 1762-1769 (1971).

Kroese, A. B. A., Van der Zalm, J. M., Van den Bercken, J.: Frequency response of the lateral-line organ of Xenopus laevis. Pflügers. Arch. 375, 167-175 (1978).

Kuiper, J. W.: The microphonic effect of the lateral line organ. Thesis, Univ. Gröningen, The Netherlands (1956).

Lowenstein, O.: The effect of galvanic polarization on the impulse discharge from sense endings in the isolated labyrinth of the thornback ray (Raja clavata). J. Physiol. 127, 104-117 (1955).

Lowenstein, O., Wersäll, J.: A functional interpretation of the electron microscopic structure of the sensory hairs in the cristae of the elasmobranch, Raja clavata, in terms of directional sensitivity. Nature 184, 1807-1810 (1959).

McGlone, F. P., Russell, I. J., Sand, O.: Measurement of calcium ion concentrations in the lateral line cupulae of Xenopus laevis. J. Exp. Biol. 83, 123-130 (1979).

Moreton, R. B.: An investigation of the electrogenic sodium pump in snail neurones, using constant field theory. J. Exp. Biol. 51, 181-201 (1969).

Moxon, E. G.: Neural and Mechanical Responses to Electric Stimulation of the Cat's Inner Ear. Thesis, Massachusetts Institute of Technology, Cambridge, 1971.

Mulroy, M. J., Altmann, D. W., Weiss, T. F., Peake, W. T.: Intracellular electric responses to sound in a vertebrate cochlea. Nature 249, 482-485 (1974).

Murray, R. W.: The lateralis organs and their innervation in Xenopus laevis. Quart. J. Micr. Sci. 96, 351-361 (1955).

Murray, R. W.: The response of the lateralis organs of Xenopus laevis to electrical stimulation by direct current. J. Physiol. Lond. 134, 408-420 (1956).

Onoda, N., Katsuki, Y.: Chemoreception of the lateral-line organ of an aquatic amphibian, Xenopus laevis. Jpn. J. Physiol. 22, 87-102 (1972).

Pabst, A.: Number and location of the sites of impulse generation in the lateral-line afferents of Xenopus laevis. J. Comp. Physiol. 114, 51-67 (1977).

Peterson, S. K., Frishkopf, L. S., Lechine, C., Oman, C. M., Weiss, T. F.: Element composition of inner ear lymph in cats, lizards and skates determined by electron probe microanalysis of liquid samples. J. Comp. Physiol. 126, 1-14 (1978).

Russell, I. J.: Influence of efferent fibers on a receptor. Nature 219, 177-178 (1968).

Russell, I. J.: Amphibian lateral line receptors. In: Frog Neurobiology. Llinás, R., Precht, W. (eds.). Berlin-Heidelberg-New York: Springer-Verlag, 1976, pp. 513-550.

Russell, I. J., Sellick, P. M.: Measurement of potassium and chloride ion concentrations in the cupulae of the lateral lines of *Xenopus laevis*. J. Physiol. Lond. 257, 245-255 (1976).

Russell, I. J., Sellick, P. M.: Intracellular studies of hair cells in the mammalian cochlea. J. Physiol. 284, 261-290 (1978).

Sand, O.: Effects of different ionic environments on the mechano-sensitivity of lateral line organs in the mudpuppy. J. Comp. Physiol. 102, 27-42 (1975).

Sand, O., Ozawa, S., Hagiwara, S.: Electrical and mechanical stimulation of hair cells in the mudpuppy. J. Comp. Physiol. 102, 13-26 (1975).

Schwartz, E.: Lateral line mechanoreceptors in fishes and amphibians. In: Handbook of Sensory Physiology, Vol. III/3. Fessard, A. (ed.). Berlin-Heidelberg-New York: Springer-Verlag, 1974, pp. 257-278.

Smith, C. A., Lowry, O. H., Wu, M. L.: The electrolytes of the laybrinthine fluids. Laryngoscope 64, 141-153 (1954).

Strelioff, D., Haas, G., Honrubia, V.: Sound-induced electrical impedance changes in the guinea pig cochlea. J. Acoust. Soc. Am. 51, 617-620 (1972).

Strelioff, D., Sitko, S. T., Honrubia, V.: Role of inner and outer hair cells in neural excitation. Trans. Am. Acad. Ophthalmol. Otolaryngol. 82, 322-327 (1976).

Strelioff, D., Honrubia, V.: Neural transduction in *Xenopus laevis* lateral line system. J. Neurophysiol. 41, 432-444 (1978).

Teas, D. C., Konishi, T., Wernick, J. S.: Effects of electrical current applied to cochlear partition on discharges in individual auditory-nerve fibers: II. Interaction of electrical polarization and acoustic stimulation. J. Acoust. Soc. Am. 47, 1527-1537 (1970).

von Békésy, G.: Experiments in Hearing. New York: McGraw-Hill, 1960, pp. 635-710.

Weiss, T. F., Mulroy, M. J., Altman, D. W.: Intracellular responses to acoustic clicks in the inner ear of the alligator lizard. J. Acoust. Soc. Am. 55, 606-621 (1974).

Wiederhold, M. L., Kiang, N. Y. S.: Effects of electric stimulation of the crossed olivocochlear bundle on single auditory nerve fibers in the cat. J. Acoust. Soc. Am. 48, 950-965 (1970).

Discussion

HAWKINS: In *Xenopus,* aren't the adjacent hair cells oriented in opposite directions?

STRELIOFF: Yes.

HAWKINS: Wouldn't that lead to rather confusing results?

STRELIOFF: I don't believe so, because the cells would alternately shunt current from one to the other, so that the cell that is being inhibited is actually providing current to the one that is being excited. If this does in fact occur, then, if you were to cross-correlate the responses of the two fibers, you would probably find a tendency for spikes not to occur simultaneously in both fibers.

HAWKINS: Would you predict what would happen in a macula, where the adjacent hair cells have a common orientation?

STRELIOFF: The mammalian cochlea is a situation we are familiar with, and there is an electrical network model to describe the properties of the cochlea. If you assume that the three rows of outer hair cells and the one row of inner cells have nearly the same electrical properties, and that the inner cells do not receive any stimulation, then mechanical stimulation of the outer cells would produce a current in the inner cells that should be less by 6 dB. There would be a great likelihood of an electrical interaction between the inner and outer cells.

CORWIN: To pursue that further, since the outer hair cells are all oriented in the same direction, then would they not be inhibiting each other?

STRELIOFF: Yes, the outer hair cells would inhibit each other in the absence of acoustic stimulation. On the other hand, when neighboring cells receive different stimuli, the differences in the resulting receptor currents should be enhanced by the electrical interaction between hair cells. At some locations along the basilar membrane there may be cancellation of such currents while at other location there may be summation.

PLATT: Hudspeth's group [Proc. Nat. Acad. Sci. 74, 2407-2411 (1977); 76, 1506-1509, (1979)] recently reported on a mechanism of adaptation to displacement in the sacculus. Do you have any data on adaptation in these cells? Furakwa and Ishii [J. Neurophysiol. 30, 1377-1403 (1967)] reported that under static pressure they did get a tonic change as they recorded efferent activity from the goldfish ear.

STRELIOFF: I did a series of experiments on the neural responses to a constant water flow. There was an initial large response when the water was first turned on, and then it decreased with a time constant of about two or three seconds. This was much longer than others have found (about 10 msec). There is a small response to a sustained water flow, and I suppose that is as it should be, since a constant flow maintains a static displacement.

BELL: What was the magnitude of your stimulus?

STRELIOFF: About 100 μV across the skin. It is really impossible to say what was the actual stimulus to the cell, except that it must have been less than that. The only way in which one can determine the fraction of the stimulus potential that actually appears across the hair cell is to do microelectrical studies and map out the potential fields.

BELL: Isn't that stimulus level within the range of sensitivity of an ampullar receptor?

BULLOCK: Sure, but some are much less sensitive. One can conceive of an acoustic detector as being, in essence, an electroreceptor with a mechanotransducer ahead of it. The sensitivity of the cochlea is in the same range. It takes a couple of millivolts to evoke cochlear microphonics, but that is well within the range of a good electroreceptor.

I wonder if there is the possibility that some acoustic receptors might have a sensitivity to an opposite polarity, as exists among photoreceptors. One group is excited by depolarization of the inner membrane, while the other is excited by depolarization of the outer membrane. In photoreceptors, there is coupling between the membranes, and transmitter release is different in the two versions of the same modality. However, maybe hearing was invented only once, and acoustic receptors are stuck with only one polarity sensitivity.

Chapter 25

Modeling the Effects of Stimulus Frequency and Intensity on Hair Cell Potentials

J. R. BOSTON*

1 Introduction

Many investigators, in attempting to describe observed properties of sensory hair cell potentials, have used the variable conductance model proposed by Davis (1965). Recent studies of hair cell potentials have shown the need to include capacitive elements in these models to match experimental results (Weiss, Mulroy, and Altman 1974). Nonlinearities in hair cell transduction have also been proposed to explain experimental findings, especially the origins of distortion components in the potentials. Nonlinear input-output functions have recently been measured in hair cells (Hudspeth and Corey 1977, Russell and Sellick 1978, Eatock, Corey, and Hudspeth 1979).

I have studied the extracellular microphonic potential recorded in the lateral line of *Fundulus heteroclitus* (killifish) and have been interested in adapting the Davis-type model to describe the experimental results. In particular, I have measured microphonic output amplitude as a function of stimulus input amplitude and have observed changes in the input-output functions as stimulus frequency is increased. This report will briefly review the experimental findings and then describe an electrical model that simulates them. Finally, the implications of this model to the hair cell transduction process will be discussed.

2 Experimental Results

Most of the experimental data were obtained with the fish wrapped in moist gauze and almost submerged in water (Boston 1976, 1980). The lateral line canals on the dorsal surface were exposed to air. A vibrating rod was inserted into one canal pore, setting the fluid within the canal into vibration and exciting the sensory organ. Response potentials were recorded with 25 μm diameter tungsten electrodes inserted into another

*Department of Anesthesia/CCM, University of Pittsburgh School of Medicine, Pittsburgh, Pennsylvania 15261.

pore of the canal. A Grass AC amplifier with gain of 1000 and band-pass .1 to 1,000 Hz was used.

The lateral line microphonic response to a sinusoidal stimulus consists primarily of second harmonic and dc distortion components. The absence of a fundamental component has been explained by the anatomical arrangement of hair cells in the sensory organ (Flock 1965). Typical microphonics generated in response to a sinusoidal vibratory burst of 200 Hz are shown in Fig. 25-1. The bottom traces show the stimulus waveform, while the upper traces show the microphonic potential recorded for increasing levels of stimulus amplitude. Computer averaging of 32 individual responses has been used to improve the signal-to-noise ratio. On the left, the 200 Hz sinusoid is phase-locked to the envelope to produce the same stimulus waveform for each response averaged. On the right, the phase between the sinusoid and the envelope is random, causing the second harmonic component to average out. The dc component is not changed. The amplitude of the dc component of the response is defined as the distance from the baseline to the maximum deflection observed in the random phase condition. The amplitude of the second harmonic is the peak-to-peak amplitude of the oscillatory component. Although the oscillatory component is very sinusoidal at low stimulus amplitudes, it becomes quite distorted at higher stimulus amplitudes.

The stimulus levels used to obtain these results were relatively large. The second harmonic amplitude was clearly saturating at the 2.2 μm stimulus level. The dc shift

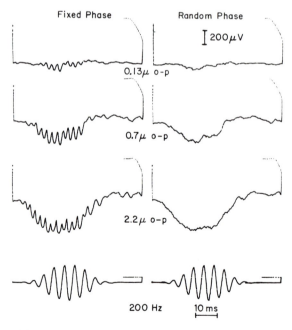

Figure 25-1. Lateral line microphonic for increasing stimulus amplitude. Bottom waveform in each column is the stimulus. Stimulus amplitude (zero to peak) is shown below each microphonic waveform. Results based on averages of 32 individual responses. (Left) Fixed phase between stimulus frequency and envelope. (Right) Random phase between stimulus frequency and envelope.

was negative, and it did not appear to be saturating as rapidly as the second harmonic. The ratio of dc to second harmonic amplitude increased considerably as stimulus level increased. Input-output functions obtained between 100 and 400 Hz showed that the increase in the dc component with respect to the second harmonic became more pronounced at higher frequencies. At constant stimulus amplitude, the dc component remained approximately constant or even increased slightly with increasing frequency, while the second harmonic amplitude decreased with frequency.

3 An Electrical Model

These results can be explained using an electrical model of the hair cell as shown in Fig. 25-2. The hair cell response is assumed to be a change in conductance, while the hair cell capacitance is assumed to remain constant. C_m and g_m represent the capacitance and conductance of the membrane of the basal region of the hair cell; C_h and g_h represent capacitance and conductance of the apical region; and C_e and g_e represent the capacitance and conductance of the external current paths between the recording electrode and ground. E_{ext} is the extracellularly recorded microphonic, and E_{int} is the potential that would be measured by an intracellular electrode. E_m is the diffusion potential assumed to be associated with the basal membrane of the hair cell; the apical membrane is assumed to be depolarized (Bracho 1977, Bracho and Budelli 1978). g_h is assumed to be a conductance that varies in response to a stimulus as $g_h = g_0(1 + \theta f)$, $|f| < 1$. f is a function of the stimulus $x(t)$. g_0 is the resting level of the conductance, and θ is the maximum change in the conductance.

The signals of interest in the model are the deviations of the extracellular and intracellular potentials from their resting values. These signals can be defined as $e = E_{ext} -$

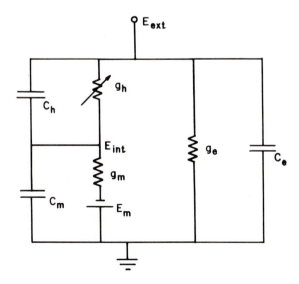

Figure 25-2. Electrical model of lateral line hair cell.

$(E_{ext})_{rest}$ and $v = E_{int} - (E_{int})_{rest}$. A pair of linear differential equations can be written for these signals:

$$(C_h + C_e)\frac{de}{dt} + [g_e + g_0(1 + \theta f)]e = C_h\frac{dv}{dt} + [g_0(1 + \theta g)]v - \frac{g_0 g_e g_m}{K_1}E_m\theta f$$

$$C_h\frac{de}{dt} + [g_0(1 + \theta f)]e = (C_h + C_m)\frac{dv}{dt} + [g_m + g_0(1 + \theta f)]v - \frac{g_0 g_e g_m}{K_1}E_m\theta f$$

where $K_1 = g_0 g_e + g_0 g_m + g_e g_m$. Using a saturating asymmetrical nonlinearity for $f(x)$ (Boston 1980), this set of equations can simulate the observed behavior of the lateral line microphonic. Specifically, the same form of input-output functions with the same dependency on frequency were obtained. The ratio of the dc to second harmonic became larger at higher frequencies, as was observed in the experimental data.

Because the function f depends on time, these equations cannot be analyzed using standard frequency transform techniques. However, if $\theta f \ll 1$, the equations can be approximated by a pair of linear, time invariant equations that can be analyzed using standard techniques. A second-order differential equation in e alone and one in v alone can be derived:

$$\frac{d^2 e}{dt^2} + (\omega_1 + \omega_2)\frac{de}{dt} + (\omega_1\omega_2)e = -\theta'(\omega_m f + \frac{df}{dt})$$

$$\frac{d^2 v}{dt^2} + (\omega_1 + \omega_2)\frac{dv}{dt} + (\omega_1\omega_2)v = \theta'\frac{C_e}{C_m}(\omega_e f + \frac{df}{dt})$$

where $(\omega_1\omega_2) = K_1/K_2$; $(\omega_1 + \omega_2) = (g_m C_h + g_m C_e + g_0 C_e + g_0 C_m + g_e C_h + g_e C_m)/K_2$; $\theta' = \theta C_m/K_2$; $\omega_m = g_m/C_m$; $\omega_e = g_e/C_e$. $K_2 = C_h C_m + C_e C_h + C_e C_m$. ω_1 and ω_2 represent the poles of the system.

The frequency dependence of e and v can be described in terms of their Fourier transforms. Denote the transforms of $e(t)$, $v(t)$, and $f[x(t)]$ by $E(j\omega)$, $V(j\omega)$, and $F(j\omega)$, respectively. Then the frequency response functions for E and V are

$$\frac{E(j\omega)}{F(j\omega)} = -\theta'\frac{j\omega + \omega_m}{(j\omega + \omega_1)(j\omega + \omega_2)}$$

$$\frac{V(j\omega)}{F(j\omega)} = \theta'\frac{C_e}{C_m}\frac{j\omega + \omega_e}{(j\omega + \omega_1)(j\omega + \omega_2)}$$

The denominators of the frequency response functions (the poles) are identical; only the numerators (the zeros) differ.

The response functions can be illustrated graphically with a Bode plot, in which the magnitude of the response is plotted as a function of frequency. Figure 25-3 shows

Bode plots for the extracellular potential frequency response for different assumed values of the capacitive elements in the model. Also shown are typical waveforms that would be recorded under these assumptions. The arrows associated with the waveforms point to the frequency of the stimulus $x(t)$. The waveforms were obtained by simulation; a sinusoidal burst such as was used in the experimental work was transformed by the function f and used to excite the differential equation for e. Because f is nonlinear, it contains distortion components as well as a fundamental component. Specifically, it contains a dc and a second harmonic component.

To simulate the effects of the oppositely oriented hair cells, two responses of single hair cells 180° out of phase were combined to form the waveforms shown. This operation cancelled the fundamentals, leaving the second harmonic and dc components. The amplitudes of the second harmonic and dc components of e depend on the form of the frequency response function of e.

For the extracellular potential, the shape of the frequency response function depends critically on the relative values of ω_m, ω_1, and ω_2. If both C_e and C_h are assumed to be zero, so that the only important capacitive element in the model is the hair cell membrane capacitance C_m, then ω_m is less than ω_1, as in Fig. 25-3, left. (The system actually becomes first order and has only one pole.) I earlier reported that this model could simulate the lateral line experimental data (Boston 1980). This conclusion resulted from an error in the formulation of the model. In fact, as shown here, the model with C_e and C_h equal to zero cannot describe the experimental data. At frequencies below ω_m the response is roughly similar to that observed experimentally at low stimulus levels. At higher frequencies, however, the dc component does not become larger than the second harmonic, and the simulated response is not

EXTRACELLULAR POTENTIAL (e)

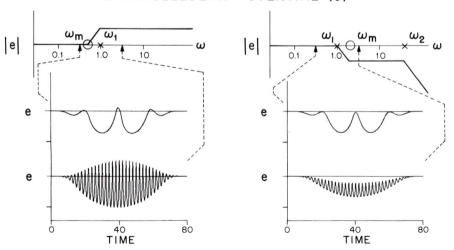

Figure 25-3. (Top) Bode plots of the magnitude of the frequency response function for the extracellular potentials. (Bottom) Extracellular potential waveforms that would be recorded for a stimulus at the indicated frequency. (Left) $C_h = C_e = 0$. (Right) C_h or $C_e \neq 0$ so that $\omega_1 < \omega_m < \omega_2$.

always negative. The cause of this result is easily explained in terms of the magnitude function of e, which shows that the high frequency response is larger than the low frequency responses, resulting in an emphasis of the second harmonic over the dc component. Clearly, these results conflict with the experimental results.

The system that actually simulates the experimental results is shown in Fig. 25-3, right. In this case, either C_e or C_h or both are assumed to be nonzero in addition to C_m being nonzero. The parameters can now be selected so that ω_m is between ω_1 and ω_2, and high frequencies are attenuated with respect to low frequencies. In this case, for higher excitation frequencies the second harmonic is attenuated with respect to the dc component, and the simulation results are similar to the experimental results. As mentioned earlier, this system was able to quantitatively simulate the experimental input-output functions. Hence, it can be concluded that it is not sufficient to include only the capacitance of the hair cell membrane in the model. One of the other capacitive elements must also be included in order to simulate the experimental results.

Sufficient data are not available to further specify the model. However, it is possible to investigate some effects of the model parameters on the behavior of the intracellular potential. Specifically, the frequency response of the intracellular potential depends on ω_e, the reciprocal of the time constant of the extracellular current path. Figure 25-4 shows Bode plots and some typical waveforms for the intracellular potentials assuming different relations between ω_e, ω_1, and ω_2. Since the intracellular potential is recorded from one cell, the fundamental is not cancelled; it is, in fact, the dominant component in the response. It can also be seen that the dc component is positive (i.e., depolarizing) instead of negative, consistent with published experimental results on intracellular potentials (Mulroy et al. 1974, Russell and Sellick 1978).

INTRACELLULAR POTENTIAL (V)

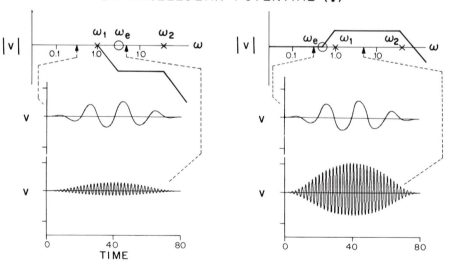

Figure 25-4. (Top) Bode plots of the magnitude of the frequency response function for the intracellular potential. (Bottom) Intracellular potential waveforms that would be recorded for a stimulus at the indicated frequency. (Left) Model parameters selected so that $\omega_1 < \omega_e < \omega_2$. (Right) Model parameters selected so that $\omega_e < \omega_1 < \omega_2$.

For very low frequencies, the response of the intracellular potential is about the same for both cases illustrated. The high frequency behavior, however, depends critically on the value of ω_e with respect to ω_1 and ω_2. If the extracellular capacitance is small, resulting in a larger value of ω_e, as shown on the left, the high frequency response is markedly attenuated, and the dc component becomes more prominent than at lower frequencies. If, on the other hand, the extracellular capacitance is larger so that ω_e is small, as shown on the right, the high frequency is enhanced and the dc component does not become very prominent.

These results can be very tentatively interpreted in terms of the hair cell response. For small extracellular capacitance (ω_e large) the hair cell sensitivity might be expected to decrease at high frequencies, and the dc component (that is, a process related to envelope detection) might become an important aspect of the transduction process. If extracellular capacitance is large, however (ω_e small), the hair cell sensitivity could remain relatively constant, even at frequencies high compared to the low pass cutoff determined by the hair cell membrane time constant. In this case, the dc component would probably not be a significant factor in transduction.

References

Boston, J. R.: The ac and dc components in lateral line microphonic potentials. J. Acoust. Soc. Am. 60, 656-664 (1976).

Boston, J. R.: A model of lateral line microphonic response to high-level stimuli. J. Acoust. Soc. Am. 67, 875-881 (1980).

Bracho, H.: Potassium dependence of membrane potentials in the *Necturus maculosus* vestibular system. Brain Res. 136, 366-370 (1977).

Bracho, H., Budelli, R.: The generation of resting membrane potentials in an inner ear hair cell system. J. Physiol. 281, 445-465 (1978).

Davis, H.: A model for transducer action in the cochlea. Cold Spring Harbor Symp. Quant. Biol. 30, 181-199 (1965).

Eatock, R. A., Corey, D. P., Hudspeth, A. J.: Adaptation in a vertebrate hair cell: stimulus induced shift of the operating range. Abstr. Soc. Neurosci. 5, 19 (1979).

Flock A.: Electron microscope and electrophysiologic studies on the lateral line canal organ. Acta Otolaryngol. Suppl. 199, 1-90 (1965).

Hudspeth, A. J., Corey, D. P.: Sensitivity, polarity, and conductance change in the response of vertebrate hair cells to controlled mechanical stimuli. Proc. Nat. Acad. Sci. 74, 2407-2411 (1977).

Mulroy, M. J., Altmann, D. W., Weiss, T. F., Peake, W. T.: Intracellular electric responses to sound in a vertebrate cochlea. Nature 249, 482-485 (1974).

Russell, I. J., Sellick, P. M.: Intracellular studies of hair cells in the mammalian cochlea. J. Physiol. 284, 261-290 (1978).

Weiss, T. F., Mulroy, M. J., Altmann, D. W.: Intracellular response to acoustic clicks in the inner ear of the alligator lizard. J. Acoust. Soc. Am. 55, 606-619 (1974).

Chapter 26

Lateral Line Function and the Internal Dynamics of Fish Schools

Brian L. Partridge*

1 Introduction

The question whether the lateral line plays a significant role in fish schooling has been posed repeatedly (Bateson 1889, Parker 1904, Parr 1927, Allen, Blaxter, and Denton 1976 and references in reviews by Shaw 1970, Radakov 1973). Until recently, how-ever, and although some studies had indicated that information from the lateral lines might be utilized (Backus 1958, Disler 1963, Cahn, Shaw, and Atz 1968, Cahn 1972, Webb 1980), most authors considered vision to be the only, or at least the primary, sensory modality employed (e.g., Parr 1927, Bowen 1931, Schlaifer 1942, Breder 1951, 1959, Atz 1953, Hunter 1969, Shaw 1962, 1970). Observation that *Pollachius virens* (saithe) could school while temporarily blindfolded (Pitcher, Partridge, and Wardle 1976) and the subsequent detailed analysis of the effects of blindfolding or cutting the posterior lateral lines on schooling (Partridge and Pitcher 1980) demon-strated that the role of the lateral lines in normal schooling is far greater than had been previously recognized. Results from those studies, coupled with analyses of the inter-nal dynamics of fish schools described below, begin to explain how sensory information from the lateral lines is used by individuals in the collective decision making process of the school.

2 Fish Schools Are Organized to Facilitate Antipredator Tactics

Of the many functions proposed for fish schools (see Cushing and Harden-Jones 1968, Breder 1959) there is a growing consensus that protection from predators is likely to be most important (Seghers 1974, Shaw 1978, Reese 1978, Hobson 1968, 1978). There is no evidence, for instance, supporting the models for hydrodynamical ad-vantages to school members which were proposed by Belyayev and Zuyev (1969) and

*Department of Biology, University of Miami, Coral Gables, Florida 33124.

Weihs (1973, 1975). Structure and dynamics of schools of herring, cod, and saithe do not match the predictions of either Belyayev and Zuyev (Partridge 1978) or Weihs (Partridge and Pitcher 1979). Thus, there must be additional selective pressures preventing fish taking advantage of possible hydrodynamical advantages. Partridge and Pitcher (1980) suggested that fish take up positions at which they are best able to monitor the changes in heading and velocity of their neighbors.

The evidence for this is quite compelling. For instance, *Phoxinus phoxinus* (European minnows) and saithe frequently take up position at 90° (i.e., alongside) their neighbors (Partridge 1980, Partridge, Pitcher, Cullen, and Wilson 1980), and this is the point at which changes in velocity would be most apparent visually. There is a significant clustering of neighbors at 90° even in normal schools, but the effect is considerably more apparent for individuals which had their lateral lines cut at the opercula (Partridge and Pitcher 1980) or were separated by a glass partition (Cahn 1972). By swimming alongside one another, sensorily deprived individuals could compensate for the lack of lateral line information.

Surprisingly, lateralis section has little effect on the degree to which individuals can monitor their neighbor's velocity, as measured by cross-correlation (Partridge and Pitcher 1980). Such results might indicate that the posterior lateral line is relatively unimportant in the determination of neighbors' velocities, but that is not the case. Studies on saithe that were temporarily blinded with opaque blinkers demonstrated that, even without vision, saithe are capable of matching changes in velocity and heading of at least their first two nearest neighbors (Partridge and Pitcher 1980). In fact, correlations between blindfolded fish and their neighbors are consistently higher than those between control fish and their neighbors. Blindfolded fish were able to maintain their position within the school by carefully matching their velocity to that of their neighbors: the posterior lateral lines provided the information necessary for this task.

The importance of the lateral line in normal schooling is further demonstrated by a series of experiments in which the entire school had their lateral lines cut at the opercula. In tests where the school was startled by a dark object suddenly being thrust overhead, the latency with which lateralis sectioned fish responded to the object was a function of the distance and direction to it (Partridge and Pitcher 1980). As found by Hunter (1969), the latency to startle was minimal for objects appearing at 90°, that is, directly alongside an individual, and increasingly greater for objects in front or behind the reference fish in a manner which approximated a sine function. Fish in control schools, on the other hand, showed no significant relationship between latency to startle and either distance or bearing (angle in the horizontal plane). In other words, with the lateral lines intact, individuals were able to respond to the startling of their schoolmates within one video frame (20 msec) so that the effects of relative positioning were no longer apparent.

These results may explain the differences between observations of Webb (1980) and Eaton (1977) who found fast-start latencies of 10-30 msec for the lateral line and Mauthner systems and those of Hunter (1969) and Partridge and Pitcher (1980) who found minimum latencies of 100-120 msec for visual responses by startled fish.

3 Integration of Information from Surrounding Fish

Further evidence that the structure of fish schools is a function of the abilities of fish to monitor changes in each other's velocities is provided by analysis of the degree to which individuals match each other's velocities as a function of their relative positions within a school. It has been shown elsewhere that fish show greater correlations to the velocities of their nearest neighbors than to other fish within the school (Cullen 1974, Partridge 1978, Partridge and Pitcher 1980). The correlation between velocities of two individuals depends upon their relative positions as well as their separation, however (Fig. 26-1). Maximum correlations are seen for fish near (less than 10 cm) and alongside one another (bearing $\approx 90°$). Correlation falls off as a function of distance. High correlations are also seen, however, for fish directly in front of or behind the reference fish. These results correspond nicely to those presented above: maximum correlation occurs for the positions at which latency is minimum and at which nearest neighbor frequency is maximum.

4 Mechanisms of Collective Decision Making

How are collective decisions made by schools? Although the rules governing choice of tactics in different contexts are still unclear, it has been possible to determine to some degree the way in which information is passed throughout the school. Observations of schools in the wild (Parr 1927, Breder 1965, 1967, Schlaifer 1942, Nursall 1973) or measurements of the latencies involved in the transmission of fright responses across the school (Keenleyside 1955, Verheijen 1956, Hunter 1969, Partridge and Pitcher 1980) demonstrate that fish must be responding to more than just their own nearest neighbor. And, since fish's velocities often show significant correlations with the velocities of both their first two nearest neighbors (Cullen 1974, Partridge 1980) one is led to ask how they integrate information from the two. Demonstration that the neighbors are unlikely to be correlated with each other (Partridge 1981) means that if a fish tries to match the velocity of both then it will reduce its correlation with each.

One possibility is that individual fish match their velocity to some average of the velocities of their neighbors, rather than to just the velocity of their NN_1. If so, then how many neighbors does a fish include in the average, and how are their velocities weighted? To examine this I looked at the correlation of the velocities of fifty randomly chosen fish from 25 film sequences of saithe schools with various averages of the velocities of their neighbors (Table 26-1) and compared this to the correlation with NN_1. Each of the four measures that include all members of the school produce higher correlations than just the nearest neighbor, but the velocity of the NN_1 is a better predictor of a fish's velocity than an average of the velocities of its two nearest neighbors. Fish do not merely match the mean velocity of the school, however, since the three measures which discount a fish's velocity as a function of distance produced higher correlations.

Individuals thus pay attention to all the rest of the fish in the school, but weight their neighbors' velocities by the square or cube of their relative distance. The measure of school velocity based on distance cubed did not produce quite as high correlations

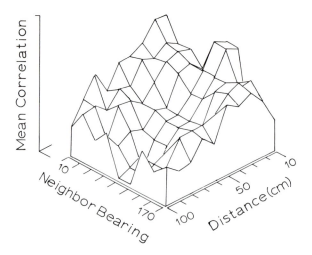

Figure 26-1. Average correlation between fish as a function of distance and angle between them. The three-dimensional surface shows average correlation as height. It can
be seen that correlation is highest for neighbors at 90° (i.e., alongside) or directly in
front and behind (0° and 180°). Correlation falls off quickly as a function of distance,
but falls off more quickly at 90° than at 180°. [The experimental methods in the
studies described here have been explained in detail elsewhere (Partridge et al. 1980,
Partridge and Pitcher 1980). Analyses described here were made of approximately 24
hours of continuous cruising data, amounting to more than 34,000 frames of video-
tape sampled at either 2.7 Hz or 50 Hz.]

Table 26-1. Mean Correlation of Fishes' Velocities by Six Different Averaging Methods

Rank	Averaging method	School speed	
		40 cm/sec	60 cm/sec
1.	School mean weighted by distance squared	.3707	.3222
2.	School mean weighted by distance cubed	.3675	.3146
3.	School mean weighted by distance	.3644	.3043
4.	School mean	.3510	.2811
5.	Velocity of NN_1	.3322	.2699
6.	Mean of NN_1 and NN_2 weighted by distance	.3269	.2479
7.	Mean of NN_1 and NN_2	.3152	.2419

as that based on distance squared, suggesting that displacement cues (such as would be perceived by the lateral lines) are not the only information being used by the fish in determining whether to speed up or slow down. In a sequence by sequence analysis, however, school averages based on distance squared are not significantly better for predicting a fish's velocity than those for distance cubed. [Both are significantly better than any of the other measures (Kendall's concordance, w = .56, p < .01).] Reliance on either distance squared or distance cubed would mean that small velocity changes by near neighbors may be tempered by the majority of the school not responding and correspondingly larger changes by fishes further away would be required in order to produce a change in a fish's velocity. Disturbances would be damped as they moved across the school (Parr 1927, Breder 1954) producing a more or less stable velocity in the absence of external stimuli, but fish could nonetheless respond quickly to sudden movements of any member of the school.

What is still unclear, is how information from vision and the lateral lines is combined. It remains for ethologists to develop experimental paradigms in which fish are presented with conflicting information from the two senses to disentangle their relative weightings and for neurophysiologists to explain the remarkable orienting capabilities exhibited by blindfolded fish.

Acknowledgments. I thank the Department of Agriculture and Fisheries, for Scotland, Marine Laboratory, Aberdeen (and especially C. S. Wardle, A. Hawkins, and R. S. Batty) for use of the gantry tank and ancilliary facilities and for help in the experiments, which were carried out jointly with T. J. Pitcher. Much of the analysis and computing was carried out at the Department of Experimental Psychology, University of Oxford, and I particularly thank R. Dawkins, J. Erichsen and A. Myrberg, Jr., for helpful comments.

References

Allen, J. W., Blaxter, J. H. S., Denton, E. J.: The functional anatomy and development of swim bladder-inner ear, and lateral line system in herring and sprat. J. Mar. Biol. Assoc. U.K. 56(2), 471-487 (1976).

Atz, J. W.: Orientation in schooling fishes. In: Proceedings of a Conference on Orientation in Animals. Schneirla, T. C. (ed.). Washington, D.C.: Office of Naval Research, 1953, pp. 115-130.

Backus, R. H.: Sound Production in Marine Animals. U.S. Navy J. Underwater Acoust. 8(2), (1958).

Bateson, W.: The sense-organs and perceptions of fish; with remarks on the supply of bait. J. Mar. Biol. Assoc. U.K. 1, 225-256 (1889).

Belyayev, N., Zuyev, G. V.: Hydrodynamic hypothesis of school formation in fishes. J. Ichthyol. 9, 589-594 (1969).

Bowen, E. S.: The role of sense organs in aggregations of *Ameiurus melas*. Ecol. Monogr. 1, 1-35 (1931).

Breder, C. M.: Studies on the structure of the fish school. Bull. Am. Mus. Nat. Hist. 98, 7-28 (1951).

Breder, C. M.: Studies on social groupings in fishes. Bull. Am. Mus. Nat. Hist. 117(6), 397-481 (1959).

Breder, C. M.: Vortices and fish schools. Zoologica 50, 97-114 (1965).

Breder, C. M.: On the survival value of fish schools. Zoologica 52(2), 25-40 (1967).

Cahn, P. H.: Sensory factors in the side-to-side spacing and positional orientation of Tuna, *Euthynnus affinis*. U.S. Fish. Bull. 70(1), 197-204 (1972).

Cahn, P., Shaw, E., Atz, J.: Lateral line nerve histology as related to the development of schooling in the atherinid fish *Menidia*. Bull. Mar. Sci. 18, 660-670 (1968).

Cullen, J. M.: Statics and dynamics of a fish school. Poster presentation, XIV International Ethological Congress, Parma, Italy (1974).

Cushing, D. H., Harden-Jones, F. R.: Why do fish school? Nature 218, 918-920 (1968).

Disler, N. N.: *Lateral line sense organs and their importance in fish behaviour*. Translated from Russian, Israel Program for Scientific Translations. N.T.I.S., Virginia, p. 328 (1963).

Eaton, R. C.: Mauthner initiated startle response in teleost fish. J. Exp. Biol. 66, 65-81 (1977).

Hobson, E.: Predatory behavior of some shore fishes in the Gulf of California. Bureau Sport Fish. Wild Life Res. Rep. 73, 1-92 (1968).

Hobson, E. S.: Aggregating as a defense against predators in aquatic and terrestrial environments. In: Contrasts in Behavior. Reese, E., Lighter, F. (eds.). New York: Wiley, 1978, pp. 219-234.

Hunter, J. R.: Communication of velocity changes in jack mackeral *Trachurus symmetricus* schools. Anim. Behav. 17, 507-514 (1969).

Keenleyside, M.: Aspects of schooling behaviour in fish. Behaviour 8, 83-248 (1955).

Nursall, J. R.: Some behavioural interactions of spottail shiner (*Notropis hudsonius*) yellow perch (*Perca flavescens*) and northern pike (*Esox lucius*). J. Fish Res. Bd. Can. 30, 1161-1178 (1973).

Parker, G. H.: The function of the lateral line organ in fishes. Bull. U.S. Bur. Fish 24, 183-207 (1904).

Parr, E. A.: A contribution to the theoretical analysis of the schooling behaviour of fishes. Occ. Pap. Bingham Oceanogr. Coll. 1, 1-32 (1927).

Partridge, B. L.: Sensory aspects of schooling. Ph.D. Thesis. University of Oxford, 1978, 520 pp.

Partridge, B. L.: The effect of school size on the structure and dynamics of minnow schools. Anim. Behav. 28, 68-77 (1980).

Partridge, B. L.: Internal dynamics and the interrelations of fish in schools. J. Comp. Physiol. (in press) (1981).

Partridge, B. L., Pitcher, T. J.: Evidence against a hydrodynamical function of fish schools. Nature 279, 418-419 (1979).

Partridge, B. L., Pitcher, T. J.: The sensory basis of fish schools: relative roles of lateral line and vision. J. Comp. Physiol. 135, 315-325 (1980).

Partridge, B. L., Pitcher, T. J., Cullen, J. M., Wilson, J.: The three-dimensional structure of fish schools. Behav. Ecol. Sociobiol. 6, 277-278 (1980).

Pitcher, T. J., Partridge, B. L., Wardle, C. S.: A blind fish can school. Science 194, 963-965 (1976).

Radakov, D. V.: Schooling in the Ecology of Fish. (Israeli Translation Series.) New York: Wiley, 1973, 173 pp.

Reese, E.: The study of space related behavior in aquatic animals: special problems and selected examples. In: Contrasts in Behavior. Reese, E., Lighter, F. (eds.). New York: Wiley, 1978, pp. 347-376.

Schlaifer, A.: The schooling behaviour of mackeral—A preliminary experimental analysis. Zoologica 27, 75-80 (1942).

Seghers, B.: Schooling behaviour in the guppy (*Poecilia reticulata*). An evolutionary response to predation. Evolution 28 (3), 488-489 (1974).

Shaw, E.: The schooling of fishes. Sci. Am. 206, 128-138 (1962).

Shaw, E.: Schooling in Fishes: Critique and Review. In: Development and Evolution of Behavior. Aronson, L. R., Tobach, E., Lehrman, D. S., Rosenblatt, J. S. (eds.). San Francisco: Freeman, 1970, pp. 452-480.

Shaw, E.: Schooling fishes. Am. Sci. 66, 166-175 (1978).

Verheijen, F. J.: Transmission of a flight reaction amongst a school of fish and the underlying sensory mechanisms. Experimentia 12, 202-204 (1956).

Webb, P. W.: Does schooling reduce the fast-start response latencies in teleosts? Comp. Biochem. Physiol. 65A, 231-234 (1980).

Weihs, D.: Hydrodynamics of fish schooling. Nature 241, 290-291 (1973).

Weihs, D.: Some hydrodynamical aspects of fish schooling. Proc. Symp. Swimming and Flying in Nature. Wu, T., et al. (eds.). Pasadena, California, July 1974, 1975.

Discussion

BLAXER: We have tried placing spectacles on herring, and we found that they became so active that they would injure themselves. How did your fish settle down after the spectacles were installed?

PARTRIDGE: We had difficulties of this sort also. However, after putting in the spectacles, we kept the fish in the dark for 24 hr or so. During this time, they settled down quite a bit. If you keep them in the dark, they seem to be less concerned about not seeing anything except that it is light or dark. The fact that so many people have failed to show schooling in blindfolded fish in the past is a function of the fish's motivation rather than visual capacity. If somebody stuck a paper bag over your head and told you to run down a highway, the fact that you *could* do it wouldn't mean that you *would*. However, I wouldn't make any claim that a totally blinded fish could survive very long in the wild.

BLAXTER: What happens if your nonoccluded fish were kept in the dark?

PARTRIDGE: They all wandered about in one particular area in a loose group, not a pod or a school. I don't know what they would do in total darkness in the wild.

CAHN: Do they have good vision?

PARTRIDGE: Yes, as far as I can see. They seem to feed visually.

CAHN: What do you think of the hydrodynamic factors, such as drag, that might play a role in schooling and school size?

PARTRIDGE: I am leaning more to the view that these factors are not important [Nature 279, 418-419 (1979)]. It seems to me that what determines the school size is the ability of the fish to keep track of each other. When you get over 15 or 20, you start seeing subgroups within the school. I think this is what keeps the school from becoming infinitely large.

PART EIGHT

Prospectives

Electrosensory systems in fishes have been the object of considerable empirical and theoretical study during the past few years. Bullock (Chapter 27) points out that there may be many similarities and parallels between electroreceptive and acoustic systems and that the rather more advanced understanding of electroreception may be of significant help in pointing the way toward new concepts and experimental paradigms for the study of auditory processing in fishes. In the final chapter of the volume, Tavolga takes the opportunity afforded him by his having participated in all earlier fish hearing conferences (and being the editor of two of those volumes) to look back on earlier studies and to speculate on the areas in which further studies are now needed.

Chapter 27

Comparisons of the Electric and Acoustic Senses and their Central Processing

THEODORE HOLMES BULLOCK*

1 Introduction: The Octavolateralis System and Its Subdivisions

Hearing is but one member of a large family of modalities that we call, collectively, the octavolateralis system, a better term than the older acousticolateralis system. This system embraces receptors for gravity, angular acceleration, vibration, water flow, turbulence and other aspects of aquatic disturbance not yet distinguished, acoustic and electric fields, indirectly magnetic fields, by the induced electric currents, not to mention the claims of chemosensitivity (Katsuki and Onada 1973, Yoshioka et al. 1978) and thermosensitivity (Hoagland 1933, 1935, Sand 1938).

Actually, the diversity of receptors is greater than is indicated by such a list. Several modalities, such as electroreception, have two or more submodalities (see Table 27-1). Some receptors are bimodal combinations having, for example, position and movement sensitivity. Some modalities such as that for angular acceleration show subdivisions for the three planes which point to a well-differentiated system of peripheral filters.

Since these VIIIth nerve and lateral line receptors and their afferent nerve fibers have a certain anatomical commonality, both peripherally and centrally, in spite of differences in adequate stimuli and transducer properties, I believe we can draw a useful parallel with the eye. As presently understood that organ serves more than six different parallel subsystems. A small number of sense cell types, perhaps four, counting rods and the cones of distinct color types, serve at least 15 types of optic nerve fibers, in the cat (Rodieck 1979), that go to at least six brain stem destinations, apparently with distinct functions. It is to be expected that acoustic reception, at any rate when well developed, will be found to have some distinct, parallel subsystems. Electroreception does, in the more derived taxa, the gymnotiforms and mormyriforms.

*Neurobiology Unit, Scripps Institution of Oceanography, and Department of Neurosciences, School of Medicine, University of California, San Diego, La Jolla, California 92093.

Table 27-1. Octavolateralis Sense Organs for Small Electric Currents and for Small Movements of the Fluid Medium[a]

ELECTRORECEPTORS

I. *Ampullary organs; low frequency sensitive receptors*
    (ampulla with patent canal to the surface and receptor cells in epithelial lining; passband ca. 0.1-50 Hz; afferent nerve fibers are frequency coders; in electric teleosts, usually not synchronized with EOD, except when that has dc component)
    A. *Ampullae of Lorenzini* (the afferent nerve fibers may be called "L" units) in elasmobranchs, holocephalans, chondrosteans, and equivalent organs of Fahrenholz of dipnoans and polypteriforms; also unknown organs of lampreys excited when ampulla goes negative, depolarizing apical membrane
        hair cell receptors with kinocilium but no microvilli
        long canal in marine species; short duct microampullae in freshwater species
    B. *Ampullae of teleosts* ("F" units)
        in siluriforms, gymnotiforms (type I of Szabo 1974, 1 of Lissman and Mullinger 1968) and mormyriforms ("small pore" mormyromast type I of Szabo 1974; gymnarchid type A of Szabo 1974)
        excited when ampulla goes positive, depolarizing basal membrane
        hair cell receptors without kinocilium but with microvilli
        long canal in marine species (sometimes, inappropriately called Lorenzinian), "small pit organs," with short duct in freshwater species
II. *Tuberous organs; high frequency sensitive receptors*
    (cavity occupied by sense cells projecting from basal attachment, usually without patent duct; sense cells with microvilli protruding or lining an invagination, or without them; excited by inward current; passband ca. 50-20,000 Hz; usually tuned to a best frequency; series capacity and active filtering contribute to tuning afferent fibers synchronized with EOD and encoding intensity in various ways not appropriately called frequency coders)
    A. *Burst duration coders* ("B" units) in gymnotiform pulse species
        small axon, unmyelinated preterminals, many sense cells per organ
        burst of spikes follows EOD, graded in number at fixed interval, little latency shift; receptor potential an active oscillation, weakly damped
        subdivided into alpha, beta, gamma, and delta types, with some subtypes, by Baker (1980)
        (part of Szabo's 1974 common tuberous, type II)
    B. *Pulse marker units* ("M" units) in gymnotiform pulse species
        large axon; myelinated preterminal branches
        high threshold in some species; low in others
        single spike follows EOD with fixed latency
        (part of Szabo's 1974 common tuberous, type I)

    A'. *Phase coders* ("T" units) in gymnotiform wave species
        medium-sized axon; myelinated preterminals
        single spike follows each EOD in physiological intensity range; shifts latency with intensity, with small jitter; receptor potential is a rapidly damped, active, tuned oscillation
        more sensitive (absolute) than P units
        some are sensitive to the difference between sinusoidal and clipped, EOD-like wave form, some are insensitive and others are intermediate (=T_c and T_s of

Table 27-1 (continued)

Scheich)

(part of Szabo's 1974 common tuberous, type I)

B'. *Probability coders* ("P" units) in gymnotiform wave species

morphology may be similar to preceding

single spike follows some EODs; probability graded with intensity; smaller, more jittery latency shift and more phasic than "T" units in same fish; difference from T in receptor potential is unknown; they may be similar

some are sensitive to the difference between sinusoidal and clipped, EOD-like wave form, some are insensitive and others are intermediate (=P_c and P_s of Scheich)

chemical transmission

A". *Knollenorgans* ("K" units) in mormyrids

"large pore" organs; myelinated preterminals; project to PLLL nucleus

more sensitive (absolute) than D units

spontaneous receptor potential in most species, rhythmic, usually spikelike, from a few hundred Hz up to several kHz under artificial conditions

useful for detecting social signals by EOD waveform

electrical transmission

B". *Mormyromasts;* burst dispersion coders ("D" units) of mormyrids

"medium-pore" organs; unmyelinated preterminals; project to PLLL cortex

two forms of sense cells, complex innervation

threshold can be high, up to 10 mV/cm

receptor potential not spontaneous; damped ringing

burst of spikes graded in latency, number and spacing (likely to be a heterogeneous category)

A'". *Gymnarchomasts type I* ("S" units) in gymnarchid

like preceding but single nerve fiber, with unmyelinated preterminals

pair of unequal sense cells, or more

like "T" units but high spontaneous firing rate, less tendency to burst below 1:1 range, small dynamic range of phase coding

B'". *Gymnarchomasts type II* ("O" units) in gymnarchid

single nerve fiber with unmyelinated preterminals

12 or more sensory units per organ; sense cells with long microvilli at bottom of invagination

like "P" units but spontaneous with more difficulty following 1:1, small dynamic range of phase part of probability coding; high intensity suppression

SMALL WATER MOVEMENT AND ACOUSTIC RECEPTORS

I. *Common lateral line receptors*

(stimulus: low frequency particle displacement or net flow)

A. Free neuromasts

B. Canal organs

II. *Otolithic receptors*

(sensitive stereocilia mechanically coupled to heavy liths; stimulus is particle displacement > displacement of fish's body = displacement of hair cell > displacement of lith; found in sacculus, utriculus, and lagena of fishes and amphibians)

Table 27-1 (continued)

A. *Sense organs without pressure transducer*
    (receptors stimulated directly by displacement wave moving them relative to
    liths; found in lagena, utricle in fishes with and without gas bladder)
        less sensitive but more directional than (B), in general
        some high, some low threshold
        variance in other properties not well known

B. *Sense organs associated with pressure transducer*
    (receptors stimulated indirectly by pressure wave; found in sacculus of ostario-
    physan fishes with Weberian ossicles, utriculus of clupeids with otic diverticu-
    lum of gas bladder)
        more sensitive but less directional than (A)
        discrete types not clear but sometimes two (goldfish); wide variance in
            properties on several dimensions:
        spontaneous firing rate varies from low to high
        spontaneous interval histogram narrower or wider, Gaussian or gamma
            function
        sensitivity at best frequency high or low
        phase-locking good or poor at medium frequency
        tonic or phasic or phasic-tonic
        excited by compression or by rarefaction
        tuning curve deep or shallow; varies in CF
        two tone suppression slight or marked
        adaptation fast or slow; frequency dependent

III. *Nonotolithic receptors of the ear*
    (cilia mechanically coupled to jelly or tectorial membrane; stimulus in aquatic
    species: particle displacement > jelly > hair cell ≥ fish's body; stimulus in birds
    and mammals: transducer displacement ≥ hair cell > tectorial membrane ≥ body;
    in amphibians and reptiles: various; found in shark macula neglecta, amphibian
    papilla, basilar papilla, amniote cochlea)

    A. *Sense organs without pressure transducer*
        (receptors are stimulated directly by propagated displacement wave moving
        cupula relative to receptor cilia; found in shark macula neglecta)
            variation in properties is not yet known
            very likely some groups, e.g., amphibians, have maculae that function as
                mixtures of (A) and (B)

    B. *Sense organs associated with pressure transducer*
        (receptors are stimulated indirectly by pressure wave; found in amphibian,
        reptile, bird, mammal)
            discrete types sometimes two (frog, lizard), sometimes controversial (mam-
            mal inner and outer hair cells); although subclasses are few or not yet
            agreed upon, variation is notable in several dimensions (these also apply to
            foregoing categories but are listed here, in lieu of types, because know-
            ledge of the variation is greater in this category namely:
                efferent terminals on hair cells or on afferent terminals, or absent
                reciprocal synapses between hair cells and afferent terminals vary in
                    extent
                presence of gap junctions between same, varies in extent
                hair cell types vary in number (4+ in frog)
                number of hair cells supplied by single afferent nerve fibers varies from

Table 27-1 (continued)

2 to 30
number of afferent fibers converging on single hair cell varies
types of acellular superstructure over cilia vary
Ca-activated outward current from hair cells may be more important in some
transmitter may vary
excited by compression or by rarefaction
tuning curve may vary in shape and CF
two tone suppression varies from strong to nil
lateral interaction varies from strong to nil
spontaneity may be high or low
interval histogram of different shapes
sensitivity may be high or low
phase locking may be good or poor
adaptation may be fast or slow

[a] Small means relative to the size of the body. Larger movements that translate the body and stimulate the vestibular acceleration and gravity sense organs are omitted from this table. Also omitted are sense organs for vibration, normally transmitted from a relatively dense stimulating mass. All of these distinctions are relative and there are continuous intergrades in amplitude and frequency. The overlap of "receptive fields" in these domains is poorly known.

We will understand hearing in fishes much better if we compare and contrast other modalities, particularly those in the same general system, the octavolateralis receptors. My purpose here is to do some of this, specifically to note, for comparison, the current state of knowledge of the electroreceptor systems central as well as peripheral. The hope is that, besides some perspective and new ways of thinking about hearing, this may suggest some new experiments, especially on a wider range of species. Both new perspective and new data are needed, for example to elucidate such elementary questions as the first evolution of hearing in the vertebrates.

It is not self-evident, given the presence of the lateral line, what the first function of hearing was. We cannot assume that the early vertebrates, who differentiated acoustic receptors among other VIIIth nerve innervated sense organs, achieved thereby a greater sensitivity, directionality, or higher frequency range than the lateral line receptors had. Maybe they did, but that is not known. If the first acoustic reception in the labyrinth was otolithic, one can argue that it was probably more sensitive than the lateral line but the most primitive vertebrates known to have hearing, the carcharhinid sharks (Corwin, Chapter 5), have both otolithic and nonotolithic receptors, so it is not self-evident what the first VIIIth nerve acoustic receptors were. In terms of adaptive significance we have yet to learn the key properties of the early auditory sense.

2 Sense Organs and Reception

2.1 Diversity of Types and Subtypes of Receptors

Table 27-1 summarizes the receptors as presently known for several of the octavolateralis modalities omitting most gravity and acceleration sense organs. Clearly there are important distinctions within each modality, so that none of them is served by a single homogeneous population of afferent fibers. There is an apparent contrast between acoustic and electric senses, which may prove to be illusory, in their respective numbers of receptor cell types. In spite of the much shorter history of investigation of the electric sense, it appears to exhibit relatively more obvious categories of afferent fibers, both major classes and subclasses of them, based on anatomy and physiology. Acoustic afferents, while varying along several dimensions, and in the better studied species of fishes, frogs, and lizards showing two types of fibers (Furukawa and Ishii 1967, Feng, Narins, and Capranica 1975, Capranica 1976, Turner 1980), are in most vertebrate literature not divided into categories in a given species. For acoustic receptors in Table 27-1, I had to resort to a list of parameters of variation among afferent fibers, which, if future work shows clustering of the variants on different dimensions, might become the basis of classes and subclasses.

Looking at Table 27-1, we see that two broad classes of electroreceptors, the ampullary organs and the tuberous organs (Figs. 27-1, 27-2) have each evolved more than once, from ancestors presumably without them. Ampullary organs, in their Lorenzinian form, we believe on recent evidence (Northcutt, Bodznick, and Bullock 1980, Bodznick, Northcutt, and Bullock, in prep., Bodznick and Northcutt 1981, Fields and Lange 1980, Teeter, Szamier, and Bennett 1980) were a general possession of almost all the nonteleost fishes, including lampreys (Petromyzontiformes) (but apparently not hagfishes = Myxiniformes), chimaeras (Holocephali), sharks, skates and rays (Elasmobranchii), sturgeons, paddlefish (Chondrostei), lungfish (Dipneusti), bichirs and reed fish (Polypteriformes), and *Latimeria* (Crossopterygii). The bowfins and garpikes (Holostei) have apparently lost ampullary organs and we believe the ancestral teleosts had likewise, because most of the 30 or so orders of teleosts apparently lack them. Of course, most orders have not even been examined by behavioral or physiological means. However, nine have (mostly unpublished from this laboratory), and only three are found to possess electroreception, in any degree that deserves the name. These three are not closely related: the world-wide Siluriformes, mostly not electric (except for *Malapterurus*, the electric catfish of the Nile) but electroreceptive, the South American Gymnotiformes, mostly weakly electric (except for *Electrophorus*, the electric eel), and the African Mormyriformes, all weakly electric. The first two are not far apart both being in the large superorder of Ostariophysi but at least twice, if not three times, ampullary organs were reinvented within the teleosts, and in quite a different form and physiology from the Lorenzinian ampulla in the nonteleost groups. I will not here take up the issue of whether these organs are homologous! The orders that have been sampled and found, in one or two species each, apparently to lack electroreception are the Osteoglossiformes (*Notopterus*), Anguilliformes (*Anguilla*), Salmoniformes (*Oncorhynchus*), Cypriniformes (*Carassius*), Synbranchiformes (*Synbranchus*), and Perciformes (*Astroscopus*). Because the sample is so small, there is still ample opportunity for further surprises, especially for intermediate cases. By that I mean species that may make use of electric detection even though they are not at present convincingly spoken of as having specialized electroreceptors; *Astroscopus* and *Anguilla* may be examples.

Tuberous organs have a much simpler and narrower distribution, but similarly puzzling in respect to homology. They are only found in the electric teleosts of the two orders Gymnotiformes and Mormyriformes (not, according to my unpublished experiments, in the electric stargazer, *Astroscopus*, which is the only electric fish belonging to a family in which not all members of the family are electric; in the Urano-

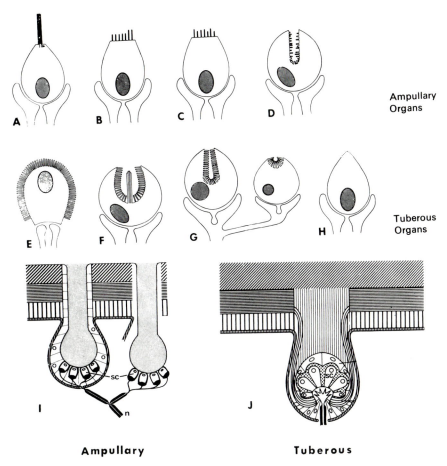

Ampullary
Organs

Tuberous
Organs

Ampullary   Tuberous

Figure 27-1. Diagrams of octavolateralis receptors specialized for electroreception. (A-D) Hair cell secondary sense cells and nerve endings of ampullary organs: (A) of elasmobranch (ampullae of Lorenzini); (B-D) of teleosts. (E-H) Hair cell secondary sense cell and nerve endings of tuberous organs of several types in mormyrid and gymnarchid electric fish. All nerve endings are afferent. (I) Scheme of two varieties of ampullary organs, in gymnotiform and mormyriform. sc, sensory cell; n, afferent nerve fiber. (J) Scheme of one variety of tuberous organ from gymnotiform. (From Szabo 1974.)

scopidae, in fact, most genera are not). Since the gymnotiforms and the mormyriforms are not even closely related, their common possession of high frequency-sensitive receptors of tuber-like morphology is interesting and presents a case of convergent evolution. However, the main point for our purposes here, is that they are not the same; beyond their common features they have differences that compel us to maintain separate subclasses and even to wonder if the class called tuberous is artificial. Note, furthermore, that each species of gymnotiform and mormyriform fish has at least two distinct kinds of tuberous organs, besides its ampullary organs (see further, Bullock 1981a).

Ampullary Receptor

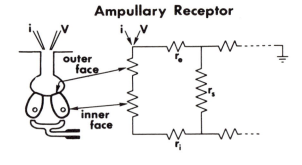

Tuberous Receptor

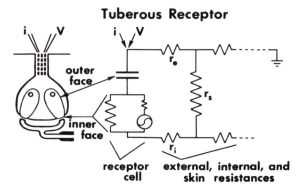

Figure 27-2. Anatomical diagrams and equivalent circuits of electroreceptors in gymnotiform and mormyriform fishes. (Above) External medium, bounded by skin and
walls of receptor cavity. (Below) Ending of afferent nerve fiber. The opening to the
external medium of the tuberous receptor cavity is shown as occluded by a porous
mass; resistance of this and of ampullary canal and of hair cell cytoplasm are assumed
to be small and are omitted. Electrodes for stimulating and recording with respect to a
distant electrode are shown at the sense organ openings. Note that generally the hair
cells of ampullary organs are largely embedded in the wall of the cavity, whereas those
of tuberous organs protrude into the cavity. (From Bennett 1967.)

Acoustic receptors can be subdivided into at least two classes, the otolithic and the
nonotolithic organs of the ear. I show also the common lateral line organs, not to take
sides on the question of hearing via the lateral line, but to underline the continuity
between sound and small disturbances of the water that can stimulate lateral line organs
by bulk flow or net flow of water relative to the fish. In fact the continuity extends to
vibration and to large, low frequency translations of the body which could stimulate
the vestibular acceleration and gravity receptors. These last have been omitted arbitrarily from Table 27-1.

Among the acoustic receptors of the VIIIth nerve, both otolithic and nonotolithic,
each has two subclasses, those stimulated directly by the particle displacement wave
and those stimulated indirectly via a pressure transducer. This does not mean that a
given fish species can have only one subclass of acoustic receptor. Sharks and some
teleost fishes seem to have otolithic displacement wave receptors in addition to their
nonotolithic acoustic organs. However, this is not yet well established; the impression

I get from most of the literature is therefore that acoustic receptors in a given fish are regarded as typically all of one class and subclass.

Homogeneity within the subclass is another matter; the better studied fishes seem to be far from homogeneous. At least two distinct types of receptors, based on clustering of several dimensions of variation are claimed for goldfish, as well as frog and lizard nonotolithic acoustic organs. There is still controversy concerning the mammals, where it has long been said that all the acoustic afferents are essentially alike, with variation along a few continua such as tuning. Today there is widespread interest in the properties of outer hair cell afferents and in the marked differences among cochlear afferents in spontaneity, interspike-interval histogram, phase-locking, adaptation, and suppression. Some suggestions of a wide range in sensitivity, under physiological conditions, have been made.

I believe it is likely that, as in electroreceptors, where distinct classes and subclasses in the same fish are associated with different functional roles, we may expect to find relatively distinct types of acoustic afferents widespread among vertebrates, at least in those groups where hearing is sufficiently developed to serve more than one behavioral role.

2.2 Properties and Mechanisms in Electroreception

In many properties both electric and acoustic receptors exhibit similar ranges of variation. *Spontaneity* may be high or low or none. The *distribution of interspike intervals* may be Gaussian or more like a gamma function or long tailed. The excitatory stimulus may in different receptors have opposite *polarities*. However, in one and the same fish this is well established only for acoustic receptors (rarefaction and compression excited units), whereas for electric receptors those sensitive to opposite polarities (excitation with ampullary side cathodal or ampullary side anodal) are in different taxa, elasmobranchs and teleosts, respectively. *Tonic as well as phasic* and phasic-tonic units are known for both modalities.

Tuning curves (threshold as function of frequency; see Fig. 27-3) have a special interest, as pointed out by Viancour (1977, 1979a, 1979b), since typical curves for tuberous electroreceptors (Scheich 1977a, Hopkins 1976, Viancour 1979c, Bastian 1977) are V-shaped and invite comparison with those for acoustic receptors. The sharpness of tuning, for tuberous units in *Eigenmannia* with a "best" frequency of ca. 300 Hz, is about the same as that for cat cochlear nerve fibers of the same best frequency. This must be regarded as support for the notion that there is filtering beyond the mechanical transducer, since in the electroreceptor there is filtering although no mechanical events are involved. The so-called second filter hypothesis for the cochlea, which is still in controversy, rests largely on the claim that the sharpness of the basilar membrane tuning is not adequate to account for that of the cochlear nerve fibers in laboratory mammals. Based on the comparison with electroreceptors, we see good reason to believe that nonmechanical tuning can be important at least for low audiofrequencies.

In view of this finding, the electroreceptors become significant objects for study of the mechanisms of nonmechanical filtering. There is a body of knowledge relating to such mechanisms, as well as opportunities for more research. Viancour (1979a, 1979b) showed that the best frequency of an afferent nerve fiber for tuberous receptors in

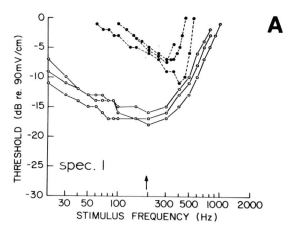

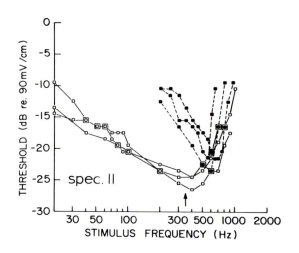

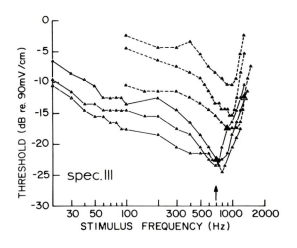

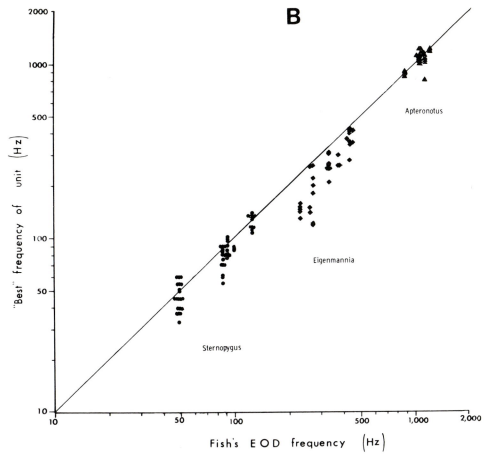

Figure 27-3. Tuning of tuberous electroreceptors. (A) Tuning curves of three species of gymnotiforms (*Hypopomus* spp.). Each graph is from a single fish of the species named; each curve from a single afferent fiber. The arrow on abscissa shows the EOD rate prior to the experiment. The low and high threshold clusters are T and P units, respectively. (From Bastian, in Heiligenberg 1977). (B) Plot of best frequencies of curves such as those in (A) as a function of the EOD rates of the fish prior to the experiment. The diagonal line represents perfect correspondence between the two quantities. (From Hopkins 1976.)

Eigenmannia predicts the frequency of a *damped ringing of the receptor,* when stimulated in the resting state by a brief pulse in the water. Individual fish of the same species have their own preferred electric organ discharge (EOD) rate between 250 and 500 Hz, commanded by a labile pacemaker unit in the medulla. This EOD rate predicts the best frequency for the whole population of afferent fibers, as well as the ringing frequency of the receptors. The power spectrum peak of the single discharge predicts it even better. What I call a ringing is an oscillation of receptor potential and of excitability following a fractional millisecond square stimulus pulse in the water outside the fish. The ringing commonly lasts from 1½ to 3 cycles, with a period of 2 to 4 msec, matched to the EOD. We believe this is an active oscillation, i.e., reflecting membrane changes, and that it contributes to but does not completely account for the tuning curve. Clusin and Bennett (1979a) find both a damped ringing and a spontaneous oscillation, each at ca. 20 Hz in the low-frequency-sensitive ampullary receptors of skates. It seems most likely to me that there is not one filter, in the sense of a single site or process, but an additive series of passive and active factors, including the impedance of the extracellular current paths, the time constants of the hair cell membrane, and of the membrane potential-to-calcium-activation step, and any other steps in the cycle that can generate receptor potential oscillation, plus perhaps the delays in transmitter release and postsynaptic potential of the afferent fiber terminal, if that is part of the response. In gymnotiform fishes oscillatory response is minimal in some units, marked but damped in others (Bennett 1967, 1971a, 1971b, Viancour 1979a, 1979b). In mormyriform species (Fessard and Szabo 1961, 1962, Szabo 1962, 1974, Bennett 1967, 1970, 1971a, Pimentel-Souza 1976) it can be a continuous, undamped rhythmic potential. There is wide variation, not only in the amplitude and damping but also in the wave form. In some mormyrids the rhythmic, spontaneous oscillation of the receptor is quite spikelike.

In this light it would seem to be a significant parallel that Fettiplace and Crawford (1978) find a damped ringing in intracellular recordings from terrapin cochlear hair cells, in response to small steps of current injected through the recording electrode; the frequency of the oscillation corresponds closely to the characteristic frequency (CF = best frequency). It is slightly higher than the CF during a depolarizing step and lower during a hyperpolarizing step; oscillation vanishes at -70 to -80 mV. They liken this ringing to the inductive power of squid axon (Cole 1968) and suppose it might involve a voltage sensitive potassium conductance. The finding that tuned ringing occurs intracellularly substantially reduces the list of candidate loci that must be considered and the observation of a systematic shift of frequency with membrane potential beautifully underlines the appropriateness of the word tuned for these filters. Viancour (1979a) observed in the electroreceptors of *Eigenmannia* a small systematic difference between the CF and the EOD rate; the CF is generally a little lower. He did not bring it under control as Fettiplace and Crawford did in the reptilian cochlea but proposed that the tuning of CF matches the power spectrum peak of the single EOD cycle. This peak lies about 15% below that of the EOD analyzed as a continuous wave, as is true for pure sine waves. If this correlation with the single EOD is not accidental, it underlines the conclusion proposed, among others by Scheich (1974) that the receptor mechanisms operate in the time domain. That is to say, the transducer may look at the instantaneous state or be influenced by the last few milliseconds so that the power spectrum of the ongoing waveform in the frequency domain is not the most relevant representation of the stimulus. The proposition is that this holds for both electro- and acoustic receptors.

Still further information about electroreceptor cellular mechanisms might well be relevant to acoustic and to other octavolateral senses. Obara and Bennett (1972), Akutsu and Obara (1974), Obara (1976), Clusin and Bennett (1977a, 1977b), Bennett (1978), Bennett and Clusin (1979), Clusin and Bennett (1979a, 1979b), and Obara and Sugawara (1979) provide a body of data from skates and *Plotosus,* a marine catfish, indicating several ionic channels. In *Plotosus* there appear to be three: a transient and a sustained voltage sensitive Ca current across the basal face of the hair cells and a late outward current, presumably a Ca-activated K current across the lumenal (outer) face. The first two both lead to transmitter release, the last one not. Bennett argues that combinations of these currents can play a role in maintaining a controlled balance between sensitivity and accommodation, adjusting the degree of spontaneous activity to achieve the lowest increment threshold.

Receptor potentials in electrosense cells are perhaps equivalent to *microphonic potentials* in the ear. If the latter are regarded as a causal link in the chain of events leading to transmitter release and afferent impulses, it is significant that the magnitude of the transmembrane voltage change for near-threshold cochlear microphonics, estimated (Honrubia, Strelioff, and Ward 1973) or measured intracellularly in lower vertebrates (Hudspeth and Corey 1977) is less than a millivolt. The reason this is significant is that it is within the range of sensitivity of specialized electroreceptors; one can look at the acoustic detector as an electroreceptor sense cell with a mechanotransducer ahead of it. Bennett (1978) points out that mammalian inner hair cells have larger cochlear microphonics at threshold (Russell and Sellick 1976), if we take present evidence at face value, but still they are only a few millivolts. Some true electroreceptors (M units = pulse markers) have similar thresholds.

Electroreceptors may be useful as model octavolateral sense organs for research on *transmitters* at sensory synapses. Besides their superficial position and wide distribution over the body (in some electrosense organs it only requires a glass pipette microelectrode inserted normal to the skin surface into one of the proper pores or pits to record unit responses close to the receptor), diverse types are available for comparison. Most electroreceptors appear to transmit chemically, and glutamate is said to be a likely candidate transmitter (Bennett 1978, but see contrary evidence, Bullock 1981a). Electrical transmission is also found in certain electroreceptors.

The *encoding* process that generates a train of spikes beginning with the postsynaptic potential in the afferent fiber terminal is a resultant of several antecedent properties, both pre- and postsynaptically. Intensity is coded in various subclasses of electroreceptors in quite different ways. Indeed it was from tuberous electroreceptor units in assorted species that we got the best evidence, so far, of diverse forms of spike code for intensity such as burst duration at fixed interval, probability of firing at fixed times, time of firing at fixed frequency and others, including as the last discovered, mean frequency coding receptors. These last are not influenced by the EOD, because the receptors are low-pass filters and the EOD power is chiefly above their best frequency. Acoustic afferents provide parallels in several degrees. Some units follow the sound more or less faithfully in a middle frequency range. Other units do not show the sound frequency in an autocorrelation or poststimulus time histogram and only change in mean rate with very irregular interspike intervals. Another way of stating this is that the tendency to phase-lock afferent spikes with the sound varies from strong to nil. The degree of jitter of intervals is widely different among electroreceptors as it is among acoustic receptors.

Efferent control is one of the most conspicuous contrasts between these modalities. Well developed in varying degree and site of termination among most acoustic receptors, efferent fibers to electroreceptors seem to be altogether absent.

Interactions between responses to two simultaneously presented stimuli, such as two tone suppression, may be another point of contrast. Not all, but many acoustic afferent fibers show some suppression. Electroreceptors so far as tested seem to show very little such nonlinearity. Viancour and Krausz (1981) got essentially the same impulse response from quasi-white noise stimulation of different power levels, bespeaking linearity in *Eigenmannia* tuberous units. The main fact is that information is so meager here, that either finding will be interesting. It is certainly unsafe to try to rule out lateral inhibition on anatomical grounds. Surprises are likely, such as the connecting strand said to be of nervous tissue found between lateral line organs, superficial to the lateral line nerve in *Tilapia* (Späth and Lehmann 1976).

3 Central Processing

In this section I select highlights from the information available on anatomy and physiology of the brain in respect to electroreception. It will be evident that this information has snowballed rapidly in recent years and in several respects is substantially more than that available for acoustic processing in fishes. A more complete review is given in Bullock (1981a).

3.1 Medulla

3.1.1 Anatomy

Two general points emerge from studies on the medullary targets in which electroreceptor afferents terminate and central processing begins. (i) One is that two quite different nuclei serve this function in the two clusters of taxa: the nonteleost groups with only ampullary input and the teleost orders with both ampullary and tuberous electroreception. It was Northcutt (Chapter 16) who assembled evidence that the well-known *dorsal nucleus* of the octavolateral region of the medulla in elasmobranchs, chondrosteans, dipnoans, and polypteriforms can be identified with electroreceptor input; apparently all the ampullary afferents end here and essentially all the lateral line and VIIIth nerve afferents end in other nuclei. The dorsal nucleus seems to be diagnostic of the possession of a specialized electric sense. In the electroreceptive teleosts, however, there is never a dorsal nucleus and the ampullary and tuberous afferents terminate in a special portion of the *medial nucleus*, whose other portions receive common lateral line afferents. That special portion is often called the posterior lateral line lobe (PLLL) (see Fig. 27-4). The medullary nuclei therefore form the best evidence that electroreception is not everywhere homologous but has been independently evolved at least twice (nonteleost and teleost). It is hard to believe that within the teleosts there have been less than two independent inventions of the whole electric apparatus, once in the osteriophysines and once in the mormyriforms. It remains to be determined whether the relevant part of the medial nucleus looks or works differently in the three teleost orders (gymnotiforms, siluriforms and mormyriforms). In contrast to the polyphyletic origin of electroreceptive systems, the first acoustic nuclei in the medulla are not, to my knowledge regarded as independently evolved more than once, but if they are, it forms another point of similarity.

(ii) Another finding of general interest about the anatomy of these nuclei is that they give the appearance of being more than simple relays; a diversity of cell types, complex connections in laminated cortical organizations with a layer of granule cells and input from other brain centers all suggest *functional complexity* and differentiation of the transactions performed. We will see in the next section that this is what we find physiologically. Electroreceptor nuclei resemble in complexity the cochlear nuclei even in mammalian species where they are well developed (Rethelyi and Szabo 1973b). The PLLL of mormyrids (Maler 1973, Maler, Karten, and Bennett 1973a, 1973b, Bell 1979, Szabo and Ravaille 1976) has a *cortex* that receives axon terminals from ampullary and mormyromast afferents in different zones and, more centrally, a *nucleus* that receives the terminals from the knollenorgans. These are quite distinct from the targets of mechanoreceptive lateral line afferents, as well as VIIIth nerve afferents. Within the cortex of PLLL there are three separate *maps* of the body surface (Fig. 27-4), with curious breaks in the continuity of the cell layers between them. One map is for ampullary afferents. Possibly the other two are for the two types of afferent fibers from mormyromasts (Table 27-1). The cortex is quite cerebellar-like, with a molecular and ganglion cell layer. A fourth map is present in the PLLL nucleus, receiving from knollenorgans though it is less precise. At least in sharks, catfish and mormyrids, if not more widely, there are presumably electric, common lateral line, and acoustic nuclei side by side in the octavolateral area of the medulla, but these projections have not yet been fully delineated from each other. One can expect, moreover, that there will be found somatotopic maps in the dorsal nucleus of elasmobranchs and their allies and in the medial nucleus of siluriforms and gymnotiforms.

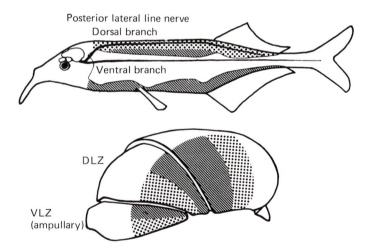

Figure 27-4. The part of the medial nucleus known as the posterior lateral line lobe in a mormyrid, *Gnathonemus petersii*, showing three of the four somatotopic maps. DLZ, dorsal lateral zone with terminals of tuberous receptors (knollenorgans and mormyromasts) in two mirror image maps; VLZ, ventral lateral zone with terminals of ampullary receptors forming their own map. (From Bell and Russell 1978.)

3.1.2 Ampullary Input Processing

Ampullary input processing was first examined by Il'inskii, Enin, and Volkova (1971) in skates (*Raja*), comparing evoked potentials to short pulse stimulation of various branches of the cranial nerves that supply ampullae of Lorenzini or lateral line canals. These authors found differences in the evoked potentials to the different nerves both in the form and in the focus of maximal activity within the octavolateral region. Platt et al. (1974) also stimulated electroreceptive nerves and mechanoreceptive nerves in the elasmobranch (*Torpedo*) and found in addition differences in the facilitation or antifacilitation to a train of stimuli. Recording from single units in the elasmobranch (*Trygon*) Andrianov, Brown, and Ilyinsky (1974), and Andrianov and Broun (1976) showed that most units are spontaneously active and modulated in rate, up or down, according to the polarity of current in the sea water or of a change in the magnetic flux. Two main populations of units were distinguished in respect to polarity: those excited by applying an anode or a south pole to the ventral side, and those excited by a cathode or a north pole applied to the ventral surface; both are inhibited by the opposite stimuli. Four main patterns of response were observed: phasic-tonic, tonic, sustained (persisting after cessation of the stimulus), and bursting. Paul and Roberts (1977) also recorded from the octavolateral area and distinguished field potentials from the dorsal nucleus, where afferent axons terminate, and from the overlying molecular layer. They distinguished units among incoming primary fibers, from second order cells in the nucleus and from efferent neurons. It is likely that they were mainly dealing with mechanoreceptive units.

In catfish (*Ictalurus*) Andrianov and Ilyinsky (1973) and Andrianov and Volkova (1975) studied second-order units and found much the same as in skates. They noted a bimodal sensitivity, to both electrical and mechanical stimulation. Roth (1975) described the properties of units (in *Kryptopterus*) located deep to the cerebellar surface in its posterolateral corner: most likely they were in the medial nucleus of the octavolateral medulla. He distinguished three main types of second-order units and compared them with primary afferents. Type a are much less active spontaneously and at 1 to 10 Hz sinusoidal stimulation, 10 to 30 times more sensitive than primary afferents. Type b have no spontaneous activity and are most responsive to stimuli below 1 Hz. Type c have a regular spontaneous activity and best frequency of 10 to 20 Hz. McCreery (1977a) also examined the second order neurons (*Ictalurus*) and found that those excited by anodal (inward) current, like all the first order afferents, receive monosynaptic excitation from those afferents. The second-order units excited by cathodal (outward) current receive disynaptic inhibitory input from the afferents. He also noted that the most effective stimulus frequency for second-order units is lower than for primary afferents. McCreery (1977b) mapped receptive fields of second-order units with a roving stimulus electrode and found that the small, round fields of primary units are replaced in the first nucleus by fields from a few to many times larger, sometimes with two, disjunct maxima, usually more rostral for caudally moving stimuli and vice versa. There is some evidence of lateral interaction. The size of the receptive fields is thought to represent a compromise between enhanced sensitivity due to spatial averaging and loss of spatial resolution, perhaps matching the spatial frequency of the analyzer properties with that of the physiological stimuli. The temporal frequency properties of the neural filter are likewise thought to be matched to those of the natural stimuli. DeWeille (in m.s.) finds a significant second kernel (Marmarelis and Marmarelis 1978) in catfish medullary units believed to be second-order neurons that receive converging input from a number of primary ampullary afferents, which themselves show only a first kernel, in response to quasi-white noise stimuli of moderate intensity. This is taken to mean that significant nonlinearities such as lateral interaction begin in the lateral lobe. He finds evidence of lateral interaction in units regarded as third-order neurons.

Ampullary units in the medulla of gymnotiform fishes have been little studied. In mormyrids, Bell and Russell (1978) found two types of neurons, much like those of McCreery (1977a, 1977b) in catfish.

3.1.3 Tuberous Input Processing

Tuberous input processing to the medullary electroreceptor nucleus has been most studied in *gymnotiform* fishes. Especially useful have been the so-called wave species (whose electric organ discharge (EOD) is about the same in duration as the interval between EODs, making a distorted sine wave), but also pulse species have been used, those whose EODs are separated by relatively long intervals. The main result is that the two types of primary afferent units (Table 27-1) diverge into half a dozen types of second order units in the medial nucleus (PLLL, or posterior lateral line lobe of some authors).

There are two broad classes of second order, high frequency sensitive units, those that follow the EOD with some degree of phase-locking and those that do not show any or only a feeble phase relation, though they are influenced by electroreceptor input. In the latter class there is considerable diversity but discrete types of units have not been categorized. To moving stimuli such as plastic or metal plates, there are tonic units in the PLLL that depend on the position but not on the direction of motion and that maintain a high rate of spontaneous discharge. There as phasic units that depend strongly on the direction of movement and others that do not. Of the direction sensitive units some respond in the same way to plastic and to metal plates, whereas others, like the primary afferents, give opposite responses. There are various degrees of sensitivity to the size of the stimulus plate; some units appear to respond mainly to the leading edge of plastic plates and, in a still more complicated way, to a silver plate (see Fig. 27-5). All these kinds of units have given some reason to be considered nuclear and not afferent but we cannot say whether they are intrinsic neurons confined to the posterior lobe or projection neurons.

Scheich (1977c) characterized four classes of posterior lobe high frequency sensitive units stimulating by introducing objects or by short circuiting between two points some distance apart in the water or by the social signal of another fish's EOD or a simulation thereof, at a frequency known (Bullock, Hamstra, and Sheich 1972a, 1972b) to elicit a "jamming avoidance response" (JAR). This is a small shift of the experimental fish's EOD rate in the presence of a neighbor with an EOD only slightly different, in the direction that will increase the difference (ΔF) between the frequencies of the two fish. One of Scheich's classes acts much like the class of afferents called T units or

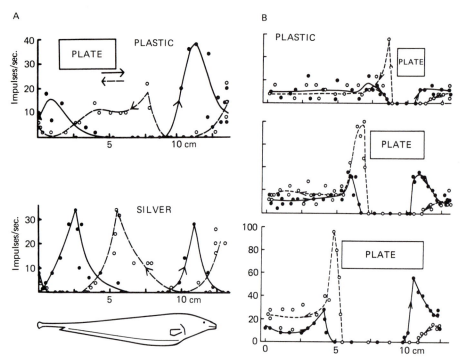

Figure 27-5. Some examples of units with integrative properties in the medial nucleus of the octavolateral area in gymnotiforms. (A) Responses in *Apteronotus* to moving 2 × 4 cm plates of plastic or silver. The posterior edge of the plate is opposite the tip of the tail at zero on the abscissa. Note that the direction of movement affects the positions of excitation and inhibition but that the plastic and silver act similarly instead of oppositely as they do for afferents. (B) Responses in *Eigenmannia* to plastic plates of 2 × 2, 2 × 4 and 2 × 6 cm. Note that the width of the zone of inhibition is proportional to the length of the plate; opposite directions of movement elicit only partly contrasting responses. (From Enger and Szabo 1965.)

phase coders though with signs of convergence of a number of peripheral units. The other three are much like the P units or probability coders in the periphery. The first of the three includes subtypes like the peripheral P_c, and P_s subtypes; they increase in the probability of firing in each EOD cycle when moving or stationary objects of high conductivity enter a receptive field, or to short circuiting, or to the beating between its own EOD and another fish's at a slightly different EOD rate. $P_c 1$ units fire earlier in the beat cycle for $-\Delta F$, $P_c 2$ units earlier for $+\Delta F$, and P_s units are not sensitive to this or any difference resulting from the actual ("slant clipped") shape of the EOD as opposed to a pure sine wave. The three types are positions along a continuous distribution of units in respect to sensitivity to harmonic content and its phase. Quantitative and timing differences distinguish the second-order units from afferents and indicate differentiation of the rate of change of firing of afferents or other processing. The second of the three classes of P units in the posterior lobe reacts in the opposite way, enhancing its firing probability to nonconducting objects, slowing down to metal plates or short circuiting between points in the bath, or to higher amplitude beating. The third class of P units hardly responds to objects or to beating but increases its firing to shorting, with a low degree of synchrony to the EOD. Tentatively it appears that T and P units are partly segregated, the T units being generally more ventral and medial. One way of looking at the evidence leads to recognizing two parallel pathways, one fast and one slow (Szabo 1974, Szabo, Sakata, and Ravaille 1975, Szabo, Enger, and Libouan 1979, Libouan and Szabo 1976a, 1976b). The fast system provides a single, electrical synapse and high velocity axons between pulse marker or phase coder receptors, in pulse and wave species, respectively, and the midbrain, such that every EOD of the fish itself causes a well-synchronized volley of spikes to arrive in the torus semicircularis (TS) within a very short time, 1.5 to 2 msec! That synapse is on the large, pear-shaped cells of the PLLL (Sotelo, Rethelyi, and Szabo 1975) which send "giant" fibers up the lateral lemniscus to a special part of the TS, the nucleus mesencephalicus magnocellularis (NMM). The slow system is all the rest, a diverse array of cells, connections and input processing in the lateral line lobe and torus semicircularis.

Little is known from the posterior lobe of the numerous pulse type gymnotiform fishes. Several studies have dealt with the physiology of *mormyrids* (Zipser 1971, Schlegel 1974, Zipser and Bennett 1976a, 1976b, Szabo et al. 1979, Bell 1979). The two broad types of afferents (K units = knollenorgans and D units = mormyromasts) end in different parts of the posterior lobe corresponding to the fast and slow systems, with a somatotopic order, and they manifest different functional properties (see Table 27-1). Most is known about the mormyromasts, themselves complex organs with two forms of hair cells and two sets of afferent nerve fibers. Intracellular recording in the cortex of the posterior lobe has shown two modes of activation of the second order cell: a direct monosynaptic input by a single primary fiber and a disynaptic and delayed input via the intermediation of granule cells, which synapse more dendritically. The monosynaptic input may be on the axon. In addition, there is inhibitory input from surrounding receptors that influences only the disynaptic input. Schlegel (1974) contrasted the cortical, mormyromast activated and the deeper, knollenorgan activated units. The former are spontaneously active and only modulated by EOD amplitude changes, normally by objects. The deeper units are driven 1:1 by adequate EODs and are unmodulated by EOD amplitude changes above threshold. Some cortical units respond differently for headward and for tailward movements of objects near the fish. Schlegel also made the interesting observation that cortical units, in comparison with primary afferents, fire impulses at a lower frequency and with more fluctuation. In contrast to high frequency gymnotiforms such as *Apteronotus* or *Eigenmannia*, the mormyrid lateral lobe cortical units do not necessarily integrate temporally over more than one EOD. Compared to the gymnotiform, *Eigenmannia*, there is notably less differentiation of types of second-order units in *Gnathonemus* and this seems not attributable to the fact that *Eigenmannia* has been more frequently studied! Knollenorgan input ends not in the cortex but in the nucleus, on cells that show low threshold, little latency change near threshold, and an ability to follow high frequency stimuli. This is the fast pathway to the midbrain. A compound evoked potential from these cells is early (1 msec latency from a stimulus in the water!), highly synchronized, confined to a small area, and saturated at intensities not yet arousing the slow, mormyromast system of the cortex. With minimal processing and delay the PLLL nucleus relays knollenorgan impulses, in their still synchronized volleys, via fast lemniscal axons to the TS. Looking over the body of knowledge, anatomical and physiological, it seems that the lesson is this: segregation occurs so that not only major modalities but submodalities distributed among each other go to distinct areas. It seems most likely that there are more subdivisions of the acoustic sense than we have so far recognized.

We come to a prominent feature of the physiology that contrasts mormyrids and gymnotiforms. Bennett and Steinbach (1969) were the first to discover a *corollary discharge* from the command signal generated in the EOD command nucleus in mormyrids. The corollary discharge does not depend on afferent input or EOD and is distributed widely, to cerebellum, midbrain and medulla. It exerts a series of powerful and complex effects; especially on the electroreceptor input nuclei (not on receptors), and on the midbrain electrosense centers. No such corollary discharge has been found in gymnotiforms, despite search; instead they use afferent input combined with special kinds of behavior to much the same end.

The influence of corollary discharge upon PLLL neurons in mormyrids is pronounced but different for the three receptor driven kinds: ampullary, knollenorgan, and mormyromast (Zipser and Bennett 1976a, 1976b, Bell and Russell unpublished). In the ampullary receptor region of PLLL the second order cells, like the afferents, show a weak tendency to follow the EOD, although the receptors are high frequency insensitive. Evidently, the small DC component between EODs is sufficient to influence them. This EOD related part of the PLLL cell discharge is altered by the corollary discharge (Bell 1979), but details are not available. In the knollenorgan region of PLLL the second order cells are completely inhibited for a period from about the 2nd to the 5th msec following the start of the EOD command signal as picked up in the fish's tail; this period corresponds to that when a volley of afferent spikes caused by the EOD is due to arrive at PLLL in the normal animal, without curare. The inhibition is not exerted on the receptors, so that they follow the animal's own EOD faithfully; however, *knollenorgan input is throttled* at the first nucleus, apparently completely. We do not know of collateral pathways or other targets of knollenorgan axon branches that might escape inhibition. The brain thus rejects this information that a self-generated EOD has occurred, instead relying on its corollary discharge. That leaves the knollenorgan region of PLLL with its high sensitivity, and disinterest in objects, free to detect the EODs of other fish at some distance.

In the mormyromast region of PLLL corollary discharge has the opposite effect: *mormyromast input is gated* such that for a brief period following each self-generated EOD it is facilitated, and for the rest of the time it hardly gets through (Zipser 1971, Zipser and Bennett 1976a, 1976b). Only the disynaptic input, mentioned above, is so affected; for about 15 ms after a corollary discharge, test stimuli via mormyromast receptors can cause a spike in the principal cells of that part of the PLLL cortex; at other times only a synaptic potential. It is not yet known whether other large cells in PLLL or later stages in the monosynaptic pathway are similarly gated. However, there is evidence from a variety of cellular and compound potential respones that a variety of effects is exerted.

3.2 Midbrain: Torus Semicircularis

3.2.1 Anatomy

In the *mesencephalon* distinct nuclei have been delineated in the catfish not only for acoustic and for electric but also for common lateral line mechanical stimuli (Knudsen 1977). They occupy the TS, homolog of the inferior colliculus of mammals. The nucleus centralis of TS in *Ictalurus* receives acoustic input; nucleus lateralis pars medialis receives lateral line mechanoreceptor input; nucleus lateralis pars lateralis receives electroreceptor input. Some details of the cell types and arrangement in these nuclei is known. The TS is relatively well developed in siluriforms but enormously hypertrophied and elaborately laminated in some gymnotiforms, apparently in association with the special development of the electric sense. In mormyrids it is intermediate.

The *gymnotiform* TS has been studied frequently (Rethelyi and Szabo 1973a, Szabo 1974, Scheich and Bullock 1974, Scheich 1977c, Scheich and Maler 1976, Szabo et al. 1975, Sotelo et al. 1975, Szabo et al. 1979) but still quite incompletely. It occupies a large volume and twelve laminae have been recognized (Fig. 27-6). Some differences between the laminae in functional significance and connections are mentioned below (Section 3.2.3). The principal input appears to be via the lateral lemniscus from the octavolateral area of the medulla particularly the electroreceptive nucleus. The outputs are more diversified and only known in part; TS projects to cerebellum, optic tectum, and probably to a diencephalic nucleus that in turn projects directly to the medullary electromotor command nucleus (see also Bullock 1981a).

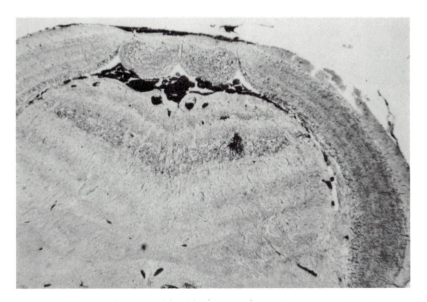

Figure 27-6. Transverse section through the midbrain of *Eigenmannia* showing tectum outside the broad flattened ventricle and torus semicircularis under it, with many layers. An Alcian blue spot is in the T cell (6th) layer; Bodian stain; the brain is ca. 4 mm wide. (From Scheich, pers. comm.)

A special part of the TS, the 6th layer of T units in *Eigenmannia* and the equivalent layer in other wave species is regarded as homologous to a structure in pulse species of gymnotiform fishes called by Rethelyi and Szabo (1973a) and Szabo et al. (1975) the nucleus mesencephalicus magnocellularis (NMM) (and in mormyriforms the nucleus mesencephalicus exterolateralis). NMM is a prominent midline structure bulging up in dorsal view and originally treated by those authors as near, but not part of the TS. Taking the evidence from all taxa into account, it is now clear that NMM is a specialized part of TS. It is the terminus of the remarkably fast pathway from one class of electro-receptors, the M or pulse marker units in pulse species, the T or phase coder units in wave species of gymnotiforms, just as the nucleus mesencephalicus exterolateralis is the terminus for fast knollenorgan units of myormyrids.

Cell types and synapses, including conspicuous gap junctions and other electron microscopic details, have been described; these correlate with the high velocity, firing rate, synchrony and electrotonic coupling (Sotelo et al. 1975, Szabo et al. 1975, Szabo et al. 1979). The output is said to be into deeper parts of TS. The other receptors, vis their diverse second-order relays, go to certain other parts of TS (lamina 9 in *Eigenmannia*) and have been collectively called the slow system.

In *mormyrids*, the nucleus mesencephalicus exterolateralis of Stendell (1914) is regarded as equivalent to the NMM on both histological and physiological grounds. As in gymnotiforms it has a dense neuropile and, besides many small neurons, a rela-tively small number of large cells remarkable in being adendritic and covered with myelin, receiving large myelinated axonal terminals with mixed synapses. These axons come directly from the lateral line lobe, mainly contralaterally, from the large pear-shaped second-order cells shown to have gap junctions, which are believed to be the terminals of knollenorgan afferents (Szabo and Ravaille 1976, Szabo et al. 1979). Thus, the fast pathway is reasonably accounted for.

New findings with the aid of HRP as well as Fink-Heimer methods (Enger et al. 1976b, Finger and Bell 1981) show massive projection from the mormyrid medullary octavolateral area into certain parts of the TS. Output from the TS is less well known, partly because it is more complex. It includes projections to the valvula of the cerebel-lum, the optic tectum, a specific thalamic nucleus that might be equivalent to ven-tralis lateralis of mammals, the nucleus preeminentialis of the medulla and the reticular formation. Thus, the control exerted by TS on the pacemaker command center is indirect.

One wonders if it would not repay the effort to study the TS of various fish groups with significantly different auditory specialization.

3.2.2 Ampullary Input Processing

Ampullary input processing is indicated by a conspicuous and moderately complex series of evoked waves in and under the tectum opticum of *elasmobranchs* (Platt et al. 1974, Bullock 1979, 1981b). Alterations in the waveform show sensitivities to electric field orientation, sign of the change of current, position of a dipole, serial number in the first few cycles of a train as well as the interval between electric events in a series of events. Separate component peaks of the evoked wave complex have different dependencies on these factors. This makes comparison of responses at different frequencies of sinusoidal stimulation not a simple matter of measuring amplitudes of waves. The "best" response by a certain criterion is at 20-30 Hz in a shark (*Carcharhinus*), and 10-15 Hz in a ray (*Potamotrygon*). The lowest threshold, with moderate averaging, was 0.015 μV/cm (=0.8 nA/cm^2) in the marine shark, and < 50 μV/cm (=0.7 nA/cm^2) in the fresh water ray. The response depends on the locus of recording, type of electrode, and state of the brain. There is evidence of topographic segregation of some stimulus parameters but mapping has not been undertaken in elasmobranchs. The best loci for electric evoked potentials are distinct from those for acoustic, although they are contiguous and appear to overlap. No interaction between them has been found. There appear to be differences between species in respect to dynamics of responses and preferred stimuli that correlate with behavior.

In *siluriform* fishes evoked potentials give similar evidence and, from their generality among all 14 genera in 9 families tested, allow us to believe that this sense is common to the very large and diversified order (ca. 30 families) of catfishes (see Fig. 27-7).

More detailed studies, including extensive single unit recording have been provided by Knudsen (1976a, 1976b, 1977, 1978) on the catfish, *Ictalurus*. The results which are particularly relevant in the present context include the following. A circumscribed part of the TS is primarily acoustic (nucleus centralis), another is mainly responsive to ampullary input (nucleus lateralis pars lateralis), and a distinct part responds mainly to lateral line input (water movement; nucleus lateralis pars medialis). In the electrorecep-

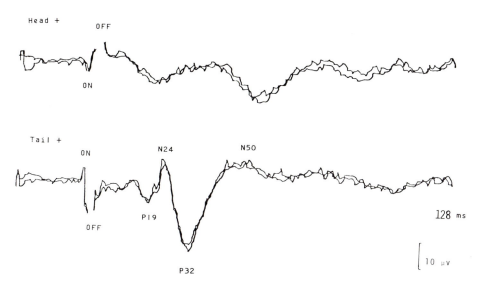

Figure 27-7. Averaged evoked potentials recorded in the torus semicircularis of catfish, *Pimelodus* sp. (Pimelodidae), following a 3.6 msec electric pulse of 3 mV/cm, longitudinally, in the water, repeated ca. every second, at ON-OFF. N, negativity; P, positivity of the brain electrode relative to an inactive reference; numbers indicate millisecond latency of peaks. Note that the population of cells sampled by the electrode at this locus in the TS "prefers" one polarity, and that the evoked potential is the most reliable way of uncovering that discrimination. (From Bullock 1979.)

tive part tactile, acoustic, and optic stimuli also influence some unit activity but electric influence dominates. Similarly the other two areas justify the designation of functional regions.

In the electrosense region five types of units based on responses to dc step stimuli are (i) phasic units with high resting discharge rate, (ii) phasic units with low resting discharge, (iii) tonic units, (iv) phase-tonic units and, (v) weakly responding erratic units. Unit latencies, as well as sensitivities at the best frequency (<0.1 to ca. 25 Hz), vary widely from 8 to 130 msec, usually 8-28 msec and a function of stimulus intensity in types i-iv. Within each type there is a spectrum of best frequencies of sinusoidal stimuli, over more than an octave and a range of sensitivity from 0.8 μV/cm to 15 μV/cm (8×10^{-11} to 15×10^{-10} A/cm^2). Four types of units based on phase locking to sinusoidal stimuli are distinguishable: monophase multispiking, monophase oligospiking, biphase multispiking, and biphase oligospiking. Another difference is that stimulus intensity increase can advance or retard or do nothing to the phase. Some units show afterdischarge following cessation of a high intensity stimulus; most do not. Some show a frequency specific inhibition and this is typically a frequency above the best excitatory frequency. Unit receptive fields show a somatotopic organization of electroreceptive input, primarily contralateral, with the main axis of the body represented by an anterior-posterior axis in the torus, anterior being anterior (see Fig. 27-8). A similar topographic representation of body regions exists in the mechanoreceptive part of the nucleus. Besides longitudinal somatotopy there is a systematic dorsoventral distribution of preferred electric field orientation; units in the dorsal 125 μm largely prefer field orientations between 180 and 360° whereas in the ventral 125 μm units

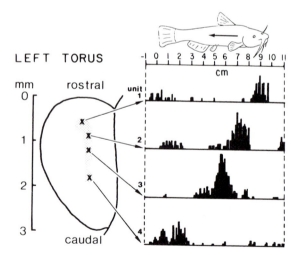

Figure 27-8. Somatotopic mapping in the catfish (*Ictalurus nebulosus*) TS, lateral nucleus, to stimulation by weak electric dipole pulses moving in the water, contralaterally, in the optimal direction and polarity for each unit. (Above) The outline of the lateral (electroreceptive) nucleus (dashed line) is projected on to the outline of the left mesencephalon and the positions of the three penetrations shown. (Below) The superimposed receptive fields (stippled) and the spatial histograms of spikes of units. (From Knudsen 1976a.)

largely prefer fields between 0 and 180°. The evoked potential recorded at the same place usually showed a somewhat different orientation preference than the units isolated. There is also a tendency for a systematic distribution of best frequencies of sinusoidal fields; dorsal units are tuned to 0.05-10 Hz, ventral units to 4-15 Hz. The more dorsal unit spikes are also broader (1 msec) compared to the ventral (0.5 msec). Little is known about ampullary input processing in *gymnotiforms* or *mormyriforms*.

3.2.3 Tuberous Input Processing

Tuberous input processing in the midbrain has been studied in *gymnotiforms* (Scheich 1974, 1977c, Scheich and Bullock 1974) chiefly in the ca. 300 Hz wave species, *Eigenmannia* sp., where the TS is both extremely large and highly differentiated, as we have seen. The two principal types of electroreceptors in this wave species (Table 27-1) are called P and T and we noted corresponding cell types in the medial nucleus of the medulla. Those cells send axons to two distinct laminae in the dorsal part of the torus where P and T units of the third order are found, the T layer being the most superficial. The P units show more precise limitation of their firing to a certain part of the beat cycle when a second fish with a different EOD rate comes within range; that is, their amplitude discrimination is increased relative to the afferents. This also means that the temporal difference is clearer in the firing phase of the beat, as between + and $-\Delta F$ (second fish or simulated fish EOD rate above or below that of the investigated fish), due to asymmetry of the beat which in turn is due to the clipped or nonsinusoidal form of the EOD (see Fig. 27-9). The T units of the torus show greater precision in their firing phase within the (ca. 3 msec) EOD cycle, i.e., the jitter is less than ¼ of the already very small jitter of afferent units; this puts the constancy of latency in the microsecond range.

I called attention above (Section 3.2.1) to the special part of the TS called the NMM in pulse species of gymnotiforms, and called the 6th layer in some wave species such as *Eigenmannia*. This is the terminus of the specialized fast system. Even with a relay in the medulla, well-synchronized impulses reach the NMM as early as 1.5-2 msec after the first peak of each EOD, one impulse volley for each EOD faithfully reporting to the midbrain that each EOD took place. In some gymnotiforms the threshold is high so that no other fish's EOD will ordinarily fire the M receptors. Blocking the EOD with curare stops the NMM firing, showing that it is not following a corollary signal from the pacemakers in the command nucleus, which continue to fire under curare. This specialized fast electrosensory system must presumably eventually interact with the rest of the system but the facts indicate that it is still quite discrete at the midbrain level, with little sign of other influence. It is not clear what the output of the NMM or the T layer of TS does or what the function of the fast system really is, in general. Notions that have been proposed are too vulnerable to be detailed here.

There are somatotopic maps of the fish's body in the T (6th) and in the P (9th) layers, mainly contralateral, and they are in register. Optimal field orientations are also congruent for P and T units near each other, but the distribution of preferred orientations has not been systematically examined. Units have been studied that are much more selectively tuned for a particular field orientation than are the afferent or the

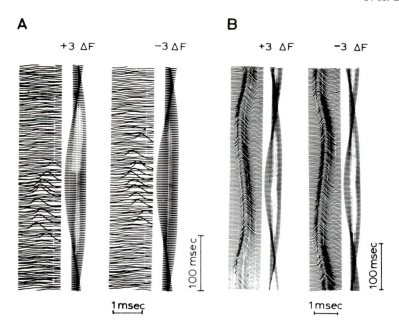

Figure 27-9. Characteristic firing patterns of the two high frequency sensitive coding types, a T unit (A), and a P unit (B), from the torus semicircularis, during the natural beat cycle of a jamming avoidance response in *Eigenmannia*. (A) The T unit spike trace is triggered by a constant phase of the EOD. For every wave the neuron fires a spike. Parallel to the beat cycle the unit exhibits a phase modulation of time of firing of the spike relative to the EOD. For $+\Delta F$ the modulation starts with a phase lag in the first half of the beat cycle. For $-\Delta F$ the unit enters the cycle with a phase lead. Note that the patterns of phase modulations for $+$ and $-\Delta F$ are not precisely mirror images but somewhat displaced along the beat cycle in the same direction as in the P unit case. (B) The spike trace on the left is triggered by a constant phase of the EOD. Parallel to the amplitude increase in the beat (next trace to the right; time runs from bottom to top) the P unit gives a burst of activity. Single spikes have a slight jittery change of phase relationship to the EOD, being usually somewhat earlier during maximum firing rate. The most significant feature, however, is that during the beat cycle the burst occurs slightly earlier than the middle of the beat for $+\Delta F$ and later for $-\Delta F$. (From Scheich and Bullock 1974.)

second or the usual third-order units; others are quite without any preferred orientation. Presumably both types represent *convergence* of selected afferents. Convergence in laminae of the ventral part of the torus occurs in various combinations. In favorable cases, Scheich (1977c) estimates 20-40 receptor units converge to one torus unit. Besides other complex units, the most sophisticated cells so far quantitatively described are the "ΔF_p" units of Scheich that fire under certain conditions normal to social encounters with a conspecific whose EOD is very close in firing rate. These units signal tonically, without obvious influence of the beat cycle that a neighbor with a different EOD rate is present, and whether the difference in EOD rate, ΔF, is $+$ or $-$; they are tuned to maximal response when the ΔF is ca. 4 Hz, just as is the behavioral JAR.

These ΔF_p units must have combined the beat amplitude signal from sign-insensitive units of the PLLL with that from sign sensitive units which Scheich found and Heiligenberg and Bastian (1980) confirmed.

The TS is relatively unconcerned with object detection but exquisitely specialized for a kind of analysis important in social interactions such as the jamming avoidance response, and probably other behaviors not yet studied. One candidate is an intermittent behavior called *active phase coupling* (APC), seen in many wave species (Langner and Scheich 1978) in which, instead of moving apart in frequency, as in the JAR, neighbors shift to lower beat frequencies and then abruptly lock into a fixed phase relation, usually in phase or in antiphase; they hold this for minutes, with microsecond precision, and then separate. A sine wave stimulus can induce APC, starting at $\Delta F =$ 10 Hz if the voltage gradient is like that of a nearby fish, much stronger than the JAR threshold. Sometimes an APC with sudden tight phase lock is seen even though EOD rates are 100 Hz or more apart, i.e., they lock into an integer ratio as high as 10:11. The social significance of APC behavior is unknown but it is clearly a real and active pattern of agonistic behavior and amenable to explanation in terms of neurons such as those known in the TS. Scheich (1974, 1977c) points out evidence from the physiology of the TS as well as first- and second-order units, that the system for analyzing these small frequency and waveform differences is working in the time domain within each EOD cycle, even in the species where that is less than 1 msec. Note also that when $\Delta F \ll 1$ Hz and beats take several seconds, the latency of a JAR can be $\ll 1$ beat cycle. To appreciate the fish's performance, it should also be noted that whereas humans can hear the difference between one beat combination and another for example in a nearly limiting case, between 300 + 303 Hz and 297 + 300 Hz, providing the intensities of the two tones are nearly equal, *Eigenmannia* detects the presence, sign, and magnitude of a sine wave added to its own field when $\Delta F \ll 1$ Hz or, at the optimal ΔF of ca. 3 Hz, when the intensity of the stimulus is 60 dB weaker than its EOD (Bullock et al. 1972a, 1972b, Scheich, Bullock, and Hamstra 1973).

Deep in the TS a few units have been reported sensitive to other modalities: acoustic, vibratory, and visual, alone or in addition to high frequency electric fields. A few units sensitive to low frequency electric fields, presumably ampullary input, have been seen.

Mormyrid midbrain physiology has been studied to some extent (Bennett and Steinbach 1969, Zipser and Bennett 1976b, Enger et al. 1976a, 1976b, Szabo et al. 1979). Szabo et al. (1979) recorded particularly in the nucleus mesencephalicus exterolateralis, the termination of the fast pathway, equivalent to the NMM of gymnotiforms. The short latency, high synchronization, and low threshold follow the properties of the knollenorgans and the nucleus of PLLL, and indicate that the postsynaptic response in the midbrain is only a third-order neuron. The corollary discharge blocks this response for a period of 1 msec occurring 3 msec after the EOD; this could be expected from the complete inhibition of the relevant part of PLLL, the source of input to nucleus mesencephalicus exterolateralis. Yet the corollary discharge signal can be recorded in the midbrain. The slow pathway from the cortex of PLLL projects to a part of the TS called in mormyrids nucleus mesencephalis lateralis (Szabo et al. 1979). Responses in this structure are facilitated, instead of blocked, for a few milliseconds after the corollary discharge; it is not clear how much this is attributable to the gating effect at PLLL. The result of the gating in the slow system is that it looks at the fish's own EOD as modulated by objects, minimally disturbed by other fish whose EODs are usually out of phase. Nucleus mesencephalicus lateralis therefore can provide suitably filtered and gated input to an analyzer of the spatial distribution of EOD current over the body, such as the valvula of the cerebellum.

3.3 Cerebellum

This organ is unusually large in many elasmobranchs, siluriforms, gymnotiforms, and mormyriforms. In the last named it is also highly specialized histologically (Fig. 27-10A). The suspicion has long been voiced but still cannot be settled that this is an hypertrophy and elaboration associated with the development of electroreception. It has induced studies and these have indeed shown an important representation of this modality. In fact we can now propose that the cerebellum is particularly involved in that part of the electrosense function that concerns electrolocation.

A

Figure 27-10. Purkinje cells in the cerebellar valvula in a mormyrid, *Gnathonemus*. (A) Golgi stain of part of a cell. Note that the dendrites are parallel, uniformly spaced, unbranched, equal in length, extending the full width of the molecular layer. (From Nieuwenhuys and Nicholson 1969). (B) Activity of one cell in the corpus cerebelli of *Apteronotus* during movement of a small stimulating dipole back and forth (separately averaged, arrows) longitudinally in the water near the fish, at different dorsoventral positions ("depth"). Each trace is the average spike firing for ten passes. Dots indicate the position of the tips of the snout and the tail. Upper and lower sets of plots are the same data viewed with a different skew to permit seeing the valleys on both sides of a peak. Irregular spontaneous activity is modulated upward in an excitatory receptive field (hatched) and downward in both anterior and posterior inhibitory fields (stippled). Direction of movement makes a substantial difference. (From Bastian 1975.)

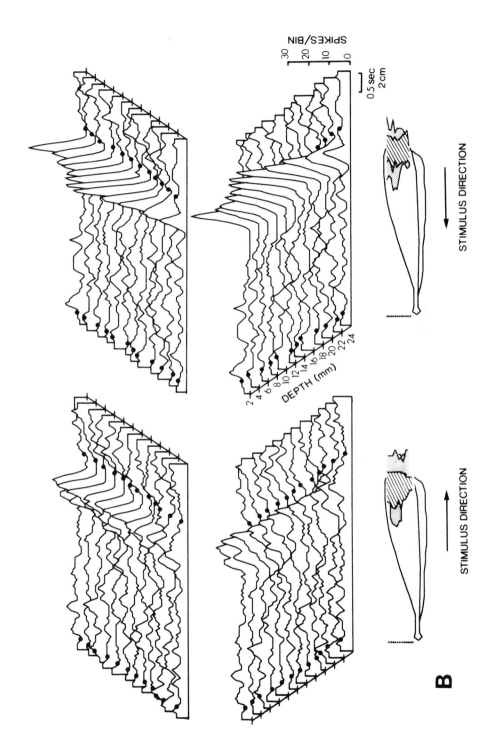

Ampullary input in sharks, rays, and catfish gives rise to evoked potentials in the corpus cerebelli (Platt et al. 1974, Bullock 1979). The wave forms, latencies, and dynamics are not essential in the present context. Localization has not been studied, nor have responses of single units, although the three types of units reported by Roth (1975) from the catfish *Kryptopterus* (see Section 3.2.1) might belong to the cerebellum.

Units have, however, been studied in gymnotiform and mormyriform fishes, in response to tuberous input from physiological stimulation. Bastian found in *Eigenmannia* (1974) and in *Apteronotus* (1975), two high EOD frequency wave species of gymnotiforms, that electroresponsive units are common in a large but circumscribed part of the caudal lobe. These units act like Purkinje cells, and show slow adapting increase or decrease from a prevailing high level of irregular firing, when adequately stimulated with electrical fields in the water. Usually motion is important and often the response is quite sensitive to the direction of motion of an object that distorts the fish's own EOD field or which is itself a source of current (Fig. 27-10B). Each unit has a receptive field, commonly with an excitatory center and inhibitory flanks. Something of a three-dimensional representation of the aquatic space around the fish is computed in the cerebellum. Some units are bimodal, responding also to tail bending proprioceptive stimuli (tested in the absence of an electric field), or to visual stimuli in a restricted receptive field. Some units show a constant response amplitude for object movement at different distances out to about 25 mm, although the electric field is markedly weaker. Other units show a weaker response and the decrement has a power function similar to that in the sensory afferents (Bastian 1976). The evidence adds up to a significant role in electrolocation of objects. This is confirmed in reversible cerebellar local cooling experiments that blocked the opercular breathing rate response to looming objects, a task involving a minimum of motor coordination but some sophisticated input processing (Bombardieri and Feng 1977). In contrast, social stimuli quite effective in causing responses elsewhere (midbrain and EOD command center) are ineffective here, indicating a minimal role of the cerebellum in electrocommunication. Behrend (1977) shed light on this in a study that revealed a low-pass filter property of the cerebellum such that jamming by a social partner with an EOD less than 10 Hz different (ΔF 10 Hz) allows the low frequency amplitude beats to interfere with Purkinje cell modulation by moving objects, but above 10 Hz the beat signal does not get through the filter to interfere. Hence, the normal JAR, mediated by the midbrain, keeps the ΔF usually above 10 Hz, thus improving electrolocation.

In *mormyriform* fishes one expects an even more elaborate involvement of the cerebellum in electroreception, because of the extreme development of that organ in both size and neuronal architecture. Bell and Russell in 1978 gave the evidence that this is so, although the proportion of the total hypertrophied valvula that is included in their map of electroreceptor projections is not massive. In unpublished reports since then, these authors say that the area is getting larger with experience—of the investigators! Most remarkable is that a separate area is found for each of the three main classes of receptors: ampullary, mormyromast, and knollenorgan; one of them receives from both of the tuberous receptor types but the other two appear to be private areas. Also special to mormyrids, and absent in gymnotiforms, is a dependence of the response to an electric event on its time in relation to a command to fire an EOD. A corollary

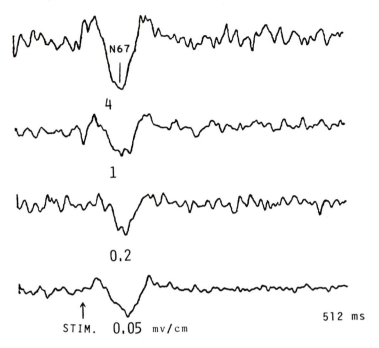

Figure 27-11. Evoked potentials recorded in the telencephalon of the ray, *Potamotrygon,* 6.3 mm deep in the posterior third, medial third, following a 1 msec stimulating electric current pulse in the water (arrow), repeated every 2 sec; the intensity of the homogeneous, transverse field indicated for each trace. Upper three traces are averages of 32, lowest trace is average of 64, at the same scale. Brain negativity down. Note that the negativity at about 67 msec shows almost no intensity effect over a range of 20:1. (From Bullock 1979.)

discharge potential that accompanies the firing of the EOD command is found in the cerebellum as well as in the brain stem (Bennett and Steinbach 1969). Studies particularly directed at its consequences in the cerebellum, and at separating the effects there from effects already exerted upon PLLL and TS have not been done. The cerebellum in these fish teaches us (1) that it can be well developed in relatively primitive vertebrates like elasmobranchs, (2) it can be neurally active continuously, while the animal is making no movements (some species at rest do not even maintain posture but lie on their sides), and (3) mostly it is a long latency sensory processor preserving local sign while mixing modalities and additional fibers. This supports the view that the cerebellum, instead of being called a motor control device, is better regarded as a sensory input processor concerned with space and time, nearby extrapersonal space and the projected position of the body in it. Of course its ultimate value is chiefly motor control and most signs if it is lesioned are motor. But they may be "visceral," for example, an altered rhythm of breathing, as in the Bombardieri et al. experiment, or motor signs may be indefinitely deferred, if no movements are initiated. The cerebellum in these lower vertebrates, where it can be quite elaborate, acts like an intelligence service for the army.

3.4 Forebrain

Ampullary input causes localized evoked activity in the telencephalon in elasmo-branchs (Platt et al. 1974, Bullock 1979) in a region in the medial third, middle third anteroposteriorly, and central third dorsoventrally (see Fig. 27-11). It appears to over-lap with the acoustic response area but the foci of maximal response are separate. Something is known of the dynamics of response but no unit or best stimulus studies have been done as yet, nor have any but preliminary explorations in siluriforms or electric teleosts. The principal finding so far is therefore that even in the telencephalon this modality is still distinct. That puts it in agreement with each of the other modali-ties so far examined in fishes, not converging into a common integrative sensorium in the midbrain, as was once believed, but projecting independently to the forebrain as in mammals. How integration between modalities takes place remains to be discovered.

No substantial study has been directed at the diencephalon as yet.

4 Inferences from Psychophysics

A considerable body of literature using behavioral performance measures establishes the abilities of fishes in electrosensory tasks, and thus permits statements about the mechanisms, especially those of central information processing.

The original experiments of Lissmann and Machin (1958) and Machin and Lissmann (1960) established the ability of *Gymnarchus* to discriminate between two objects dif-fering only in *conductivity* and by a very small amount. They had to revise their calcu-lation of sensitivity, expressed as voltage gradient in the water (Lissmann and Machin 1958, Kalmijn 1972a, 1972b), but the performance is still impressive. Other experi-ments (Lissmann 1958, Bullock 1968, Heiligenberg 1973, 1974, 1977, Harder 1972, Schlegel 1974) have shown that mormyriforms and gymnotiforms can also tell some-thing about the position, size, and form of *objects or spaces*; but our information is not adequate to say much about how well the fish can do in such tasks. We may speak of the brain creating an electrical image of discontinuities in the near surround, approxi-mately to one fish body length away, if the objects or spaces are not too small, but perhaps the limit is a good fraction of a millimeter if other parameters are favorable. I force myself to give this figure as an educated guess; it is not well founded.

Besides local discontinuities, the brain can take account of overall electric *field geometry* as in recognizing familiar from novel configurations of stimuli, walls, and large objects (Meyer, Heiligenberg, and Bullock 1976, Meyer, Becker, and Graf 1977, Fent 1977). A demonstration which has proved especially useful (Heiligenberg, Baker, and Matsubara 1978, Heiligenberg, Baker, and Bastian 1978, Heiligenberg and Bastian 1980, Partridge and Heiligenberg 1980) is the ready discrimination between a quasi-homogeneous field with stimulus electrodes at some distance on either side of the fish and that generated between such a remote electrode and one inserted into the fish's stomach. In the latter case, the field approximates in shape the fish's own EOD and both sides of the fish experience entering or exiting current at the same time, rather than out of phase. We may think of the brain creating not only an image of local dis-continuities but of large field geometry.

These findings already justify the conclusion that the brain *compares the input from different parts of the body.* Nevertheless it is a significant discovery (Heiligenberg and Bastian 1980) that a quite different problem the fish faces during a JAR, namely, which way to shift its EOD rate to increase the difference from that of its neighbor is soluble by a simple comparison of the majority of receptors taken as a reference and assumed to be due to its own EOD, with the minority whose spike probability or phase is slightly different, taken as a sign of the neighbor's EOD. More specifically, these authors find that the shift in pacemaker frequency called the JAR is controlled by operations equivalent to (i) estimating the local phase of zero crossing of the ac field mixture of two EOD-simulations ("own" and "neighbor's") relative to the phase of the majority of receptors, presumably by T units, and (ii) jointly estimating the amplitude as it rises and falls during the beats, presumably by P units. These two quantities change in opposite ways for the two stimulus situations, neighbor higher $(+\Delta F)$ and neighbor lower in EOD rate $(-\Delta F)$, i.e., phase shifts one way during amplitude increase in $+\Delta F$, the other way in $-\Delta F$. The operations could be as simple as pairwise interactions of input from two areas of the body without any necessary reference to an image of its body or of its "own" EOD and that of its neighbor. A kind of motion and direction detection in the time domain is involved. This explanation (Fig. 27-12) is not unique or even the only one for which evidence exists. Under other arrangements of stimulus geometry it was found (Scheich and Bullock 1974, Scheich 1977a, 1977b, 1977c) that asymmetry of beat envelope due to the EOD waveform in this fish (*Eigenmannia*) being like a slant-clipped sine wave, is a necessary condition and therefore a usable clue (see Fig. 27-13). Both T and P units were found to distribute into those sensitive to slant clipping and those insensitive to slant clipping (harmonic content). It seems likely that this waveform clue is not as important normally as the clue from differential contamination of areas of the skin by the neighbor's electric field. The latter explanation permits ready prediction of behavior in the face of two or more neighbors with different EOD rates and experiments by Partridge and Heiligenberg (1980) have shown the fish behaves as predicted. Neural correlates of the operations behaviorally inferred are being found (Heiligenberg and Bastian 1980). The main finding is that sign sensitive cells like those seen by Scheich and Bullock (1974) and Scheich (1977c) do not require beat asymmetry if there is geometric asymmetry available as a clue. One of the lessons I believe nature is teaching us here is that just because you find an ingenious and adequate solution requiring sophisticated brain cell processing does not mean you should stop looking or that this is the only way the animal solves the problem. There may be other solutions brought out by special conditions, even seemingly unnatural conditions.

Gating of input at certain parts of the EOD cycle, which we saw in central neurons, is evident behaviorally in pulse species, and somewhat differently in different species (Bullock 1969, MacDonald and Larimer 1970, Westby 1975).

Phase sensitivity is indicated, not only with respect to time within a beat cycle (ca. 0.2-10 Hz) and to zero-crossing time or its equivalent in a mixture of sine waves (ca. 200-500 Hz), but also in respect to the phase of harmonics. This was shown by the response specific to the sign of a slant clipped sine wave (Scheich and Bullock 1974, Scheich 1977a, 1977b), It is also demonstrated in a behavioral discrimination task in *Hypopomus artedi* given two stimuli of the same spectral amplitude composition but different harmonic phases (Heiligenberg and Altes 1978).

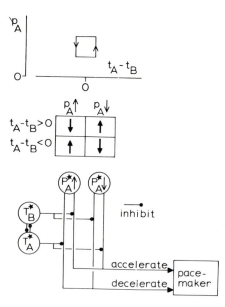

Figure 27-12. The spatial asymmetry or regional inequality scheme for detecting the sign of ΔF in the jamming avoidance response (JAR). The scheme uses known types of neurons in *Eigenmannia* and assumed connections among them and on the pacemaker. Given some smoothing to remove signs of the beat from the pacemaker, the scheme can account for the observed JAR when it is elicited in the artificially simplified condition that waveform asymmetry is removed by substituting sine waves for the natural EOD shape. The "own" EOD simulation is injected via one electrode in the stomach and the other at the tip of the tail, hence quasi-naturally and symmetrically. The "neighbor's" EOD simulation is injected via transverse or focal electrodes, hence asymmetrically and differentially contaminating the receptors of different regions (e.g., A and B). Control of PM depends on modulations of P unit activity, P_A, with beat amplitude, and differential arrival time, $t_A - t_B$, of T unit impulses from majority and minority populations of these receptors as phase (e.g., zero crossing) advances and retards during each beat cycle. Diagram at top shows how these two receptor types, P and T, can encode a succession of stimulus states defined by four combinations of amplitude and phase, at the corners of the square. These are four of the series of states repeated in each beat cycle, i.e., a normally circular or oval state plane is simplified artificially to a square, preserving the essentials of a closed figure with a sense of rotation diagnostic for the sign of the ΔF; counterclockwise means $+\Delta F$. The matrix, above, shows changes in pacemaker frequency (thick arrows) which result from increasing or decreasing P unit activity, $P_A \uparrow$ and $P_A \downarrow$, respectively and difference of arrival time, $t_A - t_B$, of T unit reponses from body surface areas A and B, respectively. The algorithm represented in this matrix is comparable to a double-pole double-throw switch mechanism in which the state of $t_A - t_B$ determines the effect of $P_A \uparrow$ and $P_A \downarrow$ on the pacemaker frequency. If ΔF ($F_{neighbor} - F_{own}$) is positive the upper left and lower right combination in the matrix are realized in alternation and the pacemaker is decelerated in both instances. If the ΔF is negative the upper right and the lower left combinations in the matrix are realized in alternation, and the pacemaker is accelerated in both instances. The diagram at the bottom suggests a minimal hardware representation of this algorithm in terms of higher order P and T unit connections. Unit $P_A \uparrow$ fires only for $P_A \uparrow$ and unit $P_A \downarrow$ fires only for $P_A \downarrow$. The mutual inhibition between T_B and T_A implies that T_B fires only for $t_A - t_B > 0$ and T_A fires only for $t_A - t_B < 0$. By virtue of the inhibitory connections, this network drives the pacemaker as indicated in the matrix above. (From Heiligenberg and Bastian 1980.)

Pulse form recognition extends to the fractional millisecond EODs of mormyrids (Hopkins 1981). The duration of the EOD pulse varies among 20 some species examined by Hopkins from approximately 0.25 to 4 msec; the peak of the power spectrum from 160 to 19 kHz and the sequence of head positive and tail positive phases varies as well (see Fig. 27-14). Behavioral experiments show that a given species can react to its species specific pulse, and discriminate against those of other species that live in the same stream. The tuberous receptors are tuned to high best frequencies —up to 18 kHz! The knollenorgans fire only with inward current and hence a complex EOD fires a distinctive sequence of left and right receptors according to the succession of phases, providing input suitable for discriminating some wave forms.

5 Summary

1. By analogy with the electric sense one is led to expect distinctions between a number of subsystems within the acoustic sense system, specialized for different roles, from hair cells to higher brain cells. Table 27-1 summarizes the known types of electro- and acoustic receptors and the variables among their properties that may be useful in distinguishing subclasses.

2. Electroreceptors are valuable as material for the study of hair cell mechanisms, including receptor potentials, ionic conductance changes, and transmission. Oscillatory receptor potentials can be more or less spontaneous, more or less damped, more or less fixed in frequency. A voltage sensitive Ca conductance and a Ca activated outward K channel have been demonstrated in certain species. These can be adjusted to control the balance between sensitivity and accommodation to achieve an optimal increment threshold. Both chemical and electrical transmission are found. Glutamate is implicated.

3. In respect to one of the measures, the tuning of the receptors to a "best" frequency (CF), tuberous receptors in some gymnotiforms can be regarded as directly relevant to the current controversy over whether there is a "second" filter, besides the mechanical one in the mammalian cochlea. Electroreceptors in these fish, without any mechanical step, have as sharp tuning (CF ca. 300 Hz) as cat cochlear fibers with the same CF. Clearly, a nonmechanical filter is available in this frequency range. It should be emphasized that the filter is likely to be the resultant of a set of parallel and sequential, passive and active properties, including an active, tuned ringing.

4. Electroreceptors manifest a variety of ways of encoding stimulus intensity; most of them should be sharply distinguished from the conventional mean-frequency-of-impulses code. Much evidence indicates that electroreceptors work in the time domain, even within a fraction of a cycle. It is suggested that the acoustic hair cells transduce and transmit and the nerve endings encode basically in the time domain.

5. Efferent control appears to be absent in all classes of electrical receptors, contrary to familiar acoustic receptors. Peripheral interaction between receptors such as lateral influence is slight or absent in electroreceptors, whereas some evidence of it is known in acoustic organs.

6. It is now clear that electroreception is a general possession of most agnathans, holocephalans, elasmobranchs, dipneustans, chondrosteans, crossopterygians, and polypteriforms, but not of holosteans or most teleosts. Among some 30 teleost orders, three have electroreception (apparently all species of siluriforms, gymnotiforms, and

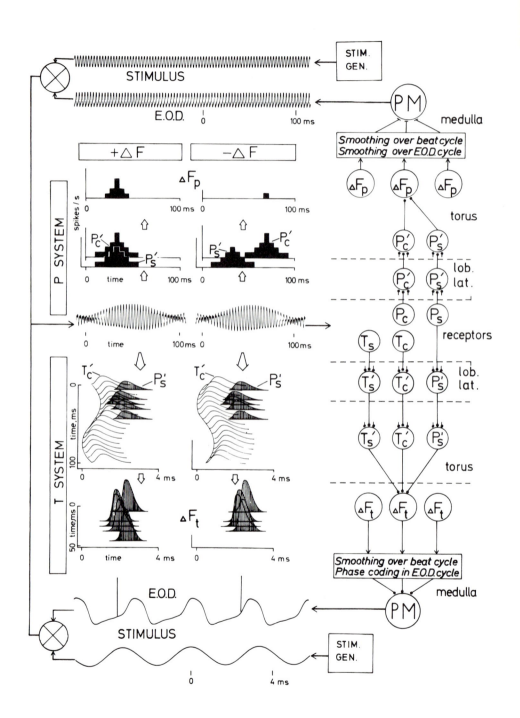

mormyriforms), and six have been shown to lack this modality (one or two species each of osteoglossiforms, anguilliforms, salmoniforms, cyrpiniforms, synbranchiforms, and perciforms). Thus, electroreception must have been independently evolved at least three if not four times. Ampullary receptors are so different in the nonteleosts and in the teleosts that they hardly deserve the same name and should be sharply distinguished as Lorenzinian and teleost ampullary organs, respectively. Tuberous organs in gymnotiforms and mormyriforms have some resemblances by convergent evolution but are unrelated and in many respects different.

7. A different nucleus in the octavolateral area of the medulla serves electroreception in the nonteleosts (the dorsal nucleus) from that in the electroreceptive teleosts (a special extension of the medial nucleus, sometimes called the posterior lateral line lobe). These nuclei with a cortex and a central nuclear portion, achieve histological differen-

◀ Figure 27-13. Scheme for detecting the sign of ΔF in the jamming avoidance response (JAR) based on the temporal asymmetry of the EOD waveform. The scheme uses known types of neurons in *Eigenmannia* and presumed connections between them. If the influence on the pacemaker (PM) is as here supposed the scheme can account for the observed JAR when that is elicited in the artifically simplified condition that geometric clues to the distinction between "own" and "neighbor's" EOD simulations are removed, by injecting them via the same electrodes, but waveform clues are available by simulating the natural shape of the fish's own EOD. This generates asymmetry of the beating and permits the neurons shown to solve the crucial problem, to recognize the sign of the ΔF. Of the seven orders of neurons known, two are omitted for simplicity, the interneurons just downstream from the PM and the electric organ motor neurons in the spinal cord. The EOD follows PM faithfully, through these two neurons and three synapses. The EOD is commanded by the PM nucleus, physiologically a single unit. The EOD, shown on a faster (bottom) and slower (top) time scale, sums with the stimulus—here an artificial sine wave for simplicity, resulting in a beat pattern in the water (center). This pattern is a mirror image for the + and $-\Delta F$ cases. The figure shows ca. 10 Hz ΔF and $> 50\%$ modulation depth for legibility of the wave forms but 3 Hz, with over a hundred EOD cycles per beat, is more effective and < 0.2 Hz is weakly effective; $< 1\%$ of the stimulus amplitude shown is still above threshold. Peripheral afferent fibers of the two coding types, T and P, fire in synchrony with the EOD. Higher order units (T' and P') preserve and enhance the difference: T' units code only by EOD phase (lower); P' code by rate of firing. T_c and P_c units show their clipped wave sensitivity by firing differently in the beat cycle for + and $-\Delta F$, as shown in rate histograms (above) over a whole beat, and time-of-firing displays (below), triggered by a fixed point in successive EOD cycles. Summation having a long time constant (seconds; P system) or a short time constant (milliseconds; T system) is assumed to occur with the known reference units P_s and T_s that do not have a ΔF sign discrimination or a sensitivity to clipping the sine wave. These convergences would explain the observed ΔF decoders, ΔF_p and ΔF_t, which are presumed to act, with others, on the PM, to excite (spherical endings) or inhibit (flat endings) its pace. The time scales are referred to arbitrary reference points in the beat cycle (above, and ordinate below) or the EOD cycle (abscissa below). The P and the T system are not equally important; they are both shown because the neuron types are found, but only the P system in principle, is essential under these conditions. The T system alone is inadequate. (From Scheich and Bullock 1974.)

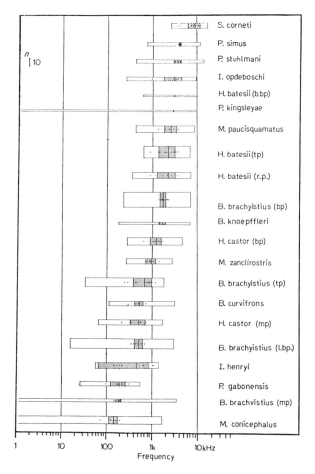

Figure 27-14. Species differences in bandwidths of EODs for mormyrids from the Ivindo region of Gabon. Each dot represents the peak spectral frequency for one individual fish's EOD, the shaded area represents the range of peak frequencies one standard deviation on either side of the mean peak frequency for the population sampled, and the vertical ends of the box show the average bandwidth 10 dB down from the peak frequency. The height of each box corresponds to the number of fish sampled. (From Hopkins 1981.)

tiation, equivalent to some mammalian cochlear nuclei. A variety of cell types, from two to five or more on each of several physiological criteria, are found in the medullary nuclei. They show evidence of somatotopic mapping, segregation of receptor types, lateral interaction, extraction of stimulus features, divergence into more categories than in the afferent fibers, and selective descending influence from higher brain levels. There are marked similarities to medullary centers for acoustic analysis in higher vertebrates but too little is known of the integrative functions of acoustic centers in fishes.

8. The principal midbrain center for electroreception is the torus semicircularis (TS) and is shared, with acoustic and lateral line mechanoreceptive input, in segregated divisions. The electrosensory TS achieves high differentiation, e.g., 12 laminae, in some

species. Afferent and efferent conditions and specializations of neurons and processing are known in some detail for the electric modality. Neurons are differentiated in respect to best position, orientation, polarity, frequency and movement of electric field stimuli, speed of afferent pathway, microsecond precision of phase of response, influence of corollary discharge in either inhibiting or facilitating particular subsystems for time slots of a few milliseconds, comparison of body regions, integrating phase sensitive and amplitude sensitive subsystems, and sensitivity to harmonics and their phase. In some parts of TS electrolocation of objects is apparently unimportant whereas analysis is devoted to parameters significant in social signals. Information on acoustic processing in teleosts for comparisons is relatively meager.

9. Evoked potentials averaged from TS or lateral line lobe have proved to be the most reliable and efficient means of estimating whether an animal, or more accurately, its brain, has electric or acoustic sensibilities, how sensitive it is, and what forms of stimuli it can discriminate. Even negative findings although inconclusive, can with experience have substantial weight.

10. Modalities and submodalities do not disappear at higher levels by converging into a general multimodal area but are preserved separately all the way to the telencephalon. Each modality indeed multiplies into many more types of units at midbrain than at medullary level and in medulla than at first-order level. Convergence and multimodal interaction occur in medulla, midbrain, and forebrain as collateral processes in special places but are difficult to find. Possibly complex, natural stimuli are more important than the simple stimuli usually used in the laboratory.

11. The cerebellum is generally large in electroreceptive species and achieves in some an unprecedented enormity and histological specialization, e.g., Purkinje cell dendrites that pass, all parallel, unbranched, equal in length, and equally spaced, in planes, the full depth of the molecular layer. This elaboration cannot on present evidence be attributed primarily to electroreception. However, electroreceptor input has a strong representation in specific regions. Purkinje units can have small receptive fields with excitatory center and inhibitory surround, require movement with preferred direction, and low-pass filter to smooth out beats above 10 Hz. Three subclasses of electroreceptors in mormyrids have three areas of special representation in the valvular cortex. Some units are multimodal, e.g., electro- and proprioceptive or electroreceptive and visual. Reversible block by cooling shows deficit in sensory processing without motor deficit. Some species habitually rest motionless, not even maintaining posture; yet the evidence suggests continued functioning of the cerebellum as an input processor of information about the nearby space, objects in it, the position of the parts of the body in it, whether movements are being commanded or not, like an army intelligence service.

Almost nothing is known of acoustic influence in the cerebellum in fishes.

12. The telencephalon shows both electric and acoustic influence, so far studied chiefly with evoked potentials. In sharks the regions concerned appear to overlap but have separate maxima. Remarks in paragraph 10, above, probably apply here, too.

13. Behavioral experiments allow the inference that the brain in electroreceptive species creates the equivalent of an electrical image of nearby space, with objects, walls and floors. In gymnotiform wave species there may be the equivalent of an electrical image of the fish's own field and those of neighbors but the psychophysics does not require this; it would be satisfied with simple comparisons of major parts of the

skin, less contaminated with a neighbor's EOD fields and minor parts more contaminated. The fish can solve the problem of whether a neighbor is above or below its own EOD rate by comparing regions with respect to the directions of simultaneous change in amplitude (beating at the difference frequency) and phase (of zero-crossing or equivalent). If regional differences are not available, it can still solve the problem using the polarity of asymmetry of beat envelopes, due to the EOD waveform with its phase-specific harmonics. Neurons have been found in the TS that perform the essential steps in the beat asymmetry solution and most of the essential steps in the spatial asymmetry solution. In mormyrids the fish's problems are recognition of its own EOD from that of its conspecifics and those from EODs of other species. The first problem is solved with a corollary discharge to the EOD command, not found in gymnotiforms. The second is solved by the EOD waveform, although the EODs are as short as 0.2 msec, with peak power above 5 kHz. Behaviorally, the fish discriminates waveforms of different species. Unit receptors are tuned to CFs up to 18 kHz, with regional differences in areas of the skin, and provide information for sequence analysis of the waveform.

14. It remains for the future to use neurothological analysis in fishes on the performance capacities of the most specialized species in acoustic sound source imaging, especially with respect to distance and movement, and on waveform discrimination when natural sounds are characterized by mixtures of frequencies or time structure.

References

Akutsu, Y., Obara, S.: Calcium dependent receptor potential of the electroreceptor of marine catfish. Proc. Jpn. Acad. 50, 247-251 (1974).

Andrianov, Yu. N., Broun, G. R.: Perception of the magnetic field by the electroreceptor system in fishes. Neurophysiology 7, 338-339 (1976).

Andrianov, G. N., Ilyinsky, O. B.: Some functional properties of central neurons connected with the lateral-line organs of the catfish (Ictalurus nebulosus). J. Comp. Physiol. 86, 365-376 (1973).

Andrianov, Yu, N., Volkova, N. K.: Some morphological and functional properties of the lateral line system of the dwarf catfish. Neurophysiology 7, 160-164 (1976).

Andrianov, G. N., Brown, H. R., Ilyinsky, O. B.: Responses of central neurons to electrical and magnetic stimuli of the ampullae of Lorenzini in the Black Sea skate. J. Comp. Physiol. 93, 287-299 (1974).

Baker, C. L., Jr.: Jamming avoidance behavior in gymnotoid electric fish with pulse-type discharges: sensory encoding for a temporal pattern discrimination. J. Comp. Physiol. 136, 165-181 (1980).

Bastian, J.: Electrosensory input to the corpus cerebelli of the high frequency electric fish Eigenmannia virescens. J. Comp. Physiol. 90, 1-24 (1974).

Bastian, J.: Receptive fields of cerebellar cells receiving exteroceptive input in a gymnotid fish. J. Neurophysiol. 38, 285-300 (1975).

Bastian, J.: The range of electrolocation: A comparison of electroreceptor responses and the responses of cerebellar neurons in a gymnotid fish. J. Comp. Physiol. 108, 193-210 (1976).

Bastian, J.: Variations in the frequency response of electroreceptors dependent on receptor location in weakly electric fish (Gymnotoidei) with a pulse discharge. J. Comp. Physiol. 121, 53-64 (1977).

Behrend, K.: Processing information carried in a high frequency wave: Properties of cerebellar units in the high frequency electric fish. J. Comp. Physiol. 118, 357-371 (1977).

Bell, C. C.: Central nervous system physiology of electroreception, A review. J. Physiol. Paris 75, 361-379 (1979).

Bell, C. C., Russell, C. J.: Termination of electroreceptor and mechanical lateral line afferents in the mormyrid acousticolateral area. J. Comp. Physiol. 182, 367-382 (1978).

Bennett, M. V. L.: Mechanisms of electroreception. In: Lateral Line Detectors. Cahn, P. (ed.). Bloomington: Indiana Univ. Press, 1967, pp. 313-393.

Bennett, M. V. L.: Comparative physiology: Electric organs. Annu. Rev. Physiol. 32, 471-528 (1970).

Bennett, M. V. L.: Electrolocation in fish. Ann. N.Y. Acad. Sci. 188, 242-269 (1971a).

Bennett, M. V. L.: Electric organs. In: Fish Physiology, Vol. 5. Hoar, W. S., Randall, D. J. (eds.). New York: Academic Press, 1971b.

Bennett, M. V. L.: Mechanism of afferent discharge from electroreceptors: implications for acoustic reception. In: Evoked Electrical Activity in the Auditory Nervous System. Naunton, R. F., Fernandez, C. (eds.). New York: Academic Press, 1978.

Bennett, M. V. L., Clusin, W. T.: Transduction at electroreceptors: Origins of sensitivity. In: Membrane Transduction Mechanisms. Cone, R. A., Dowling, J. E. (eds.). New York: Raven Press, 1979.

Bennett, M. V. L., Steinbach, A. B.: Influence of electric organ control system on electrosensory afferent pathways in mormyrids. In: Neurobiology of Cerebellar Evolution and Development. Llinas, R. (ed.). Chicago: American Medical Association (1969).

Bodznick, D. A., Northcutt, R. G.: Electroreception in lampreys. Evidence that the earliest vertebrates were electroreceptive. Science, in press.

Bombardieri, R. A., Feng, A. S.: Deficit in object detection (electrolocation) following interruption of cerebellar function in the weakly electric fish *Apteronotus albifrons*. Brain Res. 130, 343-347 (1977).

Bullock, T. H.: Biological sensors. In: Vistas in Science. Albuquerque: Univ. of New Mexico Press, 1968.

Bullock, T. H.: Species differences in effect of electroreceptor input on electric organ pacemakers and other aspects of behavior in electric fish. Brain Behav. Evol. 2, 85-118 (1969).

Bullock, T. H.: Processing of ampullary input in the brain: Comparison of sensitivity and evoked responses among elasmobranch and siluriform fishes. J. Physiol. Paris 75, 397-407 (1979).

Bullock, T. H.: Electroreception. Ann. Rev. Neurosci. Vol. 5 (1981a) (in press).

Bullock, T. H.: Physiology of the tectum mesencephali in elasmobranchs. In: Comparative Neurology of the Optic Tectum. Vanegas, H. (ed.). New York: Plenum Press (1981b) (in press).

Bullock, T. H., Hamstra, R. H., Jr., Scheich, H.: The jamming avoidance response of high frequency electric fish. I. General features. J. Comp. Physiol. 77, 1-22 (1972a).

Bullock, T. H., Hamstra, R. H., Jr., Scheich, H.: The jamming avoidance response of high frequency electric fish. II. Quantitative aspects. J. Comp. Physiol. 77, 23-48 (1972b).

Capranica, R. R.: Morphology and physiology of the auditory system. In: Frog Neurobiology. Llinas, R., Precht, W. (eds.). Berlin: Springer-Verlag, 1976, pp. 537-575.

Clusin, W. T., Bennett, M. V. L.: Calcium-activated conductance in skate electroreceptors: Current clamp experiments. J. Gen. Physiol. 69, 121-143 (1977a).

Clusin, W. T., Bennett, M. V. L.: Calcium-activated conductance in skate electrorecep-
    tors: Voltage clamp experiments. J. Gen. Physiol. 69, 145-182 (1977b).
Clusin, W. T., Bennett, M. V. L.: The oscillatory responses of skate electroreceptors to
    small voltage stimuli. J. Gen. Physiol. 73, 685-702 (1979a).
Clusin, W. T., Bennett, M. V. L.: The ionic basis of oscillatory responses of skate elec-
    troreceptors. J. Gen. Physiol. 73, 703-723 (1979b).
Cole, K. S.: Membranes, Ions and Impulses. Berkeley: Univ. of Calif. Press, 1968.
Enger, P. S., Libouban, S., Szabo, T.: Fast conducting electrosensory pathway in the
    mormyrid fish Gnathonemus petersii. Neurosci. Lett. 2, 127-133 (1976a).
Enger, P. S., Lubouban, S., Szabo, T.: Rhombo-mesencephalic connections in the fast
    conducting electrosensory system of the mormyrid fish, Gnathonemus petersii
    An HRP study. Neurosci. Lett. 3, 239-243 (1976b).
Enger, P. S., Szabo, T.: Activity of central neurons involved in electroreception in
    some weakly electric fish (Gymnotidae). J. Neurophysiol. 28, 800-818 (1965).
Feng, A. S.: The role of the electrosensory system in postural control in the weakly
    electric fish Eigenmannia virescens. J. Neurobiol. 8, 429-437 (1977).
Feng, A. S., Narins, P. M., Capranica, R. R.: Three populations of primary auditory
    fibers in the bullfrog (Rana catesbeiana): Their peripheral origins and frequency
    sensitivities. J. Comp. Physiol. 100, 221-229 (1975).
Fessard, A., Szabo, T.: Mise en évidence d'un récepteur sensible à l'électricité dans la
    peau des Mormyres. C.R. Acad. Sci. Paris 255, 1859-1860 (1961).
Fessard, A., Szabo, T.: Effets des variations de température sur l'activité de certains
    récepteurs des Mormyres. C.R. Acad. Sci. Paris 254, 2084-2085 (1962).
Fettiplace, R., Crawford, A. D.: The coding of sound pressure and frequency in
    cochlear hair cells of the terrapin. Proc. R. Soc. London Ser. B 203, 209-218 (1978).
Fields, R. D., Lange, G. D.: Electroreception in the ratfish (Hydrolagus colliei).
    Science 207, 547-548 (1980).
Finger, T. E., Bell, C. C., Russell, C. J.: Electrosensory pathways to the valvula cere-
    belli in mormyrid fish. Exp. Brain Res. (1981) (in press).
Furukawa, T., Ishii, Y.: Neurophysiological studies on hearing in goldfish. J. Neuro-
    physiol. 30, 1377-1403 (1967).
Harder, W.: Nachweis aktiver (elektrischer) Ortung bei Mormyridae (Teleostei, Pisces).
    Z. Tierpsychol. 30, 94-102 (1972).
Heiligenberg, W.: Electrolocation of objects in the electric fish Eigenmannia (Rhamphi-
    chthyidae, Gymnotoidei). J. Comp. Physiol. 87, 137-164 (1973).
Heiligenberg, W.: Electroreception and jamming avoidance in a Hypopygus (Rhamyphi-
    chthyidae, Gymnotoidei) an electric fish with pulse-type discharge. J. Comp.
    Physiol. 91, 223-240 (1974).
Heiligenberg, W.: Principles of electrolocation and jamming avoidance in electric fish.
    A neuroethological approach. In: Studies of Brain Function. Braitenberg, V., et al.
    (eds.). New York: Springer-Verlag, 1977.
Heiligenberg, W., Altes, R. A.: Phase sensitivity in electroreception. Science 199, 1001-
    1004 (1978).
Heiligenberg, W., Bastian, J.: The control of Eigenmannia's pacemaker by distributed
    evaluation of electroreceptive afferences. J. Comp. Physiol. 136, 113-133 (1980).
Heiligenberg, W., Baker, C., Bastian, J.: The jamming avoidance response in gymnotoid
    pulse-species: A mechanism to minimize the probability of pulse-train coincidence.
    J. Comp. Physiol. 124, 211-224 (1978).
Heiligenberg, W., Baker, C., Matsubara, J.: The jamming avoidance response in Eigen-
    mannia revisited: The structure of a neuronal democracy. J. Comp. Physiol. 127,
    267-286 (1978).

Hoagland, H.: Electrical responses from lateral-line nerves of fishes. III. J. Gen. Physiol. 17, 77-82 (1933).

Hoagland, H.: Pacemakers in Relation to Aspects of Behavior. New York: Macmillan, 1935.

Honrubia, V., Strelioff, D., Ward, P. H.: Quantitative studies of cochlear potentials along the scale media of the guinea pig. J. Acoust. Soc. Am. 64, 600-609 (1973).

Hopkins, C. D.: Stimulus filtering and electroreception: Tuberous electroreceptors in three species of gymnotoid fish. J. Comp. Physiol. 111, 171-207 (1976).

Hopkins, C. D.: Evolution of electric communication channels of mormyrids. Behav. Ecol. Sociobiol. 7, 1-13 (1980).

Hudspeth, A. J., Corey, D. P.: Sensitivity, polarity, and conductance change in the response of vertebrate hair cells to controlled mechanical stimuli. Proc. Nat. Acad. Sci. 74, 2407-2411 (1977).

Il'inskii, O. B., Enin, L. D., Volkova, N. K.: Evoked potentials of the medulla oblongata of the skate in response to stimulation of lateral line nerves. Neurophysiology 3, 213-218 (1972).

Kalmijn, A. J.: The second derivative or potential mode? A critical evaluation of Lissmann and Machin's theory concerning the mode of operation in the electroreceptors of *Gymnarchus niloticus* and similar fish. Scripps Inst. Oceanogr. Ref. Ser. Contr. no. 72-69, 1-32 (1972a).

Kalmijn, A. J.: Bioelectric fields in sea water and the function of the ampullae of Lorenzini in elasmobranch fishes. Scripps Inst. Oceanogr. Ref. Ser. Contr. no. 72-83, 1-21 (1972b).

Katsuki, Y., Onada, N.: The lateral-line organ of fish as a chemoreceptor. In: Responses of Fish to Environmental Changes. Chavin, W. (ed.). Springfield, Illinois: Charles C Thomas, 1973.

Knudsen, E. I.: Midbrain responses to electroreceptive input in catfish: Evidence of orientation preferences and somatotopic organization. J. Comp. Physiol. 106, 51-67 (1976a).

Knudsen, E. I.: Midbrain units in catfish. Response properties to electroreceptive input. J. Comp. Physiol. 109, 315-335 (1976b).

Knudsen, E. I.: Distinct auditory and lateral line nuclei in the midbrain of catfishes. J. Comp. Neurol. 173, 417-432 (1977).

Knudsen, E. I.: Functional organization in the electroreceptive midbrain of the catfish. J. Neurophysiol. 41, 350-364 (1978).

Langner, G., Scheich, H.: Active phase coupling in electric fish: Behavioral control with microsecond precision. J. Comp. Physiol. 128, 235-240 (1978).

Lissmann, H. W.: On the function and evolution of electric organs in fish. J. Exp. Biol. 35, 156-191 (1958).

Lissmann, H. W., Machin, K. E.: The mechanism of object location in *Gymnarchus niloticus* and similar fish. J. Exp. Biol. 35, 451-486 (1958).

Lissmann, H. W., Mullinger, A. M.: Organization of ampullary electric receptors in Gymnotidae (Pisces). Proc. Roy. Soc. London Ser. B 169, 345-378 (1968).

MacDonald, J. A., Larmier, J. L.: Phase-sensitivity of *Gymnotus carapo* in low-amplitude electrical stimuli. Z. Vergl. Physiol. 70, 322-334 (1970).

Machin, K. E., Lissmann, H. W.: The mode of operation of the electric receptors in *Gymnarchus niloticus*. J. Exp. Biol. 37, 801-811 (1960).

Maler, L.: The posterior lateral line lobe of a mormyrid fish—A Golgi study. J. Comp. Neurol. 152, 281-298 (1973).

Maler, L., Karten, H. J., Bennett, M. V. L.: The central connections of the posterior lateral line nerve of *Gnathonemus petersii*. J. Comp. Neurol. 151, 57-66 (1973a).

Maler, L., Karten, H. J., Bennett, M. V. L.: The central connections of the anterior lateral line nerve of *Gnathonemus petersii*. J. Comp. Neurol. 151, 67-84 (1973b).

Marmarelis, P. Z., Marmarelis, V. Z.: Analysis of Physiological Systems: The White Noise Approach. New York: Plenum Press, 1978.

McCreery, D. B.: Two types of electroreceptive lateral lemniscal neurons of the lateral line lobe of the catfish *Ictalurus nebulosus;* Connections from the lateral line nerve and steady-state frequency response characteristics. J. Comp. Physiol. 113, 317-340 (1977a).

McCreery, D. B.: Spatial organization of receptive fields of lateral lemniscus neurons of the lateral line lobe of catfish *Ictalurus nebulosus*. J. Comp. Physiol. 113, 341-353 (1977b).

Meyer, D. L., Heiligenberg, W., Bullock, T. H.: The ventral substrate response. A new postural control mechanism in fishes. J. Comp. Physiol. 109, 59-68 (1976).

Meyer, D. L., Becker, R., Graf, W.: The ventral substrate response of fishes. Comparative investigation of the VSR about the roll and the pitch axis. J. Comp. Physiol. 117, 209-217 (1977).

Nieuwenhuys, R., Nicholson, C.: Aspects of the histology of the cerebellum of mormyrid fishes. In: Neurobiology of Cerebellar Evolution and Development. Llinas, R. (ed.). Chicago: American Medical Association, 1969.

Northcutt, R. G., Bodznick, D. A., Bullock, T. H.: Most non-teleost fishes have electroreception. XXVIII Internat. Cong. Physiol. Sci. Budapest (Abstr.) (1980).

Obara, S.: Mechanism of electroreception in ampullae of Lorenzini of the marine catfish *Plotosus*. In: Electrobiology of Nerve, Synapse, and Muscle. Reuben, J. P., Purpura, D. P., Bennett, M. V. L., Kandel, E. R. (eds.). New York: Raven Press, 1976.

Obara, S., Bennett, M. V. L.: Mode of operation of ampullae of Lorenzini of the skate, *Raja*. J. Gen. Physiol. 60, 534-557 (1972).

Obara, S., Sugawara, Y.: Contribution of Ca to the electroreceptor mechanism in *Plotosus* ampullae. J. Physiol. Paris 75, 335-340 (1979).

Partridge, B. L., Heiligenberg, W.: Three's a crowd? Predicting *Eigenmannia*'s reponses to multiple jamming. J. Comp. Physiol. 136, 153-164 (1980).

Paul, D. H., Roberts, B. L.: Studies on a primitive cerebellar cortex. III. The projection of the anterior lateral-line nerve to the lateral-line lobes of the dogfish brain. Proc. R. Soc. London Ser. B 195, 479-496 (1977).

Pimentel-Souza, F.: Regulation of the electroreceptor potential frequency by the electric discharge of *Gnathonemus petersii*. J. Comp. Physiol. 111, 115-125 (1976).

Platt, C. J., Bullock, T. H., Cźeh, G., Kovacević, N., Konjević, Dj., Gojković, M.: Comparison of electroreceptor, mechanoreceptor and optic evoked potentials in the brain of some rays and sharks. J. Comp. Physiol. 95, 323-355 (1974).

Rethelyi, M., Szabo, T.: A particular nucleus in the mesencephalon of a weakly electric fish: *Gymnotus carapo* (Gymnotidae). Exp. Brain Res. 17, 229-241 (1973a).

Rethelyi, M., Szabo, T.: Neurohistological analysis of the lateral lobe in electric fish, *Gymnotus carapo* (Gymnotidae). Exp. Brain Res. 18, 323-339 (1973b).

Rodieck, R. W.: Visual pathways. Annu. Rev. Neurosci. 2, 193-226 (1979).

Roth, A.: Central neurons involved in the electroreception of the catfish *Kryptopterus*. J. Comp. Physiol. 100, 135-146 (1975).

Russell, I. J., Sellick, P. M.: Measurement of potassium and chloride ion concentrations in the cupulae of the lateral lines of *Xenopus laevis*. J. Physiol. 257, 245-255 (1976).

Sand, A.: The function of the ampullae of Lorenzini, with some observations on the effect of temperature on sensory rhythms. Proc. R. Soc. London Ser. B 125, 524-553 (1938).

Scheich, H.: Neuronal analysis of wave form in the time domain: Midbrain units in electric fish during social behavior. Science 185, 365-367 (1974).

Scheich, H.: Neural basis of communication in the high frequency electric fish *Eigenmannia virescens* (jamming avoidance response). I. Open loop experiments and the time domain concept of signal analysis. J. Comp. Physiol. 113, 181-206 (1977a).

Scheich, H.: Neural basis of communication in the high frequency electric fish *Eigenmannia virescens* (jamming avoidance response). II. Jammed electroreceptor neurons in the lateral line nerve. J. Comp. Physiol. 113, 207-227 (1977b).

Scheich, H.: Neural basis of communication in the high frequency electric fish *Eigenmannia virescens* (jamming avoidance response). III. Central integration in the sensory pathway and control of the pacemaker. J. Comp. Physiol. 113, 229-255 (1977c).

Scheich, H., Bullock, T. H.: The detection of electric fields from electric organs. In: Handbook of Sensory Physiology III/3. Fessard, A. (ed.). New York: Springer-Verlag, 1974.

Scheich, H., Maler, L.: Laminar organization of the torus semicircularis related to the input from two types of electroreceptors. In: Afferent and Intrinsic Organization of Laminated Structures in the Brain. Exp. Brain Research/Suppl. 1. Creutzfeldt, O. (ed.). New York: Springer-Verlag, 1976.

Scheich, H., Bullock, T. H., Hamstra, R. J., Jr.: Coding properties of two classes of afferent nerve fibers: High-frequency electroreceptors in the electric fish, *Eigenmannia*. J. Neurophysiol. 36, 39-60 (1973).

Schlegel, P. A.: Activities of rhombencephalic units in mormyrid fish. Exp. Brain Res. 19, 300-314 (1974).

Sotelo, C., Rethelyi, M., Szabo, T.: Morphological correlates for electrotonic transmission in the magnocellular mescephalic nucleus of the weakly electric fish *Gymnotus carapo*. J. Neurocytol. 4, 587-607 (1975).

Späth, M., Lehmann, B.: Inhibitory influence of the connecting strands in fish. Naturwissenschaften 9, 435-436 (1976).

Stendell, W.: Morphologische Studien an Mormyriden. Verh. Dtsch. Zool. Ges. 24, 254-261 (1914).

Szabo, T.: The activity of cutaneous sensory organs in *Gymnarchus niloticus*. Life Sci. 7, 285-286 (1962).

Szabo, T.: Anatomy of the specialized lateral line organs of electroreception. In: Handbook of Sensory Physiology III/3. Fessard, A. (ed.). New York: Springer-Verlag, 1974.

Szabo, T., Ravaille, M.: Synaptic structure of the lateral line lobe nucleus in mormyrid fish. Neurosci. Lett. 2, No. 3, 121-127 (1976).

Szabo, T., Sakata, H., Ravaille, M.: An electrotonically coupled pathway in the central neuron system of some teleost fish, Gymnotidae and Mormyridae. Brain Res. 95, 459-474 (1975).

Szabo, T., Enger, P. S., Libouban, S.: Electrosensory systems in the mormyrid fish, *Gnathonemus petersii:* Special emphasis on the fast conducting pathway. J. Physiol. Paris 75, 409-420 (1979).

Teeter, J. H., Szamier, R. B., Bennett, M. V. L.: Ampullary electroreceptors in the sturgeon *Scaphirhynchus platorynchus* (Rafinesque). J. Comp. Physiol. 138, 213-223 (1980).

Turner, R. G.: Physiology and bioacoustics in reptiles. In: Comparative Studies of Hearing in Vertebrates. Popper, A. N., Fay, R. R. (eds.). New York: Springer-Verlag, 1980, pp. 205-237.

Viancour, T. A.: Review of electroreceptor and peripheral electrosensory system physiology. Proc. XXVII Internat. Union Physiol. Sciences, Paris 13, 656 (1977).

Viancour, T. A.: Electroreceptors of a weakly electric fish. I. Characterization of tuberous receptor organ tuning. J. Comp. Physiol. 133, 317-327 (1979a).

Viancour, T. A.: Electroreceptors of a weakly electric fish. II. Individually tuned receptor oscillations. J. Comp. Physiol. 133, 328-339 (1979b).

Viancour, T. A.: Peripheral electrosense physiology: A review of recent findings. J. Physiol. Paris 75, 321-333 (1979c).

Westby, G. W. M.: Has the latency dependent response of *Gymnotus carapo* to discharge-triggered stimuli a bearing on electric fish communication? J. Comp. Physiol. 96, 307-341 (1975).

Yoshioka, T., Asanuma, A., Yanagisawa, K., Katsuki, Y.: The chemical receptive mechanism in the lateral-line organ. Jpn. J. Physiol. 28, 557-567 (1978).

Zipser, B.: The electrosensory system of mormyrids. Ph.D. Thesis, Yeshiva University, New York, 1971.

Zipser, B., Bennett, M. V. L.: Responses of cells of posterior lateral line lobe to activation of electroreceptors in a mormyrid fish. J. Neurophysiol. 39, 693-712 (1976a).

Zipser, B., Bennett, M. V. L.: Interaction of electrosensory and electromotor signals in lateral line lobe of a mormyrid fish. J. Neurophysiol. 39, 713-721 (1976b).

Discussion

PLATT: Is the electric field that an electric fish produces under normal circumstances bilaterally symmetrical? Since we have talked much about directional detection at this meeting, I wonder if there is any evidence in electric fish for something like a symmetry detector. Such a detector would give information from the two sides that could be compared.

BULLOCK: In gymnotiforms we believe the EOD is bilaterally symmetrical. A recent discovery shows that mormyriforms can discharge alternately their left and right electric organs. Even in the symmetrical case, if there is a nearby fish, not precisely coaxial with the first, the neighbor will produce a field that is asymmetrical with respect to the receptors of the first. The receiver will get a different proportion of receptors stimulated by entering currents as compared to those stimulated by exiting currents. currents.

PLATT: Are there individual neurons that can make that kind of discrimination?

BULLOCK: Yes, there are such known, especially in the midbrain.

BELL: From our anatomical studies, the abundance of coupling across the sides is striking. the posterior lobe seems to have bilateral connections to the mesencephalon.

In electroreceptive organs, there seems to be very sharp tuning in the frequency domain, yet there is evidence for considerable wave form analysis. Is there a conflict in this?

BULLOCK: Perhaps not. Tuning is not so sharp in the region of the peak, over a range of, say, 25 or 50 Hz. Sensitivity does not fall off much in this range on either side of the peak. There is much complex analysis that the fish must do in the range of from a fraction of a Hz up to 10 Hz away, so it cannot do this by the peripheral sensory filters.

BOSTON: You show different receptor cells and a lot of variability in the kinocilium and microvilli structures. Do you have any idea of the role of this variability?

BULLOCK: No, I don't. It is especially interesting when you see microvilli in these curious evaginations.

Chapter 28

Retrospect and Prospect—Listening through a Wet Filter

WILLIAM N. TAVOLGA*

1 Retrospect

The knowledge that fishes and other underwater denizens make sounds goes back to prehistory, and many reviews of early reports are available (see Tavolga 1977a for example). The logical corollary to this, i.e., that fishes can hear these sounds, is an idea of only recent origin. Weber (see Tavolga 1976a) presented his morphological evidence as far back as 1820, but it took more than a hundred years and the work of several major research biologists to show that Weber was indeed correct in his surmise and in his comparisons between fish and mammalian inner ears. As late as the 1920s, scientists were still arguing as to whether fish could hear anything at all. Some of those considered to be among the giants of modern biology have contributed to the study of this problem, e.g., Karl von Frisch, G. H. Parker, and Sven Dijkgraaf (see Tavolga 1976a).

Let me wax autobiographical for a bit. Back in 1954, I was observing the development and reproductive behavior in a goby, *Bathygobius soporator*, in Marineland, Florida. I showed some of the antics of these little fish to a colleague, Edward S. (Ted) Baylor. He was a great gadgeteer, among his other talents, and he asked if I ever considered the possibility that these fish make any sounds. This idea had never occurred to me, in spite of having seen Marie Poland Fish's recently published pioneering monograph on fish sounds (Fish 1954). We then created our first hydrophone: a cheap microphone waterproofed by a condom. An ancient amplifier (with tremendous 60 Hz hum) and a half of a pair of still more ancient earphones served as the rest of the gear. It turned out that the gobies did indeed make sounds—only the males, and only during courtship (Tavolga 1956, 1958). Subsequently, I found that several other species of tidal zone fishes (gobies and blennies) also emitted courtship sounds. I remember that I suspected the little *Gobiosoma bosci* could be a sound producer, but my equipment was not nearly sensitive enough. I was delighted with the paper by Mok (Chapter 22), confirming my long-standing suspicions.

*Mote Marine Laboratory, 1600 City Island Park, Sarasota, Florida 33577.

In the gobies the correlation between sound production and behavior was fairly clear, but still their precise role was to be investigated, and the relationship of sounds to behavior in most other fish species was only vaguely known at that time. The use of playback experiments was indicated, and immediately the problem of hearing in fishes appeared. How loud should the playback be? How hi- should the -fi be? In the 1950s information on fish hearing was scarce indeed. Actual sound level values and quantified thresholds were rarer still. A short introduction of a few pages sufficed to serve as an exhaustive review of the entire literature on hearing in fishes (Kleerekoper and Chagnon 1954). About the only paper that presented any psychophysical data was the now classic study of Poggendorf (1952) on the freshwater catfish, *Ictalurus nebulosus.*

My continuing research on fish behavior then took me to the Lerner Marine Laboratory on Bimini, in the Bahamas. That fine research institution, now extinct, was a field station of the American Museum of Natural History. It was here that I met Dr. Jerome Wodinsky, who was then investigating learning in various marine animals, and, in particular, the process of avoidance conditioning in fishes. I was fascinated by his little shuttle boxes, and the nice clear response the fish provided as they swam over the center barriers to avoid an electric shock. It occurred to us that his conditioning technique could be a fine way of getting at the problem I had. In 1962, Wodinsky and I began a productive collaboration in which we used this technique to obtain audiograms of as many species of fish as proved to be amenable to avoidance conditioning.

Even by the standards of that time, our equipment and techniques were primitive. The fish were trained and tested in aquarium tanks in which the speakers were set underwater inside the center hurdle (Tavolga and Wodinsky 1963). For observation, we used a small mirror, mounted several feet above the tank, and set so the operator could peer up and see what the fish was doing. The control equipment consisted of two telegraph keys and a stopwatch. One key turned on a sound from a sine-wave oscillator, the other would be tapped briefly and provided an electric shock to the fish from a variable line transformer. Considering the high current drawn by the sea water and the generally haywire circuitry we used, to this day, I don't understand how we survived without getting shocked ourselves. Eventually, we became automated. An electrical stopwatch with appropriate switches turned on the sound and, after a pre-selected time, produced shock pulses to the fish. All the operator had to do was to start the trial, watch the fish via the mirror, and terminate the trial. A hand stopwatch timed an intertrial period during which the operator recorded any other activities of the fish, such as intertrial crossings ("false alarms"). All data were recorded by hand in pencil.

At this time we began to appreciate the individual differences among our subjects. There was one particular squirrelfish (*Holocentrus ascensionis*) that was a most amazing performer. Each day that we tested this animal, its threshold sank lower and lower, to a level where we became suspicious. We ran some blank trials, i.e., with no sound at all. The fish responded as soon as the trial began, and acted as though it heard our non-existent stimulus. Obviously, we started thinking of the possibility of secondary cues, and eventually came to call this subject, Clever Hans. It turned out that all the operator had to do was to raise his hand to press the "start" button on the equipment. This was enough to signal the fish. Evidently while we were smugly watching the fish through the mirror, Hans was watching us with his big pink eyes.

I might add that my association with Jerry Wodinsky was educational in many ways. One of these was his insistence that all notes be taken with a simple lead pencil. The value of this procedure was proved to me when, upon returning from Bimini with six months worth of data and our allocation of liquor, a bottle of Beefeater gin broke in my suitcase. All writing not in pencil was completely erased. Data were saved (see Tavolga and Wodinsky 1963).

Gradually, the equipment evolved as we became more knowledgeable in underwater acoustics, and in electronics (Tavolga 1966). The 1962 paper by Harris and van Bergeijk came to our attention, and we visited them at Bell Laboratories in Murray Hill, New Jersey. This visit proved to have a profound influence on our entire approach to underwater acoustics. For the first time we began to appreciate the differences between the two forms of acoustic energy that affect a fish in water. Although the acoustic near field may have been familiar to physicists, it was a whole new notion to biologists.

I believe that this 1962 paper by Harris stands as a milestone in our field of study. Certainly, this paper and van Bergeijk's later contributions were to stimulate research into new paths. The death of these two men just as they were reaching the peaks of their careers was a severe loss to science. I wonder what they would think of the present papers, especially of the model for directional hearing proposed by Schuijf (Chapter 14).

To return to equipment and technology. One of the observations that forced Wodinsky and me to improve our technology was the finding that some of our subjects appeared to show a double threshold at frequencies below about 200 Hz. The second, lower, threshold was manifested after additional training. At the time, we speculated that the animals were shifting from one modality to another, i.e., from the inner ear to the lateral line. The lateral line was traditionally thought of as the more effective low frequency detector. To test this hypothesis, we built the Audio-Ichthyotron, an instrument that could test as many as six subjects at a time, with an automatic sound level attenuator. A helpful electronic engineer at City College constructed the machine, which used state-of-the-art techniques, i.e., transistors—a total of perhaps a thousand or so. With associated power supplies, relays, etc., this monster occupied a box about a cubic meter in size. It worked effectively, and we were able to apply mass production methods. There appears to be such a thing as too much data, sometimes, for these "double thresholds" were lost in the mass of statistics. Reluctantly, we gave up our hypothesis (Tavolga and Wodinsky 1965). Several years later, however, the phenomenon was replicated by Cahn, Siler, and Wodinsky (1969), by designing a device in which particle displacement and pressure energy could be varied independently and serve as separate stimuli. They even used the same species that we had tested before. Evidently, fishes are capable of discriminating between pressure and particle-displacement energy. This 1969 study used a far better system of acoustic control and measurement than Jerry and I had before, yet the threshold data came out within a decibel or two of our original figures. It's nice to be replicated.

The data on dual thresholds are still valid, however, and I noted the presence of plateaus in threshold data for *Notopterus* (Coombs, Chapter 8). Perhaps we should think not of different sensory modalities that may be involved in signal reception, rather of different *pathways* of reception.

With electronic advances, the Audio-Ichthyotron became more complex and smaller. Our last model, the Mark IV, was as small as a breadbox, but could be used for frequency and intensity discriminations (Jacobs and Tavolga 1967, 1968), and the same instrument could be used for visual studies, the Video-Ichthyotron (Tavolga 1977b). I now have a small computer that conducts such experiments, stores the data, does the statistics, plays games, calculates the laboratory payroll, and even doubles as a typewriter.

The first conference on marine bioacoustics took place at the Lerner Marine Laboratory, under the auspices of the Office of Naval Research and the American Museum of Natural History. This was in 1963, and in 1966, with additional help from the Naval Training Device Center, another conference was held at the American Museum. From a cursory look at citation frequencies of the resultant publications (Tavolga 1964, 1967a), it is evident that these meetings served as stimulation and impetus to further research in the field. In those days, the Office of Naval Research was still supporting basic research, and the Department of Animal Behavior at the American Museum was a prominent force in behavioral research. Today, the Lerner Marine Laboratory is no more, the ONR has very different interests, and I was recently informed that the Department of Animal Behavior has been liquidated. Those were halcyon days.

Over the past 10 or 15 years, the advances in this field have been truly amazing. In the area of sound detection in fishes, we have moved far from the arguments as to whether fish can hear, and from the clever demonstrations such as those of von Frisch (1923) when he trained catfish in a lake to come to him when he whistled. Fish audition has turned out to be far more precise, complex, and sophisticated than ever before suspected. Actually, we should not be very surprised at this fact. After all, the teleosts have been evolving in water for a good deal longer than any terrestrial vertebrates. Indeed, mammals are only slightly modified rhipidistians with ears that are still essentially aquatic. Structurally, the inner ears of vertebrates have been amazingly stable and conservative throughout evolution. So conservative, indeed, that just a difference in proportional size of the utriculus and sacculus is enough to excite our interest (Platt and Popper, Chapter 1; Jenkins, Chapter 4). This very conservatism of structure and function makes the fish ear a broadly useful model for the study of auditory physiology in general.

In the study of sound production and communication in fishes, we have also advanced since the postwar fascination with the simple fact that the undersea world is a noisy place. At one time, merely cataloging sound producing species was interesting and important, but we have now gone well beyond the stage where the simple report of yet another species and its sounds is useful. Questions of function, i.e., communication, are now paramount.

When Art Popper, Dick Fay and I first discussed the possibility of this conference, the thought struck each of us that perhaps the most useful function a conference could have would be to point out the most exciting and potentially fruitful directions for future research. Could some new questions be formulated? Could we, i.e., this present group assembled here, stand back and look at what we have been saying for the past few days with a wide angle lens (acoustic lens, of course)? What can we predict about future research on hearing and sound communication in fishes? Accordingly, I should like to stick my own neck out and see if I can make some predictions as to where we ought to go, influenced, of course, by the papers and discussions (formal and informal) that I have been exposed to during these meetings.

2 Prospect

2.1 Cellular Level—The Hair Cell

On the microscopic level, hair cells are particle displacement detectors, no matter where they may be. At this level of organization, we can ignore the rest of the organism, and just look at the operation of hair cells. Fishes have the advantage of carrying their hair cells on the outside, so it becomes much easier to chop away the rest of the animal than to dissect out the cochlear hair cells of a mammal. The sensory units of the lateral line in fishes (and some amphibians) are marvelously suited for sensory physiological studies. The fine structure and function of these cells was extensively described in the volume that came out of Phyllis Cahn's symposium (Cahn 1967), a conference that followed hard upon the second marine bioacoustics meetings with significant overlap of content and personnel.

Some novel ideas have appeared during this present conference. Both Strelioff and Sokolich (Chapter 24) and Boston (Chapter 25) showed that there is considerable coding, filtering, and massaging of the incoming information right at the level of the sensory cell. Furthermore, the operation of these cells supposedly specialized as motion detectors is fundamentally similar to that of other, qualitatively different sensory modalities; for example, the intriguing similarities between acoustic and electroreceptors described and dicussed by Bullock (Chapter 27).

The curious asymmetrical organizations of the hair cells has been known for some time, as well as the functional asymmetry of response. I refer, of course, to the position of the kinocilium as a determining factor of the direction of sensitivity of the hair cell to a displacement of the medium. Flock (1967), as well as other contributors to Dr. Cahn's volume on the lateral line, were able to hop from the *Xenopus* system to that of fish quite easily, since the hair cells were essentially the same. Even the cochlear hair cells in mammals could almost be interchanged with the lateral line cells in teleosts. As Flock (1967) pointed out, the kinocilium in cochlear hair cells is present in embryos and is represented by the locus of the centriole. Although the relation of the kinocilium with the active axis of the hair cell is known, the exact mechanism of this action was not clear at that time.

The functional relationships of kinocilia and stereocilia have been intensively studied recently (Hudspeth and Corey 1977, Hudspeth and Jacobs 1979). The kinocilium is a true cilium and conveys mechanical displacements to the stereocilia (large microvilli). These investigations involved some elegant microelectrophysiological techniques using inner ears of bullfrogs. However, the relatively exposed lateral line organs in fishes and some amphibians are almost ideally designed for this kind of work.

Collectively, hair cells form an array of receptors. The significance of the fact that hair cells are often segregated according to axis of polarization has been discovered only recently, and the possible relationship of these specialized groups to directional hearing is, I think, one of the most significant breakthroughs in the field of auditory physiology. If nothing else, it shows how much more interesting and challenging it is to study hearing in fish than in any so-called higher vertebrates. I shall have more to say about this in connection with directional hearing.

2.2 Organ Level—Inner Ear and Lateral Line

Functional anatomy seems to be a recurring theme in this conference. As a result of our exposure, long ago, to undergraduate comparative anatomy courses, we have all, I am sure, paid lip service on the link of form and function. Actual investigations, based on the form-function concept, are not common, however. This conference, by contrast, is replete with functional anatomical studies. For example, the curious gas-filled bullae in clupeid fishes that intrigued van Bergeijk turn out to be elegantly adapted mechanisms for sound detection and orientation to external sound sources (Blaxter, Denton, and Gray, Chapter 2). In these fishes, the swim bladder has a special role in adaptation of the inner ear to changes in pressure that result from changes in depth (Blaxter, Chapter 3). Coombs (Chapter 8) suggests that ear-swim bladder connections are related to auditory capacities in ways that may be much more complex and variable among different species than had been suspected before. Platt and Popper (Chapter 1) correlated form-function differences with phylogeny (evolutionary relationships). As appropriate in comparative zoology, they eschewed generalizations and argued against the naive notion that there exists a "typical" shark or fish. Phyletically, families of fish may be as far apart as, say, amphibians are from reptiles, and sharks could almost be a separate phylum had not someone artificially created the Phylum Chordata. Northcutt's (Chapter 16) review of the current state of fish neuroanatomy effectively pointed out the fallacies of glib homologies and other comparisons between fishes and other vertebrates, and, indeed, even between groups of teleosts. In truth, most of our neuroanatomy in fishes is based on a pitifully small representation of the perhaps 30,000 known species. How representative is our sample? The papers by Bell (Chapter 19) on momyrids and McCormick (Chapter 18) on *Amia* and other teleosts show how different and unusual are the central nervous sytem connections of the octavolateralis system. Perhaps, upon further investigation, the different and unusual may turn out to be the way of things in fish neurology.

In the case of inner ear structure, teleosts seem to be conservative, but the swim bladder-inner ear connection in the Ostariophysi is unique in the presence of the Weberian ossicles. Were it not for that feature, the several families of Ostariophysi might have been classified as far apart, phyletically. The cyprinids, catfishes, and electric eels are about as different as any teleost group could be, except for the common presence of the Weberian apparatus. Other teleosts may have an inner ear-swim bladder connection, but not as in the Ostariophysi. Although the structural distinction between the Ostariophysi and all others is clear, we are still wrestling with the functional significance of this distinction. It took over a hundred years to demonstrate, by experiment, that Weber's speculations on the function of his famous ossicles were correct. I trust it will not take another century for us to figure out just how the ostariophysine auditory system operates. Data on this point are still scarce, and the mechanism of the Weberian ossicles has not yet been precisely described.

The link of the inner ear to the central nervous sytem has, for a long time, been used just as a convenient spot for the auditory physiologist to place his electrodes. The fish acousticians are finally beginning to catch up with the other vertebrates. We are now beginning to look at how the acoustic (including lateral line) signals are projected onto the brain (Roberts, Chapter 17, for instance). However, we are still far behind the state of the art in mammals. How the fish's central nervous system can process this information, particularly with the complication of a dual, possibly redundant, acoustic modality, may well turn out to be one of the most challenging problems in the field (Fay, Chapter 10). Redundancy may be the key factor for fish, in that the lateral line input can enhance the sound detection system as a whole.

At the organ level of function, the fish ear presents some of the most exciting problems. We now know that fish are capable of fairly accurate directional hearing, and that they can discriminate frequencies almost as well as some mammals (Enger, Chapter 12; Hall, Patricoski, and Fay, Chapter 9; Horner, Hawkins, and Fraser, Chapter 11). It is clear that the question of mechanisms of directional hearing and frequency analysis have had considerable coverage at this conference. Let me summarize where I think we may be going in these two areas.

2.2.1 Frequency Analysis

Back in 1966, at the second marine bioacoustics conference, I reported a psychophysical study aimed at investigating the mechanisms by which fish are able to discriminate one frequency from another, in spite of being hampered by the lack of a cochlea. At that time (Tavolga 1967b), I made the statement that in the fish ear, there is "no place for a 'place theory'...." Willem van Bergeijk (1967) rose to the challenge and said:

> ...I believe that the saccular macula is perfectly adequate for the task. Consider a membrane, a circular one, suspended in a fairly rigid frame. Then it will be the case that the stiffness of the membrane is greatest near the frame, and least in the center; there is a *stiffness gradient* from periphery to center. The saccular macula is an approximation to this. If, as is usually the case, the nerve enters the macula near the center and fans out to the periphery, we have additionally a *mass gradient*. The combination of high-mass, low-stiffness at the center and low-mass, high-stiffness at the periphery makes this sort of membrane quite differentially sensitive to frequency, ... In fact, such a membrane shows traveling waves with frequency-dependent maxima. Any *bounded* membrane behaves this way, but most (like a bongo drum, for instance) show only a small frequency range between resonance of the centre and the periphery, and a good drummer can squeeze an octave out of a bondo drum. Refinements such as mass-loading and lengthening the membrane into a strip, rather than a circle, can greatly increase the frequency range accommodated ....

Thus was born the "bongo-drum hypothesis" to provide a home for the place mechanism. Some of van Bergeijk's points can be argued, in particular his concept of saccular structure and innervation. The idea of modes of otolith motion as accounting for a place mechanism may require some rethinking. At any rate, with all our current information on frequency discrimination in fishes, we are still faced with the inevitable comparison with a cochlea. Enger (Chapter 12) has, for the first time, presented evidence that there may really be an operational place principle in the fish ear. However,

Fay (Chapter 10) presented evidence that the goldfish ear, one of our best frequency discriminators, may be using some phase-locking mechanism. Perhaps we need to look at "place" in a new and different way. The volley principle may still be valid, and both systems could exist without being mutually exclusive (Horner, Hawkins, and Fraser, Chapter 11). Indeed, it is seldom in evolution that the baby is discarded with the bathwater. Here is another possibility for a dual, redundant sensory input, making the receptor system capable of finer discrimination than with the two principles operating separately.

This is a fine chance for the neurophysiologist to probe and to discover how the fish appears to do what its inner ear appears to be unsuited for. How is it that some species of fishes seem to be able to discriminate much better than others? Where do these differences reside? Is it in the tuning or sensitivity of the receiving structures, swimbladder included? Are there differences in neurology, peripheral or central? Or do some fishes just don't care? As I found out in looking at audition in the "spooky" bonefish, it isn't that this fish can hear sounds better than other species in the area, it is just that it *listens* better and responds more readily (Tavolga 1974a). I refer here to the signal detection theorist who makes the distinction between the physiologist's sensory threshold and the behaviorist's response threshold.

2.2.2 Directional Hearing

In 1963 there were only three published reports on directional hearing in fishes. Two of these expressed strong doubt that fishes could detect sound direction at all. Even the only positive report (Kleerekoper and Chagnon 1954) was equivocal in that the fish seemed to wander aimlessly before locating the sound source. This was called a "constrained random walk," but the term *kinesis* would apply. Willem van Bergeijk (1964) spotted the critical feature of these few studies. Under certain special conditions, fish do appear to be able to hear directionally. The special condition is that the sound be loud, low in frequency, and the source nearby. Obviously (to us with 20/20 hindsight) the animal must be within the range of significant near-field effect. Acoustic pressure is essentially a nondirectional form of energy, while particle displacement is a vector force. Van Bergeijk (1964) presented the situation with characteristic clarity:

> ... the problem is properly stated: "How can a fish with *one* pressure receptor locate and track a source of pressure waves?", then it becomes evident that a satisfactory answer is going to involve a statement of a *mechanism,* at least in the form of a postulate. The best available evidence suggests that the answer to the question is very simple: he can't. (page 297).

The logical extension of this conclusion is that the fish must have an acoustical particle-displacement detector in order to localize a sound source. Obviously, concluded van Bergeijk, it must be the lateral line. This is a fine example of a logically powerful argument. Certainly the notion that the lateral line system is the modality necessary and sufficient for directional hearing has influenced much research in this area, especially on this side of the Atlantic. My own thinking has been strongly affected by this idea. Curiously, the idea produced some semantic arguments as to what constitutes acoustic energy. For a restricted view of what acoustic energy is, I refer to comments by Dijkgraaf (1964, 1967) and Tavolga (1967a, pp. 230 and 281).

In any case, van Bergeijk was one of the first to state that the inner ear is, by itself, strictly a near-field detector, but that the presence of the nearby swimbladder allows it to receive pressure information. The nice neat idea that the swimbladder is the sole pressure transducer was very logical and influential for many years. The fact that the swimbladder does indeed act as an acoustic pressure transducer is clear, and its effect is not limited to the inner ear, but, according to Sand (Chapter 23), can also extend its range to the lateral line system.

The real pioneer in the study of lateral line physiology is Prof. Sven Dijkgraaf. He is still a profound influence in the field, and we all owe him a debt of gratitude for his inspiration and incisive thinking. It is not coincidental that we see some of the major advances coming from his institution, and his influence reflected in the contributions compiled and edited by Schuijf and Hawkins (1976). The contributions by Schuijf (Chapter 14) and by Buwalda (Chaper 7) have further elucidated their conception of the mechanics of directional hearing in fishes. It is now quite apparent that the lateral line is not the only modality that the fish can use to detect external near-field energy. The inner ear can do this directly, and can determine at least the bearing of a sound source with fair accuracy. The Schuijf and Buwalda model is not only logical but, in essence, a relatively simple explanation for the existing evidence. The accuracy of this system, however, actually has been measured and reported in only a single publication, that of Chapman and Johnstone (1974). They found directional thresholds in the order of 10 to 20° arc in the cod and the haddock. We now know something about the capacity of fish to determine the bearing of a sound source, but how about the range?

The limitations proposed by van Bergeijk still hold. Sound source localization has been studied and measured in several species since van Bergeijk's work, and the field studies on cod and related forms have been most revealing (Schuijf 1976a, 1976b, Schuijf and Buwalda 1975, Schuijf and Siemelink 1974). In every case, the adequate stimulus must be low in frequency and the sound source must be close. The rule of thumb is that the near field is an adequate stimulus within a range of about one-sixth of a wavelength, unless the sound source energy is unusually powerful. This was deduced by van Bergeijk (1964) from data on the sensitivity of lateral line hair cells to displacement. The value of this rule is now doubtful. The question of the inner ear acting as a pressure transducer independent of the swimbladder is still unanswered, but such a role appears highly unlikely.

The capacities for near-field detection without the interfering or masking effect of swimbladder vibrations can be explored in elasmobranchs. The report by Fay, Kendall, Popper, and Tester (1974) and the studies of Corwin (Chapter 5) should become fertile areas of investigation. This research provides us with a very different view of aquatic ears.

As noted earlier, the role of the swimbladder to hearing in teleost fishes has been studied in many ways, but the most elegant experiment was that of Chapman and Sand (1974). They provided a prosthetic device, a hearing aid, to a flounder, and found that if it had a swimbladder its hearing would be improved by about 10 dB. Perhaps the flounder would then hear some sounds it would rather not know about.

We can now see that the inner ear of the fish can detect near-field energy through two pathways, direct and indirect (via the swimbladder). If there is also a direct path input for pressure energy, then the requirements of the system to process all these signals must be incredibly complex. In addition to the various forms of coupling of the swimbladder to the inner ear, the vibration patterns of the otoliths and how these massive structures affect the hair cell responses are only beginning to be investigated (Hawkins and Horner, Chapter 15). By comparison, the mammalian cochlear system may turn out to be the simpler and less demanding one.

The role of the lateral line, after over one hundred years of study, still presents us with mysteries. It is beginning to look like a sort of all-purpose sensory system. Early studies claimed it to be a chemoreceptor; electroreception seems to be a derivative of the lateral line, but classically it has been considered a detector of water movements, low frequency sounds, and to serve as a basis for rheotaxis (Parker 1904). Evidence of the lateral line as an acoustic receptor has been disputed, especially in connection with the dispute of how an acoustic stimulus should be defined (see above). I think that there is still much to be learned about this sensory system, and I confess that I am still strongly influenced by Willem van Bergeijk's contention that the lateral line is perfectly adapted as an acoustic directional sense. My own studies on the echolocating capacities in the marine catfish (*Arius felis*) seem to support this notion so far (Tavolga 1971b, 1976b, 1977c). But if the lateral line is not an acoustic receptor (acoustic in the strict sense), what is it? It is clearly an ideal detector of velocity changes and particle displacement in the medium. Evidence that it is involved in schooling behavior has been reinforced by Partridge (Chapter 26). Perhaps we should return to look carefully at Dijkgraaf's paper on *Ferntastsinn* (1947). His conception of the lateral line as an essentially tactile receptor may be all that it is. The skin and hair on our own body surface can function, under special conditions, as an acoustic receptor, and this may also be true of the lateral line system.

2.3 Communication

Having looked back and tried to look ahead, I am now beginning to feel as though I have neglected my original objectives. Why did I get into the hearing business, sensory physiology, psychophysics, etc.? I thought this would be a means to an end, namely to enable me to design playback, behavioral experiments with some knowledge of what the fish could hear. We have now reached a point in our knowledge of fish hearing that should give us more data than we need for behavioral studies. However, we seem to have lost sight of the fish as a fish, and just look at it as the soft squishy stuff that surrounds the auditory organs and keeps them operating.

The essential function of the auditory system is to assist the fish to survive in its particular ecological niche. Clupeid fishes, for example, are not known to be significant sound producers, although I have recordings of the swimming sounds, i.e., hydrodynamic effects, of large schools of herring or menhaden. However, these fishes are essential prey organisms for a large variety of predators. Their auditory specializations are incredibly complex (Blaxter, Denton, and Gray, Chapter 2), and I daresay we could generate many questions and hypotheses that correlate the specializations of their auditory systems with their response to predators, as, for instance, a facilitation of a rapid startle response.

In general, it would be interesting to speculate on the functions of the fish auditory systems in the context of their normal environment. I have always wondered, for example, what is the value of the high degree of auditory acuity to a goldfish? Although today the "natural" environment of a goldfish is something created by man, the carp-like, wild ancestors of the domesticated version normally live in shallow ponds with slow moving water and relatively few fish predators. Certainly the goldfish themselves are not sound producers, yet the auditory system cannot be just an ornamental evolutionary vestige. The apparent mismatch of the auditory and sonic system spectra in the toadfish (Fine, Chapter 13) presents some questions, but the mismatch may actually serve as an adaptive mechanism for limiting the response only to nearby individuals.

We seem to have achieved some major advances in our understanding of the mechanisms of directional hearing in teleosts, yet we still know very little of the actual behavior of acoustic orientation. Under what conditions and in what behavioral context does a fish use directional information about a sound source? The apparent use of directional acoustic cues by predatory sharks is clouded by the presence, in all observations, of other directional cues. The value of acoustics in schooling behavior is still not clear, although the contribution of the lateral line system is becoming more evident (Partridge, Chapter 26).

Sound production, where the sound output clearly has some social function, has been discovered in a wide variety of teleost fishes. The groups represented are often only distantly related in phylogeny. Sonic interaction exists in such diverse groups as the sciaenids, gobies, catfishes, cichlids, toadfishes, and gouramis, to name just a few. One feature that many species have in common is the use of the swimbladder as a sound projector, and, in most of these, the sonic musculature is innervated by the spino-occipital nerves.

From the behavioral standpoint, the precise function of most fish sounds is as yet unclear. Even among the sciaenids (drum, croaker, etc.), sound production is only correlated with reproductive behavior. There are, however, a few species that have been investigated by using playback techniques both in the laboratory and in the field (see reviews by Tavolga 1971a, 1977a, Fine, Winn, and Olla 1977).

The approach by Demski (Chapter 21; Demski and Gerald 1972, 1974) represents a very different path for investigation of fish sounds. In the case of the toadfish in particular, the production of the "boatwhistle" sounds in captive individuals is rare and sporadic. A means is now available to induce the production of these sounds under controlled conditions. In addition, of course, this technique can elucidate the central nervous system mechanisms involved in sound production.

The paper by Myrberg (Chapter 20) is certainly timely. I think that the state of our data on a fair variety of species should enable us to develop some generalizations. Winn (1964) was the first to really tackle the broad aspects of acoustic communication in fishes, and his conclusion was that temporal patterning of otherwise nonspecific sounds was the key to communication in most teleosts. Myrberg has gone beyond that stage, and has tried to establish a framework for future research—a goal that is central to our present conference.

The majority of species that produce sounds with behavioral significance emit nonspecific, pulse-type sounds that provide a broad spectrum of information on the physiological state of the organism, e.g., general territorial aggressiveness or reproductive readiness. Only a few species produce sounds that are clearly related to specific activities, and only these fit the definition of *signals.* In order not to broaden the meaning of *communication* to the point of uselessness, I suggested that *communication* be restricted to levels of organization in which signal exchange is clearly identifiable. Signals are characterized by being emitted by specialized structures, and the signal energy is confined to a narrow portion of the energy spectrum. By being closely linked to the emitter's physiological condition, the signal serves as a code. The decoding by the receiver is done within the context of the signal and the receiver's condition becomes, as it were, a part of the code. At this level, the specificity of the signals and the responses to signals is generally limited to intraspecific interactions, i.e., to social behavior in the broad sense (Tavolga 1974b).

Myrberg's (Chapter 20) introduction of the idea of an *interceptor,* i.e., an individual that, in a sense, listens in to a communicatory exchange between two other organisms, is most intriguing and should prove useful. The definition and recognition of an interceptor is, of course, based on the fact that we know enough of the intercepted communication and the behavior involved to understand that the signals are not being broadcast, but are specifically directed.

The boatwhistle exchanges in the toadfish (*Opsanus*) fall into the category of signals (Fish 1972, Winn 1972). Perhaps the sonic interactions in the damselfishes (*Eupomacentrus*) may also be allocated to the level of communication (Myrberg 1972). Being a firm adherent to Lloyd Morgan's Canon and its granddaddy, Occam's razor, I prefer to assume that the particular case of interaction represents a lower level of organization, unless evidence shows it to be on a higher level.

Behavior and the problems of communication present questions where the data from different levels of research come together. Fish occupy a critical niche in phylogeny. Although highly adapted and specialized in a myriad of ways to an aquatic environment, their level of organization in behavior does not permit the complex social interactions that take place in mammals and birds. Possibly the aquatic environment is not appropriate for the evolution of higher levels of communication or, more likely, the fundamental teleost organization may not be capable of developing beyond a certain point. At any rate, teleosts appear to be stuck at a level which barely qualifies under the strictly defined term "communication," and yet their receptor systems have turned out to be marvelously sophisticated.

Myrberg's comment that he is a "token ethologist" at this meeting is more truth than poetry. With all the advances we have made in the areas of acoustic reception in fishes over the past 10-15 years, our behavioral advances have lagged far behind. Armed with our understanding of audition, we should now be able to pursue the operation of acoustics at the higher levels of organization, e.g., behavior, ethology, sociobiology, etc.

Acknowledgments. Support for the preparation of this chapter was derived from Grant No. NS 13746 from the National Institute of Neurological and Communicative Disorders and Stroke, National Institutes of Health.

References

van Bergeijk, W. A.: Directional and nondirectional hearing in fish. In: Marine Bio-
Acoustics. Tavolga, W. N. (ed.). Oxford: Pergamon Press, 1964, pp. 281-299.

van Bergeijk, W. A.: Discussion of critical bands in hearing of fishes. In: Marine Bio-
Acoustics II. Tavolga, W. N. (ed.). Oxford: Pergamon Press, 1967, p. 244.

Cahn, P. H. (ed.): Lateral Line Detectors; Bloomington: Indiana Univ. Press, 1967.

Cahn, P. H., Siler, W., Wodinsky, J.: Acoustico-lateralis system of fishes: Tests of
pressure and particle-velocity sensitivity in grunts, *Haemulon sciurus* and *Haemulon
parrai*. J. Acoust. Soc. Am. 46, 1572-1578 (1969).

Chapman, C. J., Johnstone, A. D. F.: Some auditory discrimination experiments on
marine fish. J. Exp. Biol. 61, 521-528 (1974).

Chapman, C. J., Sand, O.: Field studies of hearing in two species of flatfish *Pleuronectes platessa* (L.) and *Limanda limanda* (L.) (Family Pleuronectidae). Comp.
Biochem. Physiol. 47A, 371-385 (1974).

Demski, L. S., Gerald, J. W.: Sound production evoked by electrical stimulation of the
brain in toadfish (*Opsanus beta*). Anim. Behav. 20, 507-513 (1972).

Demski, L. S., Gerald, J. W.: Sound production and other behavioral effects of midbrain stimulation in free-swimming toadfish, *Opsanus beta*. Brain Behav. Evol. 9,
41-59 (1974).

Dijkgraaf, S.: Über die Reizung des Ferntastsinnes bei Fischen und Amphibien. Experientia 3, 206-216 (1947). (Translation in Tavolga 1976a.)

Dijkgraaf, S.: The supposed use of the lateral line as an organ of hearing in fish. Experientia 20, 586 (1964).

Dijkgraaf, S.: Biological significance of the lateral line organs. In: Lateral Line Detectors. Cahn, P. H. (ed.). Bloomington: Indiana Univ. Press, 1967, pp. 83-95.

Fay, R. R., Kendall, J. I., Popper, A. N., Tester, A. L.: Vibration detection by the
macula neglecta of sharks. Comp. Biochem. Physiol. 47A, 1235-1240 (1974).

Fine, M. L., Winn, H. E., Olla, B. L.: Communication in fishes. In: How Animals
Communicate. Sebeok, T. A. (ed.). Bloomington: Indiana Univ. Press, 1977,
pp. 472-518.

Fish, J. F.: The effect of sound playback on the toadfish. In: Behavior of Marine
Animals II. Winn, H. E., Olla, B. L. (eds.). New York: Plenum Press, 1972, pp. 386-
434.

Fish, M. P.: The character and significance of sound production among fishes of the
Western North Atlantic. Bull. Bingham Oceanogr. Coll. 14, 1-109 (1954).

Flock, Å.: Ultrastructure and function in the lateral line organs. In: Lateral Line
Detectors. Cahn, P. H. (ed.). Bloomington: Indiana Univ. Press, 1967, pp. 163-197.

Harris, G. G., van Bergeijk, W. A.: Evidence that the lateral-line organ responds to
near-field displacements of sound sources in water. J. Acoust. Soc. Am. 34, 1831-
1841 (1962). (Reprinted in Tavolga 1976a.)

Hudspeth, A. J., Corey, D. P.: Sensitivity, polarity, and conductance change in the
response of vertebrate hair cells to controlled mechanical stimuli. Proc. Nat. Acad.
Sci. 74, 2407-2411 (177).

Hudspeth, A. J., Jacobs, R.: Stereocilia mediate transduction in vertebrate hair cells.
Proc. Nat. Acad. Sci. 76, 1506-1509 (1979).

Jacobs, D. W., Tavolga, W. N.: Acoustical intensity limens in the goldfish. Anim.
Behav. 15, 324-335 (1967).

Jacobs, D. W., Tavolga, W. N.: Acoustic frequency discrimination in the goldfish.
Anim. Behav. 16, 67-71 (1968).

Kleerekoper, H., Chagnon, E. C.: Hearing in fish, with special reference to *Semotilus atromaculatus atromaculatus* (Mitchill). J. Fish. Res. Bd. Can. 11, 130-152 (1954).

Myrberg, A. A., Jr.: Ethology of the bicolor damselfish, *Eupomacentrus partitus* (Pisces: Pomacentridae): A comparative analysis of laboratory and field behaviour. Anim. Behav. Monogr. 5, 199-283 (1972).

Parker, G. H.: The function of the lateral-line organ in fishes. Bull. U.S. Bur. Fish. 24, 185-207 (1904). (Reprinted in Tavolga 1976a.)

Poggendorf, D.: Die absoluten Hörschwellen des Zwergwelses (*Amiurus nebulosus*) und Beiträge zur Physik des Weberschen Apparates der Ostariophysen. Z. Vergl. Physiol. 34, 222-257 (1952). (Translation in Tavolga 1976a.)

Schuijf, A.: The phase model of directional hearing in fish. In: Sound Reception in Fish. Schuijf, A., Hawkins, A. D. (eds.). Amsterdam: Elsevier, 1976a, pp. 63-86.

Schuijf, A.: Timing analysis and directional hearing in fish. In: Sound Reception in Fish. Schuijf, A., Hawkins, A. D. (eds.). Amsterdam: Elsevier, 1976b, pp. 87-112.

Schuijf, A., Buwalda, R. J. A.: On the mechanism of directional hearing in cod (*Gadus morhua* L.). J. Comp. Physiol. 98, 333-343 (1975).

Schuijf, A., Hawkins, A. D. (eds.): Sound Reception in Fish. Amsterdam: Elsevier, 1976.

Schuijf, A., Siemelink, M. E.: The ability of cod (*Gadus morhua*) to orient towards a sound source. Experientia 30, 773-774 (1974).

Tavolga, W. N.: Visual, chemical and sound stimuli as cues in the sex discriminatory behavior of the gobiid fish, *Bathygobius soporator*. Zoologica 41, 49-54 (1956).

Tavolga, W. N.: The significance of underwater sounds produced by males of the gobiid fish, *Bathygobius soporator*. Physiol. Zool. 31, 259-271 (1958). (Reprinted in Tavolga 1977a.)

Tavolga, W. N. (ed.): Marine Bio-Acoustics, Oxford: Pergamon Press, 1964.

Tavolga, W. N.: The Audio-Ichthyotron: The evolution of an instrument for testing the auditory capacities of fishes. Trans. N.Y. Acad. Sci. 28, 706-712 (1966).

Tavolga, W. N.: Masked auditory thresholds in teleost fishes. In: Marine Bio-Acoustics, Vol. 2. Tavolga, W. N. (ed.). Oxford: Pergamon Press, 1977b.

Tavolga, W. N.: Sound production and detection. In: Fish Physiology, Vol. 5. Hoar, W. S., Randall, D. J. (eds.). New York: Academic Press, 1971a, pp. 135-205. (Reprinted in Tavolga 1976a, 1977a.)

Tavolga, W. N.: Acoustic orientation in the sea catfish, *Galeichthys felis*. Ann. N.Y. Acad. Sci. 188, 80-97 (1971b).

Tavolga, W. N.: Sensory parameters in communication among coral reef fishes. Mt. Sinai J. Med. 41, 324-340 (1974a).

Tavolga, W. N.: Application of the concept of levels of organization to the study of animal communication. In: Non-verbal Communication. Krames, L., Pliner, P., Alloway, T. (eds.). New York: Plenum Press, 1974b, pp. 51-75.

Tavolga, W. N. (ed.): Sound Reception in Fishes. Stroudsberg, Pennsylvania: Dowden, Hutchinson and Ross, 1976a.

Tavolga, W. N.: Acoustical obstacle detection in the sea catfish (*Arius felis*). In: Sound Reception in Fish. Schuijf, A., Hawkins, A. D. (eds.). Amsterdam: Elsevier, 1976b, pp. 185-204.

Tavolga, W. N. (ed.): Sound Production in Fishes. Stroudsberg, Pennsylvania: Dowden, Hutchinson and Ross, 1977a.

Tavolga, W. N.: Behavioural thresholds for diffuse illumination in the goldfish. J. Exp. Biol. 67, 89-96 (1977b).

Tavolga, W. N.: Mechanisms for directional hearing in the sea catfish (*Arius felis*). J. Exp. Biol. 67, 97-115 (1977c).

Tavolga, W. N., Wodinsky, J.: Auditory capacities in fishes: pure tone thresholds in nine species of marine teleosts. Bull. Am. Mus. Nat. Hist. 126, 177-240 (1963). (Reprinted in Tavolga 1976a.)

Tavolga, W. N., Wodinsky, J.: Auditory capacities in fishes: threshold variability in the blue striped-grunt, *Haemulon sciurus.* Anim. Behav. 13, 301-311 (1965).

von Frisch, K.: Ein Zwergwels der kommt, wenn man ihm pfeift. Biol. Zentr. 43, 439-446 (1923). (Translation in Tavolga 1976a.)

Winn, H. E.: The biological significance of fish sounds. In: Marine Bio-Acoustics. Tavolga, W. N. (ed.). Oxford: Pergamon Press, 1964, pp. 213-231. (Reprinted in Tavolga 1977a.)

Winn, H. E.: Acoustic discrimination by the toadfish with comments on signal systems. In: Behavior of Marine Animals II. Winn. H. E., Olla, B. L. (eds.). New York: Plenum Press, 1972, pp. 361-385.

Discussion

BULLOCK: Of the many things you stimulated in my mind, I'd like to select two for consideration at our next meeting. One is the question of the ability of the central nervous system in fishes to perform computations beyond the ones we have mentioned. For instance, the ability of the brain to map a field of sensory circuits is one of the most exciting things in acoustic physiology. By historical chance, the first notion of this came in work on birds. This is not like a cochlear neuroanatomic map in a sensory cortex, but rather like a computer map.

The other point is your challenging question about why goldfish are such good hearers. Perhaps they have evolved especially good hearing to be interceptors, and use it only to avoid avian or other predators. The startle response is extraordinarily good in the wild variety, and this is essentially a Mauthner cell response—a most amazing response. With a latency of less than 10 milliseconds, the animal is already moving away from a stimulus source.

TAVOLGA: Another way of looking at it is that if the goldfish did not have such good hearing, then the species would not be nearly as widely distributed as it is today among all these auditory physiology laboratores.

ENGER: You mentioned Professor Dijkgraaf, and many of us recognize his classic contributions to this field. Perhaps it would be appropriate if we, as a group, sent him a telegram simply to show that we recognize and appreciate his pioneering work.

TAVOLGA: Hearing the general sounds of approval to this suggestion, let's you and I get together later and compose such a telegram. I, for one, am delighted with the idea.

Author Index

Abramowitz, M. *304*, 310
Adams, L. A. 14, *30*
Adrian, H. O. *501*
Ahroon, W. A. 120, *130*, 164, *177*, 180, *185*, 212, *218*
Akutsu, Y. 536, *564*
Albus, J. S. 365, *370*
Ali, M. A. *34, 421*
Allen, J. M. 41, *55*
Allen, J. W. 515, *519*
Allis, E. P. 359, *370*
Alloway, T. *421, 424, 586*
Alnaes, E. 366, 367, *370*
Altes, R. A. 557, *566*
Altmann, D. W. 485, 507, *502, 503, 513*
Anderson, D. 201, *217*
Anderson, R. A. 238, *239*
Andrianov, G. N. 367, *370, 540, 564*
Andrianov, Y. N. *564*
Anson, M. 165, *170*
Ariëns Kappers, C. U. 331, 332, *351*, 362, *371*
Arnold, A. P. 433, 440, *443*
Aronson, L. R. *423, 424, 521*
Asanuma, A. *570*
Ashcroft, R. *421*
Atz, J. W. 515, *519, 520*
Auwarter, A. 122, *129*, 142, 168, *169*, 461, *475*
Ayers, H. 331, *351*

Backus, R. H. 460, *475*, 515, *519*
Bagger-Sjöbäck, D. 19, *36*
Baird, R. 26, *30*
Baker, C. L., Jr. 526, 556, *564, 566*
Ballantyne, P. K. 406, *418*
Banbury, J. C. *101, 306, 422*
Banner, A. 13, *30*, 82, 85, *100*, 112, 114, 115, 119, 122, *128*, 135, 245, *252*, 285, *304, 305*, 403, 416, *418, 421*
Barber, V. C. 17, 28, *30*
Baretta, J. W. 110, 112, *132*, 244, *253*
Barker, J. L. 428, *443*
Barlow, G. W. 408, *419*
Bastian, J. 533, 535, 552, 554, 556, 557, 558, *564, 566*
Bateson, P. *422*
Bateson, W. 515, *519*
Bauer, B. B. 115, *128*
Bauer, D. H. 441, *443*
Bauknight, R. S. 481, 485, 489, *500*
Baylis, J. F. 397, *419*
Baylis, J. R. 261, *261*
Becker, R. 556, *568*
Behrend, E. R. 110, *133*
Behrend, K. 552, *565*
Békésy, G. von 25, *35*, 223, *241*, 485, *503*
Bell, C. C. 10, 29, 57, 103, 262, 339, 343, 344, *351*, 376, 378, 380, *380*, 382, 383, 384, 385, 387, 389, *391*, 392, 455, 504, 505, 538, 541, 543, 544, 546, 553, *565, 566*, 571, 578
Belyayev, N. 515, 516, *519*
Bennett, M. V. L. 333, 335, *351, 352*, 367,

Subject Index